LINEAR ALGEBRA

Bill Jacob	**Linear Algebra**, 1990
Larry J. Gerstein	**Discrete Mathematics and Algebraic Structures**, 1987
Daniel Finkbeiner II and Wendell Lindstrom	**A Primer of Discrete Mathematics**, 1987
Judith Gersting	**Mathematical Structures for Computer Science** (Second Edition), 1987
Branko Grünbaum and G. C. Shephard	**Tilings and Patterns**, 1987
Harley Flanders	**Calculus**, 1985
Vasek Chvatal	**Linear Programming**, 1983
Bonnie Averbach and Orin Chein	**Mathematics: Problem Solving Through Recreational Mathematics**, 1980
Michael R. Garey and David S. Johnson	**Computers and Intractability**, 1979
Michael Henle	**A Combinatorial Introduction to Topology**, 1979
Loren C. Larsen	**Algebra and Trigonometry Refresher for Calculus Students**, 1979
Underwood Dudley	**Elementary Number Theory**, 1978

LINEAR ALGEBRA

Bill Jacob

University of California, Santa Barbara

W. H. FREEMAN AND COMPANY • NEW YORK

Cover illustration by Deborah Katzburg

Cover design by Ron Ryan

Library of Congress Cataloging-in-Publication Data

Jacob, Bill.
  Linear algebra / Bill Jacob.
    p. cm. – (A Series of books in the mathematical sciences)
  Bibliography: p.
  Includes index.
  ISBN 0-7167-2031-0 :
  1. Algebras, Linear. I. Title. II. Series.
QA184.J33   1990
512'.5–dc20   89-11801
                                          CIP

Printed in the United States of America

2 3 4 5 6 7 8 9 0 V 7 6 5 4 3 2

To Debby, Michael, and Adam

CONTENTS

Part II: Advanced Topics

PREFACE

This book is a comprehensive text on linear algebra for students at the junior-senior level. Presumably, the students using this text will have some acquaintence with the basic matrix computations, although none is assumed. The book is divided into two parts. Part I treats the material which has by now become reasonably standard in most semester courses at this level. The main aim of Part I is to develop the material in as elementary a manner as possible, yet at the same time not compromise the development of the theory. Included are numerous examples and applications of the theory. In particular, applications which tie the material to other areas of mathematics are stressed.

Part II deals with advanced topics. Much of this material is more difficult and is designed for the students who continue with a second semester (or quarter). A major objective has been to write Part I at a level appropriate for students beginning a study of basic linear algebra with an emphasis on reasons why rather than on how to compute. The students who are successful with Part I should then be ready for the more advanced topics in Part II.

The book is written in the traditional mathematical style, with precisely formulated theorems followed by their proofs. Every instructor of upper-division mathematics understands the difficulties of the transition from computation-oriented classes to a class of this type. This text was written to help students over this obstacle. It offers no magic ladder to climb the barrier, nor do any technical gimmicks (computer programs, elaborate graphics, etc.) accompany this text. Instead the book begins with concrete and familiar concepts and then develops the theory at a pace that allows students time to become accustomed to the style. The foundations of linear algebra come from the study of systems of equations. For this reason this text, as much as possible, utilizes (and interprets) gaussian elimination in the proofs of the results in the earlier chapters. As the book develops the level of abstaction is increased so that by the time Part II is reached the invariant point of view is firmly entrenched.

The material in the text is developed without gaps; that is, no details of results are left to the reader or as exercises. At the same time the ideas are informally discussed and many of the common pitfalls for new students pointed out. The author believes that mathematical proof is an appropriate form of communication for upper-division university students and if we expect them to learn to speak this language, then they need to be provided with a complete and accurate model to follow. Most sections end with at least twice as many problems as need to be assigned. These problems include both the standard computational exercises as well as theoretical questions of varying difficulty. Answers and hints to most odd-numbered problems can be found at the end of the text.

There are many levels upon which one learns mathematics. Initially one learns how to perform certain computations. Then one studies examples of such calculations, searching for patterns and common themes. The conclusions of such observations are formulated as precisely as possible, and finally the mathematician manipulates the concepts into a logical order and proves results from basic principles. Students using this text will experience these different levels. In addition to learning the theory, one purpose of the "show" and "prove" problems in this text is for students to develop clear communication skills. This text acquaints students with the language of modern mathematics and at the same time exposes them to the process of transforming a computational subject into a conceptual framework with much greater applicability.

Uses of the Text

There is considerable flexibility in the design of a course using this text. Part I develops the material found in most one-semester linear algebra courses at the junior-senior level. This half of the text is designed primarily for such courses. It should be serviceable for students with a variety of backgrounds who need a course that gives more than computations. Part II develops a variety of advanced topics and applications. Depending upon students' backgrounds and interests, material from the second part can be incorporated into the latter weeks of a semester course. Taken together, both parts contain sufficient material for a rigorous full-year course.

It is not possible, nor reasonable, that the details of every proof in the text be covered in class. Instructors will need to devote time to discussing the intuitive ideas behind the results, as well as to developing the examples and discussing problems. A large amount of the feeling of this subject is best acquired verbally through classroom interaction. A major goal in writing this book has been to provide a text from which the instructor can feel comfortable about sending the students to dig out details as well as to find additional examples. The following chapter-by-chapter discussion indicates how the material is organized.

Part I

Chapter 0: Chap. 0 is devoted to matrix algebra and the basic theory of systems of linear equations. Students will have varying degrees of familiarity with these results from previous courses. Consequently, the amount of time devoted to these topics will vary anywhere from 0 to 5 or 6 (hour) lectures. The chapter develops the foundations of the subject and at the same time introduces the student to the rigor and style of the text in a familiar setting. Sec. 0.7 provides a quick review of the material needed in the sequel and is written for those who would like to cover the material of this chapter in a a single lecture.

Throughout Chap. 0, the term "field" is used to denote either the real or complex numbers. This has been done because many instructors prefer to introduce this material in as familiar a setting as possible, that is, over **R** and **C**. However, other instructors prefer to discuss general fields immediately. In this case, because the term "field" is used, all the results and their proofs in Chap. 0 can be interpreted in the general setting. This flexibility allows these instructors to give the definition of a field (Definition 1.1.1) at the beginning of Chap. 0.

Chapter 1: Chap. 1 presents the basic theory of vector spaces. Vector spaces are introduced as subspaces of $\mathbf{F}^p$ ($\mathbf{F}$ is a field). The concepts of linear independence and span are introduced concretely in this setting and the relationship between these concepts and matrix algebra is explored. The text then makes the transition to the general setting and defines arbitrary vector spaces. The traditional topics of bases, dimension, and coordinates for finite-dimensional vector spaces are developed in detail. The last (optional) section is an introduction to algebraic codes over $\mathbf{Z}/2\mathbf{Z}$, which students enjoy as an initial application of the abstract theory.

Chapter 2: Chap. 2 is devoted to a complete development of the theory of determinants. The material is independent from Chap. 1 and, if the instructor chooses, can be covered immediately after Chap. 0. All the basic results from the theory of determinants are rigorously proved. Most instructors will choose to omit many of the details and will instead explain the ideas and concentrate on the applications developed in Sec. 2.3. The last Sec. 2.4 is a listing of the key results about determinants needed for the sequel. It can be used by those instructors who desire only to give a quick review of the properties of determinants without dealing with the theory.

Chapter 3: Chap. 3 treats the basic theory of linear transformations. Matrix representations for linear transformations and matrix techniques are developed thoroughly. In addition, numerous non-matrix examples are studied in order to illustrate the need for a general theory. Eigenvalues and eigenvectors are developed in Sec. 3.4 and their connection with diagonalizability in the section that follows. As applications, Sec. 3.6 gives informal discussions of Markov chains, difference equations, and Gerschgorin's theorem.

Chapter 4: The basic theory of norms, inner products, and orthogonality are developed here. Direct sums and orthogonal projections are introduced. This chapter has a more applied flavor in that the ideas behind least-squares problems are developed in detail. In addition the basic results from spectral theory and polar decomposition are proved. The final section is devoted to optimal least-squares problems and the pseudoinverse.

Part II

Chapter 5: Chap. 5 treats the equivalence of the concepts of symmetric bilinear forms, symmetric matrices, and quadratic forms (in characteristic different from 2). The diagonalizability of quadratic forms and the signature of real quadratic forms are major topics. The connections with advanced calculus are described.

Chapter 6: The level of difficulty of the material increases throughout this chapter; the material becomes conceptually the most difficult in the text. The theory of a single linear operator on a finite-dimensional vector space is studied in detail with the main goal the development of the rational and Jordan canonical forms. The chapter begins with the detailed study of examples, so that the basic ideas behind the canonical forms can be understood before the difficult existence theorems are tackled. In fact, many users of the text may not have the time (or the desire) to give the existence proofs. It is a profitable experience for students to see some of the material of Secs. 6.1 to 6.7, even though the proofs are not given. This chapter has been written in such a way as to make this approach work. The last sections are theoretical and are tough even for the best students. The Smith normal form, which gives students an effective algorithm for computing the rational form, is developed in Sec. 6.8.

Chapter 7: In this chapter the style changes considerably. The purpose of this chapter is to expose students in an informal way to some of the infinite-dimensional topics. No attempt is made to be comprehensive, and the level of rigor varies considerably. The sections on the Wronskian and on systems of linear differential equations are important classroom topics and can be covered once the material from Chap. 3 is mastered. The remainder of this chapter requires the material from Chap. 4. It consists of an informal discussion of orthogonal polynomials, Fourier series, and the proof of the existence of a basis for an arbitrary vector space. These sections are written for students to read and increase their horizons (and are not suitable for extensive classroom development). Hopefully, the students will find reading them enjoyable and profitable.

Chapter 8: The only prerequisite for reading this chapter is the material of Chap. 3; however knowledge of multivariable calculus is necessary in order for the last section to be meaningful. The first two sections deal with multilinear algebra and tensors. The instructor who likes to discuss the double dual will find it in Sec. 8.1. As with Chap. 7, these sections are written with much less formality. The final section is basically an essay whose goal is to help the student connect the ideas developed earlier (namely tensors) with the basic concepts of multivariable calculus and advanced geometry.

The following diagram indicates the interdependence of the chapters in this book.

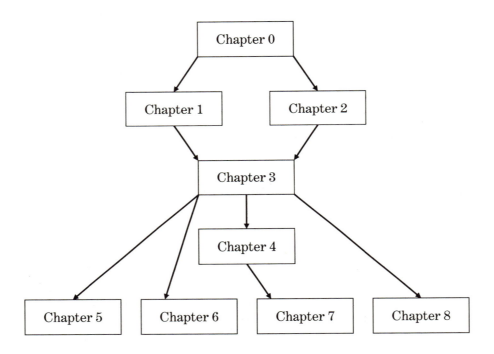

Acknowledgments

During the past four years preliminary versions of this book have been used as the text for the junior-senior level linear algebra sequence at Oregon State University. I wish to thank all my colleagues at Oregon State University for their friendly advice and many corrections throught this entire time period, especially Charles Ballantine for his extra effort and care in reading the numerous versions of the manuscript. I would also like to thank Roger Ware (The Pennsylvania State Univeristy) and Alex Rosenberg (University of California, Santa Barbara) who class tested portions of the text. I want to thank my editor, Jerry Lyons, and the staff of W. H. Freeman and Co. as well for all their help in enabling me to produce this book. My project editor, Stephen Wagley, and my copyeditor, Carol Loomis, were particularly helpful. I would also like to thank consulting editor Victor Klee for his careful reviews and advice during the early stages of writing. The many reviewers employed by W.H. Freeman provided a large number of valuable ideas and corrections. I thank them for their time and effort.

Finally I wish to thank my family for all their patience and love during the seemingly endless period in which this book was written.

Bill Jacob
Berkeley, California
June, 1989

Part I: Basic Linear Algebra

Chapter 0

Linear Equations and Matrices

0.1 Systems of Linear Equations

This chapter develops the theory of systems of linear equations. Matrices and matrix operations are introduced. So are the concepts of matrix rank and matrix inverses. The main task of this chapter is to lay rigorously the foundations for the rest of the book. While reading this chapter, students should keep in mind that it is not enough to understand merely the computations. What is crucial is *why* things work the way they do. If the reasons are carefully digested, then the student can master the basic concepts of linear algebra introduced in Chap. 1.

Linear algebra begins with the study of systems of linear equations. The foundations of this subject should be familiar from high school algebra, where it is usual to study systems of two or three equations. This first section treats a far more general situation, where the numbers of equations and variables are arbitrary and *not necessarily the same*. Such generality is critical for the development of the subject of linear algebra. In order to solve general systems of equations, one can use the computational algorithm known as gaussian elimination. The ideas behind this algorithm are extremely simple, yet its consequences are powerful. This important tool will be used explicitly (or implicitly) throughout the remainder of the book.

The symbol $\mathbf{R}$ will denote the real numbers and the symbol $\mathbf{C}$ will denote the complex numbers. In high school algebra, and again in calculus, the linear equations studied usually have real coefficients. However, in many of the important applications of linear algebra one studies linear equations whose coefficients are not real (or complex) numbers. For example, in computer science one is often interested in linear equations whose coefficients are integers (often mod p). In this introductory chapter we consider only the cases where the coefficients are real or complex numbers. In the next chapter the general situation will be described.

It turns out that in developing the theory of systems of linear equations, it is not necessary to distinguish between the real and complex cases. For this reason we shall adopt the convention of representing $\mathbf{R}$ or $\mathbf{C}$ by the single symbol $\mathbf{F}$. Every result stated in this chapter will apply to either case (namely where $\mathbf{F} = \mathbf{R}$ or $\mathbf{F} = \mathbf{C}$). To emphasize the general nature of the use of the symbol $\mathbf{F}$, we shall refer to $\mathbf{F}$ as the *field*, and we shall refer to the elements of $\mathbf{F}$ as *scalars*. This notational convention has the advantage that later, when the general notion of a field is defined, one can see that all the results of this chapter apply using this more general concept. For the discussion of different scalar fields, and their rigorous definition, we refer the reader to Sec. 1.1. If the reader desires, Definition 1.1.1 can be read now, and then this entire chapter will apply to general fields. (In fact, many instructors will prefer this approach.)

In this book capital letters like X, Y, Z or subscripted capitals like X_1, X_2, X_3 will be used to denote variables. (Variables are unknowns in equations.) Lowercase letters such as a, b, c or subscripted lowercase letters such as a_1, a_2, a_3 will be used to denote (unspecified) scalar constants.

Definition 0.1.1 A *linear equation* in the variables X_1, X_2,...,X_n is an equation of the form

$$a_1 X_1 + a_2 X_2 + \cdots + a_n X_n = b$$

where the constants a_1, a_2,..., a_n, b are scalars.

A *system* of m linear equations in the variables $X_1, X_2, \ldots, X_n$ is traditionally displayed as

$$\begin{array}{ccccccc}
a_{11}X_1 & + & a_{12}X_2 & + \cdots + & a_{1n}X_n & = & b_1 \\
a_{21}X_1 & + & a_{22}X_2 & + \cdots + & a_{2n}X_n & = & b_2 \\
\vdots & & \vdots & \ddots & \vdots & & \vdots \\
a_{m1}X_1 & + & a_{m2}X_2 & + \cdots + & a_{mn}X_n & = & b_m
\end{array}$$

Here, a_{ij} , $b_i \in \mathbf{F}$ are scalars.

Throughout this book, we will often use the phrase "a system of equations" instead of "a system of linear equations." For example, the equations $2X - 3Y + \pi Z = 7$ and $X_1 - X_4 = 0$ are both linear equations, while the equations $3X - 2Y + XY = 1$ and $(X - 2Y + Z)(X + W) = 0$ are not. The terminology "linear" comes from the fact that in the euclidean plane $\mathbf{R}^2$ the set of solutions to $aX + bY = c$ is a line (as long as a and b are not both 0). Note however, that the set of solutions in $\mathbf{R}^3$ to $X + Y + Z = 1$ is not a line (it is a plane), so do not be misled by the terminology.

As is customary, we denote by $\mathbf{R}^n$ the set of all n-tuples of real numbers. Analogously, $\mathbf{F}^n$ will denote the set of n-tuples of elements of a scalar field $\mathbf{F}$. Thus, in set theoretic notation, we have

$$\mathbf{F}^n = \{(r_1, r_2, \ldots, r_n) \mid r_1, r_2, \ldots, r_n \in \mathbf{F}\}$$

Elements of $\mathbf{F}^n$ are added (vector addition) and can be multiplied by a scalar (scalar multiplication) according to

$$(r_1, r_2, \ldots, r_n) + (s_1, s_2, \ldots, s_n) = (r_1 + s_1, r_2 + s_2, \ldots, r_n + s_n)$$

and

$$t(r_1, r_2, \ldots, r_n) = (tr_1, tr_2, \ldots, tr_n)$$

One can use $\mathbf{R}^2$ to give the coordinates of elements of the euclidean plane, and one can use $\mathbf{R}^3$ to give the coordinates of elements in 3 space. In fact it is customary to say that $\mathbf{R}^2$ "is" the plane and $\mathbf{R}^3$ "is" space. These intrepetations of $\mathbf{R}^2$ and $\mathbf{R}^3$ are crucial for initially visualizing how linear equations (more specifically their solutions) behave in one, two, or three variables. But for more variables, no such concrete geometric picture is available. One way to study equations with four or more variables is to work algebraically with the equations, using what is known from lower dimensions as a guide. If this point of view is adopted, one need not worry about any possible geometric interpretation of $\mathbf{R}^n$, for $n > 3$. This does not mean that one abandons the geometric point of view; rather it means that algebra is used as a tool to push (three-dimensional) geometric intuition to higher dimensions. Once the ideas behind linear algebra have been developed, problems can be formally solved by using only algebra. One applies geometric understanding without using formal geometry. In this setting one needs to work with $\mathbf{R}^n$ only as n-tuples of real numbers.

The next definition explicitly defines the set of solutions to a linear equation.

Definition 0.1.2 The *set of solutions* to the linear equation

$$a_1 X_1 + a_2 X_2 + \cdots + a_n X_n = b$$

is the set

$$\{(r_1, r_2, \ldots, r_n) \in \mathbf{F}^n \mid a_1 r_1 + a_2 r_2 + \cdots + a_n r_n = b\}$$

The set of solutions to the system of linear equations

$$
\begin{array}{ccccccc}
a_{11} X_1 & + & a_{12} X_2 & + \cdots + & a_{1n} X_n & = & b_1 \\
a_{21} X_1 & + & a_{22} X_2 & + \cdots + & a_{2n} X_n & = & b_2 \\
\vdots & & \vdots & \ddots & \vdots & & \vdots \\
a_{m1} X_1 & + & a_{m2} X_2 & + \cdots + & a_{mn} X_n & = & b_m
\end{array}
$$

is $\{(r_1, r_2, \ldots, r_n) \in \mathbf{F}^n \mid (r_1, r_2, \ldots, r_n)$ is a solution to each of the m equations$\}$. Specifically, the set of solutions to the system is the intersection of the solution sets of each of the m individual equations. If a system of equations has at least one solution, it is called *consistent*. If it has no solutions, it is called *inconsistent*.

Remark One common error made by students in basic linear algebra is assuming that all systems of equations have a solution. The cause of this confusion is that in many elementary treatments of systems of equations the only case considered is that where the number of equations is the same as the number of variables. As we shall see later, often (*but not always*) such a system will have a unique solution. In general however, a system of equations may have no solutions, a unique solution, or many solutions. For example, whenever a, $b \in \mathbf{R}$, the system of equations $X - Y = a$ and $X + Y = b$ has the unique real solution given by $X = (a + b)/2$ and $Y = (b - a)/2$. However the system $X - Y = a$ and $X - Y = b$ has no solutions if $a \neq b$, yet has infinitely many real solutions if $a = b$. Observe that a small change in the appearance of the system can drastically change the set of solutions. In this chapter criteria are developed for deciding, in any specific case, which of these possibilities actually occurs. Do not make conclusions until you are sure.

Solving a system of two equations in two unknowns should be familiar from high school algebra. For example, consider the system

$$
\begin{aligned}
2X + 3Y &= 5 \\
5X + 6Y &= 14
\end{aligned}
$$

This system can be solved by first subtracting twice the first equation from the second. We find

$$
\begin{array}{rcl}
5X + 6Y &=& 14 \\
- 2(2X + 3Y) &=& 2 \cdot 5 \\
\hline
X &=& 4
\end{array}
$$

Thus, necessarily $X = 4$ in any solution of the system. Substituting the X value into the first equation, we find that $5 \cdot 4 + 6Y = 14$, and thus $6Y = 14 - 20 = -6$. This shows that $Y = -1$ in any solution, and therefore the only possible solution to our system is the pair $(4, -1)$. It is readily checked that $(4, -1)$ is a solution, so we have determined all solutions.

Systems of more equations in more unknowns can be solved by a generalization of the process just illustrated. However, before outlining this general procedure, we isolate the key steps. These key steps are called *elementary (row) operations*. Elementary operations transform one system of equations into another system. After illustrating how such row operations are used, we will prove a theorem which asserts that the set of solutions to the new system is precisely the same as the set of solutions to the old system.

Definition 0.1.3 Given a system of linear equations, an *elementary row operation* applied to this system is one of the following three operations:

(i)   Multiply any single equation by a nonzero scalar, leaving all the other equations the same.

(ii)   Add a multiple of one equation to a second, leaving all but this second equation the same.

(iii) Interchange the position of any two equations on the list.

Remark 0.1.4 The third operation can, in fact, be obtained from the other two. (See Remark 0.4.2 for a proof.) For this reason, it is sometimes omitted as an elementary row operation. It should be evident that the third operation does not change the set of solutions to a system of equations, because the set of equations is exactly the same (only listed in a different order). Likewise, in Theorem 0.1.8 we prove that the other two elementary row operations do not change the set of solutions to a system of equations. This next example illustrates how to use elementary row operations in solving systems of equations.

Example 0.1.5 Consider the system of equations

$$
\begin{aligned}
3X & - Y & & + W & = 4 \\
X & + Y & - Z & + W & = 1 \\
X & & - Z & + W & = 0
\end{aligned}
$$

If we multiply the second equation by 4, we obtain

$$
\begin{aligned}
3X & - Y & & + W & = 4 \\
4X & + 4Y & - 4Z & + 4W & = 4 \\
X & & - Z & + W & = 0
\end{aligned}
$$

This is an application of elementary row operation (i). As an example of operation (ii) we multiply the first equation by -2 and add the resulting equation to the third (leaving the first and second equations unchanged):

$$
\begin{aligned}
3X & - Y & & + W & = 4 \\
4X & + 4Y & - 4Z & + 4W & = 4 \\
-5X & + 2Y & - Z & - W & = -8
\end{aligned}
$$

We now give the final definition of this section.

Definition 0.1.6 Two systems of m equations in n variables are called *row equivalent* (or simply *equivalent*) if one system can be obtained from the other system by performing a (finite) sequence of elementary row operations.

Thus for example, the three systems of equations listed in Example 0.1.5 are all row equivalent systems.

Remark 0.1.7 Observe that both row operations described in Definition 0.1.3 can be reversed. That is, whenever a single row operation is applied to a system of equations, another row operation can be applied to return the new system to the initial system. To reverse operation (i), multiply the new equation by the reciprocal of the (nonzero) scalar. To reverse operation (ii), subtract the same multiple of the first equation from the second equation. It is because of the reversibility of elementary operations that the term "equivalent" can be used in Definition 0.1.6. *Whenever a system of equations is transformed into a second system by a sequence of elementary row operations, there is another sequence of elementary row operations which can transform the second system into the original.*

We now prove the key theorem of this section. It shows why elementary row operations are important in studying the set of solutions to systems of equations. It also shows why the technique used earlier provided the solution to the system of two equations in two unknowns. Examples and applications of this basic result follow in Sec. 0.2.

Theorem 0.1.8 *If two systems of equations are row equivalent, then they have exactly the same set of solutions.*

Proof: To prove the theorem, it is sufficient to treat the case where one system has been obtained from the other by a single row operation. The general result follows by applying the single-operation case repeatedly. We now suppose that we started with the system

$$
\begin{array}{ccccccc}
a_{11}X_1 & + & a_{12}X_2 & + \cdots + & a_{1n}X_n & = & b_1 \\
a_{21}X_1 & + & a_{22}X_2 & + \cdots + & a_{2n}X_n & = & b_2 \\
\vdots & & \vdots & \ddots & \vdots & & \vdots \\
a_{m1}X_1 & + & a_{m2}X_2 & + \cdots + & a_{mn}X_n & = & b_m
\end{array}
\tag{1}
$$

It clearly does not matter in which order we write down the equations, and thus operation (iii) does not change the set of solutions to the system. Furthermore, it follows from this that we can assume our row operations have been applied to only the first and second equations, leaving the rest unchanged.

Suppose that operation (i) was used and the first equation was multiplied by the nonzero scalar k. The resulting system is:

$$
\begin{array}{cccccc}
ka_{11}X_1 & + & ka_{12}X_2 & + \cdots + & ka_{1n}X_n & = & kb_1 \\
a_{21}X_1 & + & a_{22}X_2 & + \cdots + & a_{2n}X_n & = & b_2 \\
\vdots & & \vdots & \ddots & \vdots & & \vdots \\
a_{m1}X_1 & + & a_{m2}X_2 & + \cdots + & a_{mn}X_n & = & b_m
\end{array}
\tag{2}
$$

We must see that the systems (1) and (2) have precisely the same solutions. To do this, we show that any solution to (1) is a solution to (2) and conversely that any solution to (2) is a solution to (1). Let $(r_1, r_2, \ldots, r_n)$ be a solution to system (1). By definition, $(r_1, r_2, \ldots, r_n)$ is a solution to the second, third,..., mth equations of system (1), and hence it solves these equations of system (2). Further, as it solves the first equation of (1), we know that

$$a_{11}r_1 + a_{12}r_2 + \cdots + a_{1n}r_n = b_1$$

Multiplying both sides of this equation by the scalar k and applying the distributive law, we find that

$$k(a_{11}r_1 + a_{12}r_2 + \cdots + a_{1n}r_n) = kb_1$$

so

$$ka_{11}r_1 + ka_{12}r_2 + \cdots + ka_{1n}r_n = kb_1$$

This shows that $(r_1, r_2, \ldots, r_n)$ is a solution to the first equation of system (2), and hence is a solution to the entire system (2).

We have shown that every solution of (1) is a solution to (2). For the converse, instead of repeating a similar argument, we note that as $k \neq 0$, $1/k$ exists (and is nonzero) in our field $\mathbf{F}$. Multiplying the first equation of system (2) by $1/k$ yields the original system (1). Thus, by what we just proved, any solution to system (2) is also a solution to (1). Hence, applying row operation (i) does not change the set of solutions to a system of equations. (Note: This shows why it is necessary to assume that the multiplication constant in row operation (i) is nonzero. Multiplying an equation by 0 can drastically change the set of solutions to a system.)

We next show that an application of row operation (ii) does not change the set of solutions to a system of equations. In this case we may assume that our original system (1) has been changed by adding k times the first equation to the second equation. It follows that our system of equations now looks like

$$
\begin{array}{rcl}
a_{11}X_1 + \cdots + a_{1n}X_n & = & b_1 \\
(a_{21} + ka_{11})X_1 + \cdots + (a_{2n} + ka_{1n})X_n & = & b_2 + kb_1 \\
\vdots \quad \ddots \qquad\qquad \vdots & & \vdots \\
a_{m1}X_1 + \cdots + a_{mn}X_n & = & b_m
\end{array}
\tag{3}
$$

We must show that any solution to the original system (1) is also a solution to this new system (3). Suppose that $(r_1, r_2, \ldots, r_n)$ is a solution to (1). By hypothesis we know that it is a solution to the first, third,..., mth equations in (3). Since $(r_1, r_2, \ldots, r_n)$ is a solution of the first two equations of (1), we find

$$
a_{11}r_1 + a_{12}r_2 + \cdots + a_{1n}r_n = b_1
$$

and

$$
a_{21}r_1 + a_{22}r_2 + \cdots + a_{2n}r_n = b_2
$$

Multiplying the first equation by k, we have

$$
ka_{11}r_1 + ka_{12}r_2 + \cdots + ka_{1n}r_n = kb_1
$$

Adding these last two equations and collecting terms gives

$$
(a_{21} + ka_{11})r_1 + (a_{22} + ka_{12})r_2 + \cdots + (a_{2n} + ka_{1n})r_n = b_2 + kb_1
$$

This shows that $(r_1, r_2, \ldots, r_n)$ is a solution to the second equation of (3), that is, any solution to (1) is a solution to (3). Conversely, to see that any solution of system (3) is a solution to (1), recall (from Remark 0.1.7) that the original system (1) can be obtained from the new system (3) by adding $-k$ times the first equation of (3) to the second equation of (3). Thus this result follows from what we just proved. This completes the proof of Theorem 0.1.8. //

0.2 Echelon Form Systems

This section describes how to compute the set of solutions to an arbitrary system of linear equation. The philosophy behind the procedure is this. Given a system of equations, perform row operations until a system of equations whose solutions are easily computed is obtained. Theorem 0.1.8 guarantees that the new system has precisely the same solutions as the old, so the solutions to the original system have been determined. We shall describe two specific strategies, known as *gaussian* and *Gauss-Jordan elimination*. These methods are best learned by looking at examples, so that is where we begin.

Example 0.2.1 Consider the system of equations with real coefficients

$$
\begin{aligned}
3X &- 4Y + Z = 1 \\
6X &- 8Y + 4Z = 12
\end{aligned}
$$

Subtract 2 times the first equation from the second:

$$
\begin{aligned}
3X &- 4Y + Z = 1 \\
&\quad\quad\quad 2Z = 10
\end{aligned}
$$

Multiply the second equation by $1/2$:

$$
\begin{aligned}
3X &- 4Y + Z = 1 \\
&\quad\quad\quad Z = 5
\end{aligned}
$$

Subtract the second equation from the first:

$$
\begin{aligned}
3X &- 4Y \quad\quad = -4 \\
&\quad\quad\quad Z = 5
\end{aligned}
$$

Divide the first equation by 3:

$$
\begin{aligned}
X &- (4/3)Y \quad\quad = -(4/3) \\
&\quad\quad\quad Z = 5
\end{aligned}
$$

At this point we have a system of equations whose solutions are precisely the same as the original system. In any solution of this system, the third variable Z must be 5, while the first and second variables are not uniquely determined. Representing the possible values of Y by the *parameter t* (ranging over all $\mathbf{R}$), we see that in any solution the value of X is determined by the Y value as $X = (4/3)t - 4/3$. Hence, all the solutions to the original system have the form $(X, Y, Z) = ((4/3)t - 4/3, t, 5)$ where $t \in \mathbf{R}$. Alternatively, this set of solutions can be expressed as $\{(-4/3, 0, 5) + t(4/3, 1, 0) \mid t \in \mathbf{R}\}$.

Example 0.2.2 In this example we consider a system of three equations in three unknowns, again with real coefficients:

$$
\begin{array}{rcrcrcr}
X & - & Y & + & 2Z & = & 1 \\
2X & + & Y & + & Z & = & 1 \\
4X & - & Y & + & 5Z & = & 5
\end{array}
$$

Subtract 2 times the first equation from the second:

$$
\begin{array}{rcrcrcr}
X & - & Y & + & 2Z & = & 1 \\
& & 3Y & - & 3Z & = & -1 \\
4X & - & Y & + & 5Z & = & 5
\end{array}
$$

Subtract 4 times the first equation from the third:

$$
\begin{array}{rcrcrcr}
X & - & Y & + & 2Z & = & 1 \\
& & 3Y & - & 3Z & = & -1 \\
& & 3Y & - & 3Z & = & 1
\end{array}
$$

Subtract the second equation from the third:

$$
\begin{array}{rcrcrcr}
X & - & Y & + & 2Z & = & 1 \\
& & 3Y & - & 3Z & = & -1 \\
& & & & 0 & = & 2
\end{array}
$$

At this point we can stop. The last equation is clearly nonsense; that is, this last system cannot have any solutions. It follows from Theorem 0.1.8 that the set of solutions to the original system is the empty set $\emptyset$.

Example 0.2.3 Next consider

$$
\begin{array}{rcrcrcr}
2X_1 & + & X_3 & - & X_4 & = & 0 \\
& & X_4 & - & X_3 & = & 0 \\
X_1 & - & X_2 & - & X_4 & = & 0
\end{array}
$$

First arrange the equations so that we can keep better track of all the variables:

$$
\begin{array}{rcrcrcr}
2X_1 & & & + & X_3 & - & X_4 & = & 0 \\
& & & - & X_3 & + & X_4 & = & 0 \\
X_1 & - & X_2 & & & - & X_4 & = & 0
\end{array}
$$

Multiply the first equation by 1/2:

$$
\begin{array}{rrrrl}
X_1 & & + & (1/2)X_3 & - & (1/2)X_4 & = & 0 \\
 & & - & X_3 & + & X_4 & = & 0 \\
X_1 & - & X_2 & & & - & X_4 & = & 0
\end{array}
$$

Subtract the first equation from the third:

$$
\begin{array}{rrrrl}
X_1 & & + & (1/2)X_3 & - & (1/2)X_4 & = & 0 \\
 & & - & X_3 & + & X_4 & = & 0 \\
 & - & X_2 & - & (1/2)X_3 & - & (1/2)X_4 & = & 0
\end{array}
$$

Multiply each of the second and third equations by -1 and (for reasons that will be made clear later) interchange the second and third equations:

$$
\begin{array}{rrrrl}
X_1 & & + & (1/2)X_3 & - & (1/2)X_4 & = & 0 \\
X_2 & + & (1/2)X_3 & + & (1/2)X_4 & = & 0 \\
 & & X_3 & - & X_4 & = & 0
\end{array}
$$

Subtract 1/2 the third equation from the second, and subtract 1/2 the third equation from the first:

$$
\begin{array}{rrrrl}
X_1 & & & = & 0 \\
 & X_2 & + & X_4 & = & 0 \\
 & X_3 & - & X_4 & = & 0
\end{array}
$$

This system has infinitely many solutions. However, once the value of X_4 is fixed, the remaining variables are uniquely determined. Thus, we see that the solutions to our original system of equations must all be of the form $(0, -t, t, t)$ where t is an arbitrary scalar (that is, a parameter). When describing the set of solutions in this fashion, X_4 is called a *free variable*, while X_1, X_2, and X_3 are called *determined variables*.

At this point you probably have a good idea of the strategy, so we now give it a written description. This process is known as *gaussian elimination*. In the following description, we use the phrase "variable occurs" to mean that the variable has a nonzero coefficient. We (implicitly) assume in this description that there is a fixed ordering of the variables of the system (for example a first, second, third, ... variable).

Usually this ordering is given by the alphabetical order of the variables $X, Y, Z, \ldots$ or by numerical indexing $X_1, X_2, X_3, \ldots$ (The specific ordering of the variables is not crucial, but some ordering needs to be fixed so that the definitions will make sense.) Also, the equations are listed in a specific order (the first equation on top, the second immediately below, and so forth).

Gaussian Elimination Strategy

1. Make sure the first occurring (or leading) variable occurs in the first equation, and multiply this equation by a nonzero scalar so that this first occurring variable has coefficient 1. This gives the first equation resulting from the gaussian elimination process. (If the first occurring variable does not occur in the first equation, reorder the equations to ensure that it does.)

2. Subtract an appropriate multiple of the first equation from each of the other equations to eliminate the first occurring variable from them.

3. Choose the second equation so that (if possible) the second occurring variable occurs in it. (If the second occurring variable does not occur in any equation other that the first at this point, go on to the third occurring variable, and so forth.) Multiply this second equation by a nonzero constant so this second occurring variable has coefficient 1. This gives the second equation resulting from the gaussian elimination process.

4. Subtract an appropriate multiple of the second equation from each of the equations below it to eliminate the second occurring variable from all these equations.

5. Now choose and modify the third equation so that the third (or next) occurring variable occurs with leading coefficient 1. Eliminate this variable from all the equations below the third. Continuing in this fashion, one obtains a staircaselike system of equations for which the leading variable of any equation does not occur in any of the equations below.

When this process of gaussian elimination is complete, one has obtained what is called an *echelon form* for the system of equations. An echelon form for a system of equations can be characterized as follows.

Definition 0.2.4 A system of equations (with a specified ordering for its equations and variables) is in *echelon form* (for these orderings) if
(E1)   The first occuring variable of each equation has coefficient 1. (This variable, if any, is called the *leading variable* of that equation.)
(E2)   The leading variable of any equation occurs to the right of the leading variable of any equation above it, and all equations without leading variables are listed last.

The last phrase of (E2) may seem slightly mysterious at first glance. The point is that after applying gaussian elimination, all the variables in some equations may be eliminated. In other words, an equation of the form $0X_1 + \cdots + 0X_n = k$ where $k \in \mathbf{F}$ could result. All such equations must occur at the bottom of a system in echelon form. Here are some more examples of systems of equations in echelon form.

Examples 0.2.5

(i)
$$
\begin{aligned}
X - 4Y &= 5 \\
Y &= 3
\end{aligned}
$$

(ii)
$$
\begin{aligned}
X + Y - 3Z + W &= 1 \\
Y - Z - W &= 7 \\
Z &= 4
\end{aligned}
$$

(iii)
$$
\begin{aligned}
X_1 + X_2 \qquad\qquad\quad + X_5 &= 0 \\
X_3 - X_4 - X_5 &= 2 \\
X_5 &= 1
\end{aligned}
$$

(iv)
$$
\begin{aligned}
X_1 + X_2 &= 0 \\
0 &= 1
\end{aligned}
$$

The set of solutions to a system in echelon form is reasonably easy to understand. A system in echelon form will be consistent unless an equation of the form $0 = r$ occurs where r is nonzero. (The reason for this is given below.) Suppose the system is consistent. Any variable that is not the leading variable of an equation will be called a *free variable*. The remaining variables (that is, the leading variables) are called *determined variables*. The following process, known as *back-substitution*, describes all solutions to the system by assigning parameters to the free variables.

Back-Substitution for Systems in Echelon Form

Assume that no equation of the form $0 = r$ for r nonzero occurs in the system in echelon form. Start at the bottom equation of the system. Assign a different parameter to each free variable, and express the determined variable in this bottom equation in terms of these parameters. Next, go to the second equation from the bottom, assign different parameters to any new free variables, substitute the values for all previous variables, and solve for the next determined variable. Continue moving up through all the equations. At the end of this process all free variables will be assigned a different parameter, and all the determined variables will be expressed in terms of these parameters. Evidently, all the solutions to the system are obtained by allowing the parameters to take on arbitrary scalar values. Given any assignment of the parameters, the determined variable in each equation is assigned the (unique) value necessary to make the equation true.

Note in particular that back-substitution shows that whenever an equation $0 = r$ for r nonzero, does *not* occur in a system in echelon form, the system *does* have at least one solution. This is the consistency assertion given above.

We now use back-substitution to find all solutions to the systems of equations in echelon form given in Example 0.2.5.

(i) The second equation says that $Y = 3$. So substituting this value into the first equation, we see $X - 4(3) = 5$; that is, we find that $X = 17$. Hence $(17, 3)$ is the only solution to the system.

(ii) In this system, the third equation says that $Z = 4$. The variable Y is a free variable, so to remind ourselves of this, we set $Y = t$. Then using what we know about Z, we find that $Y - (4) - (t) = 7$; that is, $Y = 11 + t$. Substituting this information into the first equation, we see that $X + (11 + t) - 3(4) + t = 1$; that is, $X = 2 - 2t$. It follows that the set of solutions is $\{(2 - 2t, 11 + t, 4, t) \mid t \in \mathbf{R}\}$.

(iii) Necessarily, $X_5 = 1$ in any solution. Assigning the free variable X_4 the parameter t and substituting $X_5 = 1$, the second equation shows that $X_3 = t + 1 + 2 = t + 3$. Now assigning the free variable X_2 the parameter u and substituting $X_5 = 1$, the first equation shows that $X_1 = -u - 1$. Consequently, the set of solutions to this system is $\{(-u - 1, u, t + 3, t, 1) \mid t, u \in \mathbf{R}\}$.

(iv) This system is inconsistent; that is, it has no solutions. In other words the solution set is $\emptyset$.

There is an extension of gaussian elimination, known as *Gauss-Jordan elimination*, which yields a more refined system of equations than does gaussian elimination. This process was illustrated in Example 0.2.1. Note that there, an echelon form for the system was obtained after the third step, but we continued applying row operations to eliminate the third variable from the first equation as well.

In Gauss-Jordan elimination, one proceeds exactly as in gaussian elimination, except that in addition one eliminates occurrences of each leading variable from the equations *above* as well as those below. In other words, the above description of the gaussian elimination strategy has additional instructions.

Further Gauss-Jordan Strategy

4. Further, subtract an appropriate multiple of the second equation from the first equation to eliminate (if necessary) the second leading variable from the first equation.

5. Further, subtract appropriate multiples of the third equation from the first and second equations to eliminate the third leading variable from these equations. Continue in this fashion so that the leading variable of each equation of the system occurs only in the single equation for which it is the leading variable.

The system of equations that results from applying Gauss-Jordan elimination is said to be in *reduced echelon* form. Systems in reduced echelon form are characterized by the following three conditions (the first two of which assert it is in echelon form).

Definition 0.2.6 A system of equations (with a specified ordering of its equations and variables) is in *reduced echelon form* if:
(E1) The leading variable of every equation has coefficient 1.
(E2) The leading variable of every equation comes after the leading

variable of each preceding equation. Equations without variables are listed last.

(RE3) The leading variable of each equation occurs in no other equation.

Whenever a system of equations has a unique solution, the reduced echelon form is exceptionally nice, for it literally tells you the solution. For example, starting with the system

$$
\begin{aligned}
X &- 4Y + Z = 2 \\
-X &+ 3Y - Z = 1 \\
X & \quad\quad + 2Z = 3
\end{aligned}
$$

we apply Gauss-Jordan elimination (recording the steps one column at a time):

$$
\begin{aligned}
X - 4Y + Z &= 2 \\
-X + 3Y - Z &= 1 \\
X \quad\quad + 2Z &= 3
\end{aligned}
\;\longmapsto\;
\begin{aligned}
X - 4Y + Z &= 2 \\
-Y \quad\quad &= 3 \\
4Y + Z &= 1
\end{aligned}
$$

$$
\longmapsto\;
\begin{aligned}
X \quad\quad + Z &= -10 \\
Y \quad\quad &= -3 \\
Z &= 13
\end{aligned}
$$

$$
\longmapsto\;
\begin{aligned}
X \quad\quad &= -23 \\
Y \quad\quad &= -3 \\
Z &= 13
\end{aligned}
$$

Consequently, $X = -23$, $Y = -3$, and $Z = 13$ is the unique solution to the original system.

The primary value of the reduced echelon form of a system of equations is that in the consistent case, it is uniquely determined (once an ordering of the variables has been fixed). This will be proved in a later section in the context of matrices. (See Theorem 0.5.5.) It is also true that systems in reduced echelon form are easier to solve than systems in (general) echelon form, since one does not have to back-substitute the values of determined variables. Of course, the work involved in such back-substitution was done earlier in the additional steps of the Gauss-Jordan elimination. In fact, as a general rule, if all that is needed is the solution set to a system, Gauss-Jordan elimination is less efficient than gaussian elimination with back-substitution.

One futher comment involves the normalization to 1 of the leading variable of equations in both gaussian and Gauss-Jordan elimination. The key reason for this normalization is that it is necessary for the uniqueness of the reduced echelon form, a result which has enormous theoretical value. However, in many instances of practical computation, it may not be at all wise to normalize the leading variable to 1. For example, this might involve a large division, which could greatly complicate the expressions giving other coefficients. Even worse, if a computer is being used, such a division could cause the machine to truncate small numbers to zero and thus drastically change the behavior of the system.

We conclude this section with some more examples.

Examples 0.2.7

(i)
$$X \qquad - \quad W = 0$$
$$ Y \quad - \quad W = 0$$
$$ Z \; - \; 5W = 0$$
is a system in reduced echelon form.

(ii) $X - Z + 8W = 3$
is a system in reduced echelon form.

(iii) $-Z + X + 8W = 3$
is *not* a system in echelon form, since the lead coefficient is not 1.

(iv)
$$X_1 \qquad - \quad X_3 \qquad\qquad = 8$$
$$ X_2 \qquad\quad + \; X_4 = 8$$
$$ X_3 \; - \; X_4 = 1$$
is a system in echelon form that is not in reduced echelon form.

(v)
$$X + Y = 1 \qquad\qquad X + Y = 0$$
$$ 0Y = 1 \quad \text{and} \quad 0Y = 1$$
are both systems in reduced echelon form. Note that these systems are row equivalent and inconsistent. Thus, the reduced echelon form is not unique in the inconsistent case.

PROBLEMS

0.2.1. Find all solutions to:

(a)
$$3X \; - \; Y \; - \; Z = 5$$
$$2X \qquad\qquad + \; Z = 5$$

(b) $\quad X + 5Y - 2Z = 0$
$\qquad X - 3Y + Z = 0$
$\qquad X + 5Y - Z = 0$

(c) $\quad X_1 - X_2 + X_3 = 1$
$\qquad X_1 + 2X_2 + X_4 = 0$
$\qquad X_1 + X_2 = 6$

(d) $\quad X_1 + X_2 = 2$
$\qquad X_1 + X_3 - X_4 = 6$
$\qquad X_2 - X_3 - X_4 = 5$

(e) $\quad X + Y + Z + W = 7$

(f) $\quad X - Z = 3$
$\qquad Y + W = 2$

(g) $\quad X + 3Y + Z = 0$
$\qquad X - Y + 3Z = 0$
$\qquad 2Y - Z = 0$

(h) $\quad X + 3Y + Z = 1$
$\qquad X - Y + 3Z = 1$
$\qquad 2Y - Z = 1$

0.2.2. Use gaussian elimination to find all solutions to the following:

(a) $\quad 2X \qquad + Z = 4$
$\qquad X + Y + Z = 3$
$\qquad X + 3Y + 2Z = 5$

(b) $\quad X + 2Y + Z = -1$
$\qquad 5X - Y = -3$
$\qquad 2X - 3Y + Z = 0$
$\qquad 2X - 2Y = 2$

(c) $\quad X - Y - Z + W = 5$
$\qquad Y - Z + 2W = 8$
$\qquad 2X - Y - 3Z + 4W = 18$

(d) $\quad 2X + 2Y - 3Z + 4W = 1$
$\qquad 2X - 4Y + 2Z - 2W = 4$
$\qquad 4X - 2Y - Z + 2W = -1$

(e) $\quad X + Y - Z - W = 1$
$\qquad X \qquad\qquad - W = 2$

(f) $\quad X + 3Y - 4Z = 0$
$\qquad 2X + 6Y - 8Z = 1$

0.2.3. For each of the systems in Prob. 0.2.1, find all possible echelon forms for the system (for the given ordering of variables). Find the reduced echelon form of the system as well.

0.2.4. For what values of u and v does the system of equations

$$\begin{aligned} X + Y &= u \\ 2X + 2Y &= v \end{aligned}$$

have a solution? For what values does it have a unique solution?

0.2.5. Show that no system of three equations in four variables has a unique solution. (Use the existence of a row-equivalent reduced echelon form system.)

0.2.6. Consider the system of equations

$$\begin{aligned} 3X - Y - Z &= 1 \\ X + 3Y + 2Z &= 3 \\ 5X + Y + 2Z &= k \end{aligned}$$

For which values of k is this system consistent? Find all solutions in this case.

0.2.7. Consider the system of equations where $a, b, c, d, e, f \in \mathbf{R}$

$$\begin{aligned} aX + bY + eZ &= 0 \\ cX + dY + fZ &= 0 \end{aligned}$$

(a) Show that this system always has a solution other than $(0,0,0)$.
(b) Show that if (r, s, t) and (u, v, w) are both solutions to the system, then $(r + u, s + v, t + w)$ is also a solution.
(c) Show that if (r, s, t) is a solution to the system and $k \in \mathbf{R}$, then (kr, ks, kt) is also a solution.

0.2.8. Find conditions on $r, s, t \in \mathbf{R}$ which guarantee that the following system has a solution:

$$\begin{aligned} X + Y + Z &= r \\ X + 6Y + 3Z &= s \\ 3X - 2Y + Z &= t \end{aligned}$$

0.2.9. Are the following systems of equations row equivalent?

$$\begin{aligned} X - Y &= 0 & \qquad 3X + Y &= 0 \\ 2X + Y &= 0 & \qquad X + Y &= 0 \end{aligned}$$

0.2.10.   What conditions must $a, b, c \in \mathbf{R}$ satisfy in order that the following system be consistent?

$$
\begin{array}{rcrcrcrcl}
3X & - & 2Y & + & 4Z & - & W & = & a \\
-2X & + & 3Y & - & 2Z & - & 2W & = & b \\
5X & & & + & 8Z & - & 7W & = & c
\end{array}
$$

0.2.11.   Prove that two consistent systems of two equations in two unknowns are row equivalent if they have the same set of solutions.

0.2.12.   Consider the system of equations over $\mathbf{R}$

$$
\begin{aligned}
a_{11}X + a_{12}Y + a_{13}Z &= b_1 \\
a_{21}X + a_{22}Y + a_{23}Z &= b_2 \\
a_{31}X + a_{32}Y + a_{33}Z &= b_3
\end{aligned}
$$

Suppose this system has two different solutions. Show it has infinitely many different solutions.

0.2.13.   Find a system of equations whose set of solutions is given by $X_1 = 3 + 2t - u$, $X_2 = t$, $X_3 = 2 - u$, $X_4 = u$, $X_5 = 2$ where $t, u \in \mathbf{R}$. (Hint: Think about the reduced echelon form.)

0.2.14.   Find three linear equations in three unknowns, any two of which form a consistent system, but all three taken together are inconsistent. Interpret your system geometrically (in $\mathbf{R}^3$).

0.3   Matrices and Matrix Operations

In this section we introduce basic matrix algebra. Our first use of matrices will be to simplify the notation involved in studying systems of equations. Shortly however, we shall see that their use is of a much broader scope. A matrix is a rectangular array of scalars, where the exact location of each scalar is crucial. In order to make this precise, we define a matrix to be a special type of function.

Definition 0.3.1 An $m \times n$ **F** matrix is a function whose domain is $\{(i, j) \mid 1 \le i \le m, 1 \le j \le n\}$ and whose range is the scalar field $\mathbf{F}$. If A is such a function, then the value of this function on the pair (i, j) is denoted by $A(i, j)$, which is called the ijth *entry* of A.

In order to visualize a matrix, instead of working with a function we represent the matrix as a rectangular array of scalars. An $m \times n$ matrix will be represented by an array with m rows and n columns. The ijth entry of the matrix is located in the ith row and jth column of the array. Here, rows go across the page (as does writing in English), and columns go up and down (like the columns on a building). We denote matrices by capital letters A, B, C,.... If A is an $m \times n$ matrix, we usually denote the ijth entry (that is, the entry in the ith row and jth column) by a_{ij}. In other words we denote $A(i, j)$ by a_{ij}. This is symbolized by writing $A = [a_{ij}]$ or $A = (a_{ij})$. For us, it makes no difference whether brackets or parentheses are used (it is only a matter of style).

Two matrices A and B are *equal* if they are the same function. Thus, two matrices A and B are equal if and only if they have the same number of rows, the same number of columns, and precisely the same entries in the same places.

For example, the 3×2 matrix represented as

$$A = \begin{pmatrix} 4 & 5 \\ 2 & 1 \\ 6 & 3 \end{pmatrix}$$

has three rows and two columns. When we write $A = (a_{ij})$, we are specifying the six real numbers $A(1, 1) = a_{11} = 4$, $A(1, 2) = a_{12} = 5$, $A(2, 1) = a_{21} = 2$, $A(2, 2) = a_{22} = 1$, $A(3, 1) = a_{31} = 6$, and $A(3, 2) = a_{32} = 3$. Whenever specifying the size of a matrix (or describing the location of an entry in a matrix), the number of rows (or row number) is listed first, and the number of columns (or column number) is listed second.

An $n \times 1$ matrix is known as a *column matrix* or *column vector* (since it consists of one column). Similarly, a $1 \times n$ matrix is known as a *row matrix* or *row vector*. Technically, there is a distinction between a row matrix like $(\begin{array}{cccc} a_1 & a_2 & \cdots & a_n \end{array})$ and the n-tuple $(a_1, a_2, \ldots, a_n) \in \mathbf{F}^n$. However, for us it will be both convenient and useful to ignore this distinction. Although we will usually use uppercase letters to denote matrices, at times we shall use lowercase letters $v, w, b, \ldots$ to denote column vectors. We shall use the symbol $\vec{0}$ to denote the column vector (of appropriate size) all of whose entries are 0.

If the scalar entries in the matrix A are all real numbers, that is, from the scalar field $\mathbf{R}$, then we say that A is a *real matrix*. In general, if the scalar entries of A lie in a specified field $\mathbf{F}$, then we say that A is a *matrix over* $\mathbf{F}$ or an $\mathbf{F}$ *matrix*. Usually, we will not mention the field and simply call A a matrix. In such cases, the field is immaterial and the results are true for all fields.

We now introduce the basic operations of matrix algebra. In the definition below we use the (formal) definition of matrices as functions. The utility of this notation will become apparent later in the proofs of the theorems of this section. After these results have been carefully proved, this notation will be dropped, and we will simply refer to the matrix with ijth entry a_{ij} as (a_{ij}).

Definition 0.3.2 (i) If $A = (a_{ij})$ is an $m \times n$ matrix and $k \in \mathbf{F}$ is a scalar, then we define the $m \times n$ matrix kA by $(kA)(i,j) = k(A(i,j))$. That is, the ijth entry of kA is ka_{ij}, where a_{ij} is the ijth entry of A. This multiplication by a scalar is called *scalar multiplication*.
(ii) If $A = (a_{ij})$ and $B = (b_{ij})$ are both $m \times n$ matrices, then we define the $m \times n$ matrix $A + B$ by $(A + B)(i,j) = A(i,j) + B(i,j)$. Thus to add two matrices of the same shape, one simply adds the corresponding entries of the matrices to obtain another matrix of the same shape.

As examples of these operations we see that

$$3 \begin{pmatrix} 1 & 1 & 1 \\ 2 & 3 & 4 \\ 0 & 1 & 9 \end{pmatrix} = \begin{pmatrix} 3 & 3 & 3 \\ 6 & 9 & 12 \\ 0 & 3 & 27 \end{pmatrix}$$

and

$$\begin{pmatrix} 2 & 1 \\ 3 & 3 \\ 2 & -1 \end{pmatrix} + \begin{pmatrix} 0 & 6 \\ 2 & 0 \\ 3 & 3 \end{pmatrix} = \begin{pmatrix} 2 & 7 \\ 5 & 3 \\ 5 & 2 \end{pmatrix}$$

Since scalar multiplication and matrix addition come from the usual multiplication and addition of scalars applied entrywise to the entries of the matrices, all of the familiar laws of algebra involving the real numbers (which are true for all fields of scalars) are also true for these two operations. This is summarized in the following:

Theorem 0.3.3 *Suppose that* A, B, C *are* $m \times n$ *matrices and* $k, r, s \in$ **F**. *Then all matrices listed below are* $m \times n$ *and*
(i) $A + B = B + A$.
(ii) $((A + B) + C) = (A + (B + C))$.
(iii) $k(A + B) = kA + kB$.
(iv) $(r + s)A = rA + sA$.
(v) $r(sA) = (rs)A$.

Proof: That every matrix is $m \times n$ is an immediate consequence of the definitions of scalar multiplication and matrix addition. We must show in each case that the corresponding entries of the matrices on both sides of the equality sign are the same.

(i) $(A + B)(i, j) = A(i, j) + B(i, j) = B(i, j) + A(i, j) = (B + A)(i, j)$ by the commutativity of addition in $\mathbf{F}$.

(ii) $((A+B)+C)(i, j) = (A+B)(i, j) + C(i, j) = (A(i, j) + B(i, j)) + C(i, j) = A(i, j) + (B(i, j) + C(i, j)) = A(i, j) + (B + C)(i, j) = (A + (B + C))(i, j)$ by the associativity of addition in $\mathbf{F}$.

(iii) $k(A + B)(i, j) = k(A(i, j) + B(i, j)) = kA(i, j) + kB(i, j) = (kA + kB)(i, j)$ by the distributive law in $\mathbf{F}$.

(iv) $((r + s)A)(i, j) = (r + s)A(i, j) = rA(i, j) + sA(i, j) = (rA + sA)(i, j)$ by the distributive law in $\mathbf{F}$.

(v) $(r(sA))(i, j) = r(sA(i, j)) = (rs)A(i, j) = ((rs)A)(i, j)$ by the associativity of multiplication in $\mathbf{F}$. This proves the theorem. $//$

We next give the definition of matrix multiplication. Unlike the definitions of addition and scalar multiplication for matrices, the definition of matrix multiplication does not (at first) seem like the "natural" thing to do. However, as we will see shortly, there are many good reasons why matrix multiplication is defined the way it is. A crucial feature to note in the next definition is the relationship between the sizes of the matrices that are being multiplied. In particular, it is *not* possible to multiply two arbitrary matrices.

Definition 0.3.4 Suppose that A is an $m \times n$ matrix and B is an $n \times p$ matrix. (Thus the number of columns of A is the same as the number of rows of B.) We define the product AB to be the $m \times p$ matrix given by

$$AB(i, k) = \sum_{j=1}^{n} A(i, j)B(j, k)$$

That is, AB is the $m \times p$ matrix whose ith row kth column entry is obtained by adding the products of the corresponding entries of the ith row of A and the kth column of B. (Note that each row of A and each column of B have precisely n entries.)

The process of matrix multiplication is illustrated in the following examples. If you have not seen matrix multiplication previously, study these examples carefully and try a number of multiplications yourself.

Again we emphasize that the matrix product AB is not defined unless the number of columns of A is the same as the number of rows of B.

Examples 0.3.5

(i) $\begin{pmatrix} 1 & 4 & 3 \\ 2 & 5 & 1 \end{pmatrix} \begin{pmatrix} 8 & 9 \\ 2 & 6 \\ 0 & 7 \end{pmatrix}$

$= \begin{pmatrix} (1 \cdot 8) + (4 \cdot 2) + (3 \cdot 0) & (1 \cdot 9) + (4 \cdot 6) + (3 \cdot 7) \\ (2 \cdot 8) + (5 \cdot 2) + (1 \cdot 0) & (2 \cdot 9) + (5 \cdot 6) + (1 \cdot 7) \end{pmatrix} = \begin{pmatrix} 16 & 54 \\ 26 & 55 \end{pmatrix}$

(ii) $\begin{pmatrix} 1 & 2 \\ 3 & 4 \end{pmatrix} \begin{pmatrix} 5 & 6 \\ 7 & 8 \end{pmatrix} = \begin{pmatrix} 5 + 14 & 6 + 16 \\ 15 + 28 & 18 + 32 \end{pmatrix} = \begin{pmatrix} 19 & 22 \\ 43 & 50 \end{pmatrix}$

(iii) $\begin{pmatrix} 5 & 6 \\ 7 & 8 \end{pmatrix} \begin{pmatrix} 1 & 2 \\ 3 & 4 \end{pmatrix} = \begin{pmatrix} 23 & 34 \\ 31 & 46 \end{pmatrix}$

(iv) $\begin{pmatrix} 1 & 2 \\ 3 & 4 \\ 2 & 2 \end{pmatrix} \begin{pmatrix} 1 & 0 & 0 \\ 0 & 1 & 0 \\ 0 & 0 & 1 \end{pmatrix}$ is *not* defined.

(v) $\begin{pmatrix} 1 & 0 & 0 \\ 0 & 1 & 0 \\ 0 & 0 & 1 \end{pmatrix} \begin{pmatrix} 1 & 2 \\ 3 & 4 \\ 2 & 2 \end{pmatrix} = \begin{pmatrix} 1 & 2 \\ 3 & 4 \\ 2 & 2 \end{pmatrix}$

(vi) $(12 \quad 5)(2) = (12 \cdot 2 + 5 \cdot 4) = (44)$

(vii) $\begin{pmatrix} 2 \\ 4 \end{pmatrix} (12 \quad 5) = \begin{pmatrix} 24 & 10 \\ 48 & 20 \end{pmatrix}$

(viii) $\begin{pmatrix} 1 & 0 \\ 0 & 1 \end{pmatrix} \begin{pmatrix} 4 & 7 & 8 \\ 1 & 0 & 3 \end{pmatrix} = \begin{pmatrix} 4 & 7 & 8 \\ 1 & 0 & 3 \end{pmatrix}$

(ix) $\begin{pmatrix} 17 & 3 \\ 22 & 4 \end{pmatrix} \begin{pmatrix} 1 & 0 \\ 0 & 1 \end{pmatrix} = \begin{pmatrix} 17 & 3 \\ 22 & 4 \end{pmatrix} = \begin{pmatrix} 1 & 0 \\ 0 & 1 \end{pmatrix} \begin{pmatrix} 17 & 3 \\ 22 & 4 \end{pmatrix}$

Note by the above examples (ii) and (iii) and also by examples (vi) and (vii) that matrix multiplication is not commutative. That is, for any two matrices A and B, AB need not be the same as BA. (In fact, they need not even be the same size, and as illustrated in (iv) and (v), one may be defined while the other is not.)

Examples (v), (viii), and (ix) illustrate an important matrix.

Definition 0.3.6 The $n \times n$ matrix with 1s on the diagonal and 0s elsewhere is called the $n \times n$ *identity matrix*. It is denoted

$$I_n = \begin{pmatrix} 1 & 0 & 0 & \cdots & 0 \\ 0 & 1 & 0 & \cdots & 0 \\ 0 & 0 & 1 & \cdots & 0 \\ \vdots & \vdots & \vdots & \ddots & \vdots \\ 0 & 0 & 0 & \cdots & 1 \end{pmatrix}$$

In other words, $I_n = (a_{ij})$ where $a_{ij} = 1$ for $i = j$ and $a_{ij} = 0$ otherwise.

Suppose that A is an $m \times n$ matrix and C is an $n \times p$ matrix. Then we show below that $AI_n = A$ and $I_nC = C$. This is exactly what happened in (v), (viii), and (ix) above. Thus the identity matrix I_n plays the same role in matrix multiplication as does $1 \in \mathbf{R}$ in the multiplication of real numbers.

There are a number of important algebraic laws involving matrix multiplication. We continue the list given above.

Theorem 0.3.3 (Continued) *Suppose that A and B are $m \times n$ matrices, C and D are $n \times p$ matrices, and $r \in \mathbf{F}$. Then*
(vi) $r(AD) = (rA)D$.
(vii) $r(AD) = A(rD)$.
(viii) $(A + B)D = AD + BD$.
(ix) $A(C + D) = AC + AD$.
(x) $AI_n = A$ *and* $I_nC = C$.

Proof: (vi) and (vii) We compute

$$r(AD)(i,k) = r(\sum_{j=1}^{n} A(i,j)D(j,k))$$

$$= \sum_{j=1}^{n} rA(i,j)D(j,k) = ((rA)D)(i,k)$$

$$= \sum_{j=1}^{n} A(i,j)rB(i,j) = (A(rD))(i,k)$$

by the distributive, associative, and commutative laws for $\mathbf{F}$. This gives (vi) and (vii).

(viii) Here,

$$
\begin{aligned}
((A+B)D)(i,k) &= \sum_{j=1}^{n}(A+B)(i,j)D(j,k) \\
&= \sum_{j=1}^{n}(A(i,j)+B(i,j))D(j,k) \\
&= \sum_{j=1}^{n}A(i,j)D(j,k) + \sum_{j=1}^{n}B(i,j)D(j,k) \\
&= (AD)(i,k) + (BD)(i,k) \;=\; (AD+BD)(i,k)
\end{aligned}
$$

(ix) In this case

$$
\begin{aligned}
(A(C+D))(i,k) &= \sum_{j=1}^{n}(A)(i,j)(C+D)(j,k) \\
&= \sum_{j=1}^{n}A(i,j)(C(j,k)+D(j,k)) \\
&= \sum_{j=1}^{n}A(i,j)C(j,k) + \sum_{j=1}^{n}A(i,j)D(j,k) \\
&= (AC)(i,k) + (AD)(i,k) \;=\; (AC+AD)(i,k)
\end{aligned}
$$

(x) For any i and k,

$$
\begin{aligned}
(I_nC)(i,k) &= I_n(i,1)C(1,k) + I_n(i,2)C(2,k) + \cdots + I_n(i,n)C(n,k) \\
&= C(i,k)
\end{aligned}
$$

since $I_n(i,j) = 0$ for $i \neq j$ and $I_n(i,i) = 1$. Consequently, $I_nC = C$. That $AI_n = A$ is proved by the analogous argument. This proves the theorem. //

By far, the most important algebraic law about matrix multiplication is the associative law. We saved this for a separate theorem.

Theorem 0.3.7 *If A is an $m \times n$ matrix, B is an $n \times p$ matrix, and C is a $p \times q$ matrix, then $(AB)C = A(BC)$.*

Proof: Using the associative and distributive laws of $\mathbf{F}$, we compute the *ir*th entry of $(AB)C$:

$$
\begin{aligned}
((AB)C)(i,r) &= \sum_{k=1}^{p}(AB)(i,k)C(k,r) \\
&= \sum_{k=1}^{p}[\sum_{j=1}^{n}A(i,j)B(j,k)]C(k,r) \\
&= \sum_{k=1}^{p}\sum_{j=1}^{n}A(i,j)B(j,k)C(k,r) \\
&= \sum_{j=1}^{n}\sum_{k=1}^{p}A(i,j)B(j,k)C(k,r) \\
&= \sum_{j=1}^{n}A(i,j)[\sum_{k=1}^{p}B(j,k)C(k,r)] \\
&= \sum_{j=1}^{n}A(i,j)(BC)(j,r) \\
&= (A(BC))(i,r)
\end{aligned}
$$

This proves the theorem. //

Two important types of matrices are the triangular and diagonal matrices.

Definition 0.3.8 Suppose A is an $n \times n$ (square) matrix. A is called *upper triangular* if $A(i,j) = 0$ whenever $i > j$. A is called *lower triangular* if $A(i,j) = 0$ whenever $i < j$. A is called *diagonal* if $A(i,j) = 0$ whenever $i \neq j$, that is, whenever A is both upper and lower triangular.

As the name suggests, upper-triangular matrices are those matrices whose nonzero entries are contained in the "triangle" that is above the diagonal. For example,

$$
\begin{pmatrix} 1 & 3 & 2 \\ 0 & 2 & 1 \\ 0 & 0 & 3 \end{pmatrix} \quad \text{and} \quad \begin{pmatrix} 4 & 0 & 0 \\ 0 & 0 & 0 \\ 0 & 0 & 2 \end{pmatrix}
$$

are both upper triangular. The second matrix is also lower triangular, so it is diagonal. The following properties of triangular and diagonal matrices will be useful in subsequent sections.

Lemma 0.3.9 (i)   *Suppose that A and B are both upper triangular $n \times n$ matrices. Then AB is an upper triangular matrix. If A and B are both lower triangular, then so is AB.*

(ii)   *If A and B are both diagonal $n \times n$ matrices, then $AB = BA$.*

Proof: (i) Suppose both A and B are upper triangular. Then $A(i, j) = 0$ whenever $i > j$ and $B(j, k) = 0$ whenever $j > k$. Suppose $i > k$. Then $(AB)(i, k) = A(i, 1)B(1, k) + \cdots + A(i, i - 1)B(i - 1, k) + A(i, i)B(i, k) + \cdots + A(i, n)B(n, k) = 0 \cdot B(1, k) + \cdots + 0 \cdot B(i - 1, k) + A(i, i) \cdot 0 + \cdots + A(i, n) \cdot 0 = 0$. Hence AB is upper triangular. The proof in the lower triangular case is the same. smallskip

(ii) By part (i), AB and BA are both upper and lower triangular, so they are diagonal. Therefore it suffices to show that AB and BA have the same diagonal entries. We find $(AB)(i, i) = A(i, 1)B(1, i) + \cdots + A(i, i - 1)B(i - 1, i) + A(i, i)B(i, i) + \cdots + A(i, n)B(n, i) = 0 \cdot 0 + \cdots + 0 \cdot 0 + A(i, i)B(i, i) + 0 \cdot 0 + \cdots + 0 \cdot 0 = A(i, i)B(i, i)$. By symmetry $(BA)(i, i) = B(i, i)A(i, i)$. We conclude that $AB = BA$. This proves the lemma. //

Digression 0.3.10 The Chain Rule for Partial Derivatives

In the first term of calculus the chain rule for the derivative of a composite function plays an important role. If $f, g : \mathbf{R} \to \mathbf{R}$ are functions with continuous derivatives, then the derivative of the composite function $f \circ g : \mathbf{R} \to \mathbf{R}$ [defined by $f \circ g(X) = f(g(X))$] has the nice expression $(f \circ g)'(x) = f'(g(X))g'(X)$. Using matrix multiplication, this convenient formula can be extended to functions of several variables.

Suppose that $F : \mathbf{R}^p \to \mathbf{R}^q$ and $G : \mathbf{R}^s \to \mathbf{R}^p$ are functions with continuous partial derivatives. We associate to F a $q \times p$ matrix of real functions, F', called the *Jacobian* of F, as follows: F can be viewed as a sequence of functions $F = (f_1, \ldots, f_q)$ where $f_i : \mathbf{R}^p \to \mathbf{R}$ is the ith coordinate function of F. In other words, for each $v \in \mathbf{R}^p$, $F(v) = (f_1(v), \ldots, f_q(v)) \in \mathbf{R}^q$. The matrix F' is then defined to be the $q \times p$ matrix whose ijth entry is the partial derivative $\partial f_i / \partial x_j$. Thus, for any $v \in \mathbf{R}^p$

$$F'(v) = (a_{ij}) \quad \text{where} \quad a_{ij} = \frac{\partial f_i}{\partial x_j}(v)$$

Similarly, we have the $p \times s$ matrix G' and the $q \times s$ matrix $(F \circ G)'$. If $G(u) = (g_1(u), g_2(u), \ldots, g_s(u))$, we define $F'(G)$ to be the $q \times p$ matrix obtained from F' by substituting $g_k(u)$ for x_k in each entry function of F'. The chain rule now reads

$$(F \circ G)' = F'(G)G'$$

where the multiplication is matrix multiplication. Note that as $F'(G)$ is a $q \times p$ matrix and G' is a $p \times s$ matrix, this multiplication makes sense, and it results in a $q \times s$ matrix, as $(F\ G)'$ should be.

We illustrate this with the values $p = 3$, $q = 1$, $s = 2$. Then $F = F(X, Y, Z) : \mathbf{R}^3 \to \mathbf{R}$, and $G = (e(U, V), g(U, V), h(U, V)) : \mathbf{R}^2 \to \mathbf{R}^3$. The Jacobian matrices are

$$F' = (\ \partial F/\partial X \quad \partial F/\partial Y \quad \partial F/\partial Z\ )$$

and

$$G' = \begin{pmatrix} \partial e/\partial U & \partial e/\partial V \\ \partial g/\partial U & \partial g/\partial V \\ \partial h/\partial U & \partial h/\partial V \end{pmatrix}$$

We find by the chain rule

$$
\begin{aligned}
(F \circ G)' &= (\ \partial(F \circ G)/\partial U \quad \partial(F \circ G)/\partial V\ ) \\
&= (\ \partial F/\partial X \quad \partial F/\partial Y \quad \partial F/\partial Z\ ) \begin{pmatrix} \partial e/\partial U & \partial e/\partial V \\ \partial g/\partial U & \partial g/\partial V \\ \partial h/\partial U & \partial h/\partial V \end{pmatrix} \\
&= (\partial F/\partial X \cdot \partial e/\partial U + \partial F/\partial Y \cdot \partial g/\partial U + \partial F/\partial Z \cdot \partial h/\partial U \\
&\qquad \partial F/\partial X \cdot \partial e/\partial V + \partial F/\partial Y \cdot \partial g/\partial V + \partial F/\partial Z \cdot \partial h/\partial V)
\end{aligned}
$$

We emphasize that all the expressions $\partial F/\partial X$ in $(F \circ G)'$ are shorthand for the partial derivative with the e, g, h substituted for X, Y, Z. For example, expanding out, our notation above really says that

$$
\begin{aligned}
\frac{\partial(F \circ G)}{\partial U}(U, V) &= \frac{\partial F}{\partial X}(e, g, h)\frac{\partial e}{\partial U}(U, V) + \frac{\partial F}{\partial Y}(e, g, h)\frac{\partial g}{\partial U}(U, V) \\
&\quad + \frac{\partial F}{\partial Z}(e, g, h)\frac{\partial h}{\partial U}(U, V)
\end{aligned}
$$

Often, when multiplying large matrices, it is useful to view the multiplication in terms of smaller component matrices which compose the larger matrices. This viewpoint is particularly useful when the larger matrices have blocks which consist entirely of zeros. To make this precise, we need the notion of a *partitioned matrix*. Consider the matrix

$$\begin{pmatrix} 1 & 9 & 7 & 0 & 0 \\ 2 & 4 & 5 & 0 & 0 \\ 0 & 0 & 0 & 3 & 3 \end{pmatrix}$$

It is convenient to think of A as being built from four smaller matrices. To illustrate we add lines and write

$$\left(\begin{array}{ccc|cc} 1 & 9 & 7 & 0 & 0 \\ 2 & 4 & 5 & 0 & 0 \\ \hline 0 & 0 & 0 & 3 & 3 \end{array}\right)$$

Thus, A can be thought of as the matrix built from the four matrices

$$A_{11} = \begin{pmatrix} 1 & 9 & 7 \\ 2 & 4 & 5 \end{pmatrix} \qquad A_{12} = \begin{pmatrix} 0 & 0 \\ 0 & 0 \end{pmatrix}$$
$$A_{21} = \begin{pmatrix} 0 & 0 & 0 \end{pmatrix} \quad \text{and} \quad A_{22} = \begin{pmatrix} 3 & 3 \end{pmatrix}$$

This is symbolized by writing

$$A = \begin{pmatrix} A_{11} & A_{12} \\ A_{21} & A_{22} \end{pmatrix}$$

In general, grouping together collections of adjacent rows and columns, one can partition a matrix into smaller matrices in various ways. (The partition with the largest number of submatrices is the one where each submatrix is 1×1!) We give this explicitly in the next definition. The formal definition of partitioning a matrix may strike the reader as being quite longwinded, compared with the simplicity of the idea it is defining. This sometimes happens in mathematics and is necessary for precision and proving theorems. It is good practice for students to work through such formal definitions and make sure that the definition describes what they (informally) understand.

Definition 0.3.11 Suppose that A is an $m \times n$ matrix. Let $r_0, \ldots, r_u$ and $s_0, \ldots, s_v$ be integers with $0 = r_0 < r_1 < r_2 < \cdots < r_u = m$ and $0 = s_0 < s_1 < s_2 < \cdots < s_v = n$, where $u, v \geq 1$. For i and j with $1 \leq i \leq u$ and $1 \leq j \leq v$, we set A_{ij} to be the $(r_i - r_{i-1}) \times (s_j - s_{j-1})$ matrix defined by $A_{ij}(k, l) = A(r_{i-1} + k, s_{j-1} + l)$ for all $1 \leq k \leq (r_i - r_{i-1})$ and $1 \leq l \leq (s_j - s_{j-1})$. We say that A has been *partitioned* by the sequences $r_0, \ldots, r_u$ and $s_0, \ldots, s_v$ and denote this by writing

$A = (A_{ij})$. The matrices A_{ij} (which are *submatrices* of A) are called the *blocks* of the partition. Evidently, the partition has uv blocks, and we say that A has been partitioned into a $u \times v$ *block matrix*.

Note that the preceding definition does NOT allow one to break up the matrix A above as

$$\begin{pmatrix} 1 & 9 & 7 & 0 & 0 \\ 2 & 4 & 5 & 0 & 0 \\ \hline 0 & 0 & 0 & 3 & 3 \end{pmatrix}$$

For, when drawn in this fashion, the upper left-hand block has more columns than the lower left-hand block. The reason the definition is restricted in this way is so that we can prove the next theorem. The result shows that when appropriately set up, the multiplication of partitioned matrices can be performed in a manner analogous to the usual multiplication of matrices. In this multiplication one multiplies and adds blocks as if they were scalars.

Theorem 0.3.12 (Block Multiplication) *Suppose that A is an $m \times n$ matrix, B is a $n \times q$ matrix, A has been partitioned into a $u \times v$ block matrix, and B has been partitioned into a $v \times w$ block matrix. Furthermore, assume that the sequence $0 = s_0 < s_1 < \cdots < s_v = n$ used in partitioning A is the same sequence used in partitioning B. Then for each i, j, k the product $A_{ij}B_{jk}$ makes sense. Moreover the product matrix AB can be viewed as a $u \times w$ block matrix, whose ik th block is the sum $\sum_{j=1}^{v} A_{ij}B_{jk}$.*

Before giving the proof, it is best to get the idea by studying an example. Consider the partitioned matrices

$$A = \begin{pmatrix} 1 & 9 & 7 & 0 & 0 \\ 2 & 4 & 5 & 0 & 0 \\ \hline 0 & 0 & 0 & 3 & 3 \end{pmatrix}$$

and

$$B = \begin{pmatrix} 1 & -1 & 7 \\ 0 & 1 & 1 \\ 1 & 2 & 0 \\ \hline 0 & 0 & 1 \\ 0 & 0 & 3 \end{pmatrix}$$

Then Theorem 0.3.12 gives that

$$AB = \begin{pmatrix} A_{11} & A_{12} \\ A_{21} & A_{22} \end{pmatrix} \begin{pmatrix} B_{11} & B_{12} \\ B_{21} & B_{22} \end{pmatrix}$$
$$= \begin{pmatrix} A_{11}B_{11} + A_{12}B_{21} & A_{11}B_{21} + A_{12}B_{22} \\ A_{21}B_{11} + A_{22}B_{21} & A_{21}B_{12} + A_{22}B_{22} \end{pmatrix}$$

We have that

$$A_{11}B_{11} + A_{12}B_{21} = \begin{pmatrix} 8 & 22 \\ 7 & 12 \end{pmatrix}$$
$$A_{11}B_{12} + A_{12}B_{22} = \begin{pmatrix} 16 \\ 6 \end{pmatrix}$$
$$A_{21}B_{11} + A_{22}B_{21} = \begin{pmatrix} 0 & 0 \end{pmatrix}$$
$$A_{21}B_{12} + A_{22}B_{22} = \begin{pmatrix} 12 \end{pmatrix}$$

Hence,

$$AB = \left(\begin{array}{cc|c} 8 & 22 & 16 \\ 7 & 12 & 6 \\ \hline 0 & 0 & 12 \end{array} \right)$$

Note for example, that if one computed the 11 entry of AB directly, one would compute the sum $(1 \cdot 1 + 9 \cdot 0 + 7 \cdot 1) + (0 \cdot 0 + 0 \cdot 0)$. The parentheses indicate how the sum decomposes according to the block multiplication. The utility of partitioning matrices and using block multiplication will become apparent later when we study the theory of linear transformations. Here is the proof of Theorem 0.3.12. The notation makes is a bit cumbersome, but it is nice to know that it can be carried out.

Proof: We assume that the sequences $0 = r_0 < r_1 < \cdots < r_u = m$ and $0 = s_0 < s_1 < \cdots < s_v = n$ are used to partition A. We assume that the sequences $0 = s_0 < s_1 < \cdots < s_v = n$ and $0 = t_0 < t_1 < \cdots < t_w = q$ are used to partition B. Note that the matrix A_{ij} has $s_j - s_{j-1}$ columns and the matrix B_{jk} has $s_j - s_{j-1}$ rows. Hence the product $A_{ij}B_{jk}$ makes sense. Moreover, each product $A_{ij}B_{jk}$ is a $(r_i - r_{i-1}) \times (t_k - t_{k-1})$ matrix. Thus the sum $\sum_{j=1}^{n} A_{ij}B_{jk}$ makes sense.

According to the definition of matrix multiplication, $AB(e, g) = \sum_{f=1}^{n} A(e, f) B(f, g)$. Using the definition of partitioning of matrices, there exist numbers h and h' for which $r_{h-1} < e \leq r_h$ and $t_{h'-1} < g \leq t_{h'}$. It follows from the definitions that

$$A_{hj}B_{jh'}(e - r_{h-1}, g - t_{h'-1})$$

$$= \sum_{p=1}^{r_h - r_{h-1}} A_{hj}(e - r_{h-1}, p)B_{jh'}(p, g - t_{h'-1})$$

$$= \sum_{f=s_{j-1}+1}^{s_j} A(e, f)B(f, g)$$

Consequently, the sum

$$\sum_{j=1}^{v} A_{hj}B_{jh'}(e - r_{h-1}, g - t_{h'-1}) \;=\; \sum_{j=1}^{v}[\sum_{f=s_{j-1}+1}^{s_j} A(e, f)B(f, g)]$$

$$= \sum_{f=1}^{n} A(e, f)B(f, g) = AB(e, g)$$

This proves the theorem. $//$

PROBLEMS

0.3.1.   Let

$$A = \begin{pmatrix} 1 & 2 & 3 \\ 1 & 2 & 3 \\ 1 & 2 & 3 \end{pmatrix} \qquad B = \begin{pmatrix} 2 & 4 & 6 \\ 2 & 2 & 2 \\ 5 & 4 & 3 \end{pmatrix} \qquad C = \begin{pmatrix} 1 & 3 \\ 2 & 3 \\ 0 & 0 \end{pmatrix}$$

$$D = \begin{pmatrix} 1 & 3 & 5 & 3 \\ 2 & 4 & 5 & 1 \\ 1 & 1 & 0 & 2 \end{pmatrix} \qquad E = \begin{pmatrix} 1 & 2 & 4 \\ 6 & 0 & 7 \end{pmatrix}$$

Find the following matrices *if* they make sense. If they do not make sense say "nonsense".
(a) A+B   (b) B+C   (c) AB   (d) BA   (e) AC   (f) CA
(g) EC   (h) CE   (i) AD   (j) ED   (k) AE   (l) AA
(m) CC   (n) 3CD   (o) B(CE) (p) (CE)B

0.3.2.   As you would expect, A^n denotes the product of A with itself n times. Suppose that

$$A = \begin{pmatrix} 0 & 1 & 0 & 0 \\ 0 & 0 & 1 & 0 \\ 0 & 0 & 0 & 1 \\ 0 & 0 & 0 & 0 \end{pmatrix}$$

Find A^n for all $n > 1$.

0.3.3. If A and B are 2×2 matrices such that

$$AB - BA = \begin{pmatrix} a & b \\ c & d \end{pmatrix}$$

show that $a + d = 0$.

0.3.4. Find all real 2×2 real matrices X and Y for which

(a) $X + Y = \begin{pmatrix} 1 & 0 \\ 1 & 0 \end{pmatrix}$ and $X - Y = \begin{pmatrix} 0 & 1 \\ 0 & 0 \end{pmatrix}$

(b) $X^2 + Y^2 = \begin{pmatrix} 0 & 0 \\ 0 & 0 \end{pmatrix}$

0.3.5. Suppose that A is an $n \times m$ matrix such that $AB = 0$ for all $m \times 1$ matrices B. (Here 0 denotes the matrix of the appropriate size with all entries zero.) Show that $A = 0$.

0.3.6. Suppose that a real 2×2 matrix A commutes with all other 2×2 matrices (that is, $AB = BA$ for all 2×2 matrices B). Show that

$$A = \begin{pmatrix} a & 0 \\ 0 & a \end{pmatrix} \quad \text{for some } a \in \mathbf{F}$$

0.3.7. Find six different 2×2 real matrices A for which

$$A^2 = \begin{pmatrix} 1 & 0 \\ 0 & 1 \end{pmatrix}$$

0.3.8. Consider the complex matrices

$$\begin{pmatrix} 0 & -1 \\ 1 & 0 \end{pmatrix} \quad \text{and} \quad \begin{pmatrix} 0 & i \\ i & 0 \end{pmatrix}$$

Show, by taking succesive products of these matrices, that a collection of eight matrices is "generated" and that this collection is closed under matrix multiplication.

0.3.9. Find all matrices M such that

$$\begin{pmatrix} 1 & 1 \\ 2 & 2 \end{pmatrix} M = \begin{pmatrix} 0 & 2 & 0 \\ 0 & 4 & 0 \end{pmatrix}$$

0.3.10. Let

$$G = \left\{ \begin{pmatrix} a & b \\ -b & a \end{pmatrix} \mid a, b \in \mathbf{R} \right\}$$

(a) Show that whenever $A, B \in G$, $AB = BA$.

(b) Show that whenever $A, B \in G$, $AB \in G$ also.

(c) Think about the matrices of the form

$$\begin{pmatrix} a & 0 \\ 0 & a \end{pmatrix} \quad \text{and} \quad \begin{pmatrix} 0 & b \\ -b & 0 \end{pmatrix}$$

in G. Explain how the set of matrices in G resemble the set of complex numbers $\mathbf{C}$.

0.3.11. Suppose that the $n \times n$ matrix $T = (t_{ij})$ is upper triangular. If $T^n = 0$, show that $t_{ii} = 0$ for all i. Is the converse true?

0.3.12. Generalize Prob. 0.3.6. Which $n \times n$ matrices A satisfy $AB = BA$ for all $n \times n$ matrices B?

0.3.13. Consider the block matrix

$$B = \left(\begin{array}{cc|c} 1 & 3 & 0 \\ 2 & 2 & 0 \\ \hline 0 & 0 & 3 \end{array} \right)$$

Compute B^2, B^3, and B^4 using block multiplication.

0.3.14. Consider the block matrix

$$C = \left(\begin{array}{cc|cc|c} 0 & 1 & 0 & 0 & 0 \\ 1 & 0 & 2 & 2 & 0 \\ \hline 0 & 2 & 0 & 0 & 2 \end{array} \right)$$

Compute BC by block multiplication, where B is the matrix of Prob. 0.3.13.

0.4 Matrices and Systems of Equations

We are now ready to use matrix multiplication to represent systems of equations. The definition of matrix multiplication was designed so that a system of linear equations can be represented by a single matrix equation in a natural way. As an example, consider the following matrix equation:

$$\begin{pmatrix} 3 & 4 & 1 & 0 \\ 1 & 2 & 6 & 7 \end{pmatrix} \begin{pmatrix} X \\ Y \\ Z \\ W \end{pmatrix} = \begin{pmatrix} 5 \\ 0 \end{pmatrix}$$

Carrying out the matrix multiplication gives

$$
\begin{array}{rrrrrr}
3X & + & 4Y & + & Z & & & = & 5 \\
X & + & 2Y & + & 6Z & + & 7W & = & 0
\end{array}
$$

which is a system of two equations in four unknowns.

As an additional example, suppose that

$$
A = \begin{pmatrix} 2 & 3 \\ 1 & 1 \\ 0 & 0 \end{pmatrix} \quad \text{and} \quad C = \begin{pmatrix} 1 \\ 2 \\ 1 \end{pmatrix}
$$

Then, the matrix equation

$$
A \begin{pmatrix} X \\ Y \end{pmatrix} = C
$$

represents the system of equations

$$
\begin{array}{rrrrr}
2X & + & 3Y & = & 1 \\
X & + & Y & = & 2 \\
0X & + & 0Y & = & 1
\end{array}
$$

Note that this system of equations has no solutions. Consequently there is no 2×1 matrix B which solves the matrix equation $AB = C$.

More generally, suppose that A in an $m \times n$ matrix, $X_1, \ldots, X_n$ are variables; and C is a $m \times 1$ (column) matrix. Then the equation

$$
A \begin{pmatrix} X_1 \\ \vdots \\ X_n \end{pmatrix} = C
$$

represents a system of m equations in n unknowns. Usually we will abbreviate the column of variables by a single X, so that the matrix equation can be written compactly as $AX = C$. The rows of A give the coefficients of the variables of the equations of the system, and the entries of C give the right hand values of each equation. For this reason, the matrix A is called the *coefficient matrix* of the system, and C is called the *value matrix*. (The terminology "coefficient matrix" is reasonably standard, but the reader is not likely to find the term "value matrix" in other texts.)

The set of solutions to the system of equations represented by $AX = B$ can be viewed as the set of column vectors which solve the matrix equation. In other words, if $X_1 = r_1$, $X_2 = r_2, \ldots, X_n = r_n$ is a solution to the system of equations, then setting

$$\vec{r} = \begin{pmatrix} r_1 \\ \vdots \\ r_n \end{pmatrix}$$

we have $A\vec{r} = C$. This viewpoint will prove to be very convenient.

When studying a system of equations, it is easier to work only with a single matrix that describes the system. This is accomplished by writing down the so-called *augmented matrix* of the system. For a system of m equations in n unknowns $AX = C$, the augmented matrix of the system is the $m \times (n+1)$ matrix $(A \mid C)$ obtained from the $m \times n$ matrix A by adding C as a new $(n+1)$st column. In the two examples above, one obtains the augmented matrices

$$\begin{pmatrix} 3 & 4 & 1 & 0 & 5 \\ 1 & 2 & 6 & 7 & 0 \end{pmatrix} \quad \text{and} \quad \begin{pmatrix} 2 & 3 & 1 \\ 1 & 1 & 2 \\ 0 & 0 & 3 \end{pmatrix}$$

It is extremely important that students learn to make the mental transition between matrix equations and the associated system of linear equations. (The notation is set up to make this especially easy; that is, the coefficients of the linear system are exactly where they should be in the augmented matrix.) This equivalence of a matrix equation with its associated system of linear equations is more that a notational convenience; it provides an important viewpoint for understanding further uses of matrices. In the sequel we shall assume that this equivalence has been mastered by the reader. We shall omit the phrase "the system of equations represented by" and simply refer to a matrix equation $AX = C$ as a system of equations.

Row operations can be applied to matrices in exactly the same fashion as they are applied to systems of equations. Theorem 0.1.8 guarantees that if row operations are applied to an augmented matrix, the resulting augmented matrix will represent a system of equations with precisely the same solutions as the original matrix. As with systems of equations, the strategies of gaussian or Gauss-Jordan elimination are followed when row-reducing matrices.

As an example, we apply Gauss-Jordan elimination to the first augmented matrix above:

$$\begin{pmatrix} 3 & 4 & 1 & 0 & | & 5 \\ 1 & 2 & 6 & 7 & | & 0 \end{pmatrix} \longmapsto \begin{pmatrix} 1 & 4/3 & 1/3 & 0 & | & 5/3 \\ 1 & 2 & 6 & 7 & | & 0 \end{pmatrix}$$

$$\longmapsto \begin{pmatrix} 1 & 4/3 & 1/3 & 0 & | & 5/3 \\ 0 & 2/3 & 17/3 & 7 & | & -5/3 \end{pmatrix}$$

$$\longmapsto \begin{pmatrix} 1 & 4/3 & 1/3 & 0 & | & 5/3 \\ 0 & 1 & 17/2 & 21/2 & | & -5/2 \end{pmatrix}$$

$$\longmapsto \begin{pmatrix} 1 & 0 & -11 & -14 & | & 5 \\ 0 & 1 & 17/2 & 21/2 & | & -5/2 \end{pmatrix}$$

The final matrix of this sequence represents a system in reduced row-echelon form and equivalent to the original system. It follows that the set of solutions to the original system is

$$\{(5 + 11z + 14w, -5/2 - (17/2)z - (21/2)w, z, w) \mid z, w \in \mathbf{F}\}$$

This set of solutions can also be expressed in the matrix form

$$\begin{pmatrix} X \\ Y \\ Z \\ W \end{pmatrix} = \begin{pmatrix} 5 \\ -5/2 \\ 0 \\ 0 \end{pmatrix} + z \begin{pmatrix} 11 \\ -17/2 \\ 1 \\ 0 \end{pmatrix} + w \begin{pmatrix} 14 \\ -21/2 \\ 0 \\ 1 \end{pmatrix}$$

where the small z, w are interpreted as arbitrary scalars (parameters).

In the sequel we shall see that there are many different applications of matrices in addition to supplying the notation needed to represent systems of equations. In these applications it turns out to be extremely important to perform row operations on an arbitrary matrix, even when not considering an associated system of equations. The row operations that are allowed are exactly the same as those for systems of equations, but we repeat the definition for emphasis.

Definition 0.4.1 Given any matrix, an *elementary row operation* applied to the matrix is one of the following:
(i) Multiply a single row by a nonzero scalar.
(ii) Add a multiple of one row to another, leaving all the other rows unchanged.
(iii) Interchange two rows.

If the matrix B can be obtained from the matrix A by a finite sequence of the above operations, then A and B are said to be *row equivalent*.

Remark 0.4.2 The third operation of interchanging rows can be obtained from (i) and (ii). For, suppose that R_i and R_j are two rows of some matrix, and consider the following sequence of row operations (i) and (ii):

$$\begin{pmatrix} R_i \\ R_j \end{pmatrix} \mapsto \begin{pmatrix} R_i - R_j \\ R_j \end{pmatrix} \mapsto \begin{pmatrix} R_i - R_j \\ R_j + (R_i - R_j) \end{pmatrix} = \begin{pmatrix} R_i - R_j \\ R_i \end{pmatrix}$$
$$\mapsto \begin{pmatrix} -R_j \\ R_i \end{pmatrix} \mapsto \begin{pmatrix} R_j \\ R_i \end{pmatrix}$$

This sequence interchanges the rows R_i and R_j. Thus, if we had desired, we could have given only the operations (i) and (ii) as elementary row operations in Definition 0.4.1, and the notion of row equivalence would have remained unchanged. We shall later use this observation to simplify the computations involved in some proofs.

Exactly as with systems of equations, there are the notions of echelon forms and reduced echelon forms for matrices.

Definition 0.4.3 A matrix is said to be a *row-echelon matrix* if it satisfies (E0), (E1), and (E2) below. A row-echelon matrix is called a *reduced row-echelon matrix* if it additionally satisfies (RE3) below. We shall often drop the word "row" while using this terminology.
(E0)   All zero rows occur below all nonzero rows.
(E1)   The first nonzero entry in any nonzero row is 1. This entry is called the *leading entry* of that row.
(E2)   The leading entry of each nonzero row occurs to the right of the leading entry of each row above (if any).
(RE3)   The leading entry of each row is the only nonzero entry in its column.

Using the techniques of gaussian elimination and Gauss-Jordan elimination, it is always possible to find for any matrix A, a row equivalent row-echelon matrix and a (uniquely determined, see Theorem 0.5.5) row equivalent reduced row-echelon matrix. The number of nonzero rows in a row equivalent reduced row-echelon matrix is an extremely useful invariant of the original matrix. In view of this we give the following definition.

Definition 0.4.4 Suppose that A and B are matrices and B is row equivalent to A. If B is a row-echelon matrix, we say that B is a row-echelon form of A. If B is a reduced row-echelon matrix, we say that B is the reduced row-echelon form of A. For any matrix A we define the *rank* of A, denoted $rk(A)$, to be the number of nonzero rows in the reduced row-echelon form of A.

Caution The rank of a matrix A is the number of nonzero rows in the reduced row-echelon form of A. In general, it is *not* the number of nonzero rows of A. We also remark that the concept of rank is well-defined only after the uniqueness of the reduced row-echelon form has been established (see Theorem 0.5.5).

Example 0.4.5 Suppose

$$A = \begin{pmatrix} 1 & -1 & 3 & 7 \\ 1 & 0 & 6 & 5 \\ 0 & 2 & 6 & -4 \end{pmatrix}$$

Applying elementary row operations we obtain

$$A \mapsto \begin{pmatrix} 1 & -1 & 3 & 7 \\ 0 & 1 & 3 & -2 \\ 0 & 2 & 6 & -4 \end{pmatrix} \mapsto \begin{pmatrix} 1 & -1 & 3 & 7 \\ 0 & 1 & 3 & -2 \\ 0 & 0 & 0 & 0 \end{pmatrix}$$

$$\mapsto \begin{pmatrix} 1 & 0 & 6 & 5 \\ 0 & 1 & 3 & -2 \\ 0 & 0 & 0 & 0 \end{pmatrix}$$

We find that both

$$\begin{pmatrix} 1 & -1 & 3 & 7 \\ 0 & 1 & 3 & -2 \\ 0 & 0 & 0 & 0 \end{pmatrix} \quad \text{and} \quad \begin{pmatrix} 1 & 0 & 6 & 5 \\ 0 & 1 & 3 & -2 \\ 0 & 0 & 0 & 0 \end{pmatrix}$$

are echelon forms of A. The latter matrix is the reduced echelon form of A. It follows that $rk(A) = 2$.

Example 0.4.6 Let

$$B = \begin{pmatrix} 3 & 4 \\ 2 & 3 \end{pmatrix}$$

Applying Gauss-Jordan elimination to B, we have

$$\begin{pmatrix} 3 & 4 \\ 2 & 3 \end{pmatrix} \mapsto \begin{pmatrix} 1 & 4/3 \\ 2 & 3 \end{pmatrix} \mapsto \begin{pmatrix} 1 & 4/3 \\ 0 & 1/3 \end{pmatrix} \mapsto \begin{pmatrix} 1 & 4/3 \\ 0 & 1 \end{pmatrix} \mapsto \begin{pmatrix} 1 & 0 \\ 0 & 1 \end{pmatrix}$$

Either of the last two matrices are echelon forms of B. The last matrix is the reduced echelon form of B. It follows that $rk(B) = 2$. Consider the second to last matrix

$$\begin{pmatrix} 1 & 4/3 \\ 0 & 1 \end{pmatrix}$$

If the second row of this matrix is multiplied by k and added to the first row, the resulting matrix is

$$\begin{pmatrix} 1 & k + 4/3 \\ 0 & 1 \end{pmatrix}$$

This is an echelon matrix which is row equivalent to B. From this it follows that any matrix of the form

$$\begin{pmatrix} 1 & r \\ 0 & 1 \end{pmatrix}$$

where $r \in \mathbf{R}$ is an echelon form of B. In fact, since the rank of B is 2, these matrices are all the possible echelon forms of B.

Example 0.4.7 Consider the matrix

$$\begin{pmatrix} 1 & 2 & 1 & 1 & 5 \\ 2 & 4 & 0 & 0 & 6 \\ 1 & 2 & 0 & 1 & 3 \\ 0 & 0 & 1 & 1 & 2 \\ 0 & 0 & 2 & -1 & 4 \end{pmatrix}$$

Gauss-Jordan elimination, applied here one column at a time, gives the reduced row-echelon form:

$$\begin{pmatrix} 1 & 2 & 1 & 1 & 5 \\ 2 & 4 & 0 & 0 & 6 \\ 1 & 2 & 0 & 1 & 3 \\ 0 & 0 & 1 & 1 & 2 \\ 0 & 0 & 2 & -1 & 4 \end{pmatrix} \mapsto \begin{pmatrix} 1 & 2 & 1 & 1 & 5 \\ 0 & 0 & -2 & -2 & -4 \\ 0 & 0 & -1 & 0 & -2 \\ 0 & 0 & 1 & 1 & 2 \\ 0 & 0 & 2 & -1 & 4 \end{pmatrix}$$

$$\mapsto \begin{pmatrix} 1 & 2 & 0 & 0 & 3 \\ 0 & 0 & 1 & 1 & 2 \\ 0 & 0 & 0 & 1 & 0 \\ 0 & 0 & 0 & 0 & 0 \\ 0 & 0 & 0 & -3 & 0 \end{pmatrix}$$

$$\mapsto \begin{pmatrix} 1 & 2 & 0 & 0 & 3 \\ 0 & 0 & 1 & 0 & 2 \\ 0 & 0 & 0 & 1 & 0 \\ 0 & 0 & 0 & 0 & 0 \\ 0 & 0 & 0 & 0 & 0 \end{pmatrix}$$

This shows the original matrix had rank 3.

We conclude this section by analyzing elementary row operations in terms of matrix multiplication. The key concept is that of an *elementary matrix*. Each of the three basic row operations given in Definition 0.4.1 can be interpreted as an appropriate matrix multiplication.

Definition 0.4.8 An $n \times n$ *elementary matrix* is a matrix obtained from the $n \times n$ identity matrix I_n by applying a single elementary row operation.

It follows that there are three types of elementary matrices. In the 3×3 case the first operation gives those that look like

$$\begin{pmatrix} a & 0 & 0 \\ 0 & 1 & 0 \\ 0 & 0 & 1 \end{pmatrix} \quad \text{or} \quad \begin{pmatrix} 1 & 0 & 0 \\ 0 & a & 0 \\ 0 & 0 & 1 \end{pmatrix} \quad \text{or} \quad \begin{pmatrix} 1 & 0 & 0 \\ 0 & 1 & 0 \\ 0 & 0 & a \end{pmatrix}$$

where $a \in \mathbf{F}$ is a nonzero scalar. The second operation gives, for any $a \in \mathbf{F}$, the six matrices (in the 3×3 case)

$$\begin{pmatrix} 1 & a & 0 \\ 0 & 1 & 0 \\ 0 & 0 & 1 \end{pmatrix}, \quad \begin{pmatrix} 1 & 0 & a \\ 0 & 1 & 0 \\ 0 & 0 & 1 \end{pmatrix}, \quad \begin{pmatrix} 1 & 0 & 0 \\ 0 & 1 & a \\ 0 & 0 & 1 \end{pmatrix}$$

$$\begin{pmatrix} 1 & 0 & 0 \\ a & 1 & 0 \\ 0 & 0 & 1 \end{pmatrix}, \quad \begin{pmatrix} 1 & 0 & 0 \\ 0 & 1 & 0 \\ a & 0 & 1 \end{pmatrix}, \quad \begin{pmatrix} 1 & 0 & 0 \\ 0 & 1 & 0 \\ 0 & a & 1 \end{pmatrix}$$

Finally, the third operation gives the three 3×3 matrices

$$\begin{pmatrix} 0 & 1 & 0 \\ 1 & 0 & 0 \\ 0 & 0 & 1 \end{pmatrix}, \quad \begin{pmatrix} 0 & 0 & 1 \\ 0 & 1 & 0 \\ 1 & 0 & 0 \end{pmatrix}, \quad \begin{pmatrix} 1 & 0 & 0 \\ 0 & 0 & 1 \\ 0 & 1 & 0 \end{pmatrix}$$

The importance of elementary matrices comes from the following lemma, which shows that row operations correspond to matrix multiplication by elementary matrices.

Lemma 0.4.9 *Suppose that E is an $n \times n$ elementary matrix and A is an $n \times m$ matrix. Then the matrix EA is the matrix obtained from A by applying the elementary row operation that was used to obtain E from I_n.*

Proof: We exploit the fact that the ith row of the matrix EA is the product of the ith row of E with the matrix A. This is a direct consequence of the definition of matrix multiplication. We denote (in this proof only) the ith row of E by $\mathcal{E}_i$ and the ith row of I_n by I_i. In case $\mathcal{E}_i = I_i$, then the ith row of EA is $\mathcal{E}_i A = I_i A$, which is the ith row of A. This is the desired result for the rows which are not affected by the elementary operation under consideration.

Suppose that the elementary matrix E was obtained from I_n by multiplying the tth row of I_n by the nonzero scalar $k \in \mathbf{F}$. Then the tth row of EA is $\mathcal{E}_t A = kI_t A$, which is k times the tth row of A. This shows that EA is the matrix obtained from A by multiplying the tth row of A by k. We are done in this case. Next suppose that the matrix E was obtained from the matrix I_n by adding k times the sth row to the tth, that is, $\mathcal{E}_t = I_t + kI_s$. Then the tth row of EA is $\mathcal{E}_t A = (I_t + kI_s)A = I_t A + kI_s A$, which is k times the sth row of A added to the tth row of A. This shows that EA is the desired matrix in this case. Finally in case E is obtained from I_n by interchanging the sth and tth rows, we have $\mathcal{E}_s = I_t$ and $\mathcal{E}_t = I_s$. Consequently, EA is the matrix obtained form A by interchanging the sth and tth rows. This proves the lemma. //

For example,

$$\begin{pmatrix} 1 & 2 \\ 0 & 1 \end{pmatrix}$$

is the elementary matrix obtained from the identity by adding 2 times the second row to the first. If we perform the multiplication

$$\begin{pmatrix} 1 & 2 \\ 0 & 1 \end{pmatrix} \begin{pmatrix} 2 & 3 & 4 \\ 1 & 2 & 0 \end{pmatrix} = \begin{pmatrix} 4 & 7 & 4 \\ 1 & 2 & 0 \end{pmatrix}$$

we obtain the same result as adding 2 times the second row of

$$\begin{pmatrix} 2 & 3 & 4 \\ 1 & 2 & 0 \end{pmatrix}$$

to the first.

The preceding lemma yields the next theorem. With this theorem, the process of row reduction (which up to now has been our main computational tool) can be understood as a matrix multiplication. In particular, we will be able to exploit theorems about matrix algebra when studying row equivalence.

Theorem 0.4.10 *Suppose that A and B are row equivalent $m \times n$ matrices. Then there exist elementary matrices $E_1, E_2, \ldots, E_s$ such that $B = E_s E_{s-1} \cdots E_1 A$. In particular, there is an $m \times m$ matrix U such that $B = UA$.*

Proof: Since A and B are row equivalent, there is a finite sequence of elementary row operations that transform A into B. We list the sequence of matrices resulting from these elementary row operations as $A = A_0, A_1, A_2, \ldots, A_{s-1}, A_s = B$. By Lemma 0.4.9 there exist elementary matrices $E_1, E_2, \ldots, E_s$ such that $A_1 = E_1 A_0 = E_1 A, A_2 = E_2 A_1, \ldots, B = A_s = E_s A_{s-1}$. Altogether this shows that $B = E_s A_{s-1} = E_s(E_{s-1} A_{s-2}) = \cdots = E_s(E_{s-1}(\cdots(E_1 A)\cdots))$. Thus the desired elementary matrices exist. By the associativity of matrix multiplication, all the parentheses can be omitted. Setting $U = E_s E_{s-1} \cdots E_1$ gives the desired result. //

For example, consider the sequence of elementary row operations

$$\begin{pmatrix} 1 & 3 & 2 \\ 1 & 5 & 4 \end{pmatrix} \longmapsto \begin{pmatrix} 1 & 3 & 2 \\ 0 & 2 & 2 \end{pmatrix}$$

$$\longmapsto \begin{pmatrix} 1 & 3 & 2 \\ 0 & 1 & 1 \end{pmatrix} \longmapsto \begin{pmatrix} 1 & 0 & -1 \\ 0 & 1 & 1 \end{pmatrix}$$

The three elementary row operations correspond to the elementary matrices

$$\begin{pmatrix} 1 & 0 \\ -1 & 1 \end{pmatrix} \begin{pmatrix} 1 & 0 \\ 0 & 1/2 \end{pmatrix} \quad \text{and} \quad \begin{pmatrix} 1 & -3 \\ 0 & 1 \end{pmatrix}$$

Consequently, the following product describes the sequence of row operations just performed

$$\begin{pmatrix} 1 & 0 & -1 \\ 0 & 1 & 1 \end{pmatrix} = \begin{pmatrix} 1 & -3 \\ 0 & 1 \end{pmatrix}\begin{pmatrix} 1 & 0 \\ 0 & 1/2 \end{pmatrix}\begin{pmatrix} 1 & 0 \\ -1 & 1 \end{pmatrix}\begin{pmatrix} 1 & 3 & 2 \\ 1 & 5 & 4 \end{pmatrix}$$

$$= \begin{pmatrix} 5/2 & -3/2 \\ -1/2 & 1/2 \end{pmatrix}\begin{pmatrix} 1 & 3 & 2 \\ 1 & 5 & 4 \end{pmatrix}$$

PROBLEMS

0.4.1. Express each of the systems of linear equations in Prob. 0.2.1. as a matrix equation.

0.4.2. For each of the following systems of equations, express the system in matrix form, row reduce the augmented matrix to reduced row-echelon form, and find all solutions.

(a) $2X - 3Y = 0$
     $X + Y = 1$

(b) $3X - Y - Z = 0$
     $X + Y + Z = 0$
     $2X + Z = 0$

(c) $X - Y - Z - W = 9$
     $X - Y - Z - W = 9$

(d) $2Y + Z = 1$
     $4X - 5Y - 7Z = 2$
     $X - Y + 2Z = 3$
     $X + Z = 1$

0.4.3. For each of the following systems of equations, express the system in matrix form, row reduce the augmented matrix to reduced row-echelon form, and find all real solutions.

(a) $2X + Y - 3Z = 3$
     $Z + W = 5$
     $X - Y - Z + W = 0$

(b)
$$
\begin{aligned}
X + Y + Z &= 1 \\
X - Y - Z &= 1 \\
5X + 3Z + 3Y &= 5
\end{aligned}
$$

(c)
$$
\begin{aligned}
2X_1 + X_2 - 2X_3 - 5X_4 + X_5 &= 0 \\
X_1 \quad\quad - \quad X_3 \quad\quad + X_5 &= 0
\end{aligned}
$$

(d)
$$
\begin{aligned}
X + Y &= 2 \\
X - Y &= 1 \\
2X + 3Y &= 4
\end{aligned}
$$

0.4.4. Find all solutions to the equation $AX = \vec{b}$ where

$$
A = \begin{pmatrix} 1 & 2 \\ 2 & 3 \\ 9 & 8 \end{pmatrix} \quad \text{and} \quad \vec{b} = \begin{pmatrix} 1 \\ 1 \\ 1 \end{pmatrix}
$$

0.4.5. Find all solutions to the equation $A\vec{X} = \vec{b}$ where

$$
A = \begin{pmatrix} 2 & -1 & -2 & -5 \\ -1 & 2 & -2 & 4 \\ 1 & 4 & -10 & 2 \end{pmatrix} \quad \text{and} \quad \vec{b} = \begin{pmatrix} -10 \\ -1 \\ -23 \end{pmatrix}
$$

0.4.6. Find the reduced row-echelon form and rank of the following matrices:

(a) $\begin{pmatrix} 2 & 1 & 3 \\ 1 & 2 & 3 \end{pmatrix}$
(b) $\begin{pmatrix} 1 & 0 & 7 \\ 2 & 4 & 6 \\ 5 & 8 & 19 \end{pmatrix}$

(c) $\begin{pmatrix} 1 & 2 & 1 & 2 \\ 3 & 3 & 3 & 3 \\ 1 & 2 & 1 & 2 \end{pmatrix}$
(d) $\begin{pmatrix} 1 & r & s \\ 0 & 1 & t \\ 0 & 0 & u \end{pmatrix}$ for arbitrary $r, s, t, u \in \mathbf{F}$

0.4.7. Find the reduced row-echelon form and the rank of each of the following matrices:

(a) $\begin{pmatrix} 1 & 2 & 3 & 4 \\ 2 & 4 & 6 & 8 \end{pmatrix}$
(b) $\begin{pmatrix} 3 & 0 & 1 & 2 \\ 6 & -2 & 1 & 0 \\ 3 & 6 & 0 & -6 \end{pmatrix}$

(c) $\begin{pmatrix} 1 \\ 2 \\ 4 \\ 2 \\ 0 \end{pmatrix}$
(d) $\begin{pmatrix} 1 & 1 & 0 & 0 \\ 1 & 1 & 0 & 0 \\ 2 & 3 & 0 & 0 \\ 2 & 3 & 0 & 0 \end{pmatrix}$

0.4.8. Find all possible echelon and reduced echelon matrices row equivalent to

(a) $\begin{pmatrix} 2 & 3 & 4 \\ 1 & 0 & 0 \end{pmatrix}$ (b) $\begin{pmatrix} 2 & 2 \\ 1 & 0 \\ 2 & 0 \end{pmatrix}$ (c) an arbitrary 2×3 matrix

0.4.9. For any $a, b, c \in \mathbf{R}$, prove the following matrices are never row equivalent

$$\begin{pmatrix} 2 & 0 & 0 \\ a & 3 & 0 \\ b & c & -8 \end{pmatrix} \quad \text{and} \quad \begin{pmatrix} 1 & 1 & 2 \\ -2 & 0 & -1 \\ 1 & 3 & 5 \end{pmatrix}$$

0.4.10. Suppose that both systems of equations $AX = \vec{b}$ and $AX = \vec{b'}$ have infinitely many solutions. Prove that the system $AX = (\vec{b} + \vec{b'})$ has infinitely many solutions.

0.4.11. If $rk(A) = 1$, show that all nonzero rows of A are multiples of one another.

0.4.12. (a) Consider the matrix

$$P = \begin{pmatrix} 0 & 0 & 1 \\ 1 & 0 & 0 \\ 0 & 1 & 0 \end{pmatrix}$$

For any $3 \times m$ matrix A describe PA in terms of A.
(b) Consider the matrix

$$Q = \begin{pmatrix} 1 & 1 & 1 \\ 0 & 1 & 1 \\ 0 & 0 & 1 \end{pmatrix}$$

For any $3 \times m$ matrix A describe QA in terms of A.

0.5   The Structure of Solutions to Systems of Linear Equations

Using the terminology and results of the preceding section, we are now ready to study carefully the nature of solutions to systems of linear equations. Throughout this section, we use matrix notation to describe systems of equations. Thus our system of equations

$$
\begin{array}{ccccccc}
a_{11}X_1 & + & a_{12}X_2 & + \cdots + & a_{1n}X_n & = & b_1 \\
a_{21}X_1 & + & a_{22}X_2 & + \cdots + & a_{2n}X_n & = & b_2 \\
\vdots & & \vdots & \ddots & \vdots & & \vdots \\
a_{m1}X_1 & + & a_{m2}X_2 & + \cdots + & a_{mn}X_n & = & b_m
\end{array}
\qquad (*)
$$

will be written as $AX = B$, where $A = (a_{ij})$ is the $m \times n$ coefficient matrix,

$$
X = \begin{pmatrix} X_1 \\ \vdots \\ X_n \end{pmatrix} \quad \text{and} \quad B = \begin{pmatrix} b_1 \\ \vdots \\ b_m \end{pmatrix}
$$

that is, X is an $n \times 1$ column matrix of variables and B is an $m \times 1$ column matrix. We shall also use matrix notation to represent the solutions to the system. A solution to $AX = B$ will be any $n \times 1$ matrix (column vector) $\vec{v}$ satisfying the equation $A\vec{v} = B$. We need some definitions.

Definition 0.5.1 The system of equations $AX = B$ is called *homogeneous* if the value matrix B is zero; that is, the above system $(*)$ is called homogeneous if $b_1 = 0, b_2 = 0, \ldots, b_m = 0$. Given any system of equations $AX = B$, we call the system of equations $AX = \vec{0}$ the *associated homogeneous* system of equations.

Observe that a homogeneous system of linear equations always has a solution, namely $X_1 = 0, X_2 = 0, \ldots, X_n = 0$. This solution is called the *trivial* solution, and any other solution is called a *nontrivial* solution. The set of solutions to a homogeneous system of equations has several important properties. The next theorem gives these properties and also describes the relationship between the set of solutions of an arbitrary system and the set of solutions of the associated homogeneous system. The reader familiar with the theory of linear differential equations will recognize that the two properties given next are also true when considering the solutions to systems of linear differential equations.

Theorem 0.5.2 (i) *Let S denote the (nonempty) set of solutions to the homogeneous system $AX = 0$. Then S is closed under addition and scalar multiplication; that is, if $\vec{v}_1, \vec{v}_2 \in S$ and $r \in \mathbf{F}$, then both*

$\vec{v}_1 + \vec{v}_2$, $r\vec{v}_1 \in S$.

(ii) *Let T denote the set of solutions to the system of equations $AX = B$ and S the set of solutions to the associated homogeneous system of equations $AX = \vec{0}$. Suppose that T is nonempty and $\vec{u} \in T$. Then every element $\vec{w} \in T$ is of the form*

$$\vec{w} = \vec{u} + \vec{v} \quad \text{for some } \vec{v} \in S$$

Proof: (i) If $\vec{v}_1$ and $\vec{v}_2$ are solutions to $AX = \vec{0}$, then by definition $A\vec{v}_1 = \vec{0}$ and $A\vec{v}_2 = \vec{0}$. But the distributive law Theorem 0.3.3 (ix) shows that $A(\vec{v}_1 + \vec{v}_2) = A\vec{v}_1 + A\vec{v}_2 = \vec{0} + \vec{0} = \vec{0}$. Thus $(\vec{v}_1 + \vec{v}_2) \in S$. Since scalars commute with matrices (Theorem 0.3.3 (vii)), we have that $A(r\vec{v}_1) = r(A\vec{v}_1) = r\vec{0} = \vec{0}$. This proves (i).

(ii) Our assumption is that $A\vec{u} = B$ and $A\vec{w} = B$. Setting $\vec{v} = \vec{w} - \vec{u}$, we find that $A\vec{v} = A(\vec{w} - \vec{u}) = A\vec{w} - A\vec{u} = B - B = \vec{0}$. Thus, $\vec{v} \in S$; this shows that $\vec{w} = \vec{u} + \vec{v}$ has the desired form. Conversely, suppose that $\vec{w} = \vec{u} + \vec{v}$ where $\vec{v} \in S$. Then $A\vec{w} = A(\vec{u} + \vec{v}) = A\vec{u} + A\vec{v} = B + \vec{0} = B$, and $\vec{w} \in T$ follows. Thus, T is precisely the set of n-tuples $\vec{u} + \vec{v}$ where $\vec{v} \in S$. //

Example 0.5.3 Consider the system of equations

$$\begin{pmatrix} 1 & 2 & 3 & 0 \\ 0 & 1 & 4 & 1 \end{pmatrix} \begin{pmatrix} X \\ Y \\ Z \\ W \end{pmatrix} = \begin{pmatrix} 6 \\ 6 \end{pmatrix}$$

The associated homogeneous system is

$$\begin{pmatrix} 1 & 2 & 3 & 0 \\ 0 & 1 & 4 & 1 \end{pmatrix} \begin{pmatrix} X \\ Y \\ Z \\ W \end{pmatrix} = \begin{pmatrix} 0 \\ 0 \end{pmatrix}$$

The reduced echelon form of

$$\begin{pmatrix} 1 & 2 & 3 & 0 \\ 0 & 1 & 4 & 1 \end{pmatrix} \quad \text{is} \quad \begin{pmatrix} 1 & 0 & -5 & -2 \\ 0 & 1 & 4 & 1 \end{pmatrix}$$

so it follows that the set of all solutions to the homogeneous system is

$$\left\{ \begin{pmatrix} 5c + 2d \\ -4c - d \\ c \\ d \end{pmatrix} \mid c, d \in \mathbf{F} \right\}$$

or alternatively

$$\left\{ c \begin{pmatrix} 5 \\ -4 \\ 1 \\ 0 \end{pmatrix} + d \begin{pmatrix} 2 \\ -1 \\ 0 \\ 1 \end{pmatrix} \mid c, d \in \mathbf{F} \right\}$$

Observe this set is closed under addition and scalar multiplication. Since $X = Y = Z = W = 1$ is a solution to the original system, it follows from Theorem 0.5.2 (ii) that

$$\left\{ \begin{pmatrix} 1 \\ 1 \\ 1 \\ 1 \end{pmatrix} + c \begin{pmatrix} 5 \\ -4 \\ 1 \\ 0 \end{pmatrix} + d \begin{pmatrix} 2 \\ -1 \\ 0 \\ 1 \end{pmatrix} \mid c, d \in \mathbf{F} \right\}$$

is the set of all solutions to the original system.

Example 0.5.4 Consider the system of equations with real coefficients

$$\begin{pmatrix} 1 & 0 & 1 \\ 0 & 2 & -1 \\ 1 & -2 & 2 \end{pmatrix} \begin{pmatrix} X \\ Y \\ Z \end{pmatrix} = \begin{pmatrix} 0 \\ 1 \\ 0 \end{pmatrix}$$

Row-reducing the augmented matrix of the associated homogeneous system, one finds

$$\begin{pmatrix} 1 & 0 & 1 & | & 0 \\ 0 & 2 & -1 & | & 0 \\ 1 & -2 & 2 & | & 0 \end{pmatrix} \longmapsto \begin{pmatrix} 1 & 0 & 1 & | & 0 \\ 0 & 2 & -1 & | & 0 \\ 0 & -2 & 1 & | & 0 \end{pmatrix}$$

$$\longmapsto \begin{pmatrix} 1 & 0 & 1 & | & 0 \\ 0 & 1 & -1/2 & | & 0 \\ 0 & 1 & -1/2 & | & 0 \end{pmatrix}$$

$$\longmapsto \begin{pmatrix} 1 & 0 & 1 & | & 0 \\ 0 & 1 & -1/2 & | & 0 \\ 0 & 0 & 0 & | & 0 \end{pmatrix}$$

It follows that all the solutions to the associated homogeneous system of equations are $(-t, (1/2)t, t)$ where $t \in \mathbf{R}$. Applying the same row operations to the augmented matrix of the original system we get

$$\left(\begin{array}{ccc|c} 1 & 0 & 1 & 0 \\ 0 & 2 & -1 & 1 \\ 1 & -2 & 2 & 0 \end{array}\right) \mapsto \left(\begin{array}{ccc|c} 1 & 0 & 1 & 0 \\ 0 & 2 & -1 & 1 \\ 0 & -2 & 1 & 0 \end{array}\right)$$

$$\mapsto \left(\begin{array}{ccc|c} 1 & 0 & 1 & 0 \\ 0 & 1 & -1/2 & 1/2 \\ 0 & 1 & -1/2 & 0 \end{array}\right)$$

$$\mapsto \left(\begin{array}{ccc|c} 1 & 0 & 1 & 0 \\ 0 & 1 & -1/2 & 1/2 \\ 0 & 0 & 0 & -1/2 \end{array}\right)$$

We see that the original system is inconsistent; that is, it has no solutions.

This example illustrates that even though the associated homogeneous system to a given system may have infinitely many solutions, the original system need not have any solutions. Note that in the hypothesis of Theorem 0.5.2 (ii) it was assumed that the system $AX = B$ had a solution.

We will shortly show how the concept of rank can be used in the study of systems of equations. However, we first must show that the concept of rank is well-defined. By definition, the rank of a matrix A is the number of nonzero rows in the reduced row-echelon form of A. But what happens if a matrix has two reduced row-echelon forms with different numbers of nonzero rows? This would cause trouble, because then our definition would not make sense. Fortunately, this cannot happen, as the next theorem shows.

Theorem 0.5.5 *The reduced row-echelon form of any matrix is unique; that is, if A is any matrix and R_1 and R_2 are both reduced row-echelon matrices that are row equivalent to A, then $R_1 = R_2$.*

Proof: Since A, R_1, and R_2 are all row equivalent, Theorem 0.1.8 guarantees that the systems of equations $AX = \vec{0}$, $R_1 X = \vec{0}$, and $R_2 X = \vec{0}$ each have precisely the same set of solutions.

Consider some row of R_1 or R_2

$$\left(\begin{array}{cccccccc} 0 & 0 & \cdots & 0 & 1 & t_{i+1} & t_{i+2} & \cdots & t_n \end{array}\right)$$

(the t_j denote unknown entries). If the leading 1 occurs in the ith column, then X_i is a determined variable, and this row represents the equation

$$X_i + t_{i+1}X_{i+1} + \cdots + t_n X_n = 0$$

in the associated homogeneous system. Such an equation *never* has a solution with $X_i = 1, X_{i+1} = 0, X_{i+2} = 0, \ldots, X_n = 0$. On the other hand, if X_i is a free variable in some system in reduced echelon form, then whenever X_i occurs in an equation in $R_1 X = \vec{0}$ or $R_2 X = \vec{0}$, that equation has the form

$$X_j + \cdots + r_i X_i + r_{i+1}X_{i+1} + \cdots + r_n X_n = 0$$

where $j < i$. Back substitution guarantees that there is some solution to the system with $X_i = 1, X_{i+1} = 0, X_{i+2} = 0, \ldots, X_n = 0$ (the values of $X_1, \ldots, X_{i-1}$ in such a solution need not be specified). As the two systems $R_1 X = \vec{0}$ and $R_2 X = \vec{0}$ have exactly the same solutions, this discussion shows that the determined variables of both systems must be the same. Of course, it then follows that the free variables are also the same.

The preceding, together with the definition of reduced echelon matrices, shows that the leading 1s in the rows of R_1 and R_2 occur in exactly the same columns. We next show that the row entries that follow these leading 1s are the same. Assume that

$$\begin{pmatrix} 0 & 0 & \cdots & 0 & 1 & t_{i+1} & \cdots & t_n \end{pmatrix}$$

is a row of R_1. Suppose that X_j is a determined variable where $j > i$. As the remaining entries in the column of a determined variable are 0, the definition of reduced row-echelon matrices gives $t_j = 0$. For the same reason the jth entry of the corresponding row of R_2 is also 0. Thus the jth entries of these rows of R_1 and R_2 are the same.

Now suppose that X_j is a free variable with $j > i$. In this case, $R_1 X = \vec{0}$ has a solution with $X_j = 1$ and $X_k = 0$ for any $k \neq j$ with $k > i$, X_k free, and $X_i = -t_j$. (The values of the other determined variables need not be specified, as their coefficients are 0 in this equation.) Since such a solution is also a solution to $R_2 X = \vec{0}$, it follows that the jth entry of the corresponding row of R_2 must also be t_j.

Altogether we have shown that the rows of R_1 and R_2 are identical, and hence $R_1 = R_2$. This proves the uniqueness of the reduced row-echelon form. $//$

We now repeat the definition of rank given in the last section, but this time, because of Theorem 0.5.5 it makes sense to talk about *the* reduced row-echelon form of a matrix A.

Definition 0.5.6 If A is any matrix, the rank of A is the number of nonzero rows in the reduced row-echelon form of A. We denote the rank of A by $rk(A)$.

Remark Observe that whenever A is an $m \times n$ matrix, $rk(A) \leq min\{m, n\}$. For clearly, the number of nonzero rows of the reduced row-echelon form of A cannot exceed the number of rows of A, that is, $rk(A) \leq m$. Yet further, any nonzero row of a reduced row-echelon matrix must begin with a leading 1, and each leading 1 lies in a distinct column. Thus the number of such rows cannot exceed the number of columns of A; that is, $rk(A) \leq n$. This simple, yet crucial, observation will be used often in the sequel.

The nature of the solutions to a system of linear equations is largely determined by the relationship between the rank of the coefficient matrix, the rank of the augmented matrix, and the number of variables. This is spelled out explicitly in the next result.

Theorem 0.5.7 *Consider a system of equations $AX = B$, and denote by $A' = (A \mid B)$ the augmented matrix of the system. We assume that there are n variables in X.*
(i)   The system $AX = B$ has a solution if and only if $rk(A) = rk(A')$.
(ii)   The system $AX = B$ has a unique solution if and only if $rk(A) = n = rk(A')$.

Proof: (i) Suppose the reduced echelon form of the (partitioned) matrix $A' = (A \mid B)$ is the matrix $R' = (R \mid S)$. Since R' is row equivalent to A', it follows that R is row equivalent to A. As R' is a reduced row-echelon matrix, it follows that the submatrix R is a reduced row-echelon matrix. Consequently, R is the reduced row-echelon form of A. Since R' cannot have fewer nonzero rows than R, we see that $rk(A') \geq rk(A)$. Suppose that $rk(A') > rk(A)$. Then the reduced echelon matrix R' has a row of the form

$$(0 \quad 0 \quad \cdots \quad 0 \quad 0 \mid 1)$$

This row corresponds to the equation $0X_1 + 0X_2 + \cdots + 0X_n = 1$, which does not have any solutions. Thus, whenever $rk(A') \neq rk(A)$, $AX = B$ does not have any solutions. In case $rk(A) = rk(A')$, then $AX = B$ has a solution by back-substitution. This proves (i).

(ii) Suppose A is an $m \times n$ matrix. Since $rk(A) \leq m$ (m is the number of rows of A), whenever $rk(A) = n$, we must have $n \leq m$. If $rk(A) = n = rk(A')$ is the number of columns of A, it follows that the reduced row-echelon form of A' must look like

$$
\begin{pmatrix}
1 & 0 & 0 & \cdots & 0 & 0 & c_1 \\
0 & 1 & 0 & \cdots & 0 & 0 & c_2 \\
0 & 0 & 1 & \cdots & 0 & 0 & c_3 \\
\vdots & \vdots & \vdots & \ddots & \vdots & \vdots & \vdots \\
0 & 0 & 0 & \cdots & 0 & 1 & c_n \\
0 & 0 & 0 & \cdots & 0 & 0 & 0 \\
\vdots & \vdots & \vdots & \ddots & \vdots & \vdots & \vdots \\
0 & 0 & 0 & \cdots & 0 & 0 & 0
\end{pmatrix}
$$

Evidently, the corresponding system of equations has the unique solution $X_1 = c_1, X_2 = c_2, \ldots, X_n = c_n$.

Now suppose the system $AX = B$ has a unique solution. Since it has a solution, part (i) shows that $rk(A) = rk(A')$. Since A is an $m \times n$ matrix, necessarily $rk(A) \leq n$. Assume that $rk(A) < n$. Then there are fewer nonzero rows in R than there are columns. In particular, there is some entry of R that is not a leading 1. The column of this entry corresponds to a free variable that gives a parameter when describing all solutions to $AX = B$ by back substitution. It follows that $AX = B$ has more than one solution, a contradiction. Hence $rk(A) = n = rk(A')$. This proves the theorem. //

The rank of the coefficient matrix of a system of equations is closely related to the number of parameters necessary to describe the solution set of the system. If A is an $m \times n$ matrix which is the coefficient matrix of a consistent system of equations, then there are $rk(A)$ determined variables and $n - rk(A)$ free variables in the reduced row-echelon form of the system. We shall see in the next chapter, using the dimension theory of vector spaces, that the number $n - rk(A)$ is precisely the smallest number of parameters that can describe the solution set. In the next corollary we record the observation that $n - rk(A)$ parameters suffice to describe all solutions.

Corollary 0.5.8 *Suppose that A is an $m \times n$ matrix. Then*
(i)   *The system $AX = \vec{0}$ always has a solution and the collection of all solutions can be described using $n - rk(A)$ parameters. In particular, if the number of variables of a homogeneous system exceeds the number of equations, the system has a nontrivial solution.*
(ii)   *If $AX = B$ has a solution, then the collection of all solutions can be described using $n - rk(A)$ parameters.*

Proof: (i) The homogeneous system $AX = \vec{0}$ always has the solution $X = \vec{0}$. The number $n - rk(A)$ is precisely the number of columns of the reduced row-echelon form R of A, which do not contain a leading 1. These columns correspond to parameters when describing the solutions to $AX = B$ by back substitution.

(ii) This follows for the same reason as does (i). //

Example 0.5.9 Consider the system of equations represented by the augmented matrix
$$\left(\begin{array}{cccc|c} 1 & 2 & 1 & 1 & 5 \\ 2 & 4 & 0 & 0 & 6 \\ 1 & 2 & 0 & 1 & 3 \\ 0 & 0 & 1 & 1 & 2 \end{array} \right)$$

Then, by the row-reduction given in Example 0.4.7, we find that the reduced row-echelon form is
$$\left(\begin{array}{cccc|c} 1 & 2 & 0 & 0 & 3 \\ 0 & 0 & 1 & 0 & 2 \\ 0 & 0 & 0 & 1 & 0 \\ 0 & 0 & 0 & 0 & 0 \end{array} \right)$$

Since the coefficient matrix and the augmented matrix have the same rank (namely 3), we know by Theorem 0.5.7 that the system of equations has a solution. Moreover, by Corollary 0.5.8 the collection of all solutions can be represented by $4 - 3 = 1$ parameter. In fact the solutions set to the system is

$$\{(3 - 2t, t, 2, 0) | t \in \mathbf{R}\}$$

PROBLEMS

0.5.1.   For each of the following systems of equations, solve the associated homogeneous system first, and then indicate how the general solution is related to the homogeneous solution.

(a) $\begin{pmatrix} 2 & 1 & 0 & 0 \\ 0 & 1 & 3 & 1 \\ 0 & 1 & 2 & 1 \end{pmatrix} \begin{pmatrix} X \\ Y \\ Z \\ W \end{pmatrix} = \begin{pmatrix} 1 \\ 1 \\ 1 \end{pmatrix}$ (b) $\begin{pmatrix} 1 & 1 \\ 1 & 2 \\ 1 & 0 \end{pmatrix} \begin{pmatrix} X \\ Y \end{pmatrix} = \begin{pmatrix} 1 \\ 1 \\ 1 \end{pmatrix}$

(c) $\begin{pmatrix} 2 & 3 & 4 \end{pmatrix} \begin{pmatrix} X \\ Y \\ Z \end{pmatrix} = \begin{pmatrix} 2 \end{pmatrix}$ (d) $\begin{pmatrix} 1 & 0 & 1 \\ 1 & 3 & 2 \\ 4 & 6 & 6 \end{pmatrix} \begin{pmatrix} X \\ Y \\ Z \end{pmatrix} = \begin{pmatrix} 1 \\ 1 \\ 1 \end{pmatrix}$

0.5.2.   For each of the following systems of equations, solve the associated homogeneous system first, and then indicate how the general solution is related to the homogeneous solution.

(a) $\begin{pmatrix} 1 & 0 & 1 \\ 2 & 3 & 6 \\ 1 & 1 & 1 \end{pmatrix} \begin{pmatrix} X \\ Y \\ Z \end{pmatrix} = \begin{pmatrix} 1 \\ 0 \\ 0 \end{pmatrix}$ (b) $\begin{pmatrix} 2 & 3 & 5 & 6 \\ 1 & 3 & 2 & 0 \end{pmatrix} \begin{pmatrix} X \\ Y \\ Z \\ W \end{pmatrix} = \begin{pmatrix} 1 \\ 2 \end{pmatrix}$

(c) $\begin{pmatrix} 1 & 0 \\ 2 & 3 \\ 0 & 1 \end{pmatrix} \begin{pmatrix} X \\ Y \end{pmatrix} = \begin{pmatrix} 1 \\ 5 \\ 1 \end{pmatrix}$ (d) $\begin{pmatrix} 2 & 0 & 2 \\ 4 & 5 & 4 \\ 0 & 0 & 0 \\ 0 & 2 & 1 \end{pmatrix} \begin{pmatrix} X \\ Y \\ Z \end{pmatrix} = \begin{pmatrix} 1 \\ 1 \\ 1 \\ 1 \end{pmatrix}$

0.5.3.   Find all real numbers a, b, c such that

$$\begin{pmatrix} 1 & 1 & 2 \\ 3 & 4 & 7 \\ 1 & 2 & 3 \end{pmatrix} \begin{pmatrix} X \\ Y \\ Z \end{pmatrix} = \begin{pmatrix} a \\ b \\ c \end{pmatrix}$$

has a solution.

0.5.4.   An upper-triangular $n \times n$ matrix has rank n if and only if ...?

0.5.5.   Suppose that A is a 3×4 rank 3 matrix. Suppose that B is any 3×2 matrix. Prove that the matrix equation $AM = B$ can always be solved for M. (Hint: Do not compute!)

0.5.6.   True or false, and why?

(a) If $n > m$ and A is an $n \times m$ matrix, then the system of equations

$AX = \vec{b}$ has infinitely many solutions.

(b) If $n = m$ and A is an $n \times m$ matrix, then the system of equations $AX = \vec{b}$ has a unique solution.

(c) If A is an $n \times m$ matrix, then the system of equations $AX = \vec{0}$ has a solution if and only if the system $AX = \vec{b}$ has a solution for all $\vec{b}$.

(d) If A is an $n \times m$ matrix of rank n, then the system of equations $AX = \vec{0}$ has a unique solution.

0.5.7.   Show that (a_1, b_1), $(a_2, b_2), \ldots, (a_n, b_n)$ are collinear in $\mathbf{R}^2$ if and only if

$$rk \begin{pmatrix} 1 & a_1 & b_1 \\ 1 & a_2 & b_2 \\ \vdots & \vdots & \vdots \\ 1 & a_n & b_n \end{pmatrix} \le 2$$

0.5.8.   Show that if u and v are solutions to a system of equations, then $u - v$ is a solution to the associated homogeneous system. Conclude that if a system of equations has at most one solution, then its associated system of homogeneous equations has only the trivial solution.

0.6   Matrix Inverses and Transposes

Whenever $a, b \in \mathbf{F}$ with $a \neq 0$, the linear equation $aX = b$ is easily solved. One can multiply on the left by the inverse (reciprocal) of a: $a^{-1}aX = a^{-1}b$, to obtain $X = a^{-1}b$. Suppose that this process could be carried out for matrices. Then the problem of solving a system of equations would reduce to a single multiplication. In fact, we shall see in this section that this often is the case.

Over any field $\mathbf{F}$, the inverse of a nonzero element a is the unique element a^{-1} defined by the condition that $aa^{-1} = a^{-1}a = 1$. In order to study matrix inversion, we need to know what the matrix version of the number 1 is. Recall from Definition 0.3.6 that the $n \times n$ identity matrix I_n is the $n \times n$ matrix with ones on the diagonal and zeros elsewhere. I_n has the important property that for any $n \times m$ matrix A, $I_n A = A$ and for any $m \times n$ matrix B, $BI_n = B$. Thus it is through the identity matrix I_n that the inverse is defined.

Example 0.6.1   Consider the 2×2 matrices

$$A = \begin{pmatrix} 2 & 3 \\ 1 & 2 \end{pmatrix} \quad B = \begin{pmatrix} 2 & -3 \\ -1 & 2 \end{pmatrix} \quad C = \begin{pmatrix} 2 & 1 \\ 2 & 1 \end{pmatrix}$$

One readily checks that

$$AB = \begin{pmatrix} 2 & 3 \\ 1 & 2 \end{pmatrix} \begin{pmatrix} 2 & -3 \\ -1 & 2 \end{pmatrix} = \begin{pmatrix} 1 & 0 \\ 0 & 1 \end{pmatrix}$$

$$= \begin{pmatrix} 2 & -3 \\ -1 & 2 \end{pmatrix} \begin{pmatrix} 2 & 3 \\ 1 & 2 \end{pmatrix} = BA$$

so that A and B are inverses. However, consider the matrix equation

$$\begin{pmatrix} 2 & 1 \\ 2 & 1 \end{pmatrix} \begin{pmatrix} X & Y \\ Z & W \end{pmatrix} = \begin{pmatrix} 1 & 0 \\ 0 & 1 \end{pmatrix}$$

In trying to solve this equation, one obtains both equations $2X + Z = 1$ and $2X + Z = 0$. It follows that the matrix C cannot have an inverse.

The matrix C in the example above is a nonzero, yet noninvertible 2×2 matrix. The determination of when a matrix has an inverse is the key problem of this section. Since matrix multiplication is not commutative, it is necessary to worry about the order of multiplication when defining matrix inverses. This is spelled out in the next definition.

Definition 0.6.2 Suppose that A is an $m \times n$ matrix. If L is an $n \times m$ matrix with the property that $LA = I_n$, we say that L is a *left inverse* of A. If R is an $n \times m$ matrix with the property that $AR = I_m$, we say that R is a *right inverse* of A. If B is both a right and a left inverse of A, we say that B is an *inverse* of A. Whenever A has an inverse, we say that A is *invertible* (or *nonsingular*).

We see in the following examples that it is possible for a matrix to have a right inverse that is not a left inverse, a left inverse that is not a right inverse, or an inverse (which is both a left and right inverse). Thus all the distinctions made in the above definition are important. Of course, with the real numbers (or the elements of any scalar field), there is no distinction between right and left inverses. (Multiplication is commutative for real numbers and consequently these two inverses are always the same.)

Examples 0.6.3 (i) Set

$$A = \begin{pmatrix} 1 & 2 \\ 1 & 3 \\ 4 & 7 \end{pmatrix} \quad \text{and} \quad B = \begin{pmatrix} 3 & -2 & 0 \\ -1 & 1 & 0 \end{pmatrix}$$

Then $BA = I_2$. It follows that B is a left inverse of A and A is a right inverse of B. However,

$$AB = \begin{pmatrix} 1 & 0 & 0 \\ 0 & 1 & 0 \\ 5 & -1 & 0 \end{pmatrix}$$

so we see that A is not a left inverse of B and B is not a right inverse of A. It is a consequence of Lemma 0.6.4 that A does not have a left inverse and B does not have a right inverse.

(ii) Set

$$B' = \begin{pmatrix} -7 & 0 & 2 \\ -11 & 3 & 2 \end{pmatrix}$$

Then for A as in (i) above we find that $B'A = I_2$. Since $B \neq B'$, we see that A has (at least) two different left inverses. Thus left (and similarly right) inverses are not necessarily unique. In fact, we shall shortly prove a result that shows the real matrix A must have infinitely many distinct left inverses.

(iii) Consider

$$C = \begin{pmatrix} 1 & 0 & 1 \\ 1 & -1 & 0 \\ 0 & 2 & 1 \end{pmatrix} \quad \text{and} \quad D = \begin{pmatrix} -1 & 2 & 1 \\ -1 & 1 & 1 \\ 2 & -2 & -1 \end{pmatrix}$$

Then one readily checks that $CD = DC = I_3$. Hence, C is an inverse of D, and D is an inverse of C. Both C and D are invertible.

The preceding examples show that a matrix need not have a unique left (or right) inverse. However, as the next lemma shows, if a matrix has an inverse, then such an inverse is unique. It is also true that if a matrix has an inverse, it must necessarily be square (that is, $n \times n$ for some n). This latter fact is given in Corollary 0.6.17.

Lemma 0.6.4 *If a matrix A has a left inverse L and a right inverse R, then $L = R$, that is, A is invertible. Further, whenever A is invertible, its inverse is unique.*

Proof: We assume that A is $m \times n$, so R and L are $n \times m$. Using the associativity of matrix multiplication, Theorem 0.3.7, we see that

$R = I_n R = (LA)R = L(AR) = LI_m = L$. This proves the first assertion. Next suppose that A has two inverses B and C. Since B is a left inverse of A and C is a right inverse of A, what we just proved shows that $B = C$. Hence the inverse of A is unique. //

Since matrix inverses are unique, we can introduce the following notation.

Definition 0.6.5 If A is an invertible matrix, we denote its unique inverse by A^{-1}.

An important property of any scalar field is that the product of any two nonzero elements is nonzero. In other words, the product of invertible scalars is invertible. This latter statement is true for matrices, but there is one peculiar twist. The inverse of a product of invertible matrices is *not* the product of the inverses *unless* you are careful about the order of multiplication. The next theorem gives this important result.

Theorem 0.6.6 *Suppose that A and B are invertible $n \times n$ matrices. Then AB is invertible and $(AB)^{-1} = B^{-1}A^{-1}$.*

Proof: One computes directly, using the associative law (Theorem 0.3.7) of matrix multiplication:

$$(AB)(B^{-1}A^{-1}) = A(BB^{-1})A^{-1} = AI_n A^{-1} = AA^{-1} = I_n$$

and

$$(B^{-1}A^{-1})(AB) = B^{-1}(A^{-1}A)B = B^{-1}I_n B = B^{-1}B = I_n$$

This proves the theorem. //

As an illustration of this result, consider the matrices

$$A = \begin{pmatrix} 1 & 1 \\ 2 & 3 \end{pmatrix} \quad \text{and} \quad B = \begin{pmatrix} 2 & 5 \\ 1 & 3 \end{pmatrix}$$

Then one can readily check that

$$A^{-1} = \begin{pmatrix} 3 & -1 \\ -2 & 1 \end{pmatrix} \quad \text{and} \quad B^{-1} = \begin{pmatrix} 3 & -5 \\ -1 & 2 \end{pmatrix}$$

Also,

$$AB = \begin{pmatrix} 3 & 8 \\ 7 & 19 \end{pmatrix} \quad \text{and} \quad (AB)^{-1} = \begin{pmatrix} 19 & -8 \\ -7 & 3 \end{pmatrix}$$

However

$$A^{-1}B^{-1} = \begin{pmatrix} 10 & -17 \\ -7 & 12 \end{pmatrix} \neq B^{-1}A^{-1} = (AB)^{-1}$$

Recall (Definition 0.4.8) that elementary matrices are those matrices that can be obtained from the identity matrix by applying a single elementary row operation. One can easily see that elementary matrices are always invertible. The invertibility of elementary matrices is nothing other than our earlier observation that row operations can always be reversed.

Lemma 0.6.7 *Every elementary matrix is invertible, and its inverse is another elementary matrix.*

Proof: It was noted in Remark 0.1.7 that whenever B is obtained from a matrix A by a single elementary operation, there is another single elementary operation which transforms B into A. Thus a single elementary row operation can transform an elementary matrix into the identity matrix. If E is an $n \times n$ elementary matrix, let E' be the elementary matrix that corresponds to the row operation that reduces E to I_n. Then Lemma 0.4.9 says that $E'E = I_n$. Since E is square, it follows that E is invertible and $E' = E^{-1}$. //

Using the above lemma, we can now refine the result given ealier as Theorem 0.4.10.

Theorem 0.6.8 *Suppose that A and B are row equivalent $m \times n$ matrices. Then there exist elementary matrices $E_1, E_2, \ldots, E_s$ such that $B = E_s E_{s-1} \cdots E_1 A$. In particular, there is an invertible matrix U such that $B = UA$.*

Proof: The existence of the representation

$$B = E_s E_{s-1} \cdots E_1 A = UA$$

has already been established in Theorem 0.4.10. It remains to check the invertibity of the matrix U. For this, each E_i, $i = 1, 2, \ldots, s$ is invertible by Lemma 0.6.7. Hence by Theorem 0.6.6, their product $U = E_s E_{s-1} \cdots E_1$ is invertible. This proves the theorem. //

In the special case where A is square, the theorem enables us to characterize invertible matrices. The nonsquare situation will be treated subsequently.

Theorem 0.6.9 *If A is an $n \times n$ (square) matrix, then the following are equivalent:*
(i)   *A is invertible.*
(ii)   *For each $\vec{y} \in \mathbf{F}^n$ the system $AX = \vec{y}$ has a unique solution (for X).*
(iii) *$rk(A) = n$.*
(iv)   *$A = E_s E_{s-1} \cdots E_1$ for some collection of elementary matrices $E_1, E_2, \ldots, E_s$.*

Proof: Assume that (i) is true. Clearly, $X = A^{-1}\vec{y}$ is a solution to $AX = \vec{y}$. Moreover, if $A\vec{b} = \vec{y}$, then multiplying this equation by A^{-1} gives $\vec{b} = A^{-1}\vec{y}$, so the equation $AX = \vec{y}$ has a unique solution. This shows that condition (ii) is a consequence of condition (i). Next we assume that (ii) is true. Then Theorem 0.5.7 (ii) shows that $rk(A) = n$. Hence (iii) is a consequence of (ii).
   Assume now that (iii) is true. Then since A is $n \times n$ and $rk(A) = n$, A can be row-reduced to the identity matrix I_n. By Theorem 0.6.8, $A = E_s E_{s-1} \cdots E_1 I_n = E_s E_{s-1} \cdots E_1$ as desired. Thus (iii) is a consequence of (ii). Finally, we assume that (iv) is true. Since A is a product of elementary matrices, each of which is invertible by Lemma 0.6.7, Theorem 0.6.6 shows that A is invertible. Hence (i) is a consequence of (iv). We have shown that the conditions (i), (ii), (iii), and (iv) are equivalent. //

The equivalence of parts (i) and (iv) of the preceeding theorem says that the group of invertible matrices is "generated" (through matrix multiplication) by the set of elementary matrices. This observation will later prove to be a useful theoretical tool.

Example 0.6.10 Consider the matrix

$$B = \begin{pmatrix} 1 & 3 \\ 2 & 4 \end{pmatrix}$$

We row-reduce B to the identity matrix in the following elementary steps:

$$\begin{pmatrix} 1 & 3 \\ 2 & 4 \end{pmatrix} \mapsto \begin{pmatrix} 1 & 3 \\ 0 & -2 \end{pmatrix} \mapsto \begin{pmatrix} 1 & 3 \\ 0 & 1 \end{pmatrix} \mapsto \begin{pmatrix} 1 & 0 \\ 0 & 1 \end{pmatrix}$$

The elementary matrices corresponding to these three row operations are:

$$\begin{pmatrix} 1 & 0 \\ -2 & 1 \end{pmatrix} \quad \begin{pmatrix} 1 & 0 \\ 0 & -1/2 \end{pmatrix} \quad \text{and} \quad \begin{pmatrix} 1 & -3 \\ 0 & 1 \end{pmatrix}$$

Lemma 0.4.9 shows that

$$\begin{pmatrix} 1 & -3 \\ 0 & 1 \end{pmatrix}\begin{pmatrix} 1 & 0 \\ 0 & -1/2 \end{pmatrix}\begin{pmatrix} 1 & 0 \\ -2 & 1 \end{pmatrix}\begin{pmatrix} 1 & 3 \\ 2 & 4 \end{pmatrix} = \begin{pmatrix} 1 & 0 \\ 0 & 1 \end{pmatrix}$$

In particular, multiplying this equation by the three inverses of these elementary matrices, we obtain

$$\begin{pmatrix} 1 & 3 \\ 2 & 4 \end{pmatrix} = \begin{pmatrix} 1 & 0 \\ 2 & 1 \end{pmatrix}\begin{pmatrix} 1 & 0 \\ 0 & -2 \end{pmatrix}\begin{pmatrix} 1 & 3 \\ 0 & 1 \end{pmatrix}$$

This shows how to express

$$\begin{pmatrix} 1 & 3 \\ 2 & 4 \end{pmatrix}$$

as a product of elementary matrices.

The next theorem gives criteria for the existence of right inverses. Its proof shows how to calculate a right inverse, if it exists. The analogous result for left inverses is given after we introduce the transpose.

Theorem 0.6.11   *Suppose that A is an $m \times n$ matrix. Then the following are equivalent:*
(i)   *A has a right inverse.*
(ii)   *For each $\vec{y} \in \mathbf{F}^m$ the system of equations $AX = \vec{y}$ can always be solved (for X).*
(iii) *$rk(A) = m$.*

Proof: If A has some right inverse B, then $A(B\vec{y}) = (AB)\vec{y} = I\vec{y} = \vec{y}$, and consequently $B\vec{y}$ is a solution to $AX = \vec{y}$. For the converse we note that if A has a right inverse B, then B must be an $n \times m$ matrix. We decompose such a possible matrix B into its columns $B_1, B_2, \ldots, B_m$ (each B_i is a $n \times 1$ matrix); that is, $B = (\, B_1 \quad B_2 \quad \cdots \quad B_m \,)$. The definition of matrix multiplication shows that the equation $AB = I_m$ is equivalent to the collection of equations

$$AB_1 = \begin{pmatrix} 1 \\ 0 \\ \vdots \\ 0 \\ 0 \end{pmatrix} \quad AB_2 = \begin{pmatrix} 0 \\ 1 \\ 0 \\ \vdots \\ 0 \end{pmatrix} \quad \cdots \quad AB_m = \begin{pmatrix} 0 \\ 0 \\ \vdots \\ 0 \\ 1 \end{pmatrix}$$

We view each of these equations as a system of equations with (unknown) variables coming from the B_i column. If we know that for each $\vec{y} \in \mathbf{F}^m$ the equation $AX = \vec{y}$ can always be solved (for X), we see that A will have a right inverse (solving for each B_i as needed). This establishes the equivalence of (i) and (ii).

If $rk(A) = m$ is the number of rows of A, the rank of the augmented matrix of any system of equations with coefficient matrix A must equal the rank of A. It follows from Theorem 0.5.7 (i) that every system $AX = \vec{y}$ can be solved. Conversely, assume $rk(A) < m$. Row-reduce A by some sequence $(*)$ of elementary row operations to its reduced echelon form R, which by assumption has a row of zeros at the bottom. Recall that all row operations can be reversed. Denote by $(*_r)$ the reversal of the sequence $(*)$. Applying $(*_r)$ to the vector $(m \times 1$ matrix)

$$e_m = \begin{pmatrix} 0 \\ 0 \\ \vdots \\ 0 \\ 1 \end{pmatrix}$$

we obtain some vector $\vec{b} \in \mathbf{F}^m$. Then, $(*)$ applied to $\vec{b}$ gives e_m back. This shows that the system of equations $AX = \vec{b}$ is row equivalent to $RX = e_m$. Clearly, the latter equation has no solutions. Thus the system $AX = \vec{y}$ cannot always be solved. This establishes the equivalence of (ii) and (iii), which proves the theorem. //

The proof of Theorem 0.6.11 gives a practical technique for finding right inverses of matrices. It shows that finding a right inverse to an $m \times n$ matrix is the same as solving m systems of m equations in n unknowns. Since each system of equations has the same coefficient matrix, namely A, all m systems can be solved at once by row reducing a larger matrix. This is illustrated in the following examples.

Example 0.6.12 (i) To find a right inverse for

$$A = \begin{pmatrix} 1 & 0 & 0 & 2 \\ 0 & 1 & 3 & 4 \\ 1 & 1 & 4 & 2 \end{pmatrix}$$

we must solve the three systems of equations in four variables each

$$A \begin{pmatrix} X_1 \\ X_2 \\ X_3 \\ X_4 \end{pmatrix} = \begin{pmatrix} 1 \\ 0 \\ 0 \end{pmatrix}, \quad A \begin{pmatrix} Y_1 \\ Y_2 \\ Y_3 \\ Y_4 \end{pmatrix} = \begin{pmatrix} 0 \\ 1 \\ 0 \end{pmatrix} \quad \text{and} \quad A \begin{pmatrix} Z_1 \\ Z_2 \\ Z_3 \\ Z_4 \end{pmatrix} = \begin{pmatrix} 0 \\ 0 \\ 1 \end{pmatrix}$$

A right inverse will then be

$$\begin{pmatrix} X_1 & Y_1 & Z_1 \\ X_2 & Y_2 & Z_2 \\ X_3 & Y_3 & Z_3 \\ X_4 & Y_4 & Z_4 \end{pmatrix}$$

Rather than solving these three systems individually, we row-reduce the single matrix where we list all three possible value matrices on the right-hand side of the line. (So, each column of the right-hand block of the augmented matrix is the value matrix of one of the three systems.)

$$\begin{pmatrix} 1 & 0 & 0 & 2 & | & 1 & 0 & 0 \\ 0 & 1 & 3 & 4 & | & 0 & 1 & 0 \\ 1 & 1 & 4 & 2 & | & 0 & 0 & 1 \end{pmatrix} \longmapsto \begin{pmatrix} 1 & 0 & 0 & 2 & | & 1 & 0 & 0 \\ 0 & 1 & 3 & 4 & | & 0 & 1 & 0 \\ 0 & 1 & 4 & 0 & | & -1 & 0 & 1 \end{pmatrix}$$

$$\longmapsto \begin{pmatrix} 1 & 0 & 0 & 2 & | & 1 & 0 & 0 \\ 0 & 1 & 3 & 4 & | & 0 & 1 & 0 \\ 0 & 0 & 1 & -4 & | & -1 & -1 & 1 \end{pmatrix}$$

$$\longmapsto \begin{pmatrix} 1 & 0 & 0 & 2 & | & 1 & 0 & 0 \\ 0 & 1 & 0 & 16 & | & 3 & 4 & -3 \\ 0 & 0 & 1 & -4 & | & -1 & -1 & 1 \end{pmatrix}$$

Now that this matrix is in reduced row-echelon form, we see that there are infinitely many possible solutions to each system, and hence infinitely many right inverses to A. In fact, the variables X_4, Y_4, and Z_4 can be taken as parameters in describing the collection of all right inverses. For example, suppose we choose $X_4 = Y_4 = Z_4 = 0$. In order to solve for X_1, X_2, X_3, we must determine the solution to the system

$$\left(\begin{array}{cccc|c} 1 & 0 & 0 & 2 & 1 \\ 0 & 1 & 0 & 16 & 3 \\ 0 & 0 & 1 & -4 & -1 \end{array} \right)$$

where $X_4 = 0$. (Here, we use the first column to the right of the line.) We find that $X_1 = 1$, $X_2 = 3$, and $X_3 = -1$. In order to solve for Y_1, Y_2, Y_3, we must determine the solution to the system

$$\left(\begin{array}{cccc|c} 1 & 0 & 0 & 2 & 0 \\ 0 & 1 & 0 & 16 & 4 \\ 0 & 0 & 1 & -4 & -1 \end{array} \right)$$

where $Y_4 = 0$. (Here, we use the second column to the right of the line.) We find that $Y_1 = 0$, $Y_2 = 4$, and $Y_3 = -1$. Analogously, we find that $Z_1 = 0$, $Z_2 = -3$, and $Z_3 = 1$. This gives the right inverse

$$\left(\begin{array}{ccc} 1 & 0 & 0 \\ 3 & 4 & -3 \\ -1 & -1 & 1 \\ 0 & 0 & 0 \end{array} \right)$$

Alternatively, if we take $X_4 = Y_4 = Z_4 = 1$, we obtain the right inverse

$$\left(\begin{array}{ccc} -1 & -2 & -2 \\ -13 & -12 & -19 \\ 3 & 3 & 5 \\ 1 & 1 & 1 \end{array} \right)$$

The general right inverse to A is

$$\left(\begin{array}{ccc} 1 - 2X_4 & -2Y_4 & -2Z_4 \\ 3 - 16X_4 & 4 - 16Y_4 & -3 - 16Z_4 \\ -1 + 4X_4 & -1 + 4Y_4 & 1 + 4Z_4 \\ X_4 & Y_4 & Z_4 \end{array} \right)$$

(ii) To find a right inverse for

$$B = \begin{pmatrix} 1 & 1 & 1 \\ 2 & -1 & 0 \\ 2 & 2 & 0 \end{pmatrix}$$

we row-reduce

$$\left(\begin{array}{ccc|ccc} 1 & 1 & 1 & 1 & 0 & 0 \\ 2 & -1 & 0 & 0 & 1 & 0 \\ 2 & 2 & 0 & 0 & 0 & 1 \end{array} \right) \longmapsto \left(\begin{array}{ccc|ccc} 1 & 1 & 1 & 1 & 0 & 0 \\ 0 & -3 & -2 & -2 & 1 & 0 \\ 0 & 0 & -2 & -2 & 0 & 1 \end{array} \right)$$

$$\longmapsto \left(\begin{array}{ccc|ccc} 1 & 1 & 0 & 0 & 0 & 1/2 \\ 0 & -3 & 0 & 0 & 1 & -1 \\ 0 & 0 & 1 & 1 & 0 & -1/2 \end{array} \right)$$

$$\longmapsto \left(\begin{array}{ccc|ccc} 1 & 0 & 0 & 0 & 1/3 & 1/6 \\ 0 & 1 & 0 & 0 & -1/3 & 1/3 \\ 0 & 0 & 1 & 1 & 0 & -1/2 \end{array} \right)$$

We find that a right inverse for B is

$$C = \begin{pmatrix} 0 & 1/3 & 1/6 \\ 0 & -1/3 & 1/3 \\ 1 & 0 & -1/2 \end{pmatrix}$$

Since B is square, Theorem 0.6.9 shows that B is invertible and $B^{-1} = C$.

(iii) Consider

$$D = \begin{pmatrix} 1 & 0 \\ 0 & 1 \\ 1 & 1 \end{pmatrix}$$

If we try to find a right inverse for D, we find that we are forced to row-reduce

$$\left(\begin{array}{cc|ccc} 1 & 0 & 1 & 0 & 0 \\ 0 & 1 & 0 & 1 & 0 \\ 1 & 1 & 0 & 0 & 1 \end{array} \right)$$

whose reduced echelon form is

$$\left(\begin{array}{cc|ccc} 1 & 0 & 1 & 0 & 0 \\ 0 & 1 & 0 & 1 & 0 \\ 0 & 0 & 1 & 1 & -1 \end{array} \right)$$

Since there are no solutions to (any of) the associated systems of equations, we find that D does not have a right inverse.

Our next definition will help us compute left inverses, and will have many other uses as well.

Definition 0.6.13 Let A be an $n \times m$ matrix. We define the *transpose* of A, denoted A^t, to be the $m \times n$ matrix with $A^t(j, i) = A(i, j)$ for all j and i where $1 \leq j \leq m$ and $1 \leq i \leq n$. In other words A^t is the matrix obtained from A by interchanging the rows and columns. If $A^t = A$, then A is called *symmetric*.

Examples 0.6.14

(i) $\begin{pmatrix} 1 & 2 & 3 \end{pmatrix}^t = \begin{pmatrix} 1 \\ 2 \\ 3 \end{pmatrix}$

(ii) $\begin{pmatrix} 3 & 4 \\ 1 & 2 \\ 5 & 7 \end{pmatrix}^t = \begin{pmatrix} 3 & 1 & 5 \\ 4 & 2 & 7 \end{pmatrix}$

(iii) $\begin{pmatrix} 1 & 0 & 0 \\ 0 & 1 & 0 \\ 0 & 0 & 1 \end{pmatrix}^t = \begin{pmatrix} 1 & 0 & 0 \\ 0 & 1 & 0 \\ 0 & 0 & 1 \end{pmatrix}$

A crucial property of the transpose is that the transpose of a product is the product of the transposes in reverse order. (Recall that the analogous result for matrix inversion was established in Theorem 0.6.6.)

Theorem 0.6.15 *Suppose that A is an $m \times n$ matrix and B is a $n \times p$ matrix. Then $(AB)^t = (B^t)(A^t)$.*

Proof: Note first that B^t is a $p \times n$ matrix and A^t is an $n \times m$ matrix, so the product $(B^t)(A^t)$ makes sense. We assume that $A = (a_{ij})$ and $B = (b_{jk})$. According to the definition of matrix multiplication, the ikth entry of AB is

$$\sum_{j=1}^{n} a_{ij} b_{jk}$$

so that this becomes the kith entry of $(AB)^t$. Using the definition of matrix multiplication, we see also that the kith entry of $(B^t)(A^t)$ is

$$\sum_{j=1}^{n} (b_{jk})(a_{ij})$$

since b_{jk} is the kjth entry of B^t and a_{ij} is the jith entry of A^t. The theorem follows since these two sums are identical. //

Using the preceding theorem, we obtain a procedure for computing the left inverses of a matrix. In the proof we make use of the obvious fact that for any matrix A, $(A^t)^t = A$.

Corollary 0.6.16 *If B is a right inverse of A^t, then B^t is a left inverse of A, and conversely.*

Proof: By hypothesis, $A^t B = I$, so $I = I^t = (A^t B)^t = B^t (A^t)^t = B^t A$. The result follows. //

As a consequence of this corollary we can apply Theorem 0.6.11 to obtain criteria for the existence of a left inverse. We also obtain the fact that invertible matrices are necessarily square.

Corollary 0.6.17 *Suppose that A is an $m \times n$ matrix. Then A has a left inverse if and only if $rk(A^t) = n$. In particular, if a matrix is invertible, it must be square, that is, $m = n$.*

Proof: By Corollary 0.6.16, A has a left inverse if and only if A^t has a right inverse. Theorem 0.6.11 shows that the $n \times m$ matrix A^t has a right inverse if and only if $rk(A^t) = n$. This gives the first statement. Now suppose that A is invertible. Since A is $m \times n$ and has a right inverse, $rk(A) = m$. This requires that $m \leq n$ (see the remark following Definition 0.5.6). Since A has a left inverse, the $n \times m$ matrix A^t has rank n. This requires that $n \leq m$. The conclusion $m = n$ follows. //

Remark In Theorem 1.4.6 it is shown for any $m \times n$ matrix A that $rk(A) = rk(A^t)$. Had we known this result, the proof of the second statement of the preceding corollary would have been quick. The existence of both left and right inverses of A would give that $m = rk(A) = n$.

The following theorem summarizes the results we have proved relating invertibility to rank.

Theorem 0.6.18 *Suppose that A is an $m \times n$ matrix. Then*
(i) $rk(A) = m$ *if and only if A has a right inverse.*
(ii) $rk(A^t) = n$ *if and only if A has a left inverse.*
(iii) *If $m < n$, then A cannot have a left inverse.*
(iv) *If $m > n$, then A cannot have a right inverse.*
(v)   *If A is $n \times n$, then A is invertible if and only if $rk(A) = n$.*
(vi) *If A is invertible, then A is square. In this case any right or left inverse of A is the inverse of A.*

In this next example we compute the left inverse of a matrix using Corollary 0.6.16.

Example 0.6.19 We wish to find a left inverse for

$$A = \begin{pmatrix} 1 & 0 \\ 1 & 2 \\ 0 & 1 \end{pmatrix}$$

To do this, we find a right inverse for its transpose A^t. We row reduce

$$\left(\begin{array}{ccc|cc} 1 & 1 & 0 & 1 & 0 \\ 0 & 2 & 1 & 0 & 1 \end{array} \right) \longmapsto \left(\begin{array}{ccc|cc} 1 & 1 & 0 & 1 & 0 \\ 0 & 1 & 1/2 & 0 & 1/2 \end{array} \right)$$

$$\longmapsto \left(\begin{array}{ccc|cc} 1 & 0 & -1/2 & 1 & -1/2 \\ 0 & 1 & 1 & 0 & 1/2 \end{array} \right)$$

and find that a right inverse for A^t is

$$\begin{pmatrix} 1 & -1/2 \\ 0 & 1/2 \\ 0 & 0 \end{pmatrix}$$

From this we see that

$$\begin{pmatrix} 1 & 0 & 0 \\ -1/2 & 1/2 & 0 \end{pmatrix}$$

is a left inverse to our original matrix A.

One could have computed the left inverse for A by solving

$$\begin{pmatrix} X_1 & X_2 & X_3 \\ Y_1 & Y_2 & Y_3 \end{pmatrix} \begin{pmatrix} 1 & 0 \\ 1 & 2 \\ 0 & 1 \end{pmatrix} = \begin{pmatrix} 1 & 0 \\ 0 & 1 \end{pmatrix}$$

for $X_1, X_2, X_3, Y_1, Y_2, Y_3$. In doing so (after some calculation), one ends up with the same system of equations that was just solved.

Example 0.6.20 We wish to compute the inverse of

$$\begin{pmatrix} -1 & 0 & 1 \\ -5 & 1 & 3 \\ 7 & -1 & -4 \end{pmatrix}$$

by row-reduction.

$$\left(\begin{array}{ccc|ccc} -1 & 0 & 1 & 1 & 0 & 0 \\ -5 & 1 & 3 & 0 & 1 & 0 \\ 7 & -1 & -4 & 0 & 0 & 1 \end{array}\right) \longmapsto \left(\begin{array}{ccc|ccc} -1 & 0 & 1 & 1 & 0 & 0 \\ 0 & 1 & -2 & -5 & 1 & 0 \\ 0 & -1 & 3 & 7 & 0 & 1 \end{array}\right)$$

$$\longmapsto \left(\begin{array}{ccc|ccc} 1 & 0 & -1 & -1 & 0 & 0 \\ 0 & 1 & -2 & -5 & 1 & 0 \\ 0 & 0 & 1 & 2 & 1 & 1 \end{array}\right)$$

$$\longmapsto \left(\begin{array}{ccc|ccc} 1 & 0 & 0 & 1 & 1 & 1 \\ 0 & 1 & 0 & -1 & 3 & 1 \\ 0 & 0 & 1 & 2 & 1 & 1 \end{array}\right)$$

Hence,

$$\begin{pmatrix} -1 & 0 & 1 \\ -5 & 1 & 3 \\ 7 & -1 & -4 \end{pmatrix}^{-1} = \begin{pmatrix} 1 & 1 & 1 \\ -1 & 3 & 2 \\ 2 & 1 & 1 \end{pmatrix}$$

This shows that the system of equations

$$\begin{pmatrix} -1 & 0 & 1 \\ -5 & 1 & 3 \\ 7 & -1 & -4 \end{pmatrix}\begin{pmatrix} X \\ Y \\ Z \end{pmatrix} = \begin{pmatrix} a \\ b \\ c \end{pmatrix}$$

always has a *unique* solution for any $a, b, c \in \mathbf{R}$. Once this is known, finding the solution becomes quite easy. Multiplying on the left by the inverse, we find

$$\begin{pmatrix} X \\ Y \\ Z \end{pmatrix} = \begin{pmatrix} 1 & 0 & 0 \\ 0 & 1 & 0 \\ 0 & 0 & 1 \end{pmatrix}\begin{pmatrix} X \\ Y \\ Z \end{pmatrix}$$

$$= \begin{pmatrix} 1 & 1 & 1 \\ -1 & 3 & 2 \\ 2 & 1 & 1 \end{pmatrix}\begin{pmatrix} -1 & 0 & 1 \\ -5 & 1 & 3 \\ 7 & -1 & -4 \end{pmatrix}\begin{pmatrix} X \\ Y \\ Z \end{pmatrix}$$

$$= \begin{pmatrix} 1 & 1 & 1 \\ -1 & 3 & 2 \\ 2 & 1 & 1 \end{pmatrix} \begin{pmatrix} a \\ b \\ c \end{pmatrix} = \begin{pmatrix} a+b+c \\ -a+3b+2c \\ 2a+b+c \end{pmatrix}$$

Hence, the solution is always given by $X = a+b+c$, $Y = -a+3b+2c$, and $Z = 2a+b+c$.

Caution: The procedure in Example 0.6.20 worked since we knew that the system of equations had a solution to begin with. It is possible to have an *inconsistent* system of equations $AX = \vec{b}$ for which the matrix A has a left inverse L. In this situation, $L\vec{b}$ cannot be a solution to the system.

PROBLEMS

0.6.1. Find all right or left inverses, if they exist, for the following matrices. (Say why they do not exist if they do not.)

(a) $\begin{pmatrix} 0 & 0 & 0 & 3 \\ 2 & 3 & 4 & 6 \\ 2 & 3 & 4 & 0 \end{pmatrix}$ (b) $\begin{pmatrix} 1 & 8 \\ 3 & 4 \\ 2 & 2 \end{pmatrix}$ (c) $\begin{pmatrix} 1 & 4 & 1 \\ 2 & 3 & 0 \\ 1 & 4 & 5 \end{pmatrix}$

(d) $\begin{pmatrix} 0 & 1 & 1 \\ 2 & 3 & 5 \\ 1 & 2 & 3 \\ 2 & 1 & 3 \end{pmatrix}$ (e) $\begin{pmatrix} 0 & 1 & 0 & 0 \\ 1 & 0 & 0 & 0 \\ 0 & 0 & 0 & 1 \\ 0 & 0 & 1 & 0 \end{pmatrix}$ (f) $\begin{pmatrix} 33 & 696 \\ 5 & 110 \end{pmatrix}$

(g) $\begin{pmatrix} 3 & 1 & -1 & 3 \\ 2 & 1 & 1 & 4 \\ 0 & 0 & 1 & 2 \\ 0 & 0 & -2 & -3 \end{pmatrix}$

0.6.2. Find A^t if A is

(a) $\begin{pmatrix} 1 & 0 & 1 & 0 \\ 2 & 3 & 4 & 6 \\ 2 & 3 & 4 & 0 \end{pmatrix}$ (b) $\begin{pmatrix} 1 & 8 \\ 3 & 4 \\ 2 & 2 \end{pmatrix}$ (c) $\begin{pmatrix} 1 & 4 & 1 \\ 2 & 3 & 0 \\ 1 & 4 & 5 \end{pmatrix}$

0.6.3. Show that for any invertible square matrix A, $(A^{-1})^{-1} = A$.

0.6.4. Suppose that A is a $m \times n$ matrix. Show that there is a nonzero $n \times n$ matrix B with AB a matrix of all zeros if and only if $rk(A) < n$.

0.6.5. A matrix U is called *skew-symmetric* if $U = -U^t$. Show that every real square matrix A can be written uniquely in the form $A = S + U$ where S is symmetric and U is skew-symmetric.

0.6.6. Find a 2×2 matrix A for which $AA^t \neq A^t A$.

0.6.7. Suppose that C is invertible and $A = CBC^{-1}$. Prove that A is invertible if and only if B is invertible.

0.6.8. If a square matrix A is invertible, show that A^n is invertible for all $n > 1$. What is $(A^n)^{-1}$?

0.6.9. Suppose there is a nonzero column matrix C such that $AC = \vec{0}$. Show that A cannot have a left inverse.

0.6.10. Show that a triangular matrix $A = (a_{ij})$ is invertible if and only if $a_{ii} \neq 0$ for all i.

0.6.11. Suppose that N is an $n \times n$ matrix such that for some $k > 0$, $N^k = 0$, the $n \times n$ matrix of all zeroes. (Such matrices are called *nilpotent.*)
(a) Show that N is not invertible.
(b) Show that $(I_n - N)$ is invertible by verifying that its inverse is $I_n + N + N^2 + \cdots + N^{k-1}$.
(c) If the matrices A and N commute, show that $I_n + AN$ is invertible.

0.6.12. An $n \times n$ matrix A is called *idempotent* if $A^2 = A$.
(a) Describe all 2×2 real idempotent matrices.
(b) Show that if A is idempotent and nonsingular, then $A = I_n$.
(c) If $rk(A) = 1$ and A is $n \times n$, show that $A^2 = kA$ for some scalar k.

0.6.13. Denote by $O_{\mathbf{F}}(2)$ the set of all 2×2 matrices U for which

$$UU^t = \begin{pmatrix} 1 & 0 \\ 0 & 1 \end{pmatrix}$$

Show that
(a) If both U and $S \in O_{\mathbf{F}}(2)$, then $US \in O_{\mathbf{F}}(2)$.
(b) If $U \in O_{\mathbf{F}}(2)$, then $U^{-1} \in O_{\mathbf{F}}(2)$.

0.6.14. Denote by L the set of all 2×2 matrices of the form

$$(1 - t^2)^{-1/2} \cdot \begin{pmatrix} 1 & t \\ t & 1 \end{pmatrix} \quad \text{where} \quad -1 < t < 1$$

Show that
(a) If both M and $N \in L$, then $MN \in L$.
(b) If $M \in L$, then $M^{-1} \in L$.

0.6.15. Suppose A and B are square commuting matrices (that is, $AB = BA$). If A and B are invertible, show that A^{-1} and B^{-1} commute.

0.6.16. Consider the collection of 2×2 complex matrices

$$H = \left\{ \begin{pmatrix} a + bi & c + di \\ -c + id & a - bi \end{pmatrix} \mid a, b, c, d \in \mathbf{R} \right\}$$

Show that H is closed under matrix addition and multiplication and every nonzero matrix in H has an inverse in H. Show that multiplication in H is not commutative. Thus, the set H is a *non-commutative field* or a *division algebra*. H is known as the *quaternions*.

0.6.17. A matrix A is called *symmetric* if $A = A^t$.
(a) If A is symmetric, show A is square.
(b) Show that for any square matrix A, the matrices AA^t and $A + A^t$ are symmetric.
(c) Assume A and B are symmetric. Prove that AB is symmetric if and only if A and B commute. (Two matrices A and B are said to *commute* if $AB = BA$.)

0.6.18. A matrix K is called *skew-symmetric* if $K^t = -K$.
(a) For any square matrix A show that $A - A^t$ is skew-symmetric.
(b) For any skew-symmetric matrix K, if $(I - K)$ is invertible, show that $B = (I + K)(I - K)^{-1}$ satisfies $B^t B = I = BB^t$.

0.6.19. Express the following matrices as products of elementary matrices:

(a) $\begin{pmatrix} 1 & 4 \\ 3 & 2 \end{pmatrix}$ (b) $\begin{pmatrix} 0 & 0 & 1 \\ 0 & 1 & 0 \\ 1 & 0 & 0 \end{pmatrix}$ (c) $\begin{pmatrix} 1 & 0 & 1 \\ 1 & 1 & 0 \\ 0 & 0 & 2 \end{pmatrix}$

0.6.20. When is the product of two elementary matrices another elementary matrix?

0.6.21. Consider the 2×2 matrices

$$A = \begin{pmatrix} 1 & 0 \\ a & 1 \end{pmatrix} \quad \text{and} \quad B = \begin{pmatrix} 2 & 0 \\ 0 & 3 \end{pmatrix}$$

where $a \in \mathbf{R}$. Suppose that R is any 2×3 matrix. Show that R and $A(BR)$ are row-equivalent matrices.

0.6.22. (a) Interpret multiplication on the right by elementary matrices as column operations.
(b) Use (a) together with the existence of the reduced row-echelon form to show that for any $m \times n$ rank m matrix A, there exist invertible matrices P and Q such that $PAQ = (I_m \mid 0)$ is a partitioned

matrix with left hand block I_m and right-hand block the $m \times (n - m)$ matrix of zeros.

0.6.23. The elementary matrices obtained by using the third elementary operation (interchanging two rows) are called *elementary permutation* matrices.

(a) Which elementary matrices E (other than I_n) satisfy $E^2 = I_n$?

(b) A *permutation* matrix is a matrix which has a single 1 in each row and in each column, and is zero elsewhere. Show that every permutation matrix is a product of elementary permutation matrices. How many $n \times n$ permutation matrices are there?

(c) Show that any permutation matrix P satisfies $P^t P = I$. (Such matrices are called *orthogonal*.)

0.6.24. Give a direct proof that if a square matrix A has a right inverse B, then B is also a left inverse for A, using the following hints as a guide: Let e_k be the column vector with all entries 0 except for a 1 in the kth row. Since $AB = I_n$, $A(BAe_k) = Ae_k$. Since the system $AX = Ae_k$ has a unique solution X, we must have $BAe_k = e_k$. Deduce that $BA = I_n$.

0.7   Chapter 0 Summary

This section is *not* intended to be a comprehensive summary of the results of Chap. 0. Instead, listed here are the results that play a critical role in the material that follows. Those readers who intend to omit Chap. 0 and start with Chap. 1 should go through this section to make sure all these results are familiar. Since row operations are the key to all linear algebra computations, and also play a major role in many of our proofs, we first list the main result of Sec. 0.1.

Theorem 0.1.8 *If two systems of equations are row equivalent, then they have exactly the same set of solutions.*

In order best to analyze the set of solutions to a system of linear equations, one uses row operations to find a row equivalent system of equations in (reduced) row-echelon form. The set of solutions to such systems of equations can be easily determined.

Definition 0.2.6 A system of equations (with a specified ordering of its equations and variables) is in *reduced echelon form* if

(E1) The leading variable of every equation has coefficient 1.

(E2) The leading variable of every equation comes after the leading variable of each preceding equation. Equations without variables are listed last.

(RE3) The leading variable of each equation occurs in no other equation.

One uses matrices and matrix algebra to represent systems of linear equations. Many familiar laws of algebra hold for matrices, but most notably *commutativity of multiplication does not hold for matrices*. The associative law for matrix multiplication is true and is extremely important.

Theorem 0.3.7 *If A is an $n \times m$ matrix, B is an $m \times p$ matrix, and C is a $p \times q$ matrix, then $(AB)C = A(BC)$.*

As with systems of equations, one has the notions of row-echelon and reduced row-echelon matrices.

Definition 0.4.3 A matrix is said to be a *row-echelon matrix* if it satisfies (E0), (E1), and (E2) below. A row-echelon matrix is called a *reduced row-echelon matrix* if it additionally satisfies (RE3) below.

(E0) All zero rows occur below all nonzero rows.

(E1) The first non-zero entry in any nonzero row is 1. This entry is called the *leading entry* of that row.

(E2) The leading entry of each nonzero row occurs to the right of the leading entry of each row above (if any).

(RE3) The leading entry of any row is the only nonzero entry in its column.

Using the techniques of gaussian elimination and Gauss-Jordan elimination, it is always possible to find, for any matrix A, row equivalent row-echelon matrices and a (uniquely determined, see Theorem 0.5.5) row equivalent reduced row-echelon matrix. Detailed descriptions of these algorithms can be found in Sec. 0.2. These notions also enable one to define the fundamental concept of the rank of a matrix.

Definition 0.4.4 Suppose A and B are matrices and B is row equivalent to A. If B is a row-echelon matrix, we say that B is a *row-echelon form of A*. If B is a reduced row-echelon form matrix, we say that B

is the *row-reduced echelon form of A*. For any matrix A we define the *rank* of A, denoted $rk(A)$, to be the number of nonzero rows in the reduced row-echelon form of A.

The structure of solutions to a system of linear equation depends heavily upon the nature of solutions to the *associated homogeneous system* of equations.

Theorem 0.5.2 (i) *Let S denote the (nonempty) set of solutions to the homogeneous system $AX = \vec{0}$. Then S is closed under addition and scalar multiplication; that is, if $\vec{v}_1, \vec{v}_2 \in S$ and $r \in \mathbf{F}$, then $\vec{v}_1 + \vec{v}_2, r\vec{v}_1 \in S$.*
(ii) *Let T denote the set of solutions to the system of equations $AX = B$ and S the set of solutions to the associated homogeneous system of equations $AX = \vec{0}$. Suppose T is nonempty and $\vec{u} \in T$. Then every element $\vec{w} \in T$ is of the form*

$$\vec{w} = \vec{u} + \vec{v} \quad for \ some \ \ \vec{v} \in S$$

In order to know that the rank of a matrix is well-defined, one needs to show that the reduced row-echelon form of a matrix is uniquely determined. This is proved by a careful analysis of the set of solutions to the associated homogeneous system of linear equations.

Theorem 0.5.5 *The reduced row echelon form of any matrix is unique, that is, if A is any matrix and R_1 and R_2 are both reduced row-echelon form matrices that are row equivalent to A, then $R_1 = R_2$.*

By studying the ranks of the matrices associated to a system of linear equations, one can determine the critical features of the set of solutions to the system.

Theorem 0.5.7 *Consider a system of equations $AX = B$, and denote by $A' = (A \mid B)$ the augmented matrix of the system. We assume that there are n variables in X.*
(i)   *The system $AX = B$ has a solution if and only if $rk(A) = rk(A')$.*
(ii) *The system $AX = B$ has a unique solution if and only if $rk(A) = n = rk(A')$.*

An important corollary of this analysis is the following.

Corollary 0.5.8 *Suppose that A is an m × n matrix. Then*
(i) *The system $AX = \vec{0}$ always has a solution and the collection of all solutions can be described using $n - rk(A)$ parameters. In particular, if the number of variables of a homogeneous system exceeds the number of equations, the system has a nontrivial solution.*
(ii) *If $AX = B$ has a solution, then the collection of all solutions can be described using $n - rk(A)$ parameters.*

When an $n \times n$ matrix is invertible, it has many nice features. A particularly important fact is the following:

Theorem 0.6.6 *Suppose that A and B are invertible $n \times n$ matrices. Then AB is invertible and $(AB)^{-1} = B^{-1}A^{-1}$.*

An important concept needed in the sequel is that of an elementary matrix.

Definition 0.4.8 An $n \times n$ *elementary matrix* is a matrix obtained from the $n \times n$ identity matrix I_n by applying a single elementary row operation.

It follows that there are three types of elementary matrices, one for each of the three types of elementary row operations. The key fact needed about elementary matrices is the following.

Lemma 0.4.9 *Suppose E is an $n \times n$ elementary matrix and A is an $n \times m$ matrix. Then the matrix EA is equal to the matrix obtained from A by applying the elementary row operation that was used to obtain E from I_n.*

Using the lemma, it is easily checked that elementary matrices are invertible. (Just reverse the row operations!) The key application of the lemma is the following.

Theorem 0.6.8 *Suppose that A and B are row equivalent $m \times n$ matrices. Then there exist elementary matrices $E_1, E_2, \ldots, E_s$ such that $B = E_s E_{s-1} \cdots E_1 A$. In particular, there is an invertible matrix U such that $B = UA$.*

It is extremely important to know when a square matrix is invertible. This next theorem gives three equivalent formulations.

Theorem 0.6.9 *If A is an $n \times n$ (square) matrix, then the following are equivalent:*
(i)   *A is invertible.*
(ii)   *For each $\vec{y} \in \mathbf{F}$ the system $AX = \vec{y}$ has a unique solution (for X).*
(iii)   *$rk(A) = n$.*
(iv)   *$A = E_s E_{s-1} \cdots E_1$ for some collection of elementary matrices $E_1, E_2, \ldots, E_s$.*

The existence of right inverses for a matix A is closely related to the systems of equations it represents.

Theorem 0.6.11 *Suppose that A is an $m \times n$ matrix. Then the following are equivalent:*
(i)   *A has a right inverse.*
(ii)   *For each $\vec{y} \in \mathbf{F}^m$ the equation $AX = \vec{y}$ can always be solved (for X).*
(iii)   *$rk(A) = m$.*

Matrix transposition satisfies a rule analogous to that satisfied by matrix inversion.

Theorem 0.6.15 *Suppose that A is an $m \times n$ matrix and B is a $n \times p$ matrix. Then $(AB)^t = (B^t)(A^t)$.*

This next corollary enables one to translate a problem about left inverses to a problem about right inverses, and vice versa.

Corollary 0.6.16 *If B is a right inverse of A^t, then B^t is a left inverse of A, and conversely.*

As a consequence of this corollary we can apply Theorem 0.6.7 to obtain criteria for the existence of a left inverse. We also obtain the fact that invertible matrices are necessarily square.

Corollary 0.6.17 *Suppose that A is an $m \times n$ matrix. Then A has a left inverse if and only if $rk(A) = n$. In particular, if a matrix is invertible, it must be square, that is $m = n$.*

Chapter 1

Vector Spaces

1.1   Fields, Spans, and Subspaces

In this chapter we depart from the computational approach of Chap. 0 and begin the study of vector spaces. Of course, we will not abandon computations, because they are necessary to illustrate and apply the theory. But our main objective is to develop a general theory which applies to vector spaces over arbitrary fields of scalars. We first study vector spaces concretely as subspaces of $\mathbf{F}^p$, where $\mathbf{F}$ is a field of scalars. (Recall that $\mathbf{F}^p = \{(a_1, a_2, \ldots, a_p) | a_i \in \mathbf{F}\}$ is the set of p-tuples of elements of $\mathbf{F}$.) The definition of arbitrary vector spaces is given in the section which follows.

It is critical that the student make the transition from a computational viewpoint to a more universal point of view at this time. This transition is difficult and does not happen overnight. While not necessarily complicated, the problems in this chapter will likely take more time to solve than those found in a traditional computation-oriented mathematics course. The reason for this is that problems involving proofs cannot be solved by some routine calculation. One must carefully analyze what the terms mean and what the question is, before it can be solved, and not expect to apply immediately some previously used algorithm. It is only by first considering examples and then struggling directly with the concepts (not manipulating numbers) that this subject can be understood.

In the preceding chapter we used the term "field" to denote either the real or the complex numbers. This was done in order to introduce the basic concepts in a familiar setting. However in many applications of linear algebra it is necessary to use scalar fields other than $\mathbf{R}$ or $\mathbf{C}$. Precisely what we mean by this is the first topic of this section. Roughly speaking, any collection of "numbers" (or "scalars") for which addition and multiplication operations are defined, and which satisfy all the same basic algebraic rules as do $\mathbf{R}$ and $\mathbf{C}$, is called a *field*. The reader should note in the definition below that all the properties required of a field in this definition are true for the real and complex numbers. The power of the concept is that there are many extremely useful fields other than the real or complex numbers.

Definition 1.1.1 A *field* is a set $\mathbf{F}$ together with two binary operations, addition which is denoted $+$ and multiplication which is denoted by a dot (or by juxtaposition), together with two distinguished elements $0, 1 \in \mathbf{F}$, such that
(i)   For all $x, y, z \in \mathbf{F}$
$$x + y \in \mathbf{F}$$
$$x + y = y + x$$
$$(x + y) + z = x + (y + z)$$
$$0 + x = x$$
(ii)   For all $x, y, z \in \mathbf{F}$
$$xy \in \mathbf{F}$$
$$xy = yx$$
$$(xy)z = x(yz)$$
$$1x = x$$
(iii)   For all $x, y, z \in F$
$$x(y + z) = xy + xz$$
(iv)   For all $x \in \mathbf{F}$ there exists a unique $z \in \mathbf{F}$ such that $x + z = 0$.
(v)   For all $x \in \mathbf{F}$ with $x \neq 0$ there exists a unique $y \in \mathbf{F}$ such that $xy = 1$.

A familiar field is the rational numbers $\mathbf{Q} = \{r/s \mid r$ and s are integers with $s \neq 0\}$ (the set of all fractions). The reader should observe that all the conditions required above are indeed satisfied by $\mathbf{Q}$. The integers $\mathbf{Z} = \{\ldots, -3, -2, -1, 0, 1, 2, 3, \ldots\}$ do *not* form a field since the multiplicative inverse of an integer does not (in general) exist as an integer. Often, specific subsets of a field are fields themselves. This will occur whenever every sum, product, and inverse of elements (except zero) in the subset are also elements of the subset. For example, the subset $\mathbf{Q}(\sqrt{2}) = \{r + s\sqrt{2} \mid r, s \in \mathbf{Q}\} \subset \mathbf{R}$ is a field having the same operations of addition and multiplication as has $\mathbf{R}$. The details are left as an exercise. Fields of this type are important in number theory.

In Chap. 3 we will need to use the fact that the set of quotients of polynomials $f(X)/g(X)$ where $g(X) \neq 0$ is a field. (Here, X is a variable, and the coefficients of $f(X)$ and $g(X)$ lie in some field $\mathbf{F}$.) The addition and multiplication of these quotients is exactly as is studied in high school algebra. Note that one must use quotients of polynomials in order that every nonzero element has a multiplicative inverse. This field is denoted by $F(X)$ and its elements are called *rational functions*. Another useful example of a field is the integers mod p, $\mathbf{Z}/p\mathbf{Z}$, where p is a prime number. mod p arithmetic is the arithmetic of addition and multiplication on a clock with p "hours." (Thus the only elements of the field are the p hours $\{1, 2, 3, \ldots, p\}$.) These fields will be described in more detail, along with an elementary but extremely nice application in Sec. 1.6.

One can perform arithmetic over any field in the same way as one does over the familiar fields $\mathbf{R}$ and $\mathbf{C}$. A point crucial for us is that the field axioms stated above are strong enough to guarantee that all the results of Chap. 0 apply over general fields. This is true because the only properties of $\mathbf{R}$ and $\mathbf{C}$ used in the proofs in Chap. 0 were consequences of the properties listed in Definition 1.1. *Throughout the remainder of the book, we will freely use this fact that the results from Chap. 0 apply to all fields.*

We now let $\mathbf{F}$ denote an arbitrary field. In this section an element of $\mathbf{F}^p$ will be called a *p-vector*. However, in most discussions, the p is unimportant, so we drop it. Hence, in this section (and *in this section only*), a *vector* refers to an element of $\mathbf{F}^p$, for some p. It proves to be convenient to represent elements of $\mathbf{F}^p$ as columns, that is, as column vectors. We shall continue the convention of referring to elements of $\mathbf{F}$ as scalars. These notational conventions have the advantage that in later sections where we work with general vector spaces, all proofs of theorems in this section will apply verbatim. In subsequent sections we will freely apply the results from this section to the case of arbitrary vector spaces over arbitrary fields.

We begin with the definition of a linear combination of a collection of vectors. This definition will be explored throughout this section and is our first fundamental definition of linear algebra. The collection of all linear combinations of a set of vectors is known as their *span* and will provide us many examples of vector spaces.

Definition 1.1.2 Suppose $v_1, v_2, \ldots, v_n$ are vectors and $r_1, r_2, \ldots, r_n$ are scalars. We say that the vector

$$w = r_1 v_1 + r_2 v_2 + \cdots + r_n v_n$$

is a *linear combination* of $v_1, v_2, \ldots, v_n$. The set of all linear combinations of $v_1, v_2, \ldots, v_n$ is called the *span* of $v_1, v_2, \ldots, v_n$ and is usually denoted $span\{v_1, v_2, \ldots, v_n\}$.

The span of a set of vectors, $span\{v_1, v_2, \ldots, v_n\} = \{r_1 v_1 + r_2 v_2 + \cdots + r_n v_n \mid r_1, r_2, \ldots, r_n \text{ are scalars }\}$, is almost always an infinite collection of vectors. These large collections of vectors are extremely important; shortly we shall show they are vector spaces. Visually, the span of a single nonzero vector in $\mathbf{R}^3$ is a line, since it is the set of all multiples of that vector. The span of two vectors in $\mathbf{R}^3$ that point in different directions is the plane that contains them.

Examples 1.1.3 (i) Since $\vec{0} = 0 v_1 + 0 v_2 + \cdots + 0 v_n$, the zero vector $\vec{0}$ is always a linear combination of any nonempty set of vectors. It proves to be convenient to adopt the convention that the zero vector is a also a linear combination of the empty set. Thus, according to this convention $span(\emptyset) = \{\vec{0}\}$.

(ii) The vector

$$\begin{pmatrix} 1 \\ 2 \\ 3 \end{pmatrix}$$

is a linear combination of

$$\begin{pmatrix} 2 \\ 2 \\ 2 \end{pmatrix} \text{ and } \begin{pmatrix} 3 \\ 0 \\ -3 \end{pmatrix}$$

over $\mathbf{R}$ since

$$\begin{pmatrix} 1 \\ 2 \\ 3 \end{pmatrix} = 1 \begin{pmatrix} 2 \\ 2 \\ 2 \end{pmatrix} + \left(-\frac{1}{3}\right) \begin{pmatrix} 3 \\ 0 \\ -3 \end{pmatrix}$$

(iii) The vector

$$\begin{pmatrix} 1 \\ 2 \\ 3 \end{pmatrix}$$

is *not* a linear combination of

$$\begin{pmatrix} 2 \\ 0 \\ 2 \end{pmatrix} \quad \text{and} \quad \begin{pmatrix} 3 \\ 0 \\ -3 \end{pmatrix}$$

since any such linear combination cannot have a nonzero second entry. In other words, if we try to solve the vector equation

$$r \begin{pmatrix} 2 \\ 0 \\ 2 \end{pmatrix} + s \begin{pmatrix} 3 \\ 0 \\ -3 \end{pmatrix} = \begin{pmatrix} 1 \\ 2 \\ 3 \end{pmatrix}$$

we obtain the system

$$
\begin{aligned}
2r + 3s &= 1 \\
0r + 0s &= 2 \\
2r - 3s &= 3
\end{aligned}
$$

which is clearly inconsistent.

(iv) Now consider the vectors

$$v_1 = \begin{pmatrix} 2 \\ 2 \\ 2 \end{pmatrix}, \quad v_2 = \begin{pmatrix} 3 \\ 0 \\ -3 \end{pmatrix} \in \mathbf{F}^3$$

The span of these vectors, $span\{v_1, v_2\}$, is the collection of all vectors of the form

$$r \begin{pmatrix} 2 \\ 2 \\ 2 \end{pmatrix} + s \begin{pmatrix} 3 \\ 0 \\ -3 \end{pmatrix} \quad \text{where} \ r, s \in \mathbf{F}$$

Alternatively expressed, this span is

$$\left\{ \begin{pmatrix} 2r + 3s \\ 2r \\ 2r - 3s \end{pmatrix} \mid r, s \in \mathbf{F} \right\}$$

(v) Consider the four vectors

$$\begin{pmatrix} 2 \\ 1 \\ 0 \end{pmatrix} , \begin{pmatrix} 1 \\ 0 \\ -1 \end{pmatrix} , \begin{pmatrix} 4 \\ 1 \\ -2 \end{pmatrix} , \begin{pmatrix} 7 \\ 2 \\ -3 \end{pmatrix} \in \mathbf{F}^3$$

The span of these vectors is the subset of $\mathbf{F}^3$ of all vectors of the form

$$\left\{ a \begin{pmatrix} 2 \\ 1 \\ 0 \end{pmatrix} + b \begin{pmatrix} 1 \\ 0 \\ -1 \end{pmatrix} + c \begin{pmatrix} 4 \\ 1 \\ -2 \end{pmatrix} + d \begin{pmatrix} 7 \\ 2 \\ -3 \end{pmatrix} \in \mathbf{F}^3 \mid a, b, c, d \in F \right\}$$

We shall see shortly that this subset is precisely the same subset as the one given in (iv).

The next result shows how to tell if a p-vector is a linear combination of a set of other p-vectors. Note that the criterion involves testing whether an associated system of equations has a solution. Thus, using gaussian elimination, one obtains an explicit algorithm to test if a given p-vector is a linear combination of a specified collection of p-vectors. It is often the case that the study of a specific vector space question can be reduced to a problem involving systems of equations.

Theorem 1.1.4   *Let $v_1, v_2, \ldots, v_m$ be a set of p-vectors. We denote by A the matrix $A = (\, v_1 \quad v_2 \quad \cdots \quad v_m \,)$, which is the $p \times m$ matrix whose columns are the v_i. Then $w \in \mathbf{F}^p$ is a linear combination of $v_1, v_2, \ldots, v_m$ if and only if the equation $AX = w$ has a solution.*

Proof: We expand $AX = w$ according to the definition of matrix multiplication to find

$$(v_1 \quad v_2 \quad \cdots \quad v_m) \begin{pmatrix} X_1 \\ X_2 \\ \vdots \\ X_m \end{pmatrix} = w$$

if and only if

$$X_1 v_1 + X_2 v_2 + \ldots + X_m v_m = w$$

Thus, the existence of a solution to $AX = w$ is equivalent to being able to express w as a linear combination of $v_1, v_2, \ldots, v_m$. This proves the result. //

The proof of Theorem 1.1.4 carries out in general terms what we observed in Example 1.1.3 (iii). There we saw that the nonexistence of a linear combination was a consequence of the inconsistency of a system of linear equations. For example, suppose we need to know if the vectors

$$w_1 = \begin{pmatrix} 2 \\ 3 \\ 4 \end{pmatrix} , \quad w_2 = \begin{pmatrix} 2 \\ 3 \\ 5 \end{pmatrix}$$

are linear combinations of the vectors

$$\begin{pmatrix} 1 \\ 2 \\ 3 \end{pmatrix} , \quad \begin{pmatrix} 1 \\ 1 \\ 1 \end{pmatrix} , \quad \begin{pmatrix} 1 \\ 3 \\ 5 \end{pmatrix} \in \mathbf{R}^3$$

To find out, we must find the solutions to the two systems of equations

$$\begin{pmatrix} 1 & 1 & 1 \\ 2 & 1 & 3 \\ 3 & 1 & 5 \end{pmatrix} \begin{pmatrix} X \\ Y \\ Z \end{pmatrix} = \begin{pmatrix} 2 \\ 3 \\ 4 \end{pmatrix} \quad \text{and} \quad \begin{pmatrix} 1 & 1 & 1 \\ 2 & 1 & 3 \\ 3 & 1 & 5 \end{pmatrix} \begin{pmatrix} X \\ Y \\ Z \end{pmatrix} = \begin{pmatrix} 2 \\ 3 \\ 5 \end{pmatrix}$$

The first system has the solution $X = 1, Y = 1, Z = 0$, while the second system has no solutions. Hence w_1 is a linear combination of the stated vectors, while w_2 is not.

We can also use Theorem 1.1.4 to show that the spans given in Examples 1.1.3 (iv) and (v) are the same. According to the theorem, a vector

$$v = \begin{pmatrix} a \\ b \\ c \end{pmatrix} \in \mathbf{F}^3$$

lies in the span of example (iv) if and only if the system

$$\begin{pmatrix} 2 & 3 \\ 2 & 0 \\ 2 & -3 \end{pmatrix} \begin{pmatrix} X \\ Y \end{pmatrix} \begin{pmatrix} a \\ b \\ c \end{pmatrix}$$

has a solution. Row-reducing the augmented matrix gives

$$\left(\begin{array}{cc|c} 2 & 3 & a \\ 2 & 0 & b \\ 2 & -3 & c \end{array} \right) \mapsto \left(\begin{array}{cc|c} 2 & 3 & a \\ 0 & -3 & b-a \\ 0 & -6 & c-a \end{array} \right) \mapsto \left(\begin{array}{cc|c} 2 & 3 & a \\ 0 & -3 & b-a \\ 0 & 0 & c-2b+a \end{array} \right)$$

The system associated with this augmented matrix has a solution if and only if the lower right entry $c - 2b + a$ is zero. Consequently,

$a - 2b + c = 0$ if and only if v lies in the span. For example (v) we row-reduce the augmented matrix:

$$\left(\begin{array}{cccc|c} 2 & 1 & 4 & 7 & a \\ 1 & 0 & 1 & 2 & b \\ 0 & -1 & -2 & -3 & c \end{array} \right) \longmapsto \left(\begin{array}{cccc|c} 1 & 0 & 1 & 2 & b \\ 0 & 1 & 2 & 3 & a - 2b \\ 0 & -1 & -2 & -3 & c \end{array} \right)$$

$$\longmapsto \left(\begin{array}{cccc|c} 1 & 0 & 1 & 2 & b \\ 0 & 1 & 2 & 3 & a - 2b \\ 0 & 0 & 0 & 0 & c + a - 2b \end{array} \right)$$

So, we again find that $a - 2b + c = 0$ is the necessary and sufficient condition for v to lie in the span. We will shortly give more efficient methods for making such calculations.

Subsets of vectors which arise as the span of a collection of vectors are extremely important in linear algebra. They are called *vector subspaces* and are defined next. This definition is a special case of the general concept of a vector space defined in the next section.

Definition 1.1.5 A *vector subspace* of $\mathbf{F}^p$ is a subset $V \subseteq \mathbf{F}^p$ of p-vectors with $\vec{0} \in V$ such that

(i)   If $v \in V$ and $a \in \mathbf{F}$, then $av \in V$.
(ii) If $v_1, v_2 \in V$, then $(v_1 + v_2) \in V$.

Equivalently, a vector subspace of $\mathbf{F}^p$ is a nonempty set of p-vectors that is closed under the operations of scalar multiplication and addition.

Of course, $\mathbf{F}^p$ is always a vector subspace of itself. So is the subset $\{\vec{0}\}$, consisting of a single vector. Thus, the familiar spaces $\mathbf{R}, \mathbf{R}^2$, and $\mathbf{R}^3$ are vector subspaces. Note that the xy plane in $\mathbf{R}^3$, namely $\{(x, y, 0) \mid x, y \in \mathbf{R}\}$, is a vector subspace of $\mathbf{R}^3$. This is because scalar multiples and sums of vectors in the xy plane still lie in the xy plane. Note that the xy plane is the span of the vectors $(1, 0, 0)$ and $(0, 1, 0)$. The same reasoning applies to show that the xz plane and the yz plane are vector subspaces of $\mathbf{R}^3$. The purpose of defining vector subspaces is to consider subcollections of vectors of exactly this form.

Throughout the remainder of this section we will usually use the terminology "vector space" instead of saying "vector subspace of $\mathbf{F}^p$." Whenever $V = span\{v_1, v_2, \ldots, v_s\}$, we say that $v_1, v_2, \ldots, v_s$ span V, or we say that $\{v_1, v_2, \ldots, v_s\}$ is a spanning set for V. The next lemma gives two key examples of subspaces of $\mathbf{F}^p$.

Lemma 1.1.6 (i) *The span of any set of vectors $\{v_1, v_2, \ldots, v_n\}$ is a vector space. Consequently, if $w_1, \ldots, w_m \in span\{v_1, v_2, \ldots, v_n\}$, then $span\{w_1, \ldots, w_m\} \subseteq span\{v_1, v_2, \ldots, v_n\}$.*
(ii) *The set of solutions to a system of homogeneous equations is a vector space.*

Proof: (i) We must see that this set is closed under scalar multiplication and vector addition. Consider $w = a_1 v_1 + a_2 v_2 + \cdots + a_n v_n$. Then for any scalar b, we have $bw = b(a_1 v_1 + a_2 v_2 + \cdots + a_n v_n) = ba_1 v_1 + ba_2 v_2 + \cdots + ba_n v_n$, which is also a linear combination of $v_1, v_2, \ldots, v_n$. This takes care of scalar multiplication. Now consider $w_1 = a_1 v_1 + a_2 v_2 + \cdots + a_n v_n$ and $w_2 = b_1 v_1 + b_2 v_2 + \cdots + b_n v_n$. Then, as $w_1 + w_2 = (a_1 + b_1)v_1 + (a_2 + b_2)v_2 + \cdots + (a_n + b_n)v_n$, we see that $w_1 + w_2$ is also a linear combination of $v_1, v_2, \ldots, v_n$. This proves that $span\{v_1, v_2, \ldots, v_n\}$ is a vector space.

Now assume that $w_1, w_2, \ldots, w_m \in span\{v_1, v_2, \ldots, v_n\}$. Any linear combination of vectors in a vector space must also lie in that vector space. Consequently, any linear combination of $w_1, w_2, \ldots, w_m$ lies inside $span\{v_1, \ldots, v_n\}$. This shows that $span\{w_1, \ldots, w_m\} \subseteq span\{v_1, \ldots, v_n\}$; this proves (i).

(ii) That the two conditions hold for this set is precisely Theorem 0.5.2 (i). This proves the lemma. //

Examples 1.1.7 (i) The set

$$V_1 = \left\{ \begin{pmatrix} -3t \\ 2t \\ t \end{pmatrix} \mid t \in \mathbf{R} \right\}$$

is a vector space. This can be checked directly, or one can observe that V_1 is the set of solutions to

$$
\begin{array}{rrcl}
X & + & 3Z & = & 0 \\
Y & - & 2Z & = & 0
\end{array}
$$

(ii) The set

$$V_2 = \left\{ \begin{pmatrix} 2r + s \\ r - t \\ 2r + 3s \\ r - s - t \\ t \end{pmatrix} \mid r, s, t \in \mathbf{R} \right\}$$

is a vector space. Again, this can easily be checked directly, or one can observe that

$$V_2 = span \left\{ \begin{pmatrix} 2 \\ 1 \\ 2 \\ 1 \\ 0 \end{pmatrix}, \begin{pmatrix} 1 \\ 0 \\ 3 \\ -1 \\ 0 \end{pmatrix}, \begin{pmatrix} 0 \\ -1 \\ 0 \\ -1 \\ 1 \end{pmatrix} \right\}$$

(iii) The set

$$V_3 = \left\{ \begin{pmatrix} 1 \\ t \\ 2t \end{pmatrix} \mid t \in \mathbf{R} \right\}$$

is *not* a vector space. Indeed, V_3 is not closed under scalar multiplication or vector addition. For example,

$$\begin{pmatrix} 1 \\ 0 \\ 0 \end{pmatrix} \in V_3 \quad \text{but} \quad 2 \begin{pmatrix} 1 \\ 0 \\ 0 \end{pmatrix} = \begin{pmatrix} 2 \\ 0 \\ 0 \end{pmatrix} \notin V_3$$

(iv) The subset of all vectors $(x, y, z) \in \mathbf{R}^3$ for which $x + y - z = 0$ is a vector space, while the collection of all (x, y, z) for which $x + y - z = 1$ is not.

The only vector subspaces of $\mathbf{R}$ are $\{0\}$ and $\mathbf{R}$. To see this, note that if $V \subseteq \mathbf{R}$ is a vector subspace and $r \in V$ is nonzero, then $kr \in V$ for all real numbers $k \in \mathbf{R}$. But kr can be any real number, so $V = \mathbf{R}$ follows. Geometrically, a vector subspace of $\mathbf{R}^2$ is either $\{(0, 0)\}$, a line through $(0, 0)$, or all of $\mathbf{R}^2$. This is because the collection of all scalar multiples of a nonzero vector in $\mathbf{R}^2$ is the line through $(0, 0)$ and that vector, and the set of linear combinations of any two noncollinear vectors in $\mathbf{R}^2$ is all of $\mathbf{R}^2$. (This will follow explicitly as a consequence of results in Sec. 1.4, but can also be seen geometrically.) By similar reasoning, a vector subspace of $\mathbf{R}^3$ must be $\{(0, 0, 0)\}$, a line through $(0, 0, 0)$, a plane through $(0, 0, 0)$, or all of $\mathbf{R}^3$.

PROBLEMS

1.1.1. Express the vector

$$\begin{pmatrix} 2 \\ 1 \\ 3 \\ 2 \end{pmatrix}$$

as a linear combination of the following (if possible):

(a) $\begin{pmatrix} 1 \\ 0 \\ 0 \\ 1 \end{pmatrix}$, $\begin{pmatrix} 0 \\ 1 \\ 2 \\ 0 \end{pmatrix}$, $\begin{pmatrix} 0 \\ 1 \\ 1 \\ 0 \end{pmatrix}$

(b) $\begin{pmatrix} 0 \\ 1 \\ 1 \\ 0 \end{pmatrix}$, $\begin{pmatrix} 2 \\ 3 \\ 1 \\ 2 \end{pmatrix}$

(c) $\begin{pmatrix} 1 \\ 0 \\ 0 \\ 1 \end{pmatrix}$, $\begin{pmatrix} 1 \\ 0 \\ 1 \\ 0 \end{pmatrix}$, $\begin{pmatrix} 1 \\ 1 \\ 0 \\ 0 \end{pmatrix}$, $\begin{pmatrix} 1 \\ 0 \\ 0 \\ 0 \end{pmatrix}$

1.1.2. Determine if the vector

$$\begin{pmatrix} 2 \\ 1 \\ 3 \\ 2 \end{pmatrix}$$

is a linear combination of the following:

(a) $\begin{pmatrix} 2 \\ 0 \\ 0 \\ 2 \end{pmatrix}$, $\begin{pmatrix} 0 \\ 3 \\ 2 \\ 0 \end{pmatrix}$, $\begin{pmatrix} 0 \\ 4 \\ 1 \\ 0 \end{pmatrix}$

(b) $\begin{pmatrix} 0 \\ 2 \\ 1 \\ 0 \end{pmatrix}$, $\begin{pmatrix} 2 \\ 3 \\ 1 \\ 2 \end{pmatrix}$

(c) $\begin{pmatrix} 1 \\ 2 \\ 0 \\ 1 \end{pmatrix}$, $\begin{pmatrix} 1 \\ 2 \\ 1 \\ 2 \end{pmatrix}$, $\begin{pmatrix} 1 \\ 4 \\ 0 \\ 1 \end{pmatrix}$, $\begin{pmatrix} 1 \\ 0 \\ 1 \\ 2 \end{pmatrix}$

1.1.3. Determine the span of the following vectors:

(a) $\begin{pmatrix} 2 \\ 2 \\ 1 \end{pmatrix}$, $\begin{pmatrix} 0 \\ 0 \\ 0 \end{pmatrix}$, $\begin{pmatrix} -6 \\ -6 \\ -3 \end{pmatrix}$

(b) $\begin{pmatrix} 1 \\ 1 \\ 1 \end{pmatrix}$, $\begin{pmatrix} 2 \\ 0 \\ 2 \end{pmatrix}$, $\begin{pmatrix} 1 \\ 1 \\ 0 \end{pmatrix}$

(c) $\begin{pmatrix} 1 \\ 2 \\ 1 \\ 0 \end{pmatrix}$, $\begin{pmatrix} 0 \\ 3 \\ 5 \\ 1 \end{pmatrix}$

(d) $\begin{pmatrix} 1 \\ 1 \end{pmatrix}$, $\begin{pmatrix} 3 \\ 4 \end{pmatrix}$, $\begin{pmatrix} 5 \\ 2 \end{pmatrix}$, $\begin{pmatrix} 7 \\ 3 \end{pmatrix}$

1.1.4. Find the span of the following sets of vectors:

(a) $\left\{ \begin{pmatrix} 2 \\ 1 \end{pmatrix}, \begin{pmatrix} 3 \\ 3 \end{pmatrix}, \begin{pmatrix} 5 \\ 4 \end{pmatrix} \right\}$

(b) $\left\{ \begin{pmatrix} 1 \\ 2 \\ 3 \end{pmatrix}, \begin{pmatrix} 3 \\ 4 \\ 6 \end{pmatrix}, \begin{pmatrix} 3 \\ 9 \\ 0 \end{pmatrix} \right\}$

(c) $\left\{ \begin{pmatrix} 0 \\ 1 \\ 0 \\ 1 \end{pmatrix}, \begin{pmatrix} 0 \\ 4 \\ 0 \\ 4 \end{pmatrix}, \begin{pmatrix} 0 \\ 5 \\ 0 \\ 5 \end{pmatrix}, \begin{pmatrix} 0 \\ 3 \\ 0 \\ 3 \end{pmatrix} \right\}$

(d) $\left\{ \begin{pmatrix} 1 \\ 2 \\ 1 \end{pmatrix}, \begin{pmatrix} 0 \\ 3 \\ 4 \end{pmatrix}, \begin{pmatrix} 0 \\ 6 \\ 7 \end{pmatrix} \right\}$

1.1.5. Which of the following are vector subspaces of $\mathbf{R}^4$?
(a) $\{(0,0,0,0)\}$
(b) $\{(1,1,1,1)\}$
(c) $\{(0,0,0,0), (1,1,1,1), (2,2,2,2), \ldots\}$
(d) $\{r(1,1,1,1) + s(2,1,4,5) \mid r, s \in \mathbf{R}\}$
(e) $\{(1,1,1,1) + s(2,1,4,5) \mid s \in \mathbf{R}\}$
(f) $\{(x,y,z,w) \mid x,y,z,w \in \mathbf{R} \text{ and } x - z + w = 0\}$
(g) $\{(x,y,z,w) \mid x,y,z,w \in \mathbf{R} \text{ and } x - z + w = 1\}$
(h) $\{(t, 1-t, 2t, 0) \mid t \in \mathbf{R}\}$
(i) $\{(t, t, 2t, 0) \mid t \in \mathbf{R}\}$

1.1.6. Which of the following are vector subspaces of $\mathbf{F}^n$?
(a) $\{(u,v,w) \in \mathbf{F}^3 \mid u + v + w = 0\}$
(b) $\{(u,v,w) \in \mathbf{F}^3 \mid u + v + w = 1\}$
(c) $\{(u,v) \in \mathbf{F}^2 \mid uv = 0\}$
(d) $\{(u,v) \in \mathbf{R}^2 \mid u^2 + v^2 = 0\}$
(e) $\{(u,v) \in \mathbf{C}^2 \mid u^2 + v^2 = 0\}$

1.1.7. Prove that if a vector v is a linear combination of $v_1, v_2, \ldots, v_n$, then $span\{v, v_1, v_2, \ldots, v_n\} = span\{v_1, v_2, \ldots, v_n\}$.

1.1.8. Show that the subset $\mathbf{Q}(\sqrt{2}) := \{r + s\sqrt{2} \mid r, s \in \mathbf{Q}\} \subset \mathbf{R}$ is a field (with addition and multiplication as in $\mathbf{R}$). Can you generalize the idea of this example?

1.1.9. Find a system of equations whose solutions are the span of

$$\begin{pmatrix} 1 \\ 0 \\ 5 \\ 2 \end{pmatrix} \quad \text{and} \quad \begin{pmatrix} 6 \\ 0 \\ 1 \\ 2 \end{pmatrix} \quad \text{inside} \quad \mathbf{R}^4$$

1.1.10. Verify that the collection of rational functions

$$F(X) = \{f(X)/g(X) \mid f(X) \text{ and } g(X) \text{ are polynomials, } g(X) \neq 0\}$$

is a field whenever $\mathbf{F}$ is a field.

1.1.11. Consider matrices A, B, C with $C = AB$. Show that the columns of C are linear combinations of the columns of A.

1.1.12. If V and W are subspaces of $\mathbf{F}^p$, is $V \cap W$ a subspace? How about $V \cup W$?

1.2 Vector Spaces

We begin this section by considering matrices and spans of vectors. For any matrix A, there are three vector subspaces associated with the matrix. They are the row space, the column space, and the nullspace. Their analysis proves to be crucial to many problems in elementary linear algebra.

Definition 1.2.1 Suppose that A is an $m \times n$ matrix. The subspace of $\mathbf{F}^n$ spanned by the rows of A is called the *row space* of A. The subspace of $\mathbf{F}^m$ spanned by the columns of A is called the *column space* of A. The subspace of $\mathbf{F}^n$ of all solutions to $AX = \vec{0}$ is called the *nullspace* of A. The row space and column space of A have the obvious abbreviations $row(A)$ and $col(A)$. We shall denote the nullspace by $ker(A)$. (*ker* is short for "kernel," terminology introduced in Chap. 3.)

For example, consider the real matrix

$$A = \begin{pmatrix} 1 & 3 & 1 \\ 1 & 0 & 2 \end{pmatrix}$$

The row space of A is the span of the (row) vectors $(1, 3, 1)$ and $(1, 0, 2)$. The column space is the span of the columns

$$\begin{pmatrix} 1 \\ 1 \end{pmatrix}, \begin{pmatrix} 3 \\ 0 \end{pmatrix}, \begin{pmatrix} 1 \\ 2 \end{pmatrix}$$

which is the same as the span of the first two columns, which happens to be $\mathbf{F}^2$. The nullspace, $ker(A)$, is the set of solutions to the system

$$\begin{pmatrix} 1 & 3 & 1 \\ 1 & 0 & 2 \end{pmatrix} \begin{pmatrix} X \\ Y \\ Z \end{pmatrix} = \begin{pmatrix} 0 \\ 0 \end{pmatrix}$$

which is span of the single vector $(6, -1, -3)$.

For an arbitrary matrix A it is usually impossible to determine any properties of these three vector spaces by just looking at the matrix. In order to find the set of solutions to a system of equations (which is finding the nullspace), one can apply gaussian elimination to obtain a row-equivalent (reduced) echelon matrix. In order to study the row

space and column space of a matrix, one can again use gaussian elimination. The next theorem shows what information can be obtained
by row-reduction. What is crucial is the relationship between the row
spaces and column spaces of row-equivalent matrices. Note carefully
the differences between parts (i) and (ii) of the theorem.

Theorem 1.2.2 *Let A be an $m \times n$ matrix. Assume B is a matrix
that is row equivalent to A. Then*
(i)   *The row space of B is the same as the row space of A.*
(ii) *Suppose that $C_1, C_2, \ldots, C_s$, and D are columns of A and let C_1',
$C_2', \ldots,$ C_s', D' denote the corresponding columns of B. Then for any
scalars $a_1, a_2, \ldots,$ a_s, $D = a_1 C_1 + a_2 C_2 + \cdots + a_s C_s$ if and only if
$D' = a_1 C_1' + a_2 C_2' + \cdots + a_s C_s'$. In particular, D is a linear combination of $C_1, C_2, \ldots, C_s$ if and only if D' is a linear combination of
$C_1', C_2', \ldots, C_s'$. This implies that the columns of A corresponding to
the spanning columns in B span $col(A)$.*

Proof: (i) It suffices to show that the row space of a matrix does not
change when an elementary row operation is applied. Suppose that the
rows of A are $R_1, R_2, \ldots, R_m$. If an elementary row operation is applied
to A, denote the resulting rows by $R_1', R_2', \ldots, R_m'$. According to the
definitions of the elementary row operations, each of $R_1', R_2', \ldots, R_m'$ is
a linear combination of the rows $R_1, R_2, \ldots, R_m$. According to Lemma
1.1.6 (i) $span\{R_1', R_2', \ldots, R_m'\} \subseteq span\{R_1, R_2, \ldots, R_m\}$. Since elementary row operations can be reversed by elementary row opera-
tions, by the same reasoning we conclude that $span\{R_1, R_2, \ldots, R_m\} \subseteq
span\{R_1', R_2', \ldots, R_m'\}$, and hence they are equal; this proves (i).

(ii) By deleting the columns of A that are not among $C_1, C_2, \ldots, C_s, D$
and performing the same sequence of row operations, we see that
we can assume that A contains only the columns $C_1, C_2, \ldots, C_s, D$.
Changing the order in which the columns appear clearly does not
change the resulting columns when row operations are applied. Thus
we can assume that $A = (\,C_1 \quad C_2 \quad \cdots \quad C_s \quad D\,)$. We now view A as
the augmented matrix of the system

$$(C_1 \quad C_2 \quad \cdots \quad C_s) \begin{pmatrix} X_1 \\ X_2 \\ \vdots \\ X_s \end{pmatrix} = D \qquad\qquad (*)$$

According to Theorem 1.1.4, $(a_1, a_2, \ldots, a_s)$ is a solution to the system
$(*)$ if and only if D is the linear combination $D = a_1 C_1 + a_2 C_2 + \cdots + a_s C_s$. By hypothesis, we know that the matrix

$$B = (\, C_1' \quad C_2' \quad \cdots \quad C_s' \quad D'\,)$$

is row equivalent to A. It follows that $(a_1, a_2, \ldots, a_s)$ is a solution to
$(*)$ if and only if $(a_1, a_2, \ldots, a_s)$ is a solution to

$$(\, C_1' \quad C_2' \quad \cdots \quad C_s'\,) \begin{pmatrix} X_1 \\ X_2 \\ \vdots \\ X_s \end{pmatrix} = D' \qquad (**)$$

However, again by Theorem 1.1.4, $(a_1, a_2, \ldots, a_s)$ is a solution to $(**)$ if
and only if D' is the linear combination $D' = a_1 C_1' + a_2 C_2' + \cdots + a_s C_s'$.
The assertions of (ii) follow from this; this proves the theorem. $/\!/$

Example 1.2.3 Consider

$$A = \begin{pmatrix} 1 & 0 & 2 & 2 & 1 \\ 0 & 1 & 1 & 2 & 5 \\ 1 & 0 & 2 & 1 & 3 \\ 0 & 1 & 1 & 3 & 3 \end{pmatrix}$$

The reduced row-echelon form of A (which we recall is unique by Theorem 0.5.5) is

$$R = \begin{pmatrix} 1 & 0 & 2 & 0 & 5 \\ 0 & 1 & 1 & 0 & 9 \\ 0 & 0 & 0 & 1 & -2 \\ 0 & 0 & 0 & 0 & 0 \end{pmatrix}$$

From this we see that $(1, 0, 2, 0, 5)$, $(0, 1, 1, 0, 9)$, $(0, 0, 0, 1, -2)$ form a
spanning set for the row space of A. The first, second, and fourth
columns of R form a spanning set for the column space of R. Consequently, the first, third, and fourth columns of A span the column
space of A. Note, however, that the first three columns of A cannot
span the column space of A, since they fail to span the column space
of R.

Next observe that it is easy to express the last column of R as a linear combination of the earlier columns:

$$\begin{pmatrix} 5 \\ 9 \\ -2 \\ 0 \end{pmatrix} = 5 \begin{pmatrix} 1 \\ 0 \\ 0 \\ 0 \end{pmatrix} + 9 \begin{pmatrix} 0 \\ 1 \\ 0 \\ 0 \end{pmatrix} + (-2) \begin{pmatrix} 0 \\ 0 \\ 1 \\ 0 \end{pmatrix}$$

Theorem 1.2.2 (ii) shows that the last column of A is precisely the same linear combination of the corresponding columns:

$$\begin{pmatrix} 1 \\ 5 \\ 3 \\ 3 \end{pmatrix} = 5 \begin{pmatrix} 1 \\ 0 \\ 1 \\ 0 \end{pmatrix} + 9 \begin{pmatrix} 0 \\ 1 \\ 0 \\ 1 \end{pmatrix} + (-2) \begin{pmatrix} 2 \\ 2 \\ 1 \\ 3 \end{pmatrix}$$

Remark   When using row operations to study the row space and column space of a matrix, one must be careful. Row operations do not change the row space, but they *do* change the column space. So, after row-reduction, if you desire a spanning set for the row space, you may take the rows of the echelon matrix. However if you want a spanning set for the column space, you *must* return to the original matrix and locate the columns which correspond to a spanning set for the echelon matrix. On the other hand, considering matrix columns and applying row operations is especially nice because you can can find precisely if (and how) one column is a linear combination of the others.

Linear algebra is a subject with far-reaching applications both in and outside mathematics. We next indicate the broader context in which linear algebra can be developed by defining general vector spaces. This subject is much more than just another topic in mathematics. It is a language, which is used in almost every branch of advanced mathematics. The main goal of any course in this subject should be to help the student understand and speak this language.

Students without previous experience with abstract mathematics may find it takes a while to adjust to this point of view. The primary reason for this is that $\mathbf{R}^1$, $\mathbf{R}^2$, and $\mathbf{R}^3$, which are familiar from basic geometry and calculus, are replaced by a general object, namely a vector space. Students should not panic if they do not immediately feel comfortable. They should push ahead to subsequent sections, and then return to this material a number of times until they master the approach. Much of the remainder of this book could be understood if the reader replaces (in his or her mind) the field **F** with the real numbers **R**, and the vector spaces V with $\mathbf{R}^n$. For many, this is a good device to learn this material, especially since intuition is largely based upon the familiar cases of **R**, $\mathbf{R}^2$, and $\mathbf{R}^3$. *However*, we must emphasize that the powerful applications of this subject will not be understood unless the student can make the transition to the general setting.

Informally, vector spaces are a set V of vectors, with some vector addition and scalar multiplication by field elements defined, for which all the familiar algebraic laws from $\mathbf{R}^n$ hold. When studying the theory of general vector spaces, one must make sure that the only algebraic laws used are consequences of those listed in the definition.

Definition 1.2.4 A *vector space V over* a field **F** is a set V, whose elements are called *vectors*, a distinguished element of V, $\vec{0}$, called the *zero vector*, together with a binary operation $+$, called *vector addition*, and a *scalar multiplication* of vectors by elements of **F** which satisfy the following conditions:

(i)   For all $u, v, w \in V$

$u + v \in V$

$\vec{0} + v = v$

$u + v = v + u$

$u + (v + w) = (u + v) + w$

(ii)   For all $u, v \in V$ and all $r, s \in \mathbf{F}$

$rv \in V$

$1v = v$

$0v = \vec{0}$

$r(u + v) = ru + rv$

$(r + s)v = rv + sv$

$(rs)v = r(sv)$

A vector space over **R** is called a *real* vector space. The vector spaces $\mathbf{R}^n$ are usually referred to as the *euclidean spaces*. The vector spaces introduced in the preceding section were the subspaces of $\mathbf{F}^n$ (including $\mathbf{F}^n$). They arose as spanning sets of collections of vectors in $\mathbf{F}^n$. Given an arbitrary vector space, its subspaces can also be considered. These are defined next.

Definition 1.2.5 A *subspace* of a vector space V is a subset $U \subseteq V$ for which U is a vector space using the same binary operations of addition and scalar multiplication that V has.

This definition of a subspace of a vector space differs from the definition of a vector subspace of $\mathbf{F}^n$ given in Definition 1.1.5. This next lemma shows that these two definitions actually define the same concept.

Lemma 1.2.6 *A nonempty subset $U \subseteq V$ of a vector space V is a subspace of V if and only if*
(i)   *Whenever $u_1, u_2 \in U$, then $u_1 + u_2 \in U$.*
(ii)   *Whenever $k \in \mathbf{F}$ and $u \in U$, then $ku \in U$.*

Proof: Suppose that W is a subspace of the vector space V. By definition a vector space is closed under the operations of vector addition and scalar multiplication. Thus (i) and (ii) hold. Conversely, if (i) and (ii) hold for some subset $U \subseteq V$, then the closure conditions for addition and scalar multiplication (in the definition of a vector space) hold in U. Since U is nonempty, there is some $u \in U$, and consequently by condition (ii) $\vec{0} = 0u \in U$. The remaining conditions for addition and scalar multiplication hold for vectors in U, since these vectors lie in V which is assumed to be a vector space. This proves the lemma. $//$

We conclude with examples. Many more will be discussed in later sections. The verification that each of these examples is actually a vector space is omitted. In each case, the addition and multiplication rules are familiar operations, and the reader should check (mentally) that all of the axioms of Definition 1.2.4 are true.

Examples 1.2.7 (i)   Let $\mathbf{F}$ be any field. For any positive integer n, the n-fold cartesian product (set of n-tuples) $\mathbf{F}^n$ is a vector space, where addition and scalar multiplication are defined entrywise (just as in $\mathbf{R}^n$). So are the subspaces $-$ the row space, column space, and nullspace of a matrix as given in Definition 1.2.1.

(ii)   For any field $\mathbf{F}$ a *polynomial* in X is an expression of the form $a_n X^n + a_{n-1} X^{n-1} + \cdots + a_1 X + a_0$ where $a_0, a_1, \ldots, a_n \in \mathbf{F}$. (Note: In this text, a polynomial is *not* a function; it is a formal expression. Of course, polynomials can be used to define functions, but it is useful to distinguish a polynomial from the function it defines.)

Let P be the set of all polynomials in the single variable X, with coefficients in $\mathbf{F}$. Usual polynomial addition, together with the usual multiplication by field elements, make P into a vector space over $\mathbf{F}$. The subset $P^n \subset P$ of all polynomials of degree less than or equal to n is closed under polynomial addition and scalar multiplication. Thus, P^n is a vector space and a subspace of P. (Sometimes we will denote P by P^∞ to emphasize that P^∞, in contrast to P^n, contains polynomials of arbitrary degree.)

(iii)   Consider some interval $[a, b] \subset \mathbf{R}$. The collection of all continuous real valued functions $\mathcal{C}([a, b], \mathbf{R})$ defined on this interval is a real vector space, with the usual addition of functions and the usual scalar multiplication of a function by a real number. The collection $\mathcal{D}([a, b], \mathbf{R})$ of differentiable functions defined on $[a, b]$ is easily checked to be a subspace of $\mathcal{C}([a, b], \mathbf{R})$. Analogously, one can consider the vector space $\mathcal{C}(\mathbf{R}, \mathbf{R})$ of continuous real valued functions defined on all of $\mathbf{R}$. This vector space has the subspace consisting of those functions $f(x) \in \mathcal{C}(\mathbf{R}, \mathbf{R})$ which satisfy

$$\int_{-\infty}^{\infty} | f(x) | \, dx < \infty$$

(iv)   Let S be any nonempty set and $\mathbf{F}$ any field. The collection of all functions, $Fun(S, \mathbf{F})$, from S to $\mathbf{F}$ is a vector space over $\mathbf{F}$, where vector addition is the usual function addition and scalar multiplication is also as usual. [In other words if $f, g \in Fun(S, \mathbf{F})$, then $(f + g)(s) = f(s) + g(s)$ and $(kf)(s) = kf(s)$ whenever $k \in \mathbf{F}$.]

(v)   The set $M_{m \times n}(\mathbf{F})$ of all $m \times n$ matrices with entries in a field $\mathbf{F}$ is a vector space over $\mathbf{F}$ with the usual definition of matrix addition and scalar multiplication. When $m = n$, $M_{n \times n}(\mathbf{F})$ has a number of important subspaces. The symmetric matrices, that is, those $A \in M_{n \times n}(\mathbf{F})$ with $A^t = A$, form a subspace. So do the skew-symmetric matrices, those for which $A^t = -A$. Note that the set of invertible matrices is *not* a subspace. (The zero matrix is not invertible!)

(vi)   The collection of infinite sequences $(r_1, r_2, r_3, \ldots)$ of real numbers can be made into a vector space with the usual entrywise addition and scalar multiplication. It is customary to view an infinite sequence of real numbers as a function $\mathbf{N} \rightarrow \mathbf{R}$. Thus from this point of view, this vector space is actually a special case of the vector spaces described in

(iv) above. The subset of sequences $(r_1, r_2, \ldots)$ for which $\sum_{i=1}^{\infty} r_i^2 < \infty$ is a useful subspace which will be studied later.

PROBLEMS

1.2.1. Find a spanning set of two vectors for:

(a) The row space of

$$
\begin{pmatrix}
1 & 0 & 0 & 3 & 2 \\
3 & 8 & 9 & 4 & 1 \\
3 & 16 & 18 & -1 & -4
\end{pmatrix}
$$

(b) The column space of

$$
\begin{pmatrix}
1 & 0 & 0 & 3 & 2 \\
3 & 8 & 9 & 4 & 1 \\
3 & 16 & 18 & -1 & -4
\end{pmatrix}
$$

(c) The nullspace of

$$
\begin{pmatrix}
1 & 0 & 0 & 3 \\
3 & 8 & 9 & 4 \\
3 & 16 & 18 & -1
\end{pmatrix}
$$

1.2.2. For each of the following matrices find spanning sets with as few vectors as possible for the row space, column space, and nullspace.

(a) $\begin{pmatrix} 1 & 1 & 1 \\ 3 & 3 & 3 \\ 2 & 2 & 2 \end{pmatrix}$ (b) $\begin{pmatrix} 1 & 0 & 1 & 0 \\ 2 & 3 & 4 & 5 \\ 4 & 3 & 6 & 5 \end{pmatrix}$ (c) $\begin{pmatrix} 1 & 3 & 4 & 5 \end{pmatrix}$

(d) $\begin{pmatrix} 2 & 2 \\ 3 & 5 \\ 1 & 1 \\ 1 & 1 \end{pmatrix}$ (e) $\begin{pmatrix} 2 & 7 & 8 & 8 & 4 \\ 1 & 3 & 5 & 7 & 1 \\ 1 & 4 & 3 & 1 & 3 \end{pmatrix}$ (f) $\begin{pmatrix} 1 & 2 & 3 & 6 & 8 \\ 2 & 4 & 5 & 3 & 1 \end{pmatrix}$

1.2.3. (a) Give an example of two 3×3 matrices with the same column space, but different row spaces.

(b) Give an example of two 3×3 matrices with the same row space, but different column spaces.

1.2.4. Show that two matrices, possibly of different size, with the same row space have the same rank. (Hint: Study their reduced echelon forms.)

1.2.5. Which of the following subsets of $\mathbf{R}^n$ are vector spaces? Give reasons.

(a) $\{(r_1, r_2, \ldots, r_n) \mid r_i \geq 0 \text{ for all } i\}$

(b) $\{(r_1, r_2, \ldots, r_n) \mid r_1^2 + r_2^2 + \cdots + r_n^2 = 1\}$

(c) $\{(r_1, r_2, \ldots, r_n) \mid r_1^2 + r_2^2 + \cdots + r_n^2 \geq 0\}$

(d) $\{(r_1, r_2, \ldots, r_n) \mid r_1 + 2r_2 + 3r_3 + \cdots + nr_n = 0\}$

(e) $\{(r_1, r_2, \ldots, r_n) \mid r_1 + 2r_2 + 3r_3 + \cdots + nr_n = 1\}$

(f) $\{\vec{0}\}$

1.2.6. Is

$$H = \left\{ \begin{pmatrix} a \\ b \\ c \end{pmatrix} \;\middle|\; \int_0^1 (aX^2 + bX + c)\, dX = 0 \right\}$$

a vector space? Why or why not?

1.2.7. Which of the following subsets are subspaces of $M_{n \times n}(F)$? Give reasons.

(a) The subset of matrices of rank 1

(b) The subset of matrices of rank 0

(c) The matrices M which solve the equation $M^2 + M = I_n$

(d) The matrices M for which $M - M^t$ is zero

1.2.8. How can one think of the complex numbers $\mathbf{C}$ as a vector space over $\mathbf{R}$? How about as a vector space over the rational numbers $\mathbf{Q}$?

1.2.9. Which of the following collections of functions are vector subspaces of $\mathcal{C}(\mathbf{R}, \mathbf{R})$? Give reasons.

(a) The even functions, those that satisfy $f(x) = f(-x)$ for all $x \in \mathbf{R}$

(b) The odd functions, those that satisfy $f(x) = -f(-x)$ for all $x \in \mathbf{R}$

(c) The positive functions, those for which $f(x) \geq 0$ for all $x \in \mathbf{R}$

(d) The constant functions

(e) Those that solve the differential equation $y'' + y' = 2y$

(f) Those that solve the differential equation $(y')^2 = y$

(g) Those that are represented by a polynomial

(h) Those that are represented by a monic poynomial (one with leading coefficient 1)

1.2.10. Verify that each example given in Examples 1.2.7 really is a vector space. Identify three different subspaces for each of these examples.

1.2.11. Consider some infinite set S and the vector space $Fun(S, \mathbf{F})$ over a field $\mathbf{F}$ as in Example 1.2.7 (iv). Let $Fun_0(S, \mathbf{F}) \subset Fun(S, \mathbf{F})$ be the subset of all functions which take the value 0 at all but a finite number of elements of S. Is $Fun_0(S, \mathbf{F})$ a subspace of $Fun(S, \mathbf{F})$? Why or why not?

1.2.12. Suppose that U and V are both $\mathbf{F}$ vector spaces. Explain how to make the set $U \times V = \{(u, v) \mid u \in U \text{ and } v \in V\}$ into a vector space. How can one view U and V as subspaces of this vector space?

1.2.13.   Let U be a $\mathbf{F}$ vector space. Suppose that V and W are subspaces of U.
(a) Show that $V \cap W$ is a subspace of U.
(b) Show that $V + W = \{v + w \mid v \in V, w \in W\}$ is a subspace of U.
(c) Suppose that $V \cap W = \{\vec{0}\}$. Show that every $u \in V + W$ can be expressed uniquely in the form $u = v + w$ where $v \in V$ and $w \in W$. When this occurs, we say that V and W are *independent subspaces* and $V + W$ is a *direct sum*. In this case we denote $V + W$ by $V \oplus W$.

1.2.14.   Suppose that U_1, U_2, and W are subspaces of a vector space V.
(a) Prove that $(U_1 \cap W) + (U_2 \cap W) \subseteq (U_1 + U_2) \cap W$.
(b) Give an example in $\mathbf{F}^2$ that shows equality need not hold in (a).

1.3   Linear Independence

We recall from Example 1.2.3 the matrix

$$\begin{pmatrix} 1 & 0 & 2 & 2 & 1 \\ 0 & 1 & 1 & 2 & 5 \\ 1 & 0 & 2 & 1 & 3 \\ 0 & 1 & 1 & 3 & 3 \end{pmatrix}$$

whose reduced row-echelon form is

$$R = \begin{pmatrix} 1 & 0 & 2 & 0 & 5 \\ 0 & 1 & 1 & 0 & 9 \\ 0 & 0 & 0 & 1 & -2 \\ 0 & 0 & 0 & 0 & 0 \end{pmatrix}$$

We observed at that time (using Theorem 1.2.2) that the first, second, and fourth columns of A formed a spanning set for $col(A)$, while the first, second, and third columns did not. Why is this? Why is it that $col(A)$ can be spanned by three vectors, when it originally is spanned by five? Such questions are answered in this section.

The key concept which enables us to answer these questions is that of the linear independence of vectors in a vector space. The definition is given next. This concept is crucial to *everything* in the remainder of this book. The reader must study the definition carefully. In particular, note the "if, then" nature of this definition. There is a great tendency among students (at least initially) to oversimplify the meaning of linear independence. Be careful, and note Remark 1.3.2.

Definition 1.3.1 A collection of vectors $v_1, v_2, \ldots, v_n$ is said to be *linearly independent* if whenever $r_1, r_2, \ldots, r_n$ are scalars and $r_1 v_1 + r_2 v_2 + \cdots + r_n v_n = \vec{0}$, then necessarily $r_1 = 0, r_2 = 0, \ldots, r_n = 0$. (In other words, the only way to express $\vec{0}$ as a linear combination of $v_1, v_2, \ldots, v_n$ is to use only 0s as scalar coefficients.) If $v_1, v_2, \ldots, v_n$ are not linearly independent, we say that $v_1, v_2, \ldots, v_n$ are *linearly dependent*.

Remark 1.3.2 The case of two vectors is quite simple to understand. By definition v_1 and v_2 are linearly dependent if one can find scalars $r_1, r_2 \in \mathbf{F}$ (not both 0) so that $r_1 v_1 + r_2 v_2 = \vec{0}$. Assuming that $r_1 \neq 0$, we find that $v_1 = -(r_2/r_1) v_2$. In particular one vector is a scalar multiple of the other. *However*, one must not incorrectly generalize this observation. For example, no one of the vectors

$$\begin{pmatrix} 1 \\ 0 \end{pmatrix}, \begin{pmatrix} 0 \\ 1 \end{pmatrix}, \begin{pmatrix} -1 \\ -1 \end{pmatrix}$$

is a scalar multiple of another. Yet these three vectors are linearly dependent since

$$\begin{pmatrix} 1 \\ 0 \end{pmatrix} + \begin{pmatrix} 0 \\ 1 \end{pmatrix} + \begin{pmatrix} -1 \\ -1 \end{pmatrix} = \begin{pmatrix} 0 \\ 0 \end{pmatrix}$$

On the other hand it is true that whenever a collection of vectors is linearly dependent, one vector is a linear combination of the others (see Lemma 1.3.8).

Examples 1.3.3 (i) The vectors

$$\begin{pmatrix} 0 \\ 0 \\ 1 \end{pmatrix}, \begin{pmatrix} 1 \\ 0 \\ 1 \end{pmatrix}, \begin{pmatrix} 0 \\ 1 \\ 0 \end{pmatrix}$$

are linearly independent in $\mathbf{R}^3$, since the equation

$$a \begin{pmatrix} 0 \\ 0 \\ 1 \end{pmatrix} + b \begin{pmatrix} 1 \\ 0 \\ 1 \end{pmatrix} + c \begin{pmatrix} 0 \\ 1 \\ 0 \end{pmatrix} = \begin{pmatrix} 0 \\ 0 \\ 0 \end{pmatrix}$$

implies that $b = 0, c = 0, a + b = 0$; hence $a = 0$ as well.

(ii) The vectors

$$\begin{pmatrix} 1 \\ 1 \end{pmatrix}, \begin{pmatrix} 3 \\ 4 \end{pmatrix}, \begin{pmatrix} 5 \\ 6 \end{pmatrix}$$

are not linearly independent since we have

$$2 \begin{pmatrix} 1 \\ 1 \end{pmatrix} + \begin{pmatrix} 3 \\ 4 \end{pmatrix} - \begin{pmatrix} 5 \\ 6 \end{pmatrix} = \begin{pmatrix} 0 \\ 0 \end{pmatrix}$$

Note that this vector equation is equivalent to the statement that the matrix equation

$$\begin{pmatrix} 1 & 3 & 5 \\ 1 & 4 & 0 \end{pmatrix} \begin{pmatrix} X \\ Y \\ Z \end{pmatrix} = \begin{pmatrix} 0 \\ 0 \end{pmatrix}$$

has the nontrivial solution

$$\begin{pmatrix} 2 \\ 1 \\ -1 \end{pmatrix}$$

(iii) The vectors

$$\begin{pmatrix} 1 \\ 1 \\ 1 \end{pmatrix}, \begin{pmatrix} 1 \\ 1 \\ 0 \end{pmatrix}, \begin{pmatrix} 1 \\ 0 \\ 0 \end{pmatrix}$$

are linearly independent in $\mathbf{F}^3$.

(iv) The vectors

$$\begin{pmatrix} 0 \\ 0 \end{pmatrix} \quad \text{and} \quad \begin{pmatrix} 1 \\ 0 \end{pmatrix}$$

are *not* linearly independent since one can express

$$\begin{pmatrix} 0 \\ 0 \end{pmatrix} = 1 \begin{pmatrix} 0 \\ 0 \end{pmatrix} + 0 \begin{pmatrix} 1 \\ 0 \end{pmatrix}$$

(v) The vectors $1, X, X^2, \ldots, X^n$ are linearly independent in the vector space P^n of all polynomials of degree at most n. This follows, since the only way to express the zero polynomial $0 = a_0 1 + a_1 X + a_2 X^2 + \cdots + a_n X^n$ is with each coefficient zero, namely $a_0 = 0, a_1 = 0, \ldots, a_n = 0$.

The next result tells precisely how to test for the linear independence of p-vectors. Note the similarity of this result with Theorem 1.1.4. Observe that in the proof we reduce the problem to a question about the nature of solutions to a system of equations (as was done in the above examples).

Theorem 1.3.4 *Let $v_1, v_2, \ldots, v_m$ be a collection of p-vectors. We denote by A the matrix $A = (\, v_1 \quad v_2 \quad \cdots \quad v_m \,)$, the $p \times m$ matrix whose columns are the v_i. Then $v_1, v_2, \ldots, v_m$ are linearly independent if and only if $rk(A) = m$.*

Proof: We expand $AX = \vec{0}$ according to the definition of matrix multiplication to find

$$(\, v_1 \quad v_2 \quad \cdots \quad v_m \,) \begin{pmatrix} X_1 \\ X_2 \\ \vdots \\ X_m \end{pmatrix} = \vec{0}$$

if and only if

$$X_1 v_1 + X_2 v_2 + \cdots + X_m v_m = \vec{0}$$

This shows that the linear independence of $v_1, v_2, \ldots, v_m$ is equivalent to the system $AX = \vec{0}$ having the unique solution $X_1 = 0, X_2 = 0, \ldots, X_m = 0$. According to Theorem 0.5.7 (ii) this occurs if and only if $rk(A) = m$. This proves the theorem. //

Example 1.3.5 Consider the vectors

$$\begin{pmatrix} 1 \\ 4 \\ 2 \\ 0 \end{pmatrix}, \begin{pmatrix} 0 \\ 0 \\ 1 \\ 0 \end{pmatrix}, \begin{pmatrix} 0 \\ 1 \\ 2 \\ -1 \end{pmatrix}, \begin{pmatrix} 1 \\ 2 \\ -1 \\ 2 \end{pmatrix} \in \mathbf{F}^4$$

Row-reducing gives

$$\begin{pmatrix} 1 & 0 & 0 & 1 \\ 4 & 0 & 1 & 2 \\ 2 & 1 & 2 & -1 \\ 0 & 0 & -1 & 2 \end{pmatrix} \mapsto \begin{pmatrix} 1 & 0 & 0 & 1 \\ 0 & 0 & 1 & -2 \\ 0 & 1 & 2 & -3 \\ 0 & 0 & -1 & 2 \end{pmatrix} \mapsto \begin{pmatrix} 1 & 0 & 0 & 1 \\ 0 & 1 & 0 & 1 \\ 0 & 0 & 1 & -2 \\ 0 & 0 & 0 & 0 \end{pmatrix}$$

Since the matrix has rank 3, this shows that the original four vectors are not linearly independent. However the first three vectors listed are linearly independent since the matrix consisting of the first three columns has rank 3. Viewing the last column above as an augmented column and using Theorem 1.1.4, we see that the fourth vector listed is a linear combination of the first three.

The preceding example showed (in the fourth column) that the failure of linear independence is closely related to the existence of linear combinations. The next lemma formulates this relationship in the abstract setting.

Lemma 1.3.6 *Suppose $v_1, v_2, \ldots, v_m$ are linearly independent vectors in a vector space V. Let $w \in V$. Then $v_1, v_2, \ldots, v_m, w$ are linearly independent if and only if w is not a linear combination of $v_1, v_2, \ldots, v_m$.*

Proof: Suppose that w is a linear combination of $v_1, v_2, \ldots, v_n$, say $w = a_1 v_1 + a_2 v_2 + \cdots + a_n v_n$. Then $a_1 v_1 + a_2 v_2 + \cdots + a_n v_n - w = \vec{0}$, which shows that $v_1, v_2, \ldots, v_n, w$ are not linearly independent. Conversely, assume that w is not a linear combination of $v_1, v_2, \ldots, v_n$. Suppose that $a_1 v_1 + a_2 v_2 + \cdots + a_n v_n + bw = \vec{0}$, where $a_1, a_2, \ldots, a_n, b$ are scalars. If $b \neq 0$, then we can divide by b, and hence $w = (-a_1/b)v_1 + (-a_2/b)v_2 + \cdots + (-a_n/b)v_n$. This contradicts our assumption that w is not a linear combination of $v_1, v_2, \ldots, v_n$ and shows that $b = 0$. But now $a_1 v_1 + \cdots + a_n v_n = \vec{0}$, so by the linear independence of $v_1, v_2, \ldots, v_n$, we see that $a_1 = 0, a_2 = 0, \ldots, a_n = 0$ as well. Altogether, this shows that $v_1, v_2, \ldots, v_n, w$ are linearly independent and proves the lemma. //

In the study of systems of linear equations, whenever an $n \times n$ matrix A has rank n, the system of equations $A\vec{x} = \vec{b}$ has a unique solution for any $\vec{b}$. (After all, such A is invertible.) By the proof of Theorem 1.1.4 this shows that $\vec{b}$ can be expressed uniquely as a linear combination of the (necessarily linearly independent) columns of A. This is a general fact about linearly independent sets of vectors and is proved next. Its importance will become apparent in the next section when we study the concept of basis.

Lemma 1.3.7 *Suppose that $v_1, v_2, \ldots, v_n$ are linearly independent and $w \in span\{v_1, v_2, \ldots, v_n\}$. Then there exists a unique sequence of* *scalars* $a_1, a_2, \ldots, a_n$ *such that* $w = a_1 v_1 + a_2 v_2 + \cdots + a_n v_n$. *In other words* w *can be expressed in exactly one way as a linear combination of the vectors* $v_1, v_2, \ldots, v_n$.

Proof: By hypothesis, since w is a linear combination of $v_1, v_2, \ldots, v_n$, we know there exist scalars $a_1, a_2, \ldots, a_n$ such that $w = a_1 v_1 + a_2 v_2 + \cdots + a_n v_n$. Suppose that $b_1, b_2, \ldots, b_n$ are also scalars such that $w = b_1 v_1 + b_2 v_2 + \cdots + b_n v_n$. We must show that $a_1 = b_1, a_2 = b_2, \ldots, a_n = b_n$. Subtracting the first expression for w from the second, we find $\vec{0} = w - w = (a_1 v_1 + \cdots + a_n v_n) - (b_1 v_1 + \cdots + b_n v_n) = (a_1 - b_1) v_1 + (a_2 - b_2) v_2 + \cdots + (a_n - b_n) v_n$. The linear independence of $v_1, v_2, \ldots, v_n$ shows that $a_1 - b_1 = 0, a_2 - b_2 = 0, \ldots, a_n - b_n = 0$; this proves the lemma. //

For example, consider the linearly independent vectors

$$\begin{pmatrix} 1 \\ 0 \end{pmatrix} \quad \text{and} \quad \begin{pmatrix} 0 \\ 1 \end{pmatrix}$$

in $\mathbf{F}^2$. These vectors clearly span $\mathbf{F}^2$. Lemma 1.3.7 asserts that every vector

$$\begin{pmatrix} a \\ b \end{pmatrix} \in \mathbf{F}^2$$

can be expressed uniquely as a linear combination

$$\begin{pmatrix} a \\ b \end{pmatrix} = a \begin{pmatrix} 1 \\ 0 \end{pmatrix} + b \begin{pmatrix} 0 \\ 1 \end{pmatrix}$$

This should be clear by inspection. As another example we look at the linearly independent set $1, X, X^2, \ldots, X^n \in P^n$. Since these vectors span P^n, we find that every polynomial in P^n can be expressed uniquely in the form $a_0 1 + a_1 X + a_2 + \cdots + a_n X^n$. But of course, by definition every polynomial in P^n is written (uniquely) in this form.

The following lemma again illustrates the interrelationship between the concepts of linear independence and linear combinations.

Lemma 1.3.8 *Suppose that* $v_1, v_2, \ldots, v_n$ *are linearly dependent vectors in some vector space* V. *Then there exists some* j *with* $1 \leq j \leq n$ *such that* v_j *is a linear combination of* $v_1, v_2, \ldots, v_{j-1}$.

Proof: If $v_1, v_2, \ldots, v_n$ are linearly dependent, then there exist scalars $a_1, a_2, \ldots, a_n$, not all zero, such that $a_1v_1 + a_2v_2 + \cdots + a_nv_n = \vec{0}$. Choose j to be the largest subscript such that $a_j \neq 0$. Then $a_1v_1 + a_2v_2 + \cdots + a_jv_j = \vec{0}$. Since $a_j \neq 0$, we find that $v_j = (-a_1/a_j)v_1 + (-a_2/a_j)v_2 + \cdots + (-a_{j-1}/a_j)v_{j-1}$. This proves the lemma. //

This next result is an important application of the theory of systems of equations to abstract vector spaces. The result is crucial for the definition of dimension, which is given in the next section.

Theorem 1.3.9 *Let V be a vector space. Suppose that $w_1, w_2, \ldots, w_m$ are linearly independent and each of $w_1, w_2, \ldots, w_m$ is a linear combination of $v_1, v_2, \ldots, v_n \in V$. Then $m \leq n$.*

Proof: We assume that $m > n$ and derive a contradiction. For $i = 1, 2, \ldots, m$ there are constants a_{ij} such that $w_i = a_{1i}v_1 + a_{2i}v_2 + \cdots + a_{ni}v_n$. Consider the equation $X_1w_1 + X_2w_2 + \cdots + X_mw_m = \vec{0}$ in the variables $X_1, X_2, \ldots, X_m$. Since $w_1, w_2, \ldots, w_m$ are linearly independent, this equation must have only the solution $X_1 = 0, X_2 = 0, \ldots, X_m = 0$. Substituting our expressions for the w_i in this equation, we obtain:

$$
\begin{aligned}
\vec{0} &= X_1(a_{11}v_1 + a_{21}v_2 + \cdots + a_{n1}v_n) \\
&\quad + X_2(a_{12}v_1 + a_{22}v_2 + \cdots + a_{n2}v_n) \\
&\quad\quad + \cdots + X_m(a_{1m}v_1 + a_{2m}v_2 + \cdots + a_{nm}v_n) \\
&= (a_{11}X_1 + a_{12}X_2 + \cdots + a_{1m}X_m)v_1 \\
&\quad + (a_{21}X_1 + a_{22}X_2 + \cdots + a_{2m}X_m)v_2 \\
&\quad\quad + \cdots + (a_{n1}X_1 + a_{n2}X_2 + \cdots + a_{nm}X_m)v_n
\end{aligned}
$$

According to Corollary 0.5.8 (i), since there are more variables than equations (that is, $m > n$), the system

$$
\begin{aligned}
a_{11}X_1 &+ a_{12}X_2 + \cdots + a_{1m}X_m = 0 \\
a_{21}X_1 &+ a_{22}X_2 + \cdots + a_{2m}X_m = 0 \\
&\quad \vdots \qquad\quad \vdots \qquad \ddots \qquad \vdots \qquad \vdots \\
a_{n1}X_1 &+ a_{n2}X_2 + \cdots + a_{nm}X_m = 0
\end{aligned}
$$

always has a nonzero solution. Since such a solution gives a nonzero solution to $X_1w_1 + X_2w_2 + \cdots + X_mw_m = \vec{0}$, we have a contradiction. This proves the theorem. //

Theorem 1.3.9 has many important consequences which will be detailed in the next section. One immediate consequence is that it is impossible to find m linearly independent vectors in $\mathbf{F}^n$ whenever $m > n$. This is because $\mathbf{F}^n$ has a set of n spanning vectors, namely $\{(1,0,\ldots,0),(0,1,0,\ldots 0),\ldots,(0,\ldots,0,1)\}$. More generally, it follows from Theorem 1.3.9 that there can never be more linearly independent vectors in a vector space than there are elements in a minimally sized spanning set. This final lemma shows how to locate minimal spanning sets for the solution space to a homogeneous system of equations.

Lemma 1.3.10 *Suppose that $RX = \vec{0}$ is a reduced echelon form system with s free variables. [Here, $s = n - rk(R)$ if R is an $m \times n$ matrix.] Denote by w_i the solution to $RX = \vec{0}$ for which the ith free variable is assigned 1 and all other free variables are assigned 0. Then the collection $\{w_1,\ldots,w_s\}$ is a spanning set for the solution space. Moreover, this collection is a minimal spanning set; that is, no proper subcollection spans.*

Proof: Let w be any solution to the system, and suppose that the value of the ith free variable in w is r_i. Then the principle of back substitution guarantees that w is the unique solution to $RX = \vec{0}$ with this assignment to free variables. As $r_1 w_1 + r_2 w_2 + \cdots + r_s w_s$ has the same assignment to the free variables as does w, we conclude that $w = r_1 w_1 + r_2 w_2 + \cdots + r_s w_s$. Hence $\{w_1,\ldots,w_s\}$ spans the solution set to $RX = \vec{0}$.

To see that the collection $\{w_1,\ldots,w_s\}$ is a minimal spanning set, observe that if w_j was left out, then any linear combination of the remaining w_i would assign the value 0 to the jth free variable. In particular, the solution w_j does not lie in the span of the remaining w_i. Thus no proper subcollection of $\{w_1,\ldots,w_s\}$ can span. This proves the lemma. //

Spanning sets of minimal size are extremely important (they are called a *basis*) and will be studied in detail in the next section. The second statement in part (i) of Lemma 1.1.6 is quite useful in this connection. For example, one often has a spanning set, say $\{v_1,\ldots,v_n\}$, of some vector space. Suppose that v_1 is a linear combination of $v_2,\ldots,v_n$. Then since $v_1,v_2,\ldots,v_n \in span\{v_2,\ldots,v_n\}$, Lemma 1.1.6 (i) says that $span\{v_1,v_2,\ldots,v_n\} \subseteq span\{v_2,\ \ldots,\ v_n\}$. But clearly,

$span\{v_2,\ldots,v_n\} \subseteq span\{v_1, v_2, \ldots, v_n\}$, so we find $span\{v_2, \ldots,v_n\} = span\{v_1, v_2, \ldots, v_n\}$. Thus we have reduced the size of our spanning set.

PROBLEMS

1.3.1. Which of the following sets of vectors are linearly independent?

(a) $\begin{pmatrix} -1 \\ 2 \\ 1 \\ 0 \\ 1 \end{pmatrix}$, $\begin{pmatrix} 2 \\ 1 \\ 1 \\ 3 \\ 2 \end{pmatrix}$, $\begin{pmatrix} 0 \\ 5 \\ 3 \\ 3 \\ 6 \end{pmatrix}$ (b) $\begin{pmatrix} 1 \\ 3 \\ 7 \end{pmatrix}$, $\begin{pmatrix} 1 \\ 2 \\ 2 \end{pmatrix}$, $\begin{pmatrix} 1 \\ 1 \\ 3 \end{pmatrix}$

(c) $\begin{pmatrix} 8 \\ 2 \\ 4 \end{pmatrix}$, $\begin{pmatrix} 4 \\ 1 \\ 2 \end{pmatrix}$

(d) $1, X, X^2 + 1, X^3 + X + 2 \in P^3$
(e) $X^3 + X + 1, X^3 - X^2 + X, 2X^2 + X + 3, 4 \in P^4$
(f) $\sin x, \cos x, e^x \in C(\mathbf{R}, \mathbf{R})$

1.3.2. Which of the following are linearly independent?

(a) $\begin{pmatrix} 2 \\ 1 \\ 0 \end{pmatrix}$, $\begin{pmatrix} 3 \\ 2 \\ 1 \end{pmatrix}$, $\begin{pmatrix} 4 \\ 3 \\ 2 \end{pmatrix}$ (b) $\begin{pmatrix} 9 \\ 1 \\ 5 \end{pmatrix}$, $\begin{pmatrix} 10 \\ 2 \\ 6 \end{pmatrix}$, $\begin{pmatrix} 11 \\ 3 \\ 7 \end{pmatrix}$, $\begin{pmatrix} 12 \\ 4 \\ 8 \end{pmatrix}$

(c) $X, X^2 + 1, X^3 + X, X^3 - 2X^2 - 2X - 2 \in P^4$
(d) $X, X^2 + 1, X^3 + X \in P^4$

1.3.3. For each of the following sets of vectors, find a subset of minimal size with the same span.

(a) $\begin{pmatrix} 2 \\ 3 \\ 1 \end{pmatrix}$, $\begin{pmatrix} 2 \\ 5 \\ 2 \end{pmatrix}$, $\begin{pmatrix} 4 \\ 8 \\ 3 \end{pmatrix}$ (b) $\begin{pmatrix} 1 \\ 1 \\ 3 \end{pmatrix}$, $\begin{pmatrix} 1 \\ 3 \\ 3 \end{pmatrix}$, $\begin{pmatrix} 2 \\ 4 \\ 3 \end{pmatrix}$, $\begin{pmatrix} 5 \\ 1 \\ 3 \end{pmatrix}$

(c) $\begin{pmatrix} 1 \\ 1 \\ 1 \end{pmatrix}$, $\begin{pmatrix} 3 \\ 2 \\ 1 \end{pmatrix}$, $\begin{pmatrix} 5 \\ 1 \\ 1 \end{pmatrix}$ (d) $\begin{pmatrix} 0 \\ 2 \\ 4 \\ 1 \end{pmatrix}$, $\begin{pmatrix} 1 \\ 3 \\ 6 \\ 2 \end{pmatrix}$, $\begin{pmatrix} 1 \\ 3 \\ 6 \\ 2 \end{pmatrix}$, $\begin{pmatrix} 0 \\ 3 \\ 6 \\ 3/2 \end{pmatrix}$, $\begin{pmatrix} 1 \\ 1 \\ 2 \\ 1 \end{pmatrix}$

1.3.4. Assume that v_1, v_2, v_3 are linearly independent in $\mathbf{F}^4$. Which of the following are linearly independent?
(a) $\{v_1 + v_2, v_2 + v_3\}$
(b) $\{\vec{0}, v_1, v_2, v_3\}$
(c) $\{v_1, v_2, v_1 + v_2\}$
(d) $\{v_1 + v_2, v_2 - v_3, v_1 + v_3\}$

1.3.5. Suppose that $v_1, v_2, \ldots, v_n$ are linearly independent. Show that $v_1, v_2, \ldots, v_{n-1}$ are linearly independent.

1.3.6. (a) Find three vectors in $\mathbf{R}^3$ which are linearly dependent, any two of which are linearly independent.
(b) Find four vectors in $\mathbf{R}^4$ which are linearly dependent, any three of which are linearly independent.

1.3.7. Suppose $v_1, v_2, \ldots, v_n$ are vectors such that $v_1 \neq \vec{0}$, $v_2 \notin span\{v_1\}$, $v_3 \notin span\{v_1, v_2\}, \ldots, v_n \notin span\{v_1, v_2, \ldots, v_{n-1}\}$. Show that $v_1, v_2, \ldots, v_n$ are linearly independent.

1.3.8. Find two linearly independent vectors such that every solution to

$$\begin{pmatrix} 1 & -1 & 2 & -2 \\ 3 & 0 & 1 & 1 \\ 2 & 1 & -1 & 3 \end{pmatrix} \begin{pmatrix} X \\ Y \\ Z \end{pmatrix} = \begin{pmatrix} 0 \\ 0 \\ 0 \end{pmatrix}$$

is a linear combination of these two vectors.

1.3.9. Suppose that $V = span\{v_1, v_2, \ldots, v_n\}$ and let $w_1, w_2, \ldots, w_m \in V$ where $m > n$. Show that $w_1, w_2, \ldots, w_m$ are linearly dependent.

1.3.10. Suppose that V and W are subspaces of $\mathbf{F}^n$ with the property that $V \cap W = \{\vec{0}\}$. Show that whenever $v_1, \ldots, v_s \in V$ are linearly independent and $w_1, \ldots, w_t \in W$ are linearly independent, then $v_1, \ldots, v_s, w_1, \ldots, w_t$ are linearly independent. Conclude that $s+t \leq n$.

1.3.11. If $v_1, v_2, \ldots, v_n$ are linearly independent, show that $v_1, v_1 + v_2, v_1 + v_2 + v_3, \ldots, v_1 + \cdots + v_n$ are linearly independent.

1.3.12. Suppose $u \in span\{v_1, v_2, \ldots, v_n, w\}$, but $u \notin span\{v_1, v_2, \ldots, v_n\}$. Show that $w \in span\{v_1, v_2, \ldots, v_n, u\}$.

1.3.13. Suppose $v_1, v_2, \ldots, v_n$ are linearly independent vectors in $\mathbf{R}^m$. Assume that A is an $m \times m$ invertible matrix. Show that $Av_1, Av_2, \ldots, Av_n$ are linearly independent.

1.3.14. Suppose u, v, w are linearly independent in some real vector space. Show that $u + v, v + w, u + w$ are linearly independent and $span\{u, v, w\} = span\{u + v, v + w, u + w\}$.

1.3.15. Find exact conditions on the real numbers $a, b, c \in \mathbf{R}$ which guarantee that the following vectors are linearly independent:

$$\begin{pmatrix} 1 \\ 1 \\ 1 \\ a \end{pmatrix}, \begin{pmatrix} 1 \\ 0 \\ 1 \\ b \end{pmatrix}, \begin{pmatrix} -2 \\ 2 \\ -2 \\ c \end{pmatrix}$$

1.3.16. Suppose that $v_1, v_2, \ldots, v_n$ are linearly independent and $v \notin span\{v_1 v_2, \ldots, v_n\}$. Show that $v_1 + v, v_2 + v, \ldots, v_n + v$ are linearly independent.

1.3.17. We say that an infinite collection of vectors $v_1, v_2, \ldots$ is linearly independent if each finite subcollection is linearly independent. Find an infinite collection of vectors in P (the vector space of all polynomials over $\mathbf{F}$) which is linearly independent.

1.3.18. Prove that two $n \times m$ matrices A and B have the same row space if and only if they are row equivalent.

1.4   Basis and Dimension

Suppose one considers the points on the plane P in $\mathbf{R}^3$ given by the equation $X + Z = 0$. All these points have the form $(x, y, -x)$ and so are described by the two coordinates x and y. By identifying a point $(x, y, -x) \in P$ with its first two coordinates $(x, y) \in \mathbf{R}^2$, it is possible to view the plane P as the plane $\mathbf{R}^2$. (This identification is a *projection*.) Clearly these planes are different, since P is a subset of 3 space $\mathbf{R}^3$, while $\mathbf{R}^2$ is abstractly a plane by itself. Why is it that these two different planes can be viewed as the same? The reason is that they have the same dimension. We begin this section by developing a precise definition of dimension. The key is to utilize simultaneously the concepts of linear independence and span.

Definition 1.4.1 A set of vectors $\{v_1, v_2, \ldots, v_n\}$ in a vector space V is called a *basis* for V, if
(i) $v_1, v_2, \ldots, v_n$ are linearly independent.
(ii) $span\{v_1, v_2, \ldots, v_n\} = V$.

The simplest example of a basis is the following: We denote by $e_i = (0, \ldots, 0, 1, 0 \ldots, 0)$ the vector in $\mathbf{F}^n$ with all zeros except for a one in the ith entry. It is easy to see that $\{e_1, e_2, \ldots, e_n\}$ is a basis for $\mathbf{F}^n$. $\{e_1, e_2, \ldots, e_n\}$ is called the *standard basis* for $\mathbf{F}^n$.
Before giving more examples, we choose to give the following simple but basic result.

Lemma 1.4.2 *Suppose that $\{v_1, v_2, \ldots, v_n\}$ and $\{w_1, w_2, \ldots, w_m\}$ are both bases of a vector space V. Then $n = m$; that is, any two (finite) bases of a vector space have the same number of elements.*

Proof: Since $v_1, v_2, \ldots, v_n$ span V, each of $w_1, w_2, \ldots, w_m$ are linear combinations of $v_1, v_2, \ldots, v_n$. As $w_1, w_2, \ldots, w_m$ are linearly independent, we conclude by Theorem 1.3.9 that $m \leq n$. By the symmetry of the hypotheses, we also find $n \leq m$. Hence $n = m$ and the lemma is proved. //

We can now define the dimension of a vector space to be the number of elements of a basis. Note that Lemma 1.4.2 shows that Definition 1.4.3 makes sense; that is, it is impossible for a vector space to have two finite bases with different numbers of elements. In view of the existence of the standard basis for $\mathbf{F}^n$, we see that (as expected!) the dimension of $\mathbf{F}^n$ is n.

Definition 1.4.3 Suppose that V is a vector space and has a basis with n elements for some natural number n. Then we say that V is *n-dimensional*. If a vector space is n-dimensional for some such n, we say that V is *finite-dimensional*. If V is not finite-dimensional, then it is called *infinite-dimensional*.

The zero vector space $\{\vec{0}\}$ by convention has as a basis the empty set $\emptyset$ which has 0 elements. Thus $\{\vec{0}\}$ is a zero-dimensional vector space. In general, a vector space has many different bases. In fact, if $\mathbf{F}$ is infinite (for example if $\mathbf{F} = \mathbf{R}$ or $\mathbf{F} = \mathbf{C}$), then any nonzero vector space has infinitely many different bases. The key, of course, is that they all have the same number of elements.

Examples 1.4.4 (i) The set of vectors

$$\left\{ \begin{pmatrix} 1 \\ 1 \end{pmatrix}, \begin{pmatrix} -1 \\ 1 \end{pmatrix} \right\}$$

form a basis for $\mathbf{R}^2$. They are linearly independent, and since

$$\begin{pmatrix} 1 & -1 \\ 1 & 1 \end{pmatrix}$$

is a rank 2 matrix, any system of equations of the form

$$X \begin{pmatrix} 1 \\ 1 \end{pmatrix} + Y \begin{pmatrix} -1 \\ 1 \end{pmatrix} = \begin{pmatrix} a \\ b \end{pmatrix}$$

can always be solved whenever $a, b \in \mathbf{R}$. Hence the two vectors span $\mathbf{R}^2$.

(ii) The set of vectors

$$\left\{ \begin{pmatrix} 1 \\ 1 \\ 0 \end{pmatrix}, \begin{pmatrix} 1 \\ 0 \\ 1 \end{pmatrix}, \begin{pmatrix} 0 \\ -1 \\ 1 \end{pmatrix} \right\}$$

do not form a basis for $\mathbf{R}^3$. They are not linearly independent since

$$\begin{pmatrix} 1 \\ 1 \\ 0 \end{pmatrix} - \begin{pmatrix} 1 \\ 0 \\ 1 \end{pmatrix} + \begin{pmatrix} 0 \\ -1 \\ 1 \end{pmatrix} = \begin{pmatrix} 0 \\ 0 \\ 0 \end{pmatrix}$$

(iii) The set of vectors

$$\left\{ \begin{pmatrix} 4 \\ 3 \\ 2 \\ 1 \\ 7 \end{pmatrix}, \begin{pmatrix} -2 \\ 0 \\ 3 \\ 0 \\ 1 \end{pmatrix}, \begin{pmatrix} 1 \\ 2 \\ 5 \\ 9 \\ 2 \end{pmatrix}, \begin{pmatrix} 4 \\ 7 \\ 6 \\ 4 \\ 7 \end{pmatrix} \right\}$$

cannot be a basis for $\mathbf{R}^5$ since $\mathbf{R}^5$ is five-dimensional and Lemma 1.4.2 shows that any basis of $\mathbf{R}^5$ must have five elements.

(iv) The vector space P^n of all polynomials with coefficients in $\mathbf{F}$ of degree at most n is an $(n+1)$-dimensional vector space. As a basis we give the set of elements $\{1, X, X^2, \ldots, X^n\}$. Since every element in P^n has a unique representation in the form $a_n X^n + \cdots + a_1 X + a_0$ where $a_n, \ldots, a_1, a_0 \in \mathbf{F}$, it is clear this set spans. Furthermore, since the zero vector is uniquely expressed as $0 = 0X^n + \cdots + 0X + 0$, the set is also linearly independent. We shall refer to this basis as the *standard basis* of P^n. There are many other bases for P^n which are useful. For example, for any $f \in \mathbf{F}$ the collection of elements $\{1, (X+f), (X+f)^2, \ldots, (X+f)^n\}$ is also a basis for P^n.

(v) The vector space P of all polynomials with coefficients in a field $\mathbf{F}$ [see Example 1.2.7 (ii)] is not a finite-dimensional $\mathbf{F}$ vector space, since any finite set of polynomials must have an upper bound to their degree. Hence, any sum of *scalar* multiples of such polynomials has degree bounded by this upper bound. Thus no finite set of polynomials can span P, and P cannot be finite-dimensional.

(vi) The vector space $M_{m \times n}(\mathbf{F})$ of $m \times n$ matrices with coefficients in $\mathbf{F}$ is an mn-dimensional vector space. One can check that the set of matrices

$$\mathcal{I} = \{E_{ij} \mid 1 \le i \le m, 1 \le j \le n\}$$

where E_{ij} is the matrix with ijth entry 1 and all other entries 0, is a basis for $M_{m \times n}(F)$.

The next result, in many cases, simplifies the work required to check if a collection of vectors is a basis for a vector space. It further illustrates the interdependence of the concepts of linear independence, span, and dimension.

Theorem 1.4.5 *Suppose V is an n-dimensional vector space.*
(i) *If $v_1, v_2, \ldots, v_n$ are linearly independent in V, then they form a basis for V.*
(ii) *If $v_1, v_2, \ldots, v_n$ span V, then they form a basis for V.*

Proof: (i) We must show that $v_1, v_2, \ldots, v_n$ span V. Let $w \in V$. If w is not a linear combination of $v_1, v_2, \ldots, v_n$, then by Lemma 1.3.6 $v_1, v_2, \ldots, v_n, w$ are linearly independent. Since V is n-dimensional, let $u_1, u_2, \ldots, u_n$ be a basis of V. Then each of $v_1, v_2, \ldots, v_n, w$ is a linear combination of $u_1, u_2, \ldots, u_n$. Since $v_1, v_2, \ldots, v_n, w$ are linearly independent, Theorem 1.3.9 shows that $n + 1 \le n$. This is nonsense; this proves (i).

(ii) We must show that $v_1, v_2, \ldots, v_n$ are linearly independent. If not, by Lemma 1.3.8 there exists some j with $1 \le j \le n$ where v_j is a linear combination of $v_1, v_2, \ldots, v_{j-1}$. It follows from Lemma 1.1.6 (i) that $v_1, v_2, \ldots, v_{j-1}, v_{j+1}, \ldots, v_n$ have the same span as the original $v_1, v_2, \ldots, v_n$. Again, let $u_1, u_2, \ldots, u_n$ be a basis for the n-dimensional vector space V. Then each of $u_1, u_2, \ldots, u_n$ are linear combinations of $v_1, v_2, \ldots, v_{j-1}, v_{j+1}, \ldots, v_n$, and the $u_1, u_2, \ldots, u_n$ are linearly independent. Theorem 1.3.9 shows that $n \le n - 1$. This contradiction proves the theorem. //

The next theorem shows how to find bases and compute the dimension of the various vector spaces associated with a matrix. It explicitly describes how to use row operations and the rank in studying possible bases for these vector spaces. Part (iii) is a nice bonus for our work.

It says that the rank of a matrix is the same as the rank of its transpose. Without the concept of a vector space, and some understanding of dimension, this result would be quite difficult to prove.

Theorem 1.4.6 *Suppose that A is an $m \times n$ matrix and R is the reduced row-echelon form of A.*
(i)   *The dimension of the row space of A is $rk(A)$. A basis for the row space of A is the collection of nonzero rows of R.*
(ii)   *The dimension of the column space of A is also $rk(A)$. A basis for the column space of A is the collection of columns of A which correspond to the columns containing leading 1s in R.*
(iii) $rk(A) = rk(A^t)$.

Proof: We know that $rk(A)$ is the number of nonzero rows of R.

(i)   According to Theorem 1.2.2 (i) the row space of A is the same as the row space of R. The row space of R is clearly spanned by the nonzero rows of R. Moreover, since each nonzero row of R contains a leading 1 in a column of R for which there are no other nonzero entries, the nonzero rows of R are clearly linearly independent. Hence, the nonzero rows of R form a basis for the row space of A. As there are $rk(A)$ such rows, this gives (i).

(ii) Suppose that $rk(A) = r$. Then the columns containing the leading 1s are the first r standard basis vectors of $\mathbf{F}^m$; that is, they are $e_1, e_2, \ldots, e_r \in \mathbf{F}^m$. These vectors are linearly independent, and by the definition of the reduced row-echelon form, they span the column space of R. (Every column of R has $m - r$ zeros at the bottom.)

Let $\mathcal{B}$ denote the collection of columns of A which correspond to these r columns of R and denote by B the matrix whose columns are the columns in $\mathcal{B}$. The row-reduction of A to R, row reduces B to the matrix whose columns are the columns consisting of $e_1, \ldots, e_r$ in R. This matrix has rank r, so Theorem 1.3.4 shows that the columns of B are linearly independent. Since the columns containing the leading 1s of R span the column space of R, by Theorem 1.2.2 (ii) every column of A must be a linear combination of the columns in $\mathcal{B}$. In particular, $\mathcal{B}$ spans the column space of A. This shows that $\mathcal{B}$ is a basis for the column space of A and proves (ii).

(iii)   Observe that the column space U of A is the same as the row space V of A^t. Consequently, $rk(A) = dim(U) = dim(V) = rk(A^t)$; this gives (iii). //

Example 1.4.7 Consider the matrix

$$M = \begin{pmatrix} 2 & 3 & 1 & 1 & 1 \\ 0 & 1 & 1 & -1 & 0 \\ 2 & 0 & -2 & 4 & 0 \end{pmatrix}$$

The reduced row-echelon form of M is

$$\begin{pmatrix} 1 & 0 & -1 & 2 & 0 \\ 0 & 1 & 1 & -1 & 0 \\ 0 & 0 & 0 & 0 & 1 \end{pmatrix}$$

Further

$$M^t = \begin{pmatrix} 2 & 0 & 2 \\ 3 & 1 & 0 \\ 1 & 1 & -2 \\ 1 & -1 & 4 \\ 1 & 0 & 0 \end{pmatrix}$$

has reduced row-echelon form

$$\begin{pmatrix} 1 & 0 & 0 \\ 0 & 1 & 0 \\ 0 & 0 & 1 \\ 0 & 0 & 0 \\ 0 & 0 & 0 \end{pmatrix}$$

Applying Theorem 1.4.6 (i), we find that the row space of M is three-dimensional with basis the rows of the reduced row-echelon form of M. If we applied part (ii) of Theorem 1.4.6 to M^t we would find that the column space of M^t (which is the row space of M) has as a basis the columns of M^t (which are the original rows of M). Note that the two bases obtained in this manner are different. Also note that the columns of the reduced row echelon form of M^t do *not* form a basis for the row space of M.

As mentioned earlier, when considering a collection of p-vectors, you may want to form the matrix where they are rows or columns, depending upon the questions you need to answer. There are advantages to either method. But be sure that you do not confuse the different conclusions (i) and (ii) of Theorem 1.4.6.

Up to this point, we have defined and studied bases of (finite-dimensional) vector spaces without considering the question of when

bases exist. It turns out that if one assumes some basic principles of
set theory (in particular the axiom of choice), then it can be proved
that every vector space (even an infinite-dimensional one) has a basis.
(For details, see Sec. 7.5.) However, in the finite dimensional setting it
is possible to give quite constructively the theorem that every finitely
spanned vector space has a basis. This section concludes with some results that guarantee the existence of bases of vector spaces with certain
properties. These results will have important theoretical uses later on.
The existence of a basis in the finitely spanned case can be found in
Corollary 1.4.9 (i).

In the next lemma we consider minimal spanning sets and maximal
linearly independent subsets. A *minimal spanning set* is a collection
of vectors which span a vector space V, but for which any proper
subset does *not* span V. (A proper subset of a set S is one different
from S.) Analogously, a *maximal linearly independent set* is a linearly
independent subset of a vector space for which any larger subset fails
to be linearly independent.

Lemma 1.4.8   *Suppose that* $v_1, v_2, \ldots, v_n$ *are elements of a vector
space* V.
(i)   *If* $\{v_1, v_2, \ldots, v_n\}$ *is a minimal spanning set for* V, *then* $\{v_1, v_2, \ldots, v_n\}$ *is a basis for* V.
(ii)   *If* $\{v_1, v_2, \ldots, v_n\}$ *is a maximal linearly independent subset of* V,
then $\{v_1, v_2, \ldots, v_n\}$ *is a basis of* V.

Proof: (i)   We must show that $v_1, v_2, \ldots, v_n$ are linearly independent. If not, then by Lemma 1.3.8 we know that some v_j is a linear combination of $v_1, \ldots, v_{j-1}$. It follows by Lemma 1.1.6 (i) that
$v_1, \ldots, v_{j-1}, v_{j+1}, \ldots, v_n$ is a spanning set for V. This contradicts the
minimality of $\{v_1, \ldots, v_n\}$, and this establishes their linear independence.

(ii) We must show that $v_1, v_2, \ldots, v_n$ span V. Assume they do not.
Choose $w \in V$ with $w \notin span\{v_1, v_2, \ldots, v_n\}$. Then by Theorem 1.3.6
we see that $v_1, v_2, \ldots, v_n, w$ are linearly independent. This contradicts
the maximality of $\{v_1, v_2, \ldots, v_n\}$ and establishes that $v_1, v_2, \ldots, v_n$
span V. //

As a consequence of this lemma we obtain some basic existence
results mentioned earlier.

Corollary 1.4.9 (i) *Every vector space V with a finite spanning set has a basis.*
(ii)   *Every subspace V of an n-dimensionsal vector space has a basis.*
(iii) *Every subspace V of $\mathbf{F}^n$ has a basis.*

Proof: (i) Take a minimal such spanning set. Such a set must exist by considering subsets of a fixed finite spanning set. By Lemma 1.4.8 (i) it is a basis for V.

(ii) By Theorem 1.3.9 a linearly independent subset of V can have at most n elements. Thus there must be a maximal (and finite) linearly independent subset of V. By Lemma 1.4.8, such a subset is a basis.

(iii) This is a special case of (ii). This proves the corollary. //

As an application we obtain the fundamental relation between the rank of a matrix A and the dimension of its nullspace $ker(A)$. This result was observed in earlier computations and in Corollary 0.5.8 (ii), where it was shown that the solutions to the equation $AX = \vec{0}$ in n variables can be described by $n - rk(A)$ parameters. In fact these parameters must be independent, because the next result shows that the dimension of $ker(A)$ is $n - rk(A)$. The dimension of $ker(A)$ is called the *nullity* of A and is denoted by $null(A)$. We shall use this result often, and we will refer to it as the rank plus nullity theorem.

Theorem 1.4.10 (Rank Plus Nullity Theorem) *If A is an $m \times n$ matrix with $rank(A) = r$, then $ker(A) = \{v \mid Av = \vec{0}\}$ is a $(n - r)$-dimensional vector space. In other words, $rk(A) + null(A) = n$.*

Proof: Apply Lemma 1.3.10 to see that $ker(A)$ has a minimal spanning set with $n - r$ vectors. Lemma 1.4.8 (i) shows $ker(A)$ is $(n - r)$-dimensional. This gives the theorem. //

The final result of this section proves to be a very convenient tool. The example that follows shows how this result can be applied.

Theorem 1.4.11 (Basis-Extension Theorem) *Suppose $v_1, v_2, \ldots, v_s$ are linearly independent vectors in an n-dimensional vector space V. Then there exist $w_{s+1}, \ldots, w_n \in V$ such that $\{v_1, v_2, \ldots, v_s, w_{s+1}, \ldots, w_n\}$ is a basis for V.*

Proof: Choose $w_{s+1}, \ldots, w_{s+t}$ so that $v_1, \ldots, v_s, w_{s+1}, \ldots, w_{s+t}$ are linearly independent and t is as large as possible. Note that such a number t must exist, as there are always at most $n = dim(V)$ vectors in any linearly independent subset of V. (If $t = 0$, no new vectors are chosen.) By Lemma 1.4.8 (ii) the maximality of the linearly independent set $\{v_1, \ldots, v_s, w_{s+1}, \ldots, w_{s+t}\}$ guarantees that it is a basis. Since $dim(V) = n$, $s + t = n$ follows. This proves the result. $//$

Example 1.4.12 The vectors $X^2 + 1$ and $X^2 - 1$ are linearly independent in P^3, the vector space of all real polynomials of degree at most 3. Suppose one needs a basis for P^3 containing these vectors. Theorem 1.4.11 says that such a basis must exist. To produce such a basis, we try adding the elements in $\{1, X, X^2, X^3\}$ one at a time until we obtain a maximal linearly independent subset of P^3. Evidently, $1 \in span\{X^2 + 1, X^2 - 1\}$, so we do not add the vector 1 to our basis. However, $X \notin span\{X^2 + 1, X^2 - 1\}$, so we add it; that is, the new set $\{X^2 + 1, X^2 - 1, X\}$ is linearly independent. Again we see that X^2 lies in the span of this set, while X^3 does not. We conclude that the set of vectors $\{X^2 + 1, X^2 - 1, X, X^3\}$ is linearly independent and consequently is a basis for P^3 containing the original two vectors.

PROBLEMS

1.4.1. Find a basis for each of the following vector spaces.
(a) The set of solutions to $3X - 4Y + 2Z = 0$ in $\mathbf{R}^3$
(b) The set of vectors of $\mathbf{R}^4$ spanned by (a), (b), (c) in Prob. 1.1.3
(c) The set of solutions to the system of equations

$$
\begin{pmatrix} 1 & 9 & 5 & 4 & 8 \\ 2 & 0 & 0 & 0 & 8 \\ 1 & 6 & 7 & 0 & 0 \end{pmatrix}
\begin{pmatrix} X \\ Y \\ Z \\ W \\ V \end{pmatrix} = \begin{pmatrix} 0 \\ 0 \\ 0 \end{pmatrix}
$$

1.4.2. Find a maximal linearly independent subset among the following vectors in $\mathbf{F}^4$, and then extend this subset to a basis for $\mathbf{F}^4$.
(a) $\{(1, 0, 1, 0), (0, 1, 0, 1), (1, 1, 1, 1), (1, 0, 2, 0)\}$
(b) $\{(1, 2, 3, 4), (4, 3, 2, 1), (1, 1, 1, 1), (0, 1, 2, 3)\}$
(c) $\{(1, 1, 1, 1), (2, 2, 3, 3), (1, 0, 1, 0), (0, 1, 0, 1), (1, 1, 0, 0)\}$

1.4.3. Let

$$A = \begin{pmatrix} 1 & 1 & 3 & 0 & 2 \\ 0 & 1 & 2 & 0 & 1 \\ 2 & 1 & 4 & 1 & 3 \\ 1 & 0 & 1 & 2 & 1 \end{pmatrix}$$

(a) Find the reduced row-echelon form of A.
(b) Find the dimension of the subspace of $\mathbf{R}^4$ spanned by the first, second, third, and fifth columns of A.
(c) Find a basis for the row space of A.
(d) Find a basis for the nullspace of A.

1.4.4. For each of the following matrices find (1) a basis for the row space, (2) a basis for the column space, and (3) a basis for the nullspace.

(a) $\begin{pmatrix} 1 & 5 & 2 \\ 2 & 3 & 1 \end{pmatrix}$ (b) $\begin{pmatrix} 1 & 4 & 8 \\ 5 & 0 & 0 \\ 1 & 0 & 0 \\ 3 & 1 & 2 \end{pmatrix}$ (c) $\begin{pmatrix} 2 & 1 & 1 \\ 3 & 1 & 2 \\ 5 & 2 & 3 \end{pmatrix}$ (d) $\begin{pmatrix} 2 & 1 \\ 0 & 0 \\ 1 & 1 \\ 2 & 0 \end{pmatrix}$

1.4.5. Find a basis for

(a) $V = span \left\{ \begin{pmatrix} 1 \\ 2 \\ 3 \\ 4 \end{pmatrix}, \begin{pmatrix} 4 \\ 3 \\ 2 \\ 1 \end{pmatrix}, \begin{pmatrix} 9 \\ 8 \\ 7 \\ 6 \end{pmatrix} \right\}$

(b) $W = span \left\{ \begin{pmatrix} 1 \\ 4 \\ 9 \end{pmatrix}, \begin{pmatrix} 2 \\ 3 \\ 8 \end{pmatrix}, \begin{pmatrix} 3 \\ 2 \\ 7 \end{pmatrix}, \begin{pmatrix} 4 \\ 1 \\ 6 \end{pmatrix} \right\}$

1.4.6. Suppose that V is a subspace of the vector space W. Assume that $dim(V) < dim(W) - 1$. Show there exists a vector subspace U of W, containing V, such that $V \neq U \neq W$.

1.4.7. Show that if V is a subspace of a finite-dimensional vector space W and $dim(V) = dim(W)$, then $V = W$.

1.4.8. Suppose that $v_1, v_2, \ldots, v_n$ are a basis of V. Let $v \in V$. Prove that $v_1 + v, v_2 + v, \ldots, v_n + v$ are a basis of V if and only if $v \neq a_1 v_1 + a_2 v_2 + \cdots + a_n v_n$ where $a_1 + a_2 + \cdots + a_n = -1$.

1.4.9. Suppose that $v_1, v_2, \ldots, v_n$ are a basis of a real vector space V. Prove that $v_1 + v_2, v_2 + v_3, \ldots, v_{n-1} + v_n, v_n + v_1$ are a basis of V if and only if n is odd.

1.4.10. Prove that whenever H is a subspace of a finite-dimensional vector space V, $dim(H) \leq dim(V)$.

1.4.11. Extend

$$\left\{\begin{pmatrix} 1 \\ 1 \\ 0 \\ 1 \end{pmatrix}, \begin{pmatrix} 2 \\ 2 \\ 3 \\ 1 \end{pmatrix}\right\}$$

to a basis of $\mathbf{R}^4$.

1.4.12. (a) What is the dimension of the *real* vector space $\mathbf{C}$?
(b) Let V be a complex vector space. Explain how to make V into a real vector space with (almost) no effort. What is the dimension of V as a real vector space (in terms of the complex dimension of V)?

1.4.13. Using the definitions from Prob. 1.2.2, show

$$dim(V) + dim(W) = dim(V \cap W) + dim(V + W)$$

1.4.14. Prove the uniqueness of the reduced echelon form R of a matrix A (Theorem 0.5.5) using Theorem 1.2.2. As an outline:
(a) Show that the columns containing leading 1s in R are characterized by being precisely those columns that are not linear combinations of the preceding columns.
(b) Show the other columns are completely determined by how they are expressed as a linear combination of the columns that precede them.

1.4.15. Prove or disprove: Two matrices of the same size and rank are row equivalent.

1.4.16. Let V be the vector space spanned by the four vectors

$$v_1 = \begin{pmatrix} 1 \\ 2 \\ 1 \\ 2 \end{pmatrix}, \quad v_2 = \begin{pmatrix} 0 \\ 1 \\ 1 \\ 0 \end{pmatrix}, \quad v_3 = \begin{pmatrix} 1 \\ 4 \\ 3 \\ 2 \end{pmatrix}, \quad v_4 = \begin{pmatrix} 1 \\ 0 \\ 2 \\ 1 \end{pmatrix} \in \mathbf{R}^4$$

(a) Find the dimension of V.
(b) Find a basis for V among the vectors v_1, v_2, v_3, and v_4.
(c) Find a second (different) basis for V among the vectors v_1, v_2, v_3, and v_4.

1.4.17. (a) Suppose that A is a 2×1 matrix and B is a 1×2 matrix. Prove that AB is not invertible.
(b) Generalize (a) to $n \times m$ and $m \times n$ matrices.

1.4.18. Assume that U and W are subspaces of a vector space V. If $dim(U) + dim(W) > dim(V)$, prove that $U \cap W \neq \{\vec{0}\}$.

1.4.19. An infinite collection S of vectors spans a vector space V if every vector in V can be written as a *finite* linear combination of elements of S. Find a basis for the infinite-dimensional vector space P of all polynomials. (See also Prob. 1.3.17.)

1.4.20. A *submatrix* of a matrix A is a matrix obtained from A by deleting some rows and/or some columns of A.
(a) Show that if B is a submatrix of A, then $rk(B) \leq rk(A)$.
(b) Suppose that $rk(A) = r$. Show that A has a $r \times r$ submatrix of rank r.

1.4.21. Suppose that A is an $m \times n$ matrix and B is an $n \times p$ matrix. Show that $rk(AB) \leq min\{rk(A), rk(B)\}$.

1.4.22. Show that $\{(1, 1, 1), (1, 1, 0), (1, 0, 0)\}$ is a basis for $\mathbf{F}^3$ by showing that each of e_1, e_2, e_3 is a linear combination of these vectors. Why does this suffice?

1.5 Coordinates with Respect to a Basis

Even though two $\mathbf{F}$ vector spaces of the same dimension may be different, from the algebraic point of view they have essentially the same structure (see Theorem 1.5.5). In this section we make this quite explicit, by showing how to identify an arbitrary n-dimensional vector space with $\mathbf{F}^n$. The next theorem, which is an easy consequence of Lemma 1.3.7, is the key. With this theorem we will be able to define the coordinates of a vector with respect to a basis.

Theorem 1.5.1 *Suppose that $\{v_1, v_2, \ldots, v_n\}$ is a basis for a vector space V. Then for every vector $w \in V$ there is a unique sequence of scalars $a_1, a_2, \ldots, a_n$ such that $w = a_1 v_1 + a_2 v_2 + \cdots + a_n v_n$.*

Proof: Since $\{v_1, v_2, \ldots, v_n\}$ is a basis of V, every element $w \in V$ is a linear combination of $v_1, v_2, \ldots, v_n$. The existence of the unique scalars $a_1, a_2, \ldots, a_n$ follows from Lemma 1.3.7. //

The uniqueness of the scalars $a_1, a_2, \ldots, a_n$ in Theorem 1.5.1 enables us to define coordinates for an arbitrary finite-dimensional vector space. We emphasize that each coordinate system depends upon a choice of a basis for the vector space and upon the order in which the basis elements are listed. Consequently, we give the next definition.

Definition 1.5.2 An *ordered basis* $\{v_1, v_2, \ldots, v_n\}$ of a vector space V is a basis of V whose elements are listed in a specific order. Suppose that $\{v_1, v_2, \ldots, v_n\}$ is an ordered basis for V and $w \in V$. We say that $(a_1, a_2, \ldots, a_n)$ are the *coordinates of w with respect to* $\{v_1, v_2, \ldots, v_n\}$ if $w = a_1 v_1 + a_2 v_2 + \cdots + a_n v_n$. Note that Theorem 1.5.1 guarantees that such $a_1, a_2, \ldots, a_n$ exist uniquely. Usually we give an ordered basis a name, say $\mathcal{B} = \{v_1, v_2, \ldots, v_n\}$, and we then denote the coordinates of a vector $w = a_1 v_1 + a_2 v_2 + \cdots + a_n v_n$ by

$$(w)_\mathcal{B} = \begin{pmatrix} a_1 \\ a_2 \\ \vdots \\ a_n \end{pmatrix}$$

Remarks (i) The preceding expression does not mean that w lies in $\mathbf{F}^n$. It means only that w lies inside some n-dimensional vector space and has the specified coordinates with respect to the basis $\mathcal{B}$. The column vector notation is used for coordinates so that matrix algebra can be applied.

(ii) Strictly speaking, the notation for an ordered basis $\{v_1, v_2, \ldots, v_n\}$ is inconsistent with common mathematical practice. Since we are using the curly brackets $\{$ and $\}$, $\{v_1, v_2, \ldots, v_n\}$ denotes only the set of n vectors and does not specify an ordering of the elements. The proper notation for an ordered set is to use parentheses (as we do for elements of $\mathbf{F}^n$), and thus one should write $(v_1, v_2, \ldots, v_n)$ for an *ordered* basis. However, in the author's experience, it is too easy to confuse this notation with elements of $\mathbf{F}^n$, so we have decided to abuse notation and use $\{v_1, v_2, \ldots, v_n\}$ when specifying an ordered basis. But do not forget about the ordering; this is extremely important in the context of coordinates!

Examples 1.5.3 (i) In the vector space $\mathbf{F}^n$ the standard basis $\mathcal{S} = \{e_1, e_2, \ldots, e_n\}$ gives the standard coordinates of a vector in $\mathbf{F}^n$. In terms of our notation, if we consider a vector $v = (a_1, a_2, \ldots, a_n) \in \mathbf{F}^n$, then since $v = a_1 e_1 + a_2 e_2 + \cdots + a_n e_n$, the standard coordinates of v are $(a_1, a_2, \ldots, a_n)$. Alternatively we can write the column notation

$$(v)_S = \begin{pmatrix} a_1 \\ a_2 \\ \vdots \\ a_n \end{pmatrix}$$

(ii) Consider the ordered basis

$$B = \left\{ \begin{pmatrix} 1 \\ 0 \\ 2 \end{pmatrix}, \begin{pmatrix} 3 \\ 2 \\ 6 \end{pmatrix}, \begin{pmatrix} 2 \\ 1 \\ 5 \end{pmatrix} \right\}$$

of $\mathbf{R}^3$. To find the coordinates of the vector $(1, 1, 1)$ with respect to B, we must express

$$\begin{pmatrix} 1 \\ 1 \\ 1 \end{pmatrix} = a_1 \begin{pmatrix} 1 \\ 0 \\ 2 \end{pmatrix} + a_2 \begin{pmatrix} 3 \\ 2 \\ 6 \end{pmatrix} + a_3 \begin{pmatrix} 2 \\ 1 \\ 5 \end{pmatrix}$$

for appropriate scalars a_1, a_2, a_3. To do this, we must solve the system of equations

$$\begin{pmatrix} 1 & 3 & 2 \\ 0 & 2 & 1 \\ 2 & 6 & 5 \end{pmatrix} \begin{pmatrix} X \\ Y \\ Z \end{pmatrix} = \begin{pmatrix} 1 \\ 1 \\ 1 \end{pmatrix} \tag{$*$}$$

It can be checked that

$$\begin{pmatrix} 1 & 3 & 2 \\ 0 & 2 & 1 \\ 2 & 6 & 5 \end{pmatrix}^{-1} = \begin{pmatrix} 2 & -3/2 & -1/2 \\ 1 & 1/2 & -1/2 \\ -2 & 0 & 1 \end{pmatrix}$$

Thus, the solution to $(*)$ is given by

$$\begin{pmatrix} X \\ Y \\ Z \end{pmatrix} = \begin{pmatrix} 2 & -3/2 & -1/2 \\ 1 & 1/2 & -1/2 \\ -2 & 0 & 1 \end{pmatrix} \begin{pmatrix} 1 \\ 1 \\ 1 \end{pmatrix} = \begin{pmatrix} 0 \\ 1 \\ -1 \end{pmatrix}$$

We obtain that

$$\begin{pmatrix} 1 \\ 1 \\ 1 \end{pmatrix}_B = \begin{pmatrix} 0 \\ 1 \\ -1 \end{pmatrix}$$

Look closely at this expression, and make sure you understand what the subscript $\mathcal{B}$ means. In general, we see that whenever $(r, s, t) \in \mathbf{R}^3$, since

$$\begin{pmatrix} 2 & -3/2 & -1/2 \\ 1 & 1/2 & -1/2 \\ -2 & 0 & 1 \end{pmatrix} \begin{pmatrix} r \\ s \\ t \end{pmatrix} = \begin{pmatrix} 2r - (3/2)s + t \\ r + (1/2)s - (1/2)t \\ -2r + t \end{pmatrix}$$

we obtain that

$$\begin{pmatrix} r \\ s \\ t \end{pmatrix}_{\mathcal{B}} = \begin{pmatrix} 2r - (3/2)s + t \\ r + (1/2)s - (1/2)t \\ -2r + t \end{pmatrix}$$

(iii) Let V be the two-dimensional subspace of $\mathbf{R}^4$ which has as an ordered basis

$$\mathcal{C} = \left\{ \begin{pmatrix} 0 \\ 1 \\ 1 \\ 2 \end{pmatrix}, \begin{pmatrix} 1 \\ 0 \\ 3 \\ 4 \end{pmatrix} \right\}$$

Consider the vector $(3, -2, 7, 8) \in V$. Since

$$\begin{pmatrix} 3 \\ -2 \\ 7 \\ 8 \end{pmatrix} = -2 \begin{pmatrix} 0 \\ 1 \\ 1 \\ 2 \end{pmatrix} + 3 \begin{pmatrix} 1 \\ 0 \\ 3 \\ 4 \end{pmatrix}$$

we find that

$$\begin{pmatrix} 3 \\ -2 \\ 7 \\ 8 \end{pmatrix}_{\mathcal{C}} = \begin{pmatrix} -2 \\ 3 \end{pmatrix}$$

(iv) Consider the standard basis for P^4, taken in the usual order, $\mathcal{S} = \{1, X, X^2, X^3, X^4\}$. Then the coordinates of a vector (which is a polynomial) pick out the coefficients. In other words

$$(aX^4 + bX^3 + cX^2 + dX + e)_{\mathcal{S}} = \begin{pmatrix} e \\ d \\ c \\ b \\ a \end{pmatrix}$$

Example 1.5.3 (ii) shows that the coordinates of a vector in $\mathbf{F}^n$ with respect to a nonstandard basis can be determined by matrix multiplication using an appropriate matrix inverse. This is spelled out in the next theorem. The proof is exactly the same as the calculation in Example 1.5.3 (ii). A more general result will be proved in Theorem 3.3.4.

Theorem 1.5.4 *Let $\mathcal{B} = \{v_1, v_2, \ldots, v_n\}$ be an ordered basis of $\mathbf{F}^n$. Set A to be the matrix $(\, v_1 \quad v_2 \quad \cdots \quad v_n \,)$; that is, A is the matrix whose columns are the coordinates of the vectors in $\mathcal{B}$. Then A is invertible and for any $v \in \mathbf{F}^n$, $(v)_\mathcal{B} = A^{-1}v$.*

Proof: Let $v \in \mathbf{F}^n$. In order to find the coordinates of v with respect to $\mathcal{B}$, one must solve the equation $v = X_1 v_1 + X_2 v_2 + \cdots + X_n v_n$ for $X_1, X_2, \ldots, X_n \in \mathbf{F}$. By the definition of matrix multiplication this is the same as solving the system of equations $A\vec{X} = v$. Since this equation always has a unique solution (as $\mathcal{B}$ is a basis of $\mathbf{F}^n$), it follows from Theorem 0.6.9 that A is invertible. The solution to the system of equations is $A^{-1}v$, which according to the definition is the column vector $(v)_\mathcal{B}$, as required. //

This final result says that the coordinates of a sum of vectors is the sum of the coordinates. This result is what we were referring to in the introduction of this section when we said "algebraically speaking" two vector spaces of the same dimension are the same.

Theorem 1.5.5 *Suppose that $\mathcal{B} = \{v_1, v_2, \ldots, v_n\}$ is an ordered basis of an n-dimensional vector space V. Let $w_1, w_2 \in V$ be vectors and suppose that $k \in \mathbf{F}$. Then*

$$(w_1 + w_2)_\mathcal{B} = (w_1)_\mathcal{B} + (w_2)_\mathcal{B} \quad \text{and} \quad (kw_1)_\mathcal{B} = k(w_1)_\mathcal{B}$$

Proof: We express $w_1 = a_1 v_1 + \cdots + a_n v_n$ and $w_2 = b_1 v_1 + \cdots b_n v_n$ so that

$$(w_1)_\mathcal{B} = \begin{pmatrix} a_1 \\ a_2 \\ \vdots \\ a_n \end{pmatrix} \quad \text{and} \quad (w_2)_\mathcal{B} = \begin{pmatrix} b_1 \\ b_2 \\ \vdots \\ b_n \end{pmatrix}$$

Since $w_1 + w_2$ is the linear combination $w_1 + w_2 = (a_1 + b_1)v_1 + \cdots + (a_n + b_n)v_n$, we obtain that

$$(w_1 + w_2)_B = \begin{pmatrix} a_1 + b_1 \\ a_2 + b_2 \\ \vdots \\ a_n + b_n \end{pmatrix}$$

Similarly we have that $kw_1 = ka_1v_1 + \cdots + ka_nv_n$ so that $(kw_1)_B = k(w_1)_B$. This proves the theorem. //

In the case where V is the vector space $\mathbf{F}^n$, the previous theorem is a simple consequence of our earlier results on matrix algebra and Theorem 1.5.4. Using the distributive law of matrix multiplication, we have, for the appropriate invertible matrix A, $(w_1 + w_2)_B = A^{-1}(w_1 + w_2) = A^{-1}(w_1 + w_2) = A^{-1}w_1 + A^{-1}w_2 = (w_1)_B + (w_2)_B$. Similarly, $(kw_1)_B = A^{-1}kw_1 = kA^{-1}w_1 = k(w_1)_B$, as required.

We shall return to coordinate systems when we study linear transformations in Chap. 3. The identification of an n-dimensional vector space V with $\mathbf{F}^n$ by using an ordered basis B can be viewed as the function $U_B : V \rightarrow \mathbf{F}^n$ defined by $U_B(v) = (v)_B$ for all $v \in V$. Theorems 1.5.1 and 1.5.5 show that such a function U_B is an example of what is called an *isomorphism*. Isomorphisms are equivalences of vector spaces and are studied in detail in Chap. 3.

PROBLEMS

1.5.1. Let V be the two-dimensional vector subspace of $\mathbf{R}^3$ with ordered basis

$$B = \left\{ \begin{pmatrix} 0 \\ 1 \\ 1 \end{pmatrix}, \begin{pmatrix} 1 \\ 0 \\ 3 \end{pmatrix} \right\}$$

Find

$$\begin{pmatrix} 3 \\ -2 \\ 7 \end{pmatrix}_B$$

1.5.2. Consider the ordered basis

$$B = \left\{ \begin{pmatrix} 0 \\ 0 \\ 1 \end{pmatrix}, \begin{pmatrix} 0 \\ 1 \\ 0 \end{pmatrix}, \begin{pmatrix} 1 \\ 0 \\ 0 \end{pmatrix} \right\}$$

of $\mathbf{F}^3$. If $v = (a, b, c) \in \mathbf{F}^3$, find $(v)_B$.

1.5.3. Consider the ordered basis $B = \{(1, 0, -1), (1, 1, 1), (0, 1, 0)\}$ of $\mathbf{R}^3$. Find the coordinates of (a, b, c) in this basis.

1.5.4. Find the coordinates of an arbitrary vector in P^4 using the ordered basis $\{1, X + 1, (X + 1)^2, (X + 1)^3, (X + 1)^4\}$.

1.5.5. Let

$$B_1 = \left\{ \begin{pmatrix} 2 \\ 3 \end{pmatrix}, \begin{pmatrix} 5 \\ 8 \end{pmatrix} \right\}$$

be an ordered basis for $\mathbf{R}^2$. Find a matrix A such that $Av = (v)_{B_1}$ for all $v \in \mathbf{R}^2$.

1.5.6. Let

$$B_2 = \left\{ \begin{pmatrix} 1 \\ 2 \\ 0 \end{pmatrix}, \begin{pmatrix} 0 \\ 1 \\ 1 \end{pmatrix}, \begin{pmatrix} 1 \\ 0 \\ 1 \end{pmatrix} \right\}$$

be an ordered basis for $\mathbf{R}^3$. Find a matrix A such that $Av = (v)_{B_2}$ for all $v \in \mathbf{R}^3$.

1.5.7. Consider the ordered basis $B = \{1, X - 1, (X - 1)^2, (X - 1)^3\}$ of P^3. Find a matrix R with the property that for any polynomial $f \in P^3$, $R(f)_S = (f)_B$ where S is the standard basis of P^3.

1.5.8. Let V be a three-dimensional $\mathbf{F}$ vector space with ordered basis $B = \{v_1, v_2, v_3\}$.
(a) Find the B coordinates of the vectors $v_1, v_1 + 4v_3$, and $v_3 - v_2$.
(b) Let B' be the ordered basis $\{v_1, v_1 + v_2, v_1 + v_2 + v_3\}$ of V. Find a matrix P such that $P(v)_B = (v)_{B'}$ for all $v \in V$.
(c) Find a matrix Q such that $Q(v)_{B'} = (v)_B$ for all $v \in V$. What is the relationship between this Q and the matrix P from part (b)?

1.5.9. Show that for each real number θ the 2×2 matrix

$$R_\theta = \begin{pmatrix} \cos \theta & \sin \theta \\ -\sin \theta & \cos \theta \end{pmatrix}$$

is invertible. Find an ordered basis B_θ of $\mathbf{R}^2$ with the property that for all vectors $v \in \mathbf{R}^2$, $R_\theta v = (v)_{B_\theta}$.

1.5.10. Suppose that B_1 and B_2 are two ordered bases for $\mathbf{F}^n$. Show there exists a matrix P with the property that for all $v \in \mathbf{F}^n$, $P(v)_{B_1} = (v)_{B_2}$. Is P necessarily invertible?

1.5.11. Suppose that $\mathcal{B}$ is an ordered basis of an n-dimensional vector space V. If $w_1, w_2, \ldots, w_s \in V$, show that $(w_1 + w_2 + \cdots + w_s)_{\mathcal{B}} = (w_1)_{\mathcal{B}} + (w_2)_{\mathcal{B}} + \cdots + (w_s)_{\mathcal{B}}$.

1.5.12. Suppose that $\mathcal{B}$ and $\mathcal{B}'$ are two ordered bases of a finite-dimensional vector space V. Assume that $(v)_{\mathcal{B}} = (v)_{\mathcal{B}'}$ for all $v \in V$. Prove that $\mathcal{B} = \mathcal{B}'$.

1.6 An Error-Correcting Code

In this section we give an application of the theory of finite fields and finite vector spaces. The computations and ideas here are quite simple, yet their utility in computer science is immense. None of the results from this section will be used subsequently. Hopefully the reader will find this section enjoyable and interesting and will be convinced of the usefulness of considering fields other than $\mathbf{R}$ or $\mathbf{C}$.

The general problem that interests us is the following. We assume that some encoded information is transmitted by electronic means. Unfortunately, there may be static or other interference so that occasional errors occur in the transmission. We would like to be able to correct these random errors (assuming they do not occur too frequently). For example, this problem could occur in transmitting information over long distances (say from the moon to the earth). A similar problem can occur transmitting data from chip to chip (or to floppy disk) in a computer.

The idea for solving this problem is to use certain sequences of digits (usually 0s and 1s) for transmitting the information. These special sequences will be called *codewords*. However, most sequences will *not* be used as codewords. Thus when an entry in a sequence is altered during transmission, the sequence will fail to be a codeword. If the code is constructed properly, one can then use linear algebra to "correct" the error.

It is customary in computer science to code information in binary, that is, with zeros and ones. The two elements $\{0, 1\}$ with the operations $0 + 0 = 0$, $0 + 1 = 1$, $1 + 1 = 0$, $0 \cdot 0 = 0$, $0 \cdot 1 = 0$, $1 \cdot 1 = 1$ form a field. This can be easily checked, or one can apply Lemma 1.6.2. The field with two elements is denoted $\mathbf{Z}/2\mathbf{Z}$. The next definition (familiar to those who have studied elementary number theory) gives a more general construction.

Definition 1.6.1 Let n be an integer. Two integers a and b are said to be congruent (mod n), denoted $a \equiv b$ (mod n), if there exists an integer d such that $(b - a) = dn$. We define $\mathbf{Z}/n\mathbf{Z}$ to be the set $\{0, 1, 2, \ldots, n-1\}$ with addition defined by $a+b = c$ in $\mathbf{Z}/n\mathbf{Z}$ whenever $a + b \equiv c$ (mod n) and multiplication defined by $ab = c$ in $\mathbf{Z}/n\mathbf{Z}$ whenever $ab \equiv c$ (mod n).

The next result shows that $\mathbf{Z}/p\mathbf{Z}$ is a field whenever p is a prime. (Recall that a natural number p is called *prime* if $p > 1$ and its only positive integer divisors are 1 and p.) Thus, whenever p is a prime, linear algebra can be applied to the $\mathbf{Z}/p\mathbf{Z}$ vector spaces $(\mathbf{Z}/p\mathbf{Z})^n$.

Lemma 1.6.2 *If p is a prime, then $\mathbf{Z}/p\mathbf{Z}$ is a field.*

Sketch of Proof: One notes that for every integer $m \in \mathbf{Z}$, there is a unique element $a \in \{0, 1, 2, \ldots, p-1\}$ such that $m \equiv a$ (mod p). (This is a consequence of the division algorithm of grade school arithmetic.) Using this, one can check that $+$ and $\cdot$ as defined in $\mathbf{Z}/p\mathbf{Z}$ actually make sense. All the required field properties of these operations, except the existence of the multiplicative inverse, follow from the corresponding properties in the integers $\mathbf{Z}$. For example, one can check the commutative property of addition: whenever $a, b \in \mathbf{Z}$, $(a+b)-(b+a) = 0 = p\cdot 0$. This shows that $a + b \equiv b + a$ (mod p) for all a, b (mod p) $\in \mathbf{Z}/p\mathbf{Z}$; that is, addition is commutative in $\mathbf{Z}/p\mathbf{Z}$.

It remains to show that there are multiplicative inverses inside $\mathbf{Z}/p\mathbf{Z}$. Let $a \in \{1, 2, \ldots, p - 1\}$. Consider the collection $\{a, 2a, 3a, \ldots, (p-1)a\}$. Form the list $b_1, b_2, \ldots, b_{p-1}$ defined by $b_i \in \{1, 2, \ldots, p-1\}$ and $b_i \equiv i \cdot a$ (mod p). We claim $\{b_1, b_2, \ldots, b_{p-1}\} = \{1, 2, \ldots, (p-1)\}$ (as sets, where the elements are not necessarily listed in the same order). Otherwise, it must happen that $b_i = b_j$ for some $i \neq j$. But this would mean that $i \cdot a \equiv b_i = b_j \equiv j \cdot a$ (mod p). From this it follows that $i \cdot a \equiv j \cdot a$ (mod p) so that there exists $d \in \mathbf{Z}$ with $d\cdot p = i\cdot a - j\cdot a = (i-j)\cdot a$. But both $a, |i-j| < p$, so since p is prime, p cannot divide the product $(i - j) \cdot a$. This contradiction completes the sketch of the proof. //

From now to the end of this section, $\mathbf{F}$ will denote the field with two elements; that is, $\mathbf{F} = \mathbf{Z}/2\mathbf{Z}$. (It is extremely useful also to consider other finite fields when constructing algebraic codes, but they will not be treated in this book.) We shall work with matrices whose coefficients lie in $\mathbf{F}$; that is, each matrix will have only entries $0, 1$. Remember that $1 + 1 = 0$ now!

Definition 1.6.3 We fix the matrix H_3 to be

$$H_3 = \begin{pmatrix} 0 & 0 & 0 & 1 & 1 & 1 & 1 \\ 0 & 1 & 1 & 0 & 0 & 1 & 1 \\ 1 & 0 & 1 & 0 & 1 & 0 & 1 \end{pmatrix}$$

The matrix H_3 is called a *Hamming matrix*. We define C_3 to be the subspace of $\mathbf{F}^7$ consisting of all solutions to the equation $H_3 X = \vec{0}$. Elements of C_3 will be called *codewords* for H_3.

The codewords of H_3 can be used for transmitting information as mentioned earlier. So, one constructs an "alphabet" with the elements of C_3 as letters and then transmits information as sequences $v_1, v_2, v_3, \ldots, v_m$ where each $v_i \in C_3$. There are exactly 16 codewords for H_3, as we show next.

Lemma 1.6.4 C_3 *is a four-dimensional* $\mathbf{F}$ *vector space. Hence, there are* 16 *codewords for* H_3.

Proof: The first, second, and fourth columns of H_3 are clearly linearly independent elements of $\mathbf{F}^3$. Since H_3 has three rows, it has rank at most 3. Hence H_3 has rank 3. It follows that the set of solutions to $H_3 X = \vec{0}$ must be a $(7 - 3)$ four-dimensional vector space. Since $\mathbf{F}$ has only two elements, any four-dimensional $\mathbf{F}$ vector space must have $2^4 = 16$ elements (as such is isomorphic to $\mathbf{F}^4$). Thus there are 16 codewords in C_3. //

There are $2^7 = 128$ vectors in $\mathbf{F}^7$. Hence, most vectors in $\mathbf{F}^7$ are not codewords. Intuitively, the idea behind using the code is that the codewords are "far apart," and whenever an error occurs in transmission the codeword moves to a nearby vector. The originally transmitted word can be found by locating the "closest" codeword. This is explicitly carried out using matrix multiplication. To tell if a given vector v is a codeword, one simply performs the multiplication $H_3 v$ to see if $\vec{0}$ results. Presumably, this occurs most of the time. When it does not, the next result is what is needed in order to correct the error.

Theorem 1.6.5 *Let $v \in \mathbf{F}^7$ be a codeword for H_3. Suppose that v' is the vector obtained from v by adding 1 to the ith entry of v. Then $H_3 v'$ is the ith column of H_3.*

Proof: We denote by $e_i \in \mathbf{F}^7$ the vector with all zeros except for a 1 in the ith row. Then $v' = v + e_i$. By matrix algebra $H_3 v' = H_3(v + e_i) = H_3 v + H_3 e_i = H_3 e_i$ which is the ith column of H_3. The theorem is proved. //

We can now explain how the matrix H_3 can be used as a code, where single errors in transmission can be corrected: Transmit all data with the 16 codewords for H_3. Whenever a word is received, multiply by H_3 to make sure it is a codeword. Presumably this will almost always be the case, but perhaps once in a while a noncodeword will be received. Assume that (since errors are rare) a single error was made in transmitting the codeword. If the error word v' is received then $v' = v + e_i$ for some i. (Note that this covers both possible cases where the ith entry of v is changed from 1 to 0 or from 0 to 1.) By Theorem 1.6.5, $H_3 v' = w$ is the i'th column of H_3. Then modify v' to $v = v' - e_i$, which is the codeword that should have been received. Codes of this kind have proved quite useful in correcting rare errors in electronic transmissions. A larger code, based exactly on the same principles as above, was used in the transmission from Mars of all data from the Mariner spacecrafts.

Example 1.6.6 Consider the vectors

$$v = \begin{pmatrix} 0 \\ 1 \\ 0 \\ 0 \\ 1 \\ 0 \\ 1 \end{pmatrix} \quad \text{and} \quad v' = \begin{pmatrix} 0 \\ 1 \\ 1 \\ 0 \\ 1 \\ 0 \\ 1 \end{pmatrix}$$

It is readily checked that the vector v is a codeword for H_3. We observe that v' would be received if v was sent but an error occured in transmitting the third entry. Multiplication (remember $\mathbf{F} = \mathbf{Z}/2\mathbf{Z}$) shows

$$H_3 v' = \begin{pmatrix} 0 \\ 1 \\ 1 \end{pmatrix}$$

This is the third column of H_3, and the code has detected the error.

PROBLEMS

1.6.1. Which of the following are linearly independent in the vector space $(\mathbf{Z}/2\mathbf{Z})^3$?

(a) $\begin{pmatrix} 1 \\ 1 \\ 0 \end{pmatrix}, \begin{pmatrix} 1 \\ 0 \\ 1 \end{pmatrix}, \begin{pmatrix} 0 \\ 1 \\ 1 \end{pmatrix}$ (b) $\begin{pmatrix} 1 \\ 1 \\ 1 \end{pmatrix}, \begin{pmatrix} 1 \\ 0 \\ 1 \end{pmatrix}, \begin{pmatrix} 0 \\ 0 \\ 1 \end{pmatrix}$ (c) $\begin{pmatrix} 0 \\ 1 \\ 1 \end{pmatrix}, \begin{pmatrix} 0 \\ 1 \\ 0 \end{pmatrix}$

1.6.2. What are the ranks of the following matrices with coefficients in $\mathbf{Z}/2\mathbf{Z}$?

(a) $\begin{pmatrix} 1 & 1 & 0 & 1 \\ 1 & 0 & 1 & 1 \\ 0 & 1 & 1 & 0 \end{pmatrix}$ (b) $\begin{pmatrix} 1 & 1 & 1 & 1 \\ 1 & 0 & 0 & 0 \\ 1 & 1 & 0 & 1 \end{pmatrix}$ (c) $\begin{pmatrix} 1 & 1 \\ 1 & 1 \\ 0 & 1 \end{pmatrix}$

1.6.3. Some of the following elements of $\mathbf{F}^7$ are codewords for H_3 (as defined in Definition 1.6.3); others were codewords, but now have a single error. Detect and correct the errors.

$$\begin{pmatrix} 1 \\ 1 \\ 0 \\ 0 \\ 1 \\ 0 \\ 1 \end{pmatrix}, \begin{pmatrix} 1 \\ 0 \\ 0 \\ 1 \\ 1 \\ 0 \\ 0 \end{pmatrix}, \begin{pmatrix} 0 \\ 0 \\ 1 \\ 0 \\ 0 \\ 1 \\ 0 \end{pmatrix}, \begin{pmatrix} 0 \\ 0 \\ 0 \\ 0 \\ 1 \\ 1 \\ 0 \end{pmatrix}, \begin{pmatrix} 1 \\ 0 \\ 1 \\ 1 \\ 0 \\ 1 \\ 0 \end{pmatrix}$$

1.6.4. Consider the following matrix with coefficients in $\mathbf{F} = \mathbf{Z}/2\mathbf{Z}$.

$$H_4 = \begin{pmatrix} 1 & 1 & 1 & 1 & 1 & 1 & 1 \\ 0 & 0 & 0 & 1 & 1 & 1 & 1 \\ 0 & 1 & 1 & 0 & 0 & 1 & 1 \\ 1 & 0 & 1 & 0 & 1 & 0 & 1 \end{pmatrix}$$

(a) What are the codewords for H_4? Find a basis for the codeword space.

(b) This code can correct single errors and tell you when a double error occurs. Figure out why.

(c) Give some examples to illustrate your answer to (b).

Chapter 2

Determinants

2.1 Determinant Functions

In this chapter we study the determinant function $det : M_{n \times n}(\mathbf{F}) \to \mathbf{F}$. Most students are acquainted with the definition of the 2×2 and 3×3 determinants. The $n \times n$ determinant generalizes these familiar cases. The purpose of this chapter is to explain why the determinant function has the important properties it does. Unfortunately, one cannot develop the theory behind the $n \times n$ determinant merely by manipulating the formula which defines it (as is often done in the 2×2 and 3×3 cases). Instead one must carefully analyze the relationship between the determinant and row operations.

In this section we define general *determinant functions*. All the important properties of the determinant are proved for these functions. We then show that there is only one such determinant function on $M_{m \times n}(\mathbf{F})$. This approach also yields the most practical method for computing determinants; see Example 2.1.9.

Definition 2.1.1 For any field $\mathbf{F}$, we denote by $M_{n \times n}(\mathbf{F})$ the set of $n \times n$ matrices with entries from $\mathbf{F}$. Suppose

$$\mathcal{D} : M_{n \times n}(\mathbf{F}) \to \mathbf{F}$$

is a scalar-valued function on $M_{n \times n}(\mathbf{F})$. We say that D is *n-linear* if
(i) Whenever $A \in M_{n \times n}(\mathbf{F})$ and B is obtained from A by multiplying a single row of A by $k \in \mathbf{F}$, $D(B) = kD(A)$.
(ii) Whenever $A, B, C \in M_{n \times n}(\mathbf{F})$ are identical except for the ith rows and the ith row of C is the sum of the ith rows of A and B, then $D(C) = D(A) + D(B)$.
In addition, an n-linear function D is called *alternating* if whenever $A \in M_{n \times n}(\mathbf{F})$ has two identical rows, $D(A) = 0$. Any function $D : M_{n \times n} \to \mathbf{F}$ which is n-linear and alternating and for which $D(I_n) = 1$ is called a *determinant function*.

Remark Misreading condition (ii) in the definition of n-linearity can cause a great deal of grief. It is NOT true that $D(A + B) = D(A) + D(B)$. Read condition (ii) carefully!

Theorem 2.1.8 shows that the determinant function is uniquely determined by Definition 2.1.1. The sections which follow give two different formulations of the determinant function. Throughout this section we shall assume that a determinant function exists and derive its basic properties.

Condition (i) of n-linearity shows how the determinant changes when the first row operation (multiplication by a nonzero scalar) is applied to a matrix. If a row is multiplied by a scalar $k \in \mathbf{F}$, then the determinant changes by multiplication by k. This also shows that whenever a matrix A has an entire row of zeros, $D(A) = 0$. This is because the row of zeros can be multiplied by 0 without changing the matrix, and the value of the determinant becomes 0. How the other two row operations affect the determinant is taken up in the next sequence of lemmas.

This first lemma describes what happens to the determinant when the third row operation of interchanging rows is applied.

Lemma 2.1.2 *Suppose D is a determinant function. If the square matrix B is obtained from the matrix A by interchanging two rows, then $D(B) = -D(A)$.*

Proof: Suppose that the ith and jth rows are interchanged. Let R_i denote the ith row of A, and let R_j denote the jth row of A. We consider the three matrices

$$A = \begin{pmatrix} \vdots \\ R_i \\ \vdots \\ R_j \\ \vdots \end{pmatrix}, \quad B = \begin{pmatrix} \vdots \\ R_j \\ \vdots \\ R_i \\ \vdots \end{pmatrix}, \quad C = \begin{pmatrix} \vdots \\ R_i + R_j \\ \vdots \\ R_i + R_j \\ \vdots \end{pmatrix}$$

Since $\mathcal{D}$ is alternating, Definition 2.1.1 gives that $\mathcal{D}(C) = 0$. We now repeatedly apply the second condition of n-linearity to obtain that

$$0 = \mathcal{D}(C) = \mathcal{D}\begin{pmatrix} \vdots \\ R_i + R_j \\ \vdots \\ R_i + R_j \\ \vdots \end{pmatrix} = \mathcal{D}\begin{pmatrix} \vdots \\ R_i \\ \vdots \\ R_i + R_j \\ \vdots \end{pmatrix} + \mathcal{D}\begin{pmatrix} \vdots \\ R_j \\ \vdots \\ R_i + R_j \\ \vdots \end{pmatrix}$$

$$= \mathcal{D}\begin{pmatrix} \vdots \\ R_i \\ \vdots \\ R_i \\ \vdots \end{pmatrix} + \mathcal{D}\begin{pmatrix} \vdots \\ R_i \\ \vdots \\ R_j \\ \vdots \end{pmatrix} + \mathcal{D}\begin{pmatrix} \vdots \\ R_j \\ \vdots \\ R_i \\ \vdots \end{pmatrix} + \mathcal{D}\begin{pmatrix} \vdots \\ R_j \\ \vdots \\ R_j \\ \vdots \end{pmatrix}$$

$$= \mathcal{D}\begin{pmatrix} \vdots \\ R_i \\ \vdots \\ R_j \\ \vdots \end{pmatrix} + \mathcal{D}\begin{pmatrix} \vdots \\ R_j \\ \vdots \\ R_i \\ \vdots \end{pmatrix} = \mathcal{D}(A) + \mathcal{D}(B)$$

It follows that $\mathcal{D}(B) = -\mathcal{D}(A)$; this proves the lemma. //

We next treat the case of the second elementary row operation. The amazing thing is that this row operation does not change the determinant!

Lemma 2.1.3 *Suppose $\mathcal{D}$ is a determinant function. Assume that the matrix B is obtained from the matrix A by adding a multiple of one row to another. Then $\mathcal{D}(B) = \mathcal{D}(A)$.*

Proof: Suppose that $A = (a_{ij})$ and $B = (b_{ij})$ where $a_{ij} = b_{ij}$ for all j with $i \neq t$ and $b_{tj} = a_{tj} + ka_{sj}$ where $t \neq s$. If we denote by A' the matrix obtained from A by replacing the tth row of A by the sth row of A, then A' has two identical rows. It follows from the alternating property of $\mathcal{D}$ that $\mathcal{D}(A') = 0$. Since $\mathcal{D}$ is n-linear, condition (i) of Definition 2.1.1 shows that the matrix A'' obtained from A' by multiplying the tth row by the scalar k also satisfies $\mathcal{D}(A'') = 0$. Since the matrix B is identical to the matrices A and A'' at each row except the tth, which is the sum of the tth rows of A and A'', condition (ii) of n-linearity gives $\mathcal{D}(B) = \mathcal{D}(A) + \mathcal{D}(A'') = \mathcal{D}(A)$, as desired. $//$

We summarize the fundamental relationship between the determinant functions and row operations in the following theorem.

Theorem 2.1.4 *Suppose that* $\mathcal{D} : M_{n \times n}(\mathbf{F}) \to \mathbf{F}$ *is a determinant function. Let* $A \in M_{n \times n}(\mathbf{F})$. *Then*
(i) *If* B *is obtained from* A *by multiplying a single row of* A *by* $k \in \mathbf{F}$, *then* $\mathcal{D}(B) = k\mathcal{D}(A)$.
(ii) *If* B *is obtained from* A *by adding a multiple of one row to another, then* $\mathcal{D}(B) = \mathcal{D}(A)$.
(iii) *If* B *is obtained from the matrix* A *by interchanging two rows, then* $\mathcal{D}(B) = -\mathcal{D}(A)$.

Proof: (i)   This is true by definition (it is the first condition in Definition 2.1.1).
(ii) This is Lemma 2.1.3.
(iii) This is Lemma 2.1.2. $//$

Let $\mathcal{D}$ be a (the) determinant function. We can use row operations to compute the value of $\mathcal{D}(A)$ for any matrix A. The point is that if a matrix A is row-reduced to a matrix B where $\mathcal{D}(B)$ is known, then $\mathcal{D}(A)$ is a (determined) scalar multiple of $\mathcal{D}(B)$. We illustrate this in the following example.

Example 2.1.5 Consider

$$A = \begin{pmatrix} 2 & 1 & 4 & 7 \\ 3 & 0 & 1 & 1 \\ 7 & 2 & 9 & 15 \\ 0 & 1 & 1 & 1 \end{pmatrix}$$

We apply row operations. Eliminating the nonzero entries from the first column gives

$$B = \begin{pmatrix} 2 & 1 & 4 & 7 \\ 0 & -3/2 & -5 & -19/2 \\ 0 & -3/2 & -5 & -19/2 \\ 0 & 1 & 1 & 1 \end{pmatrix}$$

Applying Lemma 2.1.3 twice (once for each row operation used in reducing A to B), we find that $\mathcal{D}(A) = \mathcal{D}(B)$. The alternating property of $\mathcal{D}$ gives $\mathcal{D}(B) = 0$, so $\mathcal{D}(A) = 0$.

We next expand upon the the idea of the preceding example. In order to apply row operations effectively in the computation of determinants, we need to know the determinants of a reasonably large class of matrices. A convenient such class is the triangular matrices, whose definition we recall.

Definition 2.1.6 A square matrix $A = (a_{ij})$ is called *upper triangular* if $a_{ij} = 0$ whenever $i > j$. Analogously, A is called *lower triangular* if $a_{ij} = 0$ whenever $j > i$. A is called *diagonal* if $a_{ij} = 0$ whenever $i \neq j$ (in other words A is both upper and lower triangular).

The determinants of triangular (and diagonal) matrices are very easy to compute, as the next lemma shows.

Lemma 2.1.7 *Suppose that $\mathcal{D}$ is a determinant function. If $A = (a_{ij})$ is triangular (upper or lower), then $\mathcal{D}(A) = a_{11}a_{22}\cdots a_{nn}$.*

Proof: Assume that A is triangular (upper or lower) and each diagonal entry a_{ii} of A is nonzero. Applying condition (i) of n-linearity n times shows that $\mathcal{D}(A) = a_{11}a_{22}\cdots a_{nn}\mathcal{D}(B)$ where B is a triangular matrix with all diagonal entries 1. (To produce the matrix B, simply multiply the ith row of B by a_{ii}^{-1}.) Repeated applications of the second row operation, together with Lemma 2.1.3, shows that $\mathcal{D}(B) = \mathcal{D}(R)$ where R is the reduced row-echelon form of B. Since B has rank n, necessarily $R = I_n$. We conclude that $\mathcal{D}(A) = a_{11}a_{22}\cdots a_{nn}\mathcal{D}(I_n) = a_{11}a_{22}\cdots a_{nn}$ since $\mathcal{D}(I_n) = 1$ for any determinant function $\mathcal{D}$.

In case some diagonal entry a_{ii} of A is zero, then the reduced row-echelon form R of A has an entire row zero. We find, therefore, that

$D(R) = 0$. Since $D(A)$ differs from $D(R)$ by a scalar multiple, it follows that $D(A) = 0 = a_{11}a_{22} \cdots a_{nn}$. This gives the lemma. //

As an immediate application of the preceding lemma we obtain the uniqueness of the determinant function.

Theorem 2.1.8 (The Characterization of the Determinant) *There is a unique function* $D : M_{n \times n}(\mathbf{F}) \to \mathbf{F}$ *which is n-linear and alternating and for which* $D(I_n) = 1$.

Proof: Suppose that D and D' are both determinant functions on $M_{n \times n}(\mathbf{F})$. Let $A \in M_{n \times n}(\mathbf{F})$ and suppose that R is the reduced row-echelon form of A. The matrix R is upper triangular, so applying Lemma 2.1.7, we find that $D(R) = D'(R)$. However, R can be obtained from A by a sequence of elementary row operations, so repeatedly applying Theorem 2.1.4 for this sequence of elementary operations shows that $D(A) = D'(A)$ as well. From this the theorem follows. //

We have now established the uniqueness of the determinant function. Two proofs of its existence will be given in the sections which follow. For the remainder of this section we shall use the standard notation $det(A)$ for *the* determinant of an $n \times n$ matrix A. As remarked earlier, our characterization of the determinant function via row operations is very effective in computations.

Example 2.1.9 Theorem 2.1.4 together with Lemma 2.1.7 shows how to compute a particular determinant by row operations. The point is to row-reduce a given matrix to a triangular matrix, while keeping track of how the determinant changes with each row operation. For example consider

$$A = \begin{pmatrix} 2 & 0 & 2 & 4 & 8 \\ 0 & 3 & 2 & 4 & 0 \\ 1 & 2 & 3 & 0 & 1 \\ 0 & 1 & 4 & 6 & 7 \\ 1 & 3 & 6 & 7 & 8 \end{pmatrix}$$

Then, repeatedly apply row operations (together with Theorem 2.1.4):

$$det(A) = 2det \begin{pmatrix} 1 & 0 & 1 & 2 & 4 \\ 0 & 3 & 2 & 4 & 0 \\ 1 & 2 & 3 & 0 & 1 \\ 0 & 1 & 4 & 6 & 7 \\ 1 & 3 & 6 & 7 & 8 \end{pmatrix}$$

$$= \ 2det\begin{pmatrix} 1 & 0 & 1 & 2 & 4 \\ 0 & 3 & 2 & 4 & 0 \\ 0 & 2 & 2 & -2 & -3 \\ 0 & 1 & 4 & 6 & 7 \\ 1 & 3 & 6 & 7 & 8 \end{pmatrix}$$

$$= \ 2det\begin{pmatrix} 1 & 0 & 1 & 2 & 4 \\ 0 & 3 & 2 & 4 & 0 \\ 0 & 2 & 2 & -2 & -3 \\ 0 & 1 & 4 & 6 & 7 \\ 0 & 3 & 5 & 5 & 4 \end{pmatrix}$$

$$= \ (3)(2)det\begin{pmatrix} 1 & 0 & 1 & 2 & 4 \\ 0 & 1 & 4/3 & 2/3 & 0 \\ 0 & 2 & 2 & -2 & -3 \\ 0 & 1 & 4 & 6 & 7 \\ 0 & 3 & 5 & 5 & 4 \end{pmatrix}$$

$$= \ 6det\begin{pmatrix} 1 & 0 & 1 & 2 & 4 \\ 0 & 1 & 4/3 & 2/3 & 0 \\ 0 & 0 & -2/3 & -10/3 & -3 \\ 0 & 0 & 8/3 & 16/3 & 7 \\ 0 & 0 & 1 & 3 & 4 \end{pmatrix}$$

$$= \ (-2/3)(6)det\begin{pmatrix} 1 & 0 & 1 & 2 & 4 \\ 0 & 1 & 4/3 & 2/3 & 0 \\ 0 & 0 & 1 & 5 & 9/2 \\ 0 & 0 & 8/3 & 16/3 & 7 \\ 0 & 0 & 1 & 3 & 4 \end{pmatrix}$$

$$= \ -4det\begin{pmatrix} 1 & 0 & 1 & 2 & 4 \\ 0 & 1 & 4/3 & 2/3 & 0 \\ 0 & 0 & 1 & 5 & 9/2 \\ 0 & 0 & 0 & -24/3 & -5 \\ 0 & 0 & 0 & -2 & -1/2 \end{pmatrix}$$

$$= \ (-8)(-4)det\begin{pmatrix} 1 & 0 & 1 & 2 & 4 \\ 0 & 1 & 4/3 & 2/3 & 0 \\ 0 & 0 & 1 & 5 & 9/2 \\ 0 & 0 & 0 & 1 & 5/8 \\ 0 & 0 & 0 & -2 & -1/2 \end{pmatrix}$$

$$
\begin{aligned}
&= 32 det \begin{pmatrix}
1 & 0 & 1 & 2 & 4 \\
0 & 1 & 4/3 & 2/3 & 0 \\
0 & 0 & 1 & 5 & 9/2 \\
0 & 0 & 0 & 1 & 5/8 \\
0 & 0 & 0 & 0 & (5/4 - 1/2)
\end{pmatrix} \\
&= (32)(3/4) \\
&= 24
\end{aligned}
$$

One word of caution: Be sure to keep track of the changes in the determinant which arise whenever a row is multiplied by a constant. Since the second row operation is the most frequently used and since it does not change the determinant, it is easy to become sloppy and forget this.

The practical calculation of determinants is not the only reason we have characterized the determinant via row operations; there are important theoretical applications as well. Recall that row operations are the same as multiplication (on the left) by elementary matrices. An immediate use of Theorem 2.1.4 is the following.

Lemma 2.1.10   *If E is an $n \times n$ elementary matrix and A is an $n \times n$ matrix, then $det(EA) = det(A)$.*

Proof: There are three cases to consider, but the proof is the same in each case. If the elementary matrix E is obtained from the identity matrix by multiplying the jth row by a scalar k, then Theorem 2.1.4 (i) shows that $det(E) = k \cdot det(I_n) = k$. For any matrix A, Lemma 0.4.9 observes that the matrix EA can be obtained from A by multiplying the jth row of A by the scalar k. Hence, again applying Theorem 2.1.4 (i) shows that $det(EA) = k \cdot det(A) = det(E)det(A)$, as required. By a similar argument, Theorem 2.1.4 (ii) and (iii) gives the result for the other two row operations. $//$

Lemma 2.2.10 can now be iteratively applied together with Theorem 0.6.9 to obtain the following.

Lemma 2.1.11   *Suppose that A is an invertible $n \times n$ matrix and B is any $n \times n$ matrix. Then $det(AB) = det(A)det(B)$.*

Proof: Since A is invertible, by Theorem 0.6.9 $A = E_s E_{s-1} \cdots E_1$ where $E_1, E_2, \ldots, E_s$ are elementary matrices. Applying Lemma 2.1.10 repeatedly gives

$$
\begin{aligned}
det(AB) &= det(E_s E_{s-1} \cdots E_1 B) \\
&= det(E_s)det(E_{s-1} \cdots E_1 B) \\
&= \cdots = det(E_s)det(E_{s-1}) \cdots det(E_1)det(B) \\
&= det(E_s E_{s-1})det(E_{s-2}) \cdots det(E_1)det(B) \\
&= \cdots = det(E_s E_{s-1} \cdots E_1)det(B) \\
&= det(A)det(B)
\end{aligned}
$$

This proves the lemma. //

The determinant provides a nice characterization of invertibility; in fact this may be its best-known applicaton. The result should not be surprising, given all our work up to this point. We already know that invertibility corresponds to full rank and the determinant of full-rank triangular matrices is nonzero by Lemma 2.1.8. We remark that as a practical test for invertibility, row operations are still the best bet. (After all, in order to compute the determinant effectively, row operations will probably be involved!)

Theorem 2.1.12 *An $n \times n$ matrix A is invertible if and only if $det(A) \neq 0$.*

Proof: If A is invertible, then by Theorem 2.1.11, $1 = det(I_n) = det(AA^{-1}) = det(A)det(A^{-1})$, so $det(A) \neq 0$. If A is not invertible, then as $rk(A) < n$, A is row equivalent to a triangular matrix R with a row of zeros at the bottom (the reduced echelon form of A, for example). Then $A = UR$ for some invertible matrix U by Theorem 0.6.8. Hence $det(A) = det(U)det(R) = det(U)0 = 0$ by Lemmas 2.1.11 and 2.1.8. This proves the theorem. //

Using the two preceding theorems, we obtain the most important property of the determinant function. This result is precisely Lemma 2.1.11, without the invertibility assumption on the matrix A.

Theorem 2.1.13 *If A and B are arbitrary $n \times n$ matrices, then $det(AB) = det(A)det(B)$.*

Proof: If A is invertible, then this result is precisely Lemma 2.1.11. If A is not invertible, then by Theorem 2.1.12, $det(A) = 0$. Suppose that $UA = R$ is the reduced echelon form of A. Then as R has a row of zeros, the definition of matrix multiplication shows that RB has a row of zeros. Since $RB = U(AB)$, we see that AB is row equivalent to a matrix with a row of zeros. It follows that AB cannot be invertible either. Hence, $0 = det(AB) = 0det(B) = det(A)det(B)$, as desired. **//**

The determinant function ignores transposes, as we show in the next theorem. This result could be proved directly by using the definition of the determinant given in the next section. However in this proof we take advantage of the theory of elementary matrices.

Theorem 2.1.14 *For any $n \times n$ matrix A, $det(A) = det(A^t)$.*

Proof: If A is not invertible, then by Corollary 0.6.13 A^t is not invertible either. Hence $det(A) = 0 = det(A^t)$. So we can assume A is invertible. We see easily that whenever E is an elementary matrix, $det(E) = det(E^t)$. Expressing $A = E_s E_{s-1} \cdots E_1$ with each E_i elementary, we see by Theorem 0.6.12 that $A^t = E_1^t E_2^t \cdots E_s^t$. Putting all this together, $det(A) = det(E_s)det(E_{s-1}) \cdots det(E_1) = det(E_1^t)det(E_2^t) \cdots det(E_s^t) = det(A^t)$, as required. **//**

Using Theorem 2.1.14, we obtain the next corollary. This corollary is really a general remark, which says that as far as the computation of determinants is concerned, we can use column operations in the same ways that row operations are used.

Corollary 2.1.15 *Each of the three results in* Theorem 2.1.4 *remain true when the word "row" is replaced by the word "column." In other words, column operations can be performed on square matrices and they effect the determinant in exactly the same ways as row opertions.*

Proof: In each case apply Theorem 2.1.4 to the transpose, and then apply Theorem 2.1.14. **//**

PROBLEMS

2.1.1. Compute the the determinants of the following matrices using row operations.

(a) $\begin{pmatrix} 1 & 1 & 1 \\ 2 & 0 & 4 \\ 2 & 3 & 4 \end{pmatrix}$ (b) $\begin{pmatrix} 1 & 0 & 1 & 2 \\ 2 & 3 & 4 & 5 \\ 1 & 2 & 3 & 4 \\ 4 & 5 & 8 & 11 \end{pmatrix}$ (c) $\begin{pmatrix} 1 & 1 & 1 & 1 & 2 \\ 1 & 1 & 1 & 1 & 1 \\ 0 & 0 & 1 & 2 & 3 \\ 0 & 0 & 0 & 3 & 5 \\ 2 & 3 & 3 & 1 & 1 \end{pmatrix}$

(d) $\begin{pmatrix} 2 & 1 & 3 \\ 3 & 5 & 1 \\ 5 & 6 & 5 \end{pmatrix}$ (e) $\begin{pmatrix} 2 & 3 & -1 & 0 \\ 3 & 1 & 7 & 1 \\ 5 & 4 & 6 & 2 \\ 7 & 7 & 6 & 1 \end{pmatrix}$ (f) $\begin{pmatrix} 2 & 3 & 4 & 5 \\ 1 & 1 & 1 & 1 \\ 2 & 2 & 2 & 2 \\ 9 & 8 & 7 & 6 \end{pmatrix}$

2.1.2. Compute the the determinants of the following matrices using row operations.

(a) $\begin{pmatrix} 1 & 0 & 1 \\ 2 & 0 & 0 \\ 2 & 1 & 4 \end{pmatrix}$ (b) $\begin{pmatrix} 1 & 1 & 1 & 2 \\ 2 & 2 & 2 & 5 \\ 1 & 2 & 1 & 4 \\ 4 & 4 & 4 & 11 \end{pmatrix}$ (c) $\begin{pmatrix} 1 & 1 & 1 & 1 \\ 1 & 1 & 1 & 1 \\ 0 & 0 & 1 & 2 \\ 0 & 0 & 0 & 3 \end{pmatrix}$

(d) $\begin{pmatrix} 2 & 2 & 3 \\ 3 & 3 & 1 \\ 5 & 6 & 5 \end{pmatrix}$ (e) $\begin{pmatrix} 2 & -1 & -1 & 0 \\ 3 & 1 & 1 & -1 \\ 5 & 0 & 0 & 2 \\ 7 & 0 & 0 & -1 \end{pmatrix}$ (f) $\begin{pmatrix} 2 & 3 & 4 & 5 \\ 1 & 1 & 1 & 1 \\ 2 & 2 & 2 & 0 \\ 1 & 1 & 0 & 0 \end{pmatrix}$

2.1.3. A *nilpotent* matrix N is a matrix for which $N^k = 0$ for some k. Show that if N is a nilpotent matrix, then $det(N) = 0$.

2.1.4. Find two 2×2 matrices A and B such that $det(A + B) \neq det(A) + det(B)$.

2.1.5. Two matrices A and B are said to *anticommute* if $AB = -BA$. Suppose that n is odd and A and B are $n \times n$ real matrices which anticommute. Show that one of A or B is not invertible. Show this can fail if n is even.

2.1.6. If A is an $n \times n$ symmetric matrix (that is, $A = A^t$), show that for any $n \times n$ matrix B, $det(A + B) = det(A + B^t)$.

2.1.7. Show that for any two $n \times n$ matrices A and B $det(AB) = det(BA)$.

2.1.8. Show that three points $(a, b), (c, d), (e, f) \in \mathbf{R}^2$ are collinear if and only if

$$det \begin{pmatrix} a & b & 1 \\ c & d & 1 \\ e & f & 1 \end{pmatrix} = 0$$

2.1.9. Compute $det(B^t A^{-1} BA)$ where

$$A = \begin{pmatrix} 1 & 3 & 2 \\ -1 & 4 & 1 \\ 5 & -2 & 6 \end{pmatrix} \quad \text{and} \quad B = \begin{pmatrix} 3 & 1 & 6 \\ 3 & 2 & 7 \\ 3 & 1 & 8 \end{pmatrix}$$

2.1.10. For any real $n \times n$ matrix Q show that
(a) The diagonal entries of QQ^t are nonnegative.
(b) $det(QQ^t) \geq 0$

2.1.11. (a) Compute the determinant of

$$\begin{pmatrix} 1 & 1 & 2 & 1 \\ a & (a+1) & 2a & a \\ 0 & b & 2 & 1 \\ c & c & 2d & d+1 \end{pmatrix}$$

(b) If a 4×4 matrix A satisfies $det(A) = 1$, prove that the row space of A is $\mathbf{F}^4$.

2.1.12. If A is an $n \times n$ matrix and $det(A) = 0$, prove that $ker(A) \neq \{\vec{0}\}$.

2.1.13. If

$$C = \begin{pmatrix} 2 & 0 & 0 \\ 2 & 9 & 3 \\ 5 & 7 & 8 \end{pmatrix} \quad \text{and} \quad D = \begin{pmatrix} 1 & 2 & 5 \\ 0 & 1 & 9 \\ 0 & 0 & 2 \end{pmatrix}$$

find $det(C^t D C^{-1})$.

2.2   Permutations and the Determinant

We now give an explicit construction of the determinant function. So, we shall know that the function discussed at length in the preceding section really exists! There are several equivalent ways to construct the determinant function. The two standard methods are the *permutation definition* (given in this section) and the *cofactor expansion*, which is an inductive definition described in detail in Sec. 2.3. An advantage to giving the permutation definition is that far fewer inductions are required in the proofs of the main results. Also, the permutation definition is clearly well-defined, while (as we shall see) the cofactor method requires some effort to check that the various cofactor expansions all give the same function. The cofactor definition is popular because it gives an easily learned procedure for determinant calculation. Unfortunately, this procedure is (usually) not efficient for matrices larger than 3×3. A disadvantage with the permutation definition is that additional algebraic terminology is required (namely permutations and their signs). We begin with these preliminary definitions.

Definition 2.2.1 A *permutation of n* is an n-tuple $\sigma = (i_1, i_2, \ldots, i_n)$ where $i_1, i_2, \ldots, i_n$ are the n distinct integers $1, 2, \ldots, n$ (but not necessarily listed in the usual order). We denote the jth entry in a permutation σ by σ_j. The collection of all permutations of n is denoted by S_n.

For example, suppose that $n = 3$. Then $S_3 = \{(1, 2, 3), (1, 3, 2),$ $(2, 1, 3), (2, 3, 1), (3, 1, 2), (3, 2, 1)\}$ consists of six permutations. Note that the sequence $(3, 1, 3)$ is *not* considered to be a permutation of 3, since the entries in this three-tuple are not distinct. For the permutation $\sigma = (3, 2, 1)$, our notation gives that $\sigma_1 = 3$, $\sigma_2 = 2$, and $\sigma_3 = 1$. For general n, S_n always has $n! = n(n-1)(n-2)\cdots 2 \cdot 1$ elements. (Many texts define the permutations of n to be the set of all functions from $\{1, 2, \ldots, n\}$ to itself which are one-to-one. In this view one considers $\sigma = (i_1, i_2, \ldots, i_n)$ to be the function whose value at j is i_j. This point of view is extremely important in abstract algebra. However, for our purposes of defining and studying determinants, it suffices to consider a permutation as an n-tuple.)

Associated with every permutation is a sign, which is either $+1$ or -1. This is defined next.

Definition 2.2.2 Suppose that $\sigma = (i_1, i_2, \ldots, i_n) \in S_n$. If $1 \le j < k \le n$ and $i_j > i_k$, we say that $i_j > i_k$ is an *inversion of* σ. For any permutation σ we define the *sign* of σ, denoted $sg(\sigma)$, by $sg(\sigma) = +1$ if the total number of inversions of σ is even and by $sg(\sigma) = -1$ if the total number of inversions of σ is odd. When $sg(\sigma) = +1$, σ is called *even*, and when $sg(\sigma) = -1$, σ is called *odd*.

For example, consider the permutations of 4: $(1, 2, 4, 3)$, $(4, 3, 2, 1)$ and $(4, 1, 3, 2)$. The permutation $(1, 2, 4, 3)$ has the single inversion $4 > 3$; so $sg(1, 2, 4, 3) = -1$. $(4, 3, 2, 1)$ has the six inversions $4 > 3$, $4 > 2$, $4 > 1$, $3 > 2$, $3 > 1$, $2 > 1$. Hence $sg(4, 3, 2, 1) = +1$. Similarly, $sg(4, 1, 3, 2) = +1$.

We now can give a definition of the determinant function. In the examples that follow we show that this definition coincides with the familiar definition in the 2×2 and 3×3 cases.

Definition 2.2.3 Suppose that $A = (a_{ij})$ is an $n \times n$ matrix. We define the *determinant* of A, denoted $det(A)$, by

$$dct(A) = \sum_{\sigma \in S_n} sg(\sigma) a_{1\sigma_1} a_{2\sigma_2} \cdots a_{n\sigma_n}$$

Examples 2.2.4 (i) If $A = (a)$ is a 1×1 matrix, then since S_1 has only one element, $det(A) = a$.

(ii) Suppose that

$$A = \begin{pmatrix} a_{11} & a_{12} \\ a_{21} & a_{22} \end{pmatrix}$$

Then, since S_2 has only the two elements $(1,2)$, $(2,1)$ where the first has sign $+1$ and the second has sign -1, we find that $det(A) = (1)a_{11}a_{22} + (-1)a_{12}a_{21}$. In more compact notation we obtain the familiar expression

$$det \begin{pmatrix} a & b \\ c & d \end{pmatrix} = ad - bc$$

(iii) Now suppose that $A = (a_{ij})$ is a 3×3 matrix. The six elements of S_3 are $(1,2,3)$, $(1,3,2)$, $(2,3,1)$, $(2,1,3)$, $(3,1,2)$, $(3,2,1)$. The respective signs are $+1, -1, +1, -1, +1, -1$. It follows that the 3×3 determinant is given by

$$det \begin{pmatrix} a_{11} & a_{12} & a_{13} \\ a_{21} & a_{22} & a_{23} \\ a_{31} & a_{32} & a_{33} \end{pmatrix} = \begin{aligned} & a_{11}a_{22}a_{33} - a_{11}a_{23}a_{32} + a_{12}a_{23}a_{31} \\ & -a_{12}a_{21}a_{33} + a_{13}a_{21}a_{32} - a_{13}a_{22}a_{31} \end{aligned}$$

In general, since there are $n!$ elements in S_n, using Definition 2.2.3 to compytate the determinant of (even not very large) square matrices makes the computation extremely laborious. In the last section, however, we saw that the value of the determinant function can be more easily computed using row operations. The next three results show that the determinant, defined as in Definition 2.2.3, is a determinant function as defined in Definition 2.1.1. We begin with the first condition of n-linearity.

Lemma 2.2.5 *Let det be as in* Definition 2.2.3. *Suppose that A is a square matrix and B is obtained from A by multiplying a single row of A by a scalar k. Then* $det(B) = k \cdot det(A)$.

Proof: We assume that $A = (a_{ij})$ and $B = (b_{ij})$. Suppose that B is obtained from A by multiplying the sth row by k. Then $b_{ij} = a_{ij}$ for all $i \neq s$ and all j, while $b_{sj} = ka_{sj}$ for all j. By Definition 2.2.3

$$
\begin{aligned}
det(B) &= \sum_{\sigma \in S_n} sg(\sigma) b_{1\sigma_1} b_{2\sigma_2} \cdots b_{s\sigma_s} \cdots b_{n\sigma_n} \\
&= \sum_{\sigma \in S_n} sg(\sigma) a_{1\sigma_1} a_{2\sigma_2} \cdots ka_{s\sigma_s} \cdots a_{n\sigma_n} \\
&= k \sum_{\sigma \in S_n} sg(\sigma) a_{1\sigma_1} a_{2\sigma_2} \cdots a_{s\sigma_s} \cdots a_{n\sigma_n} \\
&= k \cdot det(A)
\end{aligned}
$$

This proves the lemma. //

The second half of n-linearity is treated in this next lemma.

Lemma 2.2.6 *Let det be as in* Definition 2.2.3. *Suppose that the three matrices $A = (a_{ij})$, $B = (b_{ij})$, and $C = (c_{ij})$ are identical except for the kth row, where we assume for all j that $a_{kj} + b_{kj} = c_{kj}$. Then* $det(A) + det(B) = det(C)$.

Proof: Our hypotheses give that for all j $a_{ij} = b_{ij} = c_{ij}$ whenever $i \neq k$ and $a_{kj} + b_{kj} = c_{kj}$ otherwise. It then follows from Definition 2.2.3 that

$$
\begin{aligned}
det(C) &= \sum_{\sigma \in S_n} sg(\sigma) a_{1\sigma_1} a_{2\sigma_2} \cdots (a_{k\sigma_k} + b_{k\sigma_k}) \cdots a_{n\sigma_n} \\
&= \sum_{\sigma \in S_n} sg(\sigma) a_{1\sigma_1} a_{2\sigma_2} \cdots a_{k\sigma_k} \cdots a_{n\sigma_n} \\
&\quad + \sum_{\sigma \in S_n} sg(\sigma) b_{1\sigma_1} b_{2\sigma_2} \cdots b_{k\sigma_k} \cdots b_{n\sigma_n} \\
&= det(A) + det(B)
\end{aligned}
$$

This proves the lemma. //

Before we can establish the alternating property of the determinant, we need a technical result about permutations.

Lemma 2.2.7 *Suppose that $1 \leq s < t \leq n$, $\sigma = (i_1, i_2, \ldots, i_n) \in S_n$, and σ' is obtained from σ by interchanging i_s and i_t. Then $sg(\sigma') = -sg(\sigma)$.*

Proof: Consider the permutation $\sigma = (\ldots, i_s, \ldots, i_t, \ldots)$ where $s < t$. Without loss of generality we can assume $i_s < i_t$. (One of σ or σ' must have this property.) Then σ' is the permutation $(\ldots, i_t, \ldots, i_s, \ldots)$ which is identical to σ at every place, except that the tth and sth entries are reversed. We must show $sg(\sigma') = -sg(\sigma)$. Consider the collection of entries $i_{s+1}, i_{s+2}, \ldots, i_{t-1}$ of σ. Assume that among this collection there are e numbers less than i_s, f numbers larger than i_s but less than i_t, and g numbers larger than i_t. When i_s and i_t are interchanged, the number of inversions passing from σ to σ' is:

(i) Decreased by e, and increased by $f + g$ (those inversions involving i_s but not i_t)

(ii) Decreased by g and increased by $e + f$ (those inversions involving i_t but not i_s)

(iii) Increased by 1 (the inversion involving both i_t and i_s)

The total change in the number of inversions is

$$-e + (f + g) + -g + (e + f) + 1 = 2f + 1$$

Since $2f + 1$ is always odd, $sg(\sigma') = -sg(\sigma)$. This proves the lemma.
//

Lemma 2.2.7 was proved precisely so that we could establish the alternating property of the determinant function as defined in Definition 2.2.3.

Lemma 2.2.8 *Let det be as in* Definition 2.2.3. *If a square matrix A has two equal rows, then $det(A) = 0$.*

Proof: Suppose that $A = (a_{ij})$ and $a_{sj} = a_{tj}$ for all j where $1 \leq s < t \leq n$. We partition S_n into two subsets C and C', where C consists of those permutations in S_n with $i_s < i_t$ and C' consists of those permutations in S_n with $i_t < i_s$. Note that $C \cup C' = S_n$ and $C \cap C' = \emptyset$. (This is what "partition" means.) It follows from the definitions that whenever $\sigma \in C$, then $\sigma' \in C'$. Moreover, each element of C' is σ' for some $\sigma \in C$. (Here, σ' is as described in Lemma 2.2.7.)

Since $a_{sj} = a_{tj}$ for all j, we find that $a_{s\sigma_s} = a_{t\sigma'_t}$ and $a_{t\sigma_t} = a_{s\sigma'_s}$. Using Lemma 2.2.7 together with the fact that $a_{i\sigma_i} = a_{i\sigma'_i}$ for all $i \neq s, t$, we see

$$
\begin{aligned}
det(A) &= \sum_{\sigma \in S_n} sg(\sigma)a_{1\sigma_1} \cdots a_{s\sigma_s} \cdots a_{t\sigma_t} \cdots a_{n\sigma_n} \\
&= \sum_{\sigma \in C} sg(\sigma)a_{1\sigma_1} \cdots a_{s\sigma_s} \cdots a_{t\sigma_t} \cdots a_{n\sigma_n} \\
&\quad + \sum_{\sigma \in C'} sg(\sigma)a_{1\sigma_1} \cdots a_{s\sigma_s} \cdots a_{t\sigma_t} \cdots a_{n\sigma_n} \\
&= \sum_{\sigma \in C} sg(\sigma)a_{1\sigma_1} \cdots a_{s\sigma_s} \cdots a_{t\sigma_t} \cdots a_{n\sigma_n} \\
&\quad + \sum_{\sigma \in C} sg(\sigma')a_{1\sigma'_1} \cdots a_{s\sigma'_s} \cdots a_{t\sigma'_t} \cdots a_{n\sigma'_n} \\
&= \sum_{\sigma \in C} sg(\sigma)a_{1\sigma_1} \cdots a_{s\sigma_s} \cdots a_{t\sigma_t} \cdots a_{n\sigma_n} \\
&\quad + \sum_{\sigma \in C} -sg(\sigma)a_{1\sigma_1} \cdots a_{s\sigma_s} \cdots a_{t\sigma_t} \cdots a_{n\sigma_n} \\
&= 0
\end{aligned}
$$

This proves the lemma. //

As an immediate consequence of Lemmas 2.2.5, 2.2.6, and 2.2.8 we obtain the following theorem.

Theorem 2.2.9 *The function* det $: M_{n \times n}(\mathbf{F}) \to \mathbf{F}$ *defined for* $A = (a_{ij})$ *by*

$$ det(A) = \sum_{\sigma \in S_n} sg(\sigma)a_{1\sigma_1}a_{2\sigma_2} \cdots a_{n\sigma_n} $$

is a determinant function in the sense of Definition 2.1.1.

This section concludes with a result which computes the determinant of certain block matrices. The result can be proved by considering row operations on partitioned matrices, but we give the result here since it nicely illustrates our permutation definition of the determinant. This result will be quite convenient to have later when we study linear operators, but will not be subsequently used in this chapter.

Theorem 2.2.10 *Suppose that* $1 \leq r < n$ *and*

$$A = \begin{pmatrix} A_{11} & 0 \\ A_{21} & A_{22} \end{pmatrix}$$

is a partitioned $n \times n$ *matrix, where* A_{11} *is* $r \times r$, A_{22} *is* $(n-r) \times (n-r)$, *and the upper right* $r \times (n - r)$ *matrix* A_{12} *consists entirely of zeros. Then* $det(A) = det(A_{11})det(A_{22})$.

Proof: We must consider a special type of permutation in S_n. We shall say that $\sigma = (i_1, \ldots, i_n)$ is *partitioned* by r if $i_1, i_2, \ldots, i_r \in \{1, 2, \ldots, r\}$. Whenever this occurs, note also that $i_{r+1}, i_{r+2}, \ldots, i_n \in \{r + 1, r + 2, \ldots, n\}$. Consequently, for such σ we obtain two smaller permutations: The first is $\sigma' = (i_1, i_2, \ldots, i_r) \in S_r$, and the second is $\sigma^* = (i_{r+1} - r, i_{r-2} - r, \ldots, i_n - r) \in S_{n-r}$. (Note that our conditions guarantee that σ^* makes sense as defined.) Whenever σ is partitioned by r, the product arising in the definition of the determinant factors as

$$a_{1\sigma_1} a_{2\sigma_2} \cdots a_{n\sigma_n} = \left(a_{1\sigma'_1} \cdots a_{r\sigma'_r}\right)\left(a_{r+1\sigma^*_1+r} \cdots a_{n\sigma^*_{n-r}+r}\right)$$

Observe that the first factor in this product is the product arising in the definition of $det(A_{11})$ corresponding to the permutation σ' and the second factor is the product arising in the definition of $det(A_{22})$ corresponding to the permutation σ^*. Finally observe that $sg(\sigma) = sg(\sigma')sg(\sigma^*)$ since the number of transpositions in σ is the sum of the number of transpositons in σ' and σ^*. (This is because a transposition $i_t < i_s$ of σ with $1 \leq t < s \leq n$ must occur with $s \leq r$ or $t > r$.)

Now suppose that $\sigma \in S_n$ is not partitioned by r. For such a permutation there is some $k \leq r$ with $\sigma_k > r$. Denoting $A = (a_{ij})$, we see from our hypotheses on A that $a_{k\sigma_k} = 0$ for this k. In particular, whenever σ is not partitioned by r, we have $a_{1\sigma_1} a_{2\sigma_2} \cdots a_{n\sigma_n} = 0$. From this we conclude that the only products contributing to the determinant of A come from the permutations partitioned by r.

We now can compute $det(A)$ by Definition 2.1.3:

$$det(A) = \sum_{\sigma \in S_n} sg(\sigma) a_{1\sigma_1} a_{2\sigma_2} \cdots a_{n\sigma_n}$$

$$= \sum_{\sigma \text{ partitioned by } r} sg(\sigma) \left(a_{1\sigma'_1} \cdots a_{r\sigma'_r}\right)\left(a_{r+1\sigma^*_1+r} \cdots a_{n\sigma^*_{n-r}+r}\right)$$

$$= \left(\sum_{\sigma' \in S_r} sg(\sigma')a_{1\sigma'_1} \cdots a_{r\sigma'_r}\right)\left(\sum_{\sigma^* \in S_{n-r}} sg(\sigma^*)a_{r+1\sigma^*_1+r} \cdots a_{n\sigma^*_{n-r}+r}\right)$$

$$= det(A_{11})det(A_{22})$$

This proves the theorem. //

PROBLEMS

2.2.1. Find the sign of the following permutations:
(a) $(5, 4, 3, 2, 1)$ (b) $(4, 5, 3, 2, 1)$ (c) $(1, 2, 3, 5, 4)$

2.2.2. Compute the determinants of the following matrices using Definition 2.2.3.

(a) $\begin{pmatrix} 2 & 3 \\ 5 & 7 \end{pmatrix}$ (b) $\begin{pmatrix} 1 & 2 & 0 \\ 5 & 1 & 0 \\ 3 & 3 & 1 \end{pmatrix}$ (c) $\begin{pmatrix} 3 & 1 & 0 & 1 \\ 2 & 0 & 0 & 0 \\ 2 & 2 & 1 & 0 \\ 0 & 1 & 0 & 1 \end{pmatrix}$

2.2.3. Suppose that A is an $n \times n$ matrix with more than $n^2 - n$ entries which are 0. Prove that $det(A) = 0$.

2.2.4. Using only Definition 2.2.3, prove:
(a) Lemma 2.1.3
(b) Lemma 2.1.7
(c) Theorem 2.1.14

2.2.5. An *orthogonal matrix* is an $n \times n$ matrix A for which $AA^t = I_n$. Show that if A is orthogonal, then $det(A) = \pm 1$.

2.2.6. Verify by direct calculation that $det(AB) = det(A)det(B)$ for 2×2 and 3×3 matrices.

2.2.7. A *skew-symmetric matrix* is a matrix A for which $A^t = -A$. Suppose that A is an $n \times n$ real skew-symmetric matrix where n is odd. Show $det(A) = 0$.

2.2.8. Show the 3×3 Vandermonde determinant

$$det \begin{pmatrix} 1 & a & a^2 \\ 1 & b & b^2 \\ 1 & c & c^2 \end{pmatrix}$$

is given by $(b - a)(c - a)(c - b)$. When is the Vandermonde matrix invertible?

2.2.9. Show that half the permutations in S_n have sign $+1$ and half have sign -1 ($n > 1$ of course).

2.2.10. Recall that a permutation matrix (see Prob. 0.6.23) is a matrix that has a single 1 in each row and each column and is 0 elsewhere. Relate permutation matrices to permutations in a natural way. How can one interpret the determinant of a permutation matrix?

2.3   The Laplace Expansion of the Determinant; Applications

In this section we give the inductive description of the determinant, known as the *cofactor* or *Laplace expansion*. In many texts, this inductive description is used to define the determinant. We also give two applications of the inductive characterization of the determinant. The first is a computation of the inverse matrix in terms of determinants. While not always practical from a computational point of view, this description of the inverse has important theoretical uses. The second application, Cramer's rule, describes the unique solution to $AX = \vec{b}$ whenever A is invertible. This next definition is needed to give the inductive description of the determinant.

Definition 2.3.1 Suppose that $A = (a_{ij})$ is an $n \times n$ matrix. We define $A(i \mid j)$ to be the $(n-1) \times (n-1)$ matrix obtained from A by deleting the ith row and jth column. $A(i \mid j)$ is called the ijth *maximal submatrix* of A.

Example 2.3.2 Suppose that

$$A = \begin{pmatrix} 2 & 1 & 5 \\ 3 & 6 & 7 \\ 4 & 8 & 9 \end{pmatrix}$$

Then we have that

$$A(1 \mid 1) = \begin{pmatrix} 6 & 7 \\ 8 & 9 \end{pmatrix} \quad A(2 \mid 3) = \begin{pmatrix} 2 & 1 \\ 4 & 8 \end{pmatrix} \quad \text{and} \quad A(3 \mid 1) = \begin{pmatrix} 1 & 5 \\ 6 & 7 \end{pmatrix}$$

We now give the Laplace (or cofactor) expansion for determinant function. This description gives the determinant $det(A)$ in terms of the determinant of certain maximal submatrices of A. Applying this same description to these smaller submatrices, we obtain an inductive characterization of the determinant.

There are different ways in which these cofactor expansions can be computed. Specifically, we shall introduce functions denoted $\mathcal{D}_{n,i}$ and $\mathcal{D}_j^n$, each of which will be shown to be the determinant function on $n \times n$ matrices. (See Theorem 2.3.5.) The function $\mathcal{D}_{n,i}$ is the cofactor expansion along the ith row, and the function $\mathcal{D}_j^n$ is the cofactor expansion along the jth column. We remark that this notation for the cofactor expansion in *not* standard and is introduced merely as a technical convenience. Once Theorem 2.3.5 has been proved, we will abandon this notation and continue to use *det* to mean *the* determinant.

Definition 2.3.3 If $A = (a)$, we define $\mathcal{D}_{1,1}(A) = \mathcal{D}_1^1(A) = a = det(A)$. If $A = (a_{ij})$ is an $n \times n$ matrix with $n > 1$, for i with $1 \le i \le n$ we define

$$\mathcal{D}_{n,i}(A) = \sum_{j=1}^{n}(-1)^{i+j}a_{ij}det(A(i \mid j))$$

which is called the ith *row cofactor* or *Laplace expansion* of *det*. Similarly, for j with $1 \le j \le n$ we define

$$\mathcal{D}_j^n(A) = \sum_{i=1}^{n}(-1)^{i+j}a_{ij}det(A(i \mid j))$$

which is called the jth *column cofactor* or *Laplace expansion* of *det*.

Remark The reader may be curious why the definitions of $\mathcal{D}_{n,i}$ and $\mathcal{D}_j^n$ are given in terms of *det*, rather than inductively in terms of $\mathcal{D}_{n-1,i}$ and $\mathcal{D}_j^{n-1}$. Of course, this could have been done (in view of Theorem 2.3.5). The reason we chose to use *det* instead is that it simplifies the proof of Theorem 2.3.5 and, after all, we end up at the same place!

Example 2.3.4 Suppose that

$$A = \begin{pmatrix} 3 & 4 & 6 \\ 0 & 1 & 2 \\ 2 & 5 & 0 \end{pmatrix}$$

Then by the definition we find that

$$\mathcal{D}_{3,1}(A) = (-1)^2 3 det\begin{pmatrix} 1 & 2 \\ 5 & 0 \end{pmatrix} + (-1)^3 4 det\begin{pmatrix} 0 & 2 \\ 2 & 0 \end{pmatrix}$$

$$+ (-1)^4 6 det\begin{pmatrix} 0 & 1 \\ 2 & 5 \end{pmatrix}$$

$$= 3(-10) - 4(-4) + 6(-2) = -26$$

Note that this expansion takes place along the first row of A. Analogously we have

$$D_{3,2}(A) = (-1)^3 0 det \begin{pmatrix} 4 & 6 \\ 5 & 0 \end{pmatrix} + (-1)^4 1 det \begin{pmatrix} 3 & 6 \\ 2 & 0 \end{pmatrix}$$

$$+ (-1)^5 2 det \begin{pmatrix} 3 & 4 \\ 2 & 5 \end{pmatrix}$$

$$= 0(-30) + 1(-12) - 2(7) = -26$$

and

$$D_{3,3}(A) = (-1)^4 2 det \begin{pmatrix} 4 & 6 \\ 1 & 2 \end{pmatrix} + (-1)^5 5 det \begin{pmatrix} 3 & 6 \\ 0 & 2 \end{pmatrix}$$

$$+ (-1)^6 0 det \begin{pmatrix} 3 & 4 \\ 0 & 1 \end{pmatrix}$$

$$= 2(2) - 5(6) + 0(3) = -26$$

The column cofactor expansions also give the same result. For example,

$$D_1^3(A) = (-1)^4 3 det \begin{pmatrix} 1 & 2 \\ 5 & 0 \end{pmatrix} + (-1)^5 0 det \begin{pmatrix} 4 & 6 \\ 5 & 0 \end{pmatrix}$$

$$+ (-1)^6 2 det \begin{pmatrix} 4 & 6 \\ 1 & 2 \end{pmatrix}$$

$$= 3(-10) + (-1)0(-30) + 2(2) = -26$$

The most common error in applying cofactor expansions is failing to keep track of the $+$ and $-$ signs appropriately. The sign is $+1$ if $i + j$ (the sum of the row and column numbers) is even, and it is -1 if $i + j$ is odd. We now give the theorem that has been promised, that all the cofactor expansions agree with the determinant.

Theorem 2.3.5 *For any $n \times n$ matrix A and any i, j with $1 \le i, j \le n$,*

$$D_{n,i}(A) = D_j^n(A) = det(A)$$

Proof: We show that for each j, D_j^n is an n-linear, alternating function with $D_j^n(I_n) = 1$. The result then follows by Theorem 2.1.8. In this proof we shall make use of the fact that the $(n-1) \times (n-1)$ determinant function det is $(n-1)$-linear and alternating. First assume that A' has been obtained from A by multiplying the sth row of $A = (a_{ij})$ by k. Then $A'(s \mid j) = A(s \mid j)$, and for $i \ne s$ we have $det(A(i \mid k)) =$

$k \cdot det(A(i \mid j))$ since det is $(n-1)$-linear. Applying the defintion of $\mathcal{D}_j^n$, we get

$$\mathcal{D}_j^n(A') = \sum_{i=1,i\neq s}^{n}(-1)^{i+j}a_{ij}det(A'(i \mid j)) + ka_{sj}(-1)^{s+j}det(A'(s \mid j))$$

$$= \sum_{i=1,i\neq s}^{n}k(-1)^{i+j}a_{ij}det(A(i \mid j)) + ka_{sj}(-1)^{s+j}det(A(s \mid j))$$

$$= k\sum_{i=1}^{n}(-1)^{i+j}a_{ij}det(A(i \mid j)) = k\mathcal{D}_j^n(A)$$

Next we assume that A, B, C are identical except for the kth rows, where the kth row of C is the sum of the kth rows of A and B. Assume $C = (c_{ij})$, the kth row of A is (a_{kj}), and the kth row of B is (b_{kj}). Observe that in this case $A(k \mid j) = B(k \mid j) = C(k \mid j)$. It follows from the definitions and the fact that det is n-linear that

$$\mathcal{D}_j^n(C) = \sum_{i=1}^{n}c_{ij}(-1)^{i+j}det(C(i \mid j))$$

$$= \sum_{i=1,i\neq k}^{n}c_{ij}(-1)^{i+j}det(C(i \mid j))$$

$$+(a_{kj}+b_{kj})(-1)^{k+j}det(C(k \mid j))$$

$$= \sum_{i=1,i\neq k}^{n}c_{ij}(-1)^{i+j}[det(A(i \mid j)) + det(B(i \mid j))]$$

$$+(a_{kj}+b_{kj})(-1)^{k+j}det(C(k \mid j))$$

$$= \sum_{j}^{n}a_{ij}(-1)^{i+j}det(A(i \mid j)) + \sum_{i=1}^{n}b_{ij}(-1)^{i+j}det(B(i \mid j))$$

$$= \mathcal{D}_j^n(A) + \mathcal{D}_j^n(B)$$

This shows that $\mathcal{D}_j^n$ is n-linear.

We now show that $\mathcal{D}_j^n$ is alternating. Suppose that the sth and tth rows of $A = (a_{ij})$ are the same. It follows from this that the matrix $A(s \mid j)$ can be transformed into the matrix $A(t \mid j)$ by a sequence of interchanges of adjacent rows. If $s < t$, then starting with $A(s \mid j)$, we interchange (A's) tth row with (A's) $(t-1)$th row, then we interchange the $(t-1)$th with the $(t-2)$th, and so forth, until we finally interchange the $(s+2)$th with the $(s+1)$th. As the sth row of A is missing from $A(s \mid j)$, we have now moved the tth row to the sth. Thus $(t-s-1)$ adjacent row interchanges transform $A(s \mid j)$ into $A(t \mid j)$. Consequently we find that $det(A(s \mid j)) = (-1)^{(t-s-1)}det(A(t \mid j))$. We note also that whenever $i \neq s,t$, $det(A(i \mid j)) = 0$ since $A(i \mid j)$ has two identical rows. Expanding according to the definition we find

$$
\begin{aligned}
\mathcal{D}_j^n(A) &= \sum_{i=1,i\neq s,t}^{n} a_{ij}(-1)^{i+j}det(A(i \mid j)) + \\
&\quad +a_{sj}(-1)^{s+j}det(A(s \mid j)) + a_{tj}(-1)^{t+j}det(A(t \mid j)) \\
&= 0 + a_{sj}(-1)^{s+j}det(A(s \mid j)) \\
&\quad +a_{sj}(-1)^{t+j}(-1)^{(t-s-1)}det(A(s \mid j)) \\
&= 0
\end{aligned}
$$

Finally, $\mathcal{D}_j^n(I_n) = 1$ is easily checked by induction, so it follows from Theorem 2.1.8 that $\mathcal{D}_j^n = det$. To conclude the proof, we observe that $\mathcal{D}_{n,j}(A) = \mathcal{D}_j^n(A^t) = det(A^t) = det(A)$ by Theorem 2.1.14. This completes the proof of the theorem. //

We now turn to the applications mentioned earlier. First we give a technical definition needed in these applications.

Definition 2.3.6 Suppose that $A = (a_{ij})$ is an $n \times n$ matrix. We define the *adjoint* of A, denoted $adj(A)$, to be the $n \times n$ matrix whose jith entry is $(-1)^{i+j}det(A(i \mid j))$. In other words

$$adj(A)(j,i) = (-1)^{i+j}det(A(i \mid j))$$

Using the adjoint, we have the following elegant description of the inverse of an invertible matrix.

Theorem 2.3.7 *For any invertible matrix A, $A^{-1} = det(A)^{-1}adj(A)$.*

Proof: We compute $A(det(A)^{-1} \cdot adj(A)) = det(A)^{-1} \cdot A \cdot adj(A)$. Since the ith row of A is $(\, a_{i1} \quad a_{i2} \quad \cdots \quad a_{in}\,)$ and the jth column of $adj(A)$ (do not forget the transpose in the definition) is

$$
\begin{pmatrix}
(-1)^{j+1}det(A(j \mid 1)) \\
(-1)^{j+2}det(A(j \mid 2)) \\
\vdots \\
(-1)^{j+n}det(A(j \mid n))
\end{pmatrix}
$$

we see from the definition of matrix multiplication that the ijth entry of $A \cdot adj(A)$ is

$$\sum_{k=1}^{n} a_{ik}(-1)^{i+k} det(A(j \mid k))$$

When $i = j$, this is precisely $\mathcal{D}_{n,j}(A) = det(A)$. When $i \neq j$, this equals $\mathcal{D}_{n,i}(A^{*})$ where A^{*} is the matrix obtained from A by replacing the jth row of A with the ith row of A. [Note that the entries in the jth row of A do not effect the matrix $A(j \mid k)$.] Since A^{*} has two identical rows, $\mathcal{D}_{n,i}(A^{*}) = det(A^{*}) = 0$. It follows that $A \cdot adj(A) = det(A)I_{n}$, and from this the theorem follows. //

For example, we consider the matrix

$$\begin{pmatrix} 2 & 0 & 0 \\ 1 & 1 & 3 \\ 4 & 0 & 1 \end{pmatrix}$$

Computing the adjoint of A, we find

$$adj(A) = \begin{pmatrix} 1 & 11 & -4 \\ 0 & 2 & 0 \\ 0 & -6 & 2 \end{pmatrix}^{t} = \begin{pmatrix} 1 & 0 & 0 \\ 11 & 2 & -6 \\ -4 & 0 & 2 \end{pmatrix}$$

Since $det(A) = 2$, Theorem 2.3.7 gives

$$A^{-1} = \frac{1}{2} \begin{pmatrix} 1 & 0 & 0 \\ 11 & 2 & -6 \\ -4 & 0 & 2 \end{pmatrix}$$

This next theorem uses the preceding inversion formula to describe the unique solution to a system of n equations in n unknowns *whenever* such a unique solution exists. If the system does not have a unique solution, then this result *cannot* be applied.

Theorem 2.3.8 (Cramer's Rule) *Suppose that $A = (a_{ij})$ is an invertible matrix, $\vec{X}$ is a column of n variables, and B is a column of n constants. We denote by $A(j)$ the $n \times n$ matrix obtained from A by replacing the jth column of A by B. Then the system of equations $A\vec{X} = B$ has the unique solution given by*

$$X_{i} = \frac{det(A(i))}{det(A)}$$

Proof: The system of equations $A\vec{X} = B$ must have the solution $\vec{X} = A^{-1}B = det(A)^{-1}adj(A)B$. So it suffices to show that the ith row entry in the column matrix $adj(A)B$ is precisely $det(A(i))$. The ith row of $adj(A)$ is

$$((-1)^{i+1}det(A(1 \mid i)) \quad (-1)^{i+2}det(A(2 \mid i)) \quad \cdots \quad (-1)^{i+n}det(A(n \mid i)))$$

It follows that the ith entry of $adj(A)B$ is

$$\sum_{k=1}^{n}(-1)^{i+k}det(A(k \mid i))b_k = \sum_{k=1}^{n}(-1)^{i+k}b_k det(A(k \mid i))$$

But this is precisely $\mathcal{D}_i^n(A(i))$, which equals $det(A(i))$ by Theorem 2.3.5. This proves Cramer's rule. //

As an example, according to Cramer's rule the matrix equation

$$\begin{pmatrix} 1 & 2 \\ 3 & 4 \end{pmatrix}\begin{pmatrix} X \\ Y \end{pmatrix} = \begin{pmatrix} 5 \\ 6 \end{pmatrix}$$

has the unique solution given by

$$X = \frac{det\begin{pmatrix} 5 & 2 \\ 6 & 4 \end{pmatrix}}{det\begin{pmatrix} 1 & 2 \\ 3 & 4 \end{pmatrix}} \quad Y = \frac{det\begin{pmatrix} 51 & 5 \\ 3 & 6 \end{pmatrix}}{det\begin{pmatrix} 1 & 2 \\ 3 & 4 \end{pmatrix}}$$

We find that $X = 8/{-2} = -4$ and $Y = -9/{-2} = 9/2$.

Discussion 2.3.9 Determinants and Volume

The determinant of an $n \times n$ real matrix has an important geometric interpretation. Let A be an $n \times n$ real matrix, and suppose the columns of A are given by the vectors $v_1, v_2, \ldots, v_n \in \mathbf{R}^n$. The *parallelotope spanned by* $v_1, v_2, \ldots, v_n$ is defined to be the set of all linear combinations $r_1v_1 + r_2v_2 + \cdots + r_nv_n$ where $0 \leq r_i \leq 1$ for each i. When $n = 3$, the parallelotope spanned by three linearly independent vectors in $\mathbf{R}^3$ is the solid region pictured on the next page.

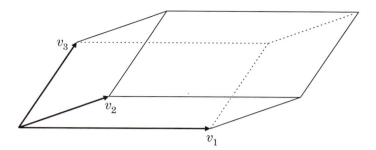

Observe that the vectors v_1, v_2, v_3 are mutually adjacent edges of the parallelotope.

For the purposes of this discussion, we denote by $Vol(A)$ the volume (n-dimensional) of the parallelotope spanned by the columns $v_1, v_2, \ldots, v_n$ of A. It turns out that $Vol(A) = | \, det(A) \, |$. We give an informal proof of this result. First observe that the parallelotope spanned by $e_1, e_2, \ldots, e_n$ (the standard basis of $\mathbf{R}^n$) is the unit n cube in $\mathbf{R}^n$. This n cube has volume 1, which is precisely $det(A) = det(I_n)$ in this case.

We next compare the volume of the parallelotopes which are the spanned by $v_1, v_2, \ldots, v_n$, by $v_1', v_2, \ldots, v_n$, and by $v_1 + v_1', v_2, \ldots, v_n$. By studying the picture below (of the $n = 2$ case), the reader should be convinced that the sum of the areas of the first two parallelotopes is the area of the third.

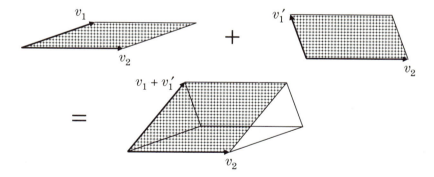

Considering the special case of this where $v_1' = v_1$ shows that the volume of the parallelotope spanned by $2v_1, v_2, \ldots, v_n$ is twice the volume of the parallelotope spanned by $v_1, v_2, \ldots, v_n$. Iterating this argument shows that for any natural number p, the volume of the parallelotope spanned by $pv_1, v_2, \ldots, v_n$ is p times the volume of the paralleleotope spanned by $v_1, v_2, \ldots, v_n$. It follows from this by division that for any rational number p/q the volume of the parallelotope spanned by $(p/q)v_1, v_2, \ldots, v_n$ is $|p/q|$ times the volume of the parallelotope spanned by $v_1, v_2, \ldots, v_n$. The continuity of the volume function shows that for any $r \in \mathbf{R}$, the volume of the parallelotope spanned by $rv_1, v_2, \ldots, v_n$ is $|r|$ times the volume of the parallelotope spanned by $v_1, v_2, \ldots, v_n$.

The arguments just given show that parallelotope volume, considered as a function of n vectors $v_1, v_2, \ldots, v_n$, is n-linear (modulo absolute values) and that $Vol(I_n) = 1$. Furthermore, this function is alternating, since the paralleotope spanned by $v_1, v_2, \ldots, v_n$ with $v_i = v_j$ and $i \neq j$ has volume 0. Specifically, the function $V^*(A) = [det(A)/ |det(A)|]Vol(A)$ is n-linear and alternating, with $V^*(I_n) = 1$. We conclude that $V^*(A) = det(A)$, and consequently $Vol(A) = |det(A)|$ for all A. The reader may recall that this result surfaces in multivariable calculus when the change of variable formula for the integration of functions of several variables is derived.

PROBLEMS

2.3.1. Compute the determinants of the following matrices by cofactor expansions.

(a) $\begin{pmatrix} 1 & 4 & 1 \\ 5 & 0 & 2 \\ 1 & 1 & 1 \end{pmatrix}$ (b) $\begin{pmatrix} 1 & 1 & 1 \\ 4 & 2 & 1 \\ 5 & 1 & 1 \end{pmatrix}$ (c) $\begin{pmatrix} 1 & 0 & 2 & 0 \\ 0 & 1 & 0 & 1 \\ 2 & 2 & 1 & 1 \\ 1 & 0 & 0 & 1 \end{pmatrix}$

(d) $\begin{pmatrix} 2 & 3 & 1 & 3 \\ 0 & 2 & 0 & 1 \\ 0 & 2 & 0 & 2 \\ 0 & 2 & 1 & 0 \end{pmatrix}$ (e) $\begin{pmatrix} 1 & 0 & 0 & 0 \\ 9 & 1 & 3 & 5 \\ 8 & 0 & 2 & 3 \\ 3 & 0 & 1 & 2 \end{pmatrix}$ (f) $\begin{pmatrix} 3 & 3 & 3 \\ 3 & 3 & 3 \\ 3 & 3 & 3 \end{pmatrix}$

2.3.2. Compute the determinants of the following matrices by cofactor expansions.

(a) $\begin{pmatrix} 2 & 4 & 0 \\ 5 & 0 & 2 \\ 1 & 2 & 1 \end{pmatrix}$ (b) $\begin{pmatrix} 1 & 0 & 1 \\ 4 & 2 & 0 \\ 5 & 0 & 1 \end{pmatrix}$ (c) $\begin{pmatrix} 1 & 1 & 2 & 1 \\ 0 & 1 & 0 & 1 \\ 0 & 2 & 0 & 1 \\ 0 & 0 & 0 & 1 \end{pmatrix}$

(d) $\begin{pmatrix} 2 & 0 & 1 & 3 \\ 0 & 2 & 0 & 0 \\ 1 & 2 & 1 & 2 \\ 1 & 0 & 1 & 0 \end{pmatrix}$ (e) $\begin{pmatrix} 1 & 0 & 1 & 0 \\ 1 & 1 & 3 & 5 \\ 1 & 0 & 2 & 3 \\ 3 & 0 & 1 & 0 \end{pmatrix}$ (f) $\begin{pmatrix} 1 & 2 & 3 \\ 4 & 5 & 6 \\ 7 & 8 & 9 \end{pmatrix}$

2.3.3.   Use the adjoint formula to invert

$$\begin{pmatrix} -2 & 0 & 1 \\ 3 & 0 & 1 \\ 0 & 1 & -1 \end{pmatrix}$$

2.3.4.   Use the adjoint formula to invert

$$\begin{pmatrix} 1 & 1 & 1 \\ 2 & 0 & -1 \\ 3 & 4 & 2 \end{pmatrix}$$

2.3.5.   Use Cramer's rule to solve

$$\begin{pmatrix} 1 & 1 & 1 \\ 2 & 0 & -1 \\ 3 & 4 & 2 \end{pmatrix} \begin{pmatrix} X \\ Y \\ Z \end{pmatrix} = \begin{pmatrix} 1 \\ 1 \\ 1 \end{pmatrix}$$

2.3.6.   Use Cramer's rule to solve

$$\begin{pmatrix} -2 & 0 & 1 \\ 3 & 0 & 1 \\ 0 & 1 & -1 \end{pmatrix} \begin{pmatrix} X \\ Y \\ Z \end{pmatrix} = \begin{pmatrix} 1 \\ 0 \\ 2 \end{pmatrix}$$

2.3.7.   Compute the determinant of the following matrices by any manner you like:

(a) $\begin{pmatrix} 2 & 3 & 4 & 1 \\ 1 & 0 & 2 & 3 \\ 2 & 0 & 0 & 4 \\ 1 & 0 & 0 & 9 \end{pmatrix}$ (b) $\begin{pmatrix} 1 & 2 & 3 & 4 \\ 5 & 6 & 7 & 8 \\ 9 & 10 & 11 & 12 \\ 13 & 14 & 15 & 16 \end{pmatrix}$

(c) $\begin{pmatrix} -1 & 2 & -3 \\ 1 & 2 & 3 \\ 0 & 3 & 2 \end{pmatrix}$ (d) $\begin{pmatrix} 0 & 1 & 1 & 1 \\ 1 & 0 & 1 & 1 \\ 1 & 1 & 0 & 1 \\ 1 & 1 & 1 & 0 \end{pmatrix}$

2.3.8.   (a) Suppose that A is an $n \times n$ matrix and $rk(A) < n - 1$. Show that $adj(A)$ is the matrix of all zeros.
(b) If A is an $n \times n$ matrix and $rk(A) = n - 1$, show that $\mathrm{rk}(\mathrm{adj}(A)) = 1$.

2.3.9. Show that if A is an $n \times n$ matrix, then $det(adj(A)) = det(A)^{n-1}$.

2.3.10.   (a) Show that for any $n \times n$ matrix A, $adj(A)^t = adj(A^t)$.
(b) Show that for any $n \times n$ matrix A, $adj(adj(A)) = det(A)^{n-2}A$.

2.4   Determinants: Summary

In this section we summarize the properties of the determinant function which are necessary for the understanding of the remainder of the book. Those readers who omitted any of the previous sections from this chapter should check to make sure that the results highlighted here are familiar.

The determinant function $det : M_{n \times n}(\mathbf{F}) \to \mathbf{F}$ is a scalar-valued function defined on $n \times n$ (square) matrices. A central property (and probably the best known) of determinants is that an $n \times n$ matrix A is invertible if and only if $det(A) \neq 0$. This result (Theorem 2.1.12) plays a major role in the succeeding chapters, especially in understanding the characteristic polynomial (Sec. 3.4). There are two customary approaches to defining the determinant function. One is the definition via permutations (given in Sec. 2.2), and the other is the Laplace (or cofactor) expansion definition (given in Sec. 2.3). These are summarized next.

The set of all permutations of n is denoted by S_n. In case the permutation σ is $(i_1, i_2, \ldots, i_n)$, we abbreviate the jth entry of σ by σ_j. The sign of a permutation $\sigma \in S_n$ is denoted by $sg(\sigma)$ (which is either $+1$ if σ is even or -1 if σ is odd). See Definitions 2.2.1 and 2.2.2 for details. Using this notation, we give the permutation definition of the determinant.

Definition 2.2.3 Suppose that $A = (a_{ij})$ is an $n \times n$ matrix. We define the *determinant* of A, denoted $det(A)$, by

$$det(A) = \sum_{\sigma \in S_n} sg(\sigma) a_{1\sigma_1} a_{2\sigma_2} \cdots a_{n\sigma_n}$$

Many texts introduce the determinant in a different fashion. In the following definition, $A(i \mid j)$ denotes the matrix obtained from A by deleting the ith row and the jth column. This definition inductively gives the determinant of a matrix in terms of the submatrices $A(i \mid j)$ by expanding either upon a row or upon a column. (We remark that in Sec. 3, where this definition is given, we denote these definitions of $det(A)$ as $\mathcal{D}_{n,i}(A)$ and $\mathcal{D}_j^n(A)$ since we must later *prove* (see Theorem 2.3.5) that these definitions give the same function as that defined in Definition 2.2.3.)

Definition 2.3.3 If $A = (a)$, we define $\mathcal{D}_{1,1}(A) = \mathcal{D}_1^1(A) = a = det(A)$. If $A = (a_{ij})$ is an $n \times n$ matrix with $n > 1$, for i with $1 \leq i \leq n$ we define

$$\mathcal{D}_{n,i}(A) = \sum_{j=1}^{n} (-1)^{i+j} a_{ij} det(A(i \mid j))$$

which is called the ith *row cofactor* or *Laplace expansion* of *det*. Similarly, for j with $1 \leq j \leq n$ we define

$$\mathcal{D}_j^n(A) = \sum_{i=1}^{n} (-1)^{i+j} a_{ij} det(A(i \mid j))$$

which is called the jth *column cofactor* or *Laplace expansion* of *det*.

While both of the definitions above give a method for computing determinants, once matrices become large the most effective means of computation is row operations. In order to compute a determinant using row operations, one needs to know how the determinant changes when row operations are applied to a matrix. The next theorem summarizes these relationships.

Theorem 2.1.4 *Suppose that* $\mathcal{D} : M_{n \times n}(\mathbf{F}) \rightarrow \mathbf{F}$ *is a determinant function. Let* $A \in M_{n \times n}(\mathbf{F})$. *Then*
(i) *If B is obtained from A by multiplying a single row of A by $k \in \mathbf{F}$, then $\mathcal{D}(B) = k\mathcal{D}(B)$.*
(ii) *If B is obtained from A by adding a multiple of one row to another, then $\mathcal{D}(B) = \mathcal{D}(A)$.*
(iii) *If B is obtained from the matrix A by interchanging two rows, then $\mathcal{D}(B) = -\mathcal{D}(A)$.*

Once a matrix has been reduced to triangular form, its determinant is easily computed:

Lemma 2.2.8 *If* $A = (a_{ij})$ *is triangular (upper or lower), then* $det(A) = a_{11}a_{22} \cdots a_{nn}$.

Two important properties of the determinant are its relationship to invertibility (mentioned above) and the multiplicativity property.

Theorem 2.2.12 *An* $n \times n$ *matrix* A *is invertible if and only if* $det(A) \neq 0$.

Theorem 2.1.13 *If* A *and* B *are arbitrary* $n \times n$ *matrices, then* $det(AB) = det(A)det(B)$.

There are two important formulas which enable one to invert matrices and solve $(n \times n)$ linear systems. These use the adjoint of a matrix.

Definition 2.3.6 Suppose that $A = (a_{ij})$ is an $n \times n$ matrix. We define the *adjoint* of A, denoted $adj(A)$, to be the $n \times n$ matrix whose jith entry is $(-1)^{i+j}det(A(i \mid j))$. In other words

$$adj(A)(j,i) = (-1)^{i+j}det(A(i \mid j))$$

Using the adjoint, we have the following elegant description of the inverse of an invertible matrix.

Theorem 2.3.7 *For any invertible matrix* A, $A^{-1} = det(A)^{-1}adj(A)$.

Cramer's rule enables one to determine the unique solution to a system of n equations in n unknowns, *whenever* such a unique solution exists.

Theorem 2.3.8 (Cramer's Rule) *Suppose that* $A = (a_{ij})$ *is an invertible matrix,* $\vec{X}$ *is a column of* n *variables, and* B *is a column of* n *constants. We denote by* $A(j)$ *the* $n \times n$ *matrix obtained from* A *by replacing the* j*th column of* A *by* B. *Then the system of equations* $A\vec{X} = B$ *has the unique solution given by*

$$X_i = \frac{det(A(i))}{det(A)}$$

Chapter 3

Linear Transformations and Eigenvalues

3.1 Linear Transformations

Throughout mathematics, in addition to the study of objects such as numbers, shapes, or sets, one studies functions between collections of such objects. In each case, the functions considered have special properties, depending upon the type of questions that need to be answered. For example, in grade school a main class of objects of study was the real numbers. They arose in counting and measuring. Later, in high school algebra, polynomial and trigonometric functions on the real numbers were introduced. The reason for studying such functions was their usefulness in solving algebraic and geometric problems. By introducing these functions, one increased the utility of the real number system beyond its original use of counting and measuring.

In calculus, the continuous functions play a special role. The key concept of calculus is the notion of a limit, and continuous functions are precisely those functions which "respect" limits. More precisely, suppose that $a_i \in \mathbf{R}$, $i = 1, 2, 3, \ldots$, is a convergent sequence of real numbers with $lim_{n \to \infty} a_i = a \in \mathbf{R}$. If $f : \mathbf{R} \to \mathbf{R}$ is a continuous function, then the sequence $f(a_i)$ converges and $lim_{n \to \infty} f(a_i) = f(lim_{n \to \infty} a_i) = f(a)$. In fact, this property of a continuous function can be used to define the concept of continuous functions on the real line.

The subject of linear algebra has as objects vectors, which are collected together in groups called *vector spaces*. In linear algebra we study special types of functions $T : V \rightarrow W$, between two vector spaces V and W, called *linear transformations*. The important property of linear transformations is that they preserve the basic vector operations of addition and scalar multiplication (see Lemma 3.1.3). This means that various relationships between vectors in one vector space can be compared with and transformed to relationships between vectors in another.

Linear transformations are the key to the main applications of linear algebra. In this chapter we develop their basic properties and see how they can be described by matrix algebra. This process will require a change in our point of view. We must change from the computational and matrix perspective to the functional perspective. In other words, the functions, not the vectors, will be our primary concern. With this different point of view, new doors open. Many problems are easily understood when thinking about functions, yet are nearly impossible to solve using matrix computations.

We now give the definition of a linear transformation. Note that this definition is similar to the defining property of continuous functions discussed above in that linear transformations "respect" vector operations just as continuous functions "respect" convergent sequences.

Definition 3.1.1 Suppose that V and W are vector spaces over the same field **F**. A *linear transformation* from V to W is a function $T : V \rightarrow W$ which satisfies:
(i)   If $u, v \in V$, then $T(u + v) = T(u) + T(v)$.
(ii) If $v \in V$ and $k \in \mathbf{F}$, then $T(kv) = kT(v)$.

When $T : V \rightarrow W$ is a linear transformation, we refer to V as the *domain of definition*, for short *domain*, of the linear transformation T. We shall refer to W as the *value space* of T (some authors use "codomain"). In general function theory, it is common to refer to the set where a function takes its values as the "image" or "range" of the function. We shall avoid this terminology, since we want to use the terms "image" and "range" in a slightly different way in the next section.

Often, instead of saying that a function is a linear transformation, we shall simply say that it is linear. Likewise, we shall refer to a linear transformation as a transformation, the linearity property being implicit from the context. Sometimes the terminology vector space homomorphism is used synonymously with "linear transformation." A linear transformation $T : V \to V$ is often called a *linear operator* on V.

Examples 3.1.2 (i) Consider $F : \mathbf{R} \to \mathbf{R}$ defined by $F(x) = 3x$ and $G : \mathbf{R} \to \mathbf{R}$ defined by $G(x) = 5x + 8$. Then as $F(a + b) = 3(a+b) = 3a + 3a = F(a) + F(b)$ and $F(ka) = 3(ka) = k(3a) = kF(a)$, we see that F is linear. On the other hand, $G(a + b) = 5(a + b) + 8 \neq (5a + 8) + (5b + 8) = G(a) + G(b)$, so $G(x)$ is not linear. [Note: Even though the graph of G is a line, this does *not* mean $G(x)$ is linear in the sense of linear algebra. The use of the word "linear" in the definition of a linear transformation is more restrictive than the use of the word "linear" in, say, elementary geometry.]

(ii) Define $R, S : \mathbf{F}^2 \to \mathbf{F}^4$ by $R(x, y) = (x - y, x, x, 2y)$ and $S(x, y) = (x, y, 0, |x|)$. Then R is linear, by a calculation similar to (i) above. But S is not linear because $S((1, 0) + (-1, 0)) = S(0, 0) = (0, 0, 0, 0)$ while $S(1, 0) + S(-1, 0) = (1, 0, 0, 1) + (-1, 0, 0, 1) = (0, 0, 0, 2) \neq (0, 0, 0, 0)$.

(iii) This example begins with some notation that we will use throughout the remainder of this book. Suppose that A is an $m \times n$ matrix. Then multiplication on the left by A can be viewed as a function from $\mathbf{F}^n$ to $\mathbf{F}^m$. We denote this function by $T_A : \mathbf{F}^n \to \mathbf{F}^n$. In other words

$$T_A(\vec{v}) = A\vec{v} \quad \text{for all} \quad n \times 1 \text{ matrices } \vec{v} \ (\in \mathbf{F}^n)$$

It follows from Theorem 0.3.3 (vii) and (ix) that T_A is a linear transformation. Note that in this example we distinguish between the function T_A and the matrix A, even though the two are closely related. At this point we are beginning to make the transition to the functional point of view, mentioned in the introduction to this section.

(iv) Suppose that

$$A = \begin{pmatrix} 1 & 2 \\ 3 & 4 \\ 5 & 6 \end{pmatrix}$$

Then in the notation of (iii) above, $T_A : \mathbf{F}^2 \to \mathbf{F}^3$ is the linear function defined by

$$T_A \begin{pmatrix} x \\ y \end{pmatrix} = \begin{pmatrix} 1 & 2 \\ 3 & 4 \\ 5 & 6 \end{pmatrix} \begin{pmatrix} x \\ y \end{pmatrix} = \begin{pmatrix} x + 2y \\ 3x + 4y \\ 5x + 6y \end{pmatrix}.$$

(v) Let $V = C^\infty(\mathbf{R}, \mathbf{R})$ denote the real vector space of all infinitely differentiable continuous functions on $\mathbf{R}$, that is, those functions $f : \mathbf{R} \to \mathbf{R}$ which have continuous nth-order derivatives $f^{(n)}$ for all n. Define a function $D : V \to V$ by $D(f) = f'$ for all functions $f \in V$. According to the linearity of the derivative, for any functions $f_1, f_2 \in V$ and any $r \in \mathbf{R}$, $(f_1 + f_2)' = f_1' + f_2'$, and $(rf_1)' = rf_1'$. This shows that $D : V \to V$ is a linear transformation.

(vi) We can generalize (v) as follows. Consider

$$S(f) = a_n f^{(n)} + a_{n-1} f^{(n-1)} + \cdots + a_1 f' + a_0 f$$

where $a_0, a_1, \ldots, a_n \in \mathbf{R}$. [Then the equation $S(f) = g(x)$ is an nth-order linear differential equation.] The expression $S(f)$ defines a function $S : V \to V$. By the same reasoning as in (v), S is a linear transformation. In fact, this is why such a differential equation is called linear. The set of solutions to the differential equation $S(f) = g(x)$ is precisely the set of vectors in V whose value under S is the vector $g(x) \in V$.

(vii) Let $V = C^\infty(\mathbf{R}, \mathbf{R})$ be as in (v), and define $Int : V \to \mathbf{R}$ by $Int(f) = \int_0^1 f(x)\, dx$. By the linearity of the integral, we see that Int is a linear transformation.

(viii) Expanding upon the preceding idea, define $INT_a : C^\infty(\mathbf{R}, \mathbf{R}) \to C^\infty(\mathbf{R}, \mathbf{R})$ by $(INT_a(f))(x) = \int_a^x f(y)\, dy$. Then again by the properties of the integral INT_a is a linear transformation.

(ix) Recall that $M_{m \times n}$ denotes the vector space of all $m \times n$ matrices. If B is a fixed $n \times p$ matrix, then $G : M_{m \times n} \to M_{m \times p}$ defined by $G(X) = XB$ is a linear transformation because of the properties of matix multiplication [see Theorem 0.3.3 (vii) and (ix)].

(x) The transpose operation is a linear transformation; that is, the function $Tps : M_{m \times n} \to M_{n \times m}$ defined by $Tps(A) = A^t$ is a linear transformation.

(xi) Let V be any vector space. Then the functions $I : V \to V$ and $0 : V \to V$ defined by $I(v) = v$ and $0(v) = \vec{0}$ are both linear transformations. I is called the *identity transformation*, and 0 is called the *zero transformation*.

(xii) Let $r(X) \in P$ be a polynomial. The distributive law for polynomial multiplication shows that the function $R : P \to P$ defined by $R(f(X)) = r(X)f(X)$ is a linear transformation.

(xiii) Let W be the vector space of all piecewise continuous real valued functions $f(x)$ on $[0, \infty)$ for which there exist real constants $a, K, M \in$ **R** such that $|f(t)| < Ke^{at}$ for all $t > M$. For any $f(x) \in W$ the *Laplace transform* $\mathcal{L}(f(x)) = F(s)$ is defined by $F(s) = \int_0^\infty e^{-st} f(t)\, dt$. The Laplace transform is a linear transformation $\mathcal{L} : W \to W$ since $\mathcal{L}(a_1 f_1(x) + a_2 f_2(x)) = a_1 \mathcal{L}(f_1(x)) + a_2 \mathcal{L}(f_2(t))$. [The remarkable property $\mathcal{L}(f'(x)) = s\mathcal{L}(f(x)) - f(0)$ is a seperate issue, unrelated to the linearity of $\mathcal{L}$.]

The examples above show that linear transformations arise in many situations in mathematics. For this reason it is important to study their properties in the setting of general vector spaces and not merely in $\mathbf{R}^n$. This first result is a direct consequence of the definition.

Lemma 3.1.3 *Suppose that $T : V \to W$ is a linear transformation.*
(i) $T(\vec{0}) = \vec{0}$.
(ii) *Let $v_1, v_2, \ldots, v_n \in V$. Then for all scalars $a_1, a_2, \ldots, a_n$,*

$$T(a_1 v_1 + a_2 v_2 + \cdots + a_n v_n) = a_1 T(v_1) + a_2 T(v_2) + \cdots + a_n T(v_n)$$

Proof: (i) Part (ii) of Definition 3.1.1 shows that $T(\vec{0}) = T(0 \cdot \vec{0}) = 0 \cdot T(\vec{0}) = \vec{0}$.

(ii) We repeatedly apply both (i) and (ii) of Definition 3.1.1:

$$\begin{aligned} T(a_1 v_1 + \cdots + a_n v_n) &= T(a_1 v_1) + T(a_2 v_2 + \cdots + a_n v_n) \\ &= \cdots = T(a_1 v_1) + T(a_2 v_2) + \cdots + T(a_n v_n) \\ &= a_1 T(v_1) + a_2 T(v_2) + \cdots + a_n T(v_n) \end{aligned}$$

as required. //

Lemma 3.1.3 will be used so often throughout the rest of this book, we will not refer to it by number. When using this lemma, we shall simply say "by linearity." This is a common practice throughout mathematics, and this phrase is never used differently. The next result is a key application of linearity. It says that a linear transformation on a finite-dimensional vector space is uniquely determined by how it acts on a basis.

Theorem 3.1.4 *Assume that V and W are vector spaces. Suppose that $\{v_1, v_2, \ldots, v_n\}$ is a basis of V and $w_1, w_2, \ldots, w_n \in W$ are arbitrary (not necessarily distinct). Then there is a unique linear transformation $T : V \to W$ such that $T(v_1) = w_1$, $T(v_2) = w_2, \ldots, T(v_n) = w_n$.*

Proof: Since $\{v_1, v_2, \ldots, v_n\}$ is a basis of V, every element of V can be written uniquely as a linear combination $a_1 v_1 + a_2 v_2 + \cdots + a_n v_n$. We define the value of a function T on such a linear combination by

$$
\begin{aligned}
T(a_1 v_1 + a_2 v_2 + \cdots + a_n v_n) &= a_1 T(v_1) + a_2 T(v_2) + \cdots + a_n T(v_n) \\
&= a_1 w_1 + a_2 w_2 + \cdots + a_n w_n
\end{aligned}
$$

Note that according to Lemma 3.1.3, we have defined T in the only way possible. Thus, if such a linear transformation exists, it must be this function T. The uniqueness assertion of the theorem follows. We must check that T defined in this manner is linear. Consider two elements of V, $u = a_1 v_1 + a_2 v_2 + \cdots + a_n v_n$ and $u' = b_1 v_1 + b_2 v_2 + \cdots + b_n v_n$. Then $u + u' = (a_1 + b_1)v_1 + (a_2 + b_2)v_2 + \cdots + (a_n + b_n)v_n$, so that

$$
\begin{aligned}
T(u + u') &= (a_1 + b_1)w_1 + (a_2 + b_2)w_2 + \cdots + (a_n + b_n)w_n \\
&= (a_1 w_1 + a_2 w_2 + \cdots + a_n w_n) \\
&\quad + (b_1 w_1 + b_2 w_2 + \cdots + b_n w_n) \\
&= T(u) + T(u')
\end{aligned}
$$

Similarly, for any scalar k,

$$
\begin{aligned}
T(ku) &= T(ka_1 v_1 + ka_2 v_2 + \cdots + ka_n v_n) \\
&= ka_1 w_1 + ka_2 w_2 + \cdots + ka_n w_n \\
&= k(a_1 w_1 + a_2 w_2 + \cdots + a_n w_n) = kT(u)
\end{aligned}
$$

This shows that T is linear, and the theorem is proved. //

To illustrate Theorem 3.1.4, consider the vector space P^n of all polynomials of degree at most n and let $S = \{1, X, X^2, \ldots, X^n\}$ denote the standard basis. The field $\mathbf{F}$ can be viewed as a one-dimensional $\mathbf{F}$ vector space, and for each $i = 1, 2, \ldots, n$ we let the w_i of Theorem 3.1.4 be $1 \in \mathbf{F}$. It then follows there is a unique linear transformation $T : P^n \to \mathbf{F}$ with the property that $T(1) = 1$, $T(X) = 1, \ldots, T(X^n) = 1$. The linear transformation T is precisely the linear transformation, which when applied to a polynomial, sums the coefficients; that is, $T(a_n X^n + a_{n-1} X^{n-1} + \cdots + a_0) = a_n + a_{n-1} + \cdots + a_0$.

A key application of Theorem 3.1.4 is the fact that every linear transformation from $\mathbf{F}^n$ to $\mathbf{F}^m$ arises from matrix multiplication as described in Example 3.1.2 (iii). This result will enable us to translate questions about such linear transformations into questions about matrices, and vice versa. This next theorem uses the T_A notation defined in Example 3.1.2 (iii).

Theorem 3.1.5 *Suppose that* $G : \mathbf{F}^n \to \mathbf{F}^m$ *is a linear transformation. Then there exists a unique* $m \times n$ *matrix* A *such that* $G = T_A$.

Proof: Denote by $\{e_1, e_2, \ldots, e_n\}$ the standard basis of $\mathbf{F}^n$, and denote by $\{\varepsilon_1, \varepsilon_2, \ldots, \varepsilon_m\}$ the standard basis of $\mathbf{F}^m$. We define scalars a_{ij} for $1 \leq i \leq m$, $1 \leq j \leq n$ using the expression $G(e_i) = a_{1i}\varepsilon_1 + a_{2i}\varepsilon_2 + \cdots + a_{mi}\varepsilon_m$. Such a_{ij} are uniquely determined since $\{\varepsilon_1, \varepsilon_2, \ldots, \varepsilon_m\}$ is a basis for $\mathbf{F}^m$. Then $A = (a_{ij})$ is an $m \times n$ matrix. We claim that $G = T_A$. By Example 3.1.2 (iii) multiplication on the left by A defines a linear transformation $T_A : \mathbf{F}^n \to \mathbf{F}^m$. Thus, to see that $G = T_A$ (as functions), it suffices by Theorem 3.1.4 to show that $G(e_1) = T_A(e_1), G(e_2) = T_A(e_2), \ldots, G(e_n) = T_A(e_n)$.

Note that according to the definition of A, the ith column of A is nothing other than (the coordinates of) the vector $G(e_i)$. As Ae_i is (according to the definition of matrix multiplication) precisely the ith column of A, we see that $T_A(e_i) = Ae_i = G(e_i)$. The theorem is proved. //

The unique $m \times n$ matrix A associated with a linear transformation $G : \mathbf{F}^n \to \mathbf{F}^m$ given in Theorem 3.1.5 is called the *standard matrix of* G. We often denote this matrix A by $[G]$, so we have $G = T_{[G]}$. Also note that with this notation $[T_A] = A$ for any matrix A. The proof of the above theorem gives an algorithm for finding the matrix $A = [G]$ that corresponds to the linear transformation $G : \mathbf{F}^n \to \mathbf{F}^m$. The algorithm is the following:

Compute the values of the linear transformation G applied to the standard basis vectors $e_1, e_2, \ldots, e_n$ of $\mathbf{F}^n$. The ith column of the desired matrix A is precisely the (coordinates of) the vector $G(e_i)$.

This algorithm is illustrated in the following examples.

Examples 3.1.6 (i) Suppose that $G : \mathbf{F}^2 \to \mathbf{F}^2$ is defined by

$$G \begin{pmatrix} x \\ y \end{pmatrix} = \begin{pmatrix} x - y \\ x + 2y \end{pmatrix}$$

Then we compute that

$$G\begin{pmatrix} 1 \\ 0 \end{pmatrix} = \begin{pmatrix} 1 \\ 1 \end{pmatrix} \quad \text{and} \quad G\begin{pmatrix} 0 \\ 1 \end{pmatrix} = \begin{pmatrix} -1 \\ 2 \end{pmatrix}$$

From this we find that $G = T_A$ where

$$A = \begin{pmatrix} 1 & -1 \\ 1 & 2 \end{pmatrix}$$

(ii) Suppose that $G : \mathbf{F}^3 \to \mathbf{F}^4$ is defined by

$$G\begin{pmatrix} x \\ y \\ z \end{pmatrix} = \begin{pmatrix} 2x - y \\ z \\ x - z + y \\ x \end{pmatrix}$$

then we compute the images of the standard basis under G:

$$G\begin{pmatrix} 1 \\ 0 \\ 0 \end{pmatrix} = \begin{pmatrix} 2 \\ 0 \\ 1 \\ 1 \end{pmatrix}, \ G\begin{pmatrix} 0 \\ 1 \\ 0 \end{pmatrix} = \begin{pmatrix} -1 \\ 0 \\ 1 \\ 0 \end{pmatrix}, \ G\begin{pmatrix} 0 \\ 0 \\ 1 \end{pmatrix} = \begin{pmatrix} 0 \\ 1 \\ -1 \\ 0 \end{pmatrix}$$

so it follows that $G = T_A$ where

$$A = \begin{pmatrix} 2 & -1 & 0 \\ 0 & 0 & 1 \\ 1 & 1 & -1 \\ 1 & 0 & 0 \end{pmatrix}$$

Recall that if $H : U \to V$ and $G : V \to W$ are functions, then the composition of these functions $G \circ H$ is the function $G \circ H : U \to W$ defined by $G \circ H(x) = G(H(x))$. Composing functions, together with understanding the composition in terms of its components, is an important problem in mathematics. Recall for example, the chain rule from calculus, and how useful it was. The next task is to determine how to find the matrix of a composition of linear transformations in terms of the associated matrices of the component functions.

Consider an example where we compose two linear transformations: Suppose $H : \mathbf{F}^2 \to \mathbf{F}^3$ is defined by $H(x, y) = (x - y, y, 0)$ and $G :$

$\mathbf{F}^3 \to \mathbf{F}^1$ is defined by $G(r, s, t) = 2r - s + t$. Then the composition $G \circ H(x, y) = 2(x - y) - 1(y) + 1(0)$. This expression is the same as matrix product

$$(2 \quad -1 \quad 1) \begin{pmatrix} x - y \\ y \\ 0 \end{pmatrix}$$

However, $(2 \quad -1 \quad 1)$ is the matrix that represents G, and

$$\begin{pmatrix} x - y \\ y \\ 0 \end{pmatrix} = \begin{pmatrix} 1 & -1 \\ 0 & 1 \\ 0 & 0 \end{pmatrix} \begin{pmatrix} x \\ y \end{pmatrix} \quad \text{where} \quad \begin{pmatrix} 1 & -1 \\ 0 & 1 \\ 0 & 0 \end{pmatrix}$$

is the matrix that represents H. Altogether this shows that

$$G \circ H = (2 \quad -1 \quad 1) \begin{pmatrix} 1 & -1 \\ 0 & 1 \\ 0 & 0 \end{pmatrix} \begin{pmatrix} x \\ y \end{pmatrix}$$

In other words, $G \circ H = T_A$ where A is the product $[G][H]$ of the matrices that represent the transformations G and H.

The next result generalizes the preceding observation. The point is that matrix multiplication, interpreted in terms of linear transformations, corresponds to functional composition. The proof of this fact is nothing more that the associativity of matrix multiplication, together with the definition of T_A.

Theorem 3.1.7 *Suppose that $H : U \to V$ and $G : V \to W$ are both linear transformations.*
(i) *Then the composition $G \circ H : U \to W$ is a linear transformation.*
(ii) *If $H = T_A : \mathbf{F}^m \to \mathbf{F}^n$ and $G = T_B : \mathbf{F}^n \to \mathbf{F}^s$ where A is an $n \times m$ matrix and B is a $s \times n$ matrix, then $G \circ H = T_{BA}$.*
Hence, $[G \circ H] = [G][H]$ and $T_{BA} = T_B \circ T_A$.

Proof: (i) Suppose that $u, v \in U$ and k is a scalar. Then $G \circ H(u+v) = G(H(u + v)) = G(H(u) + H(v)) = G(H(u)) + G(H(v)) = G \circ H(u) + G \circ H(v)$. Also $G \circ H(ku) = G(H(ku)) = G(kH(u)) = kG(H(u)) = k[G \circ H(u)]$. This proves (i).

(ii) Observe that for any $v \in \mathbf{F}^m$ (expressed as a column vector) $G \circ H(v) = G(H(v)) = G(Av) = B(Av) = (BA)v = T_{BA}(v)$, where the second to last equality is the associativity of matrix multiplication (Theorem 0.3.7). This proves the theorem. //

In the next section we describe how various properties of linear transformations (as functions) can be related to properties of the associated matrix. We conclude this section with an informal discussion of how the linear transformation associated with a matrix arises in advanced calculus.

Digression 3.1.8 Recall from calculus that if the curve $Y = f(X)$ has derivative $f'(a)$ at a point (a, b), then the tangent line to the curve is described by the equation $(Y - b) = f'(a)(X - a)$. The real number $f'(a)$ is the slope of this line. The equation of this tangent line gives the best linear approximation to the function $Y = f(X)$.

Functions of several variables can also be approximated by linear functions. Suppose that $Y = F(X_1, X_2, \ldots, X_n)$ is a function of n variables, with all partial derivatives defined at a point $(a_1, a_2, \ldots, a_n) \in \mathbf{R}^n$. If $b = F(a_1, a_2, \ldots, a_n)$, then the tangent hyperplane in $\mathbf{R}^{n+1}$ to the graph of $Y = F(X_1, X_2, \ldots, X_n)$ is given by

$$Y - b = \left(\frac{\partial F}{\partial X_1}(a_1, \ldots, a_n) \quad \cdots \quad \frac{\partial F}{\partial X_n}(a_1, \ldots, a_n) \right) \begin{pmatrix} X_1 - a_1 \\ X_2 - a_2 \\ \vdots \\ X_n - a_n \end{pmatrix}$$

$$= F'(\vec{a})(\vec{X} - \vec{a})$$

where F' is the Jacobian of F defined in Digression 0.3.10. Recall that the vector $F'(\vec{a})$ of partial derivatives has the property that for any unit vector u, $F'(\vec{a}) \cdot u = \partial F/\partial u(\vec{a})$, the directional derivative of F in the direction of u. Thus, the single vector $F'(\vec{a})$ captures all the information regarding rates of increase and decrease of the function $Y = F(\vec{X})$.

Now suppose that $G : \mathbf{R}^n \to \mathbf{R}^m$ and the Jacobian matrix $G'(\vec{a})$ is defined for some $\vec{a} \in \mathbf{R}^n$. Then, as above, the function $G(\vec{X})$ has as a linear approximation $G'(\vec{a})(\vec{X} - \vec{a})$. In multivariable calculus, it is shown that the linear transformation associated to $G'(\vec{a})$ determines the local behavior of the function G. For example, if $n = m$ and $G'(\vec{a})$ is invertible, then under appropriate continuity assumptions, G is locally invertible as a function (inverse function theorem). This occurs because the invertibility of the matrix $G'(\bar{a})$ gives the invertibility of the linear operator $T_{G'(\bar{a})}$.

PROBLEMS

3.1.1. Indicate whether each of the following functions is linear and if it is, find the standard matrix that represents the linear transformation.
(a) $T : \mathbf{R}^2 \to \mathbf{R}^2$ given by $T(x, y) = (1 + x, y)$
(b) $T : \mathbf{R}^2 \to \mathbf{R}^2$ given by $T(x, y) = (x - y, 0)$
(c) $T : \mathbf{R}^2 \to \mathbf{R}^2$ given by $T(x, y) = (0, -xy)$
(d) $T : \mathbf{R}^3 \to \mathbf{R}^2$ given by $T(x, y, z) = (0, 0)$
(e) $T : \mathbf{R}^3 \to \mathbf{R}^2$ given by $T(x, y, z) = (x, y)$
(f) $T : \mathbf{R}^3 \to \mathbf{R}^2$ given by $T(x, y, z) = (\sin x, \cos y)$
(g) $T : \mathbf{R}^3 \to \mathbf{R}^4$ given by $T(x, y, z) = (x, x, y, x + \pi z)$
(h) $T : \mathbf{R}^3 \to \mathbf{R}^4$ given by $T(x, y, z) = (z^2, x, y)$

3.1.2. Find two linear transformations $T, U : \mathbf{R}^3 \to \mathbf{R}^3$ such that $T \circ U \neq U \circ T$.

3.1.3. Answer the question (and give reasons): Is there a linear transformation $T : \mathbf{R}^3 \to \mathbf{R}^3$ such that
(a) $T(1, 0, 0) = (1, 2, 3)$, $T(0, 1, 0) = (3, 2, 1)$, $T(0, 0, 1) = (4, 4, 4)$
(b) $T(1, 0, 0) = (1, 2, 3)$, $T(0, 1, 0) = (3, 2, 1)$, $T(1, 2, 0) = (4, 4, 4)$

3.1.4. Suppose that $f : \mathbf{R}^3 \to \mathbf{R}^2$ and $g : \mathbf{R}^2 \to \mathbf{R}^2$ are linear transformations. Find the standard matrix for $g \circ f$, $[g \circ f]$, if
(a) $g(x, y) = (x - y, 0)$ and $f(x, y, z) = (z - x, z - y)$
(b) $f(x, y, z) = (0, 0)$
(c) $f(x, y, z) = (x, x)$ and $[g] = \begin{pmatrix} 2 & 3 \\ 2 & 3 \end{pmatrix}$

3.1.5. Find the standard matrix of the linear transformation $F : \mathbf{R}^2 \to \mathbf{R}^2$ where F is geometrically described by
(a) The reflection across the line $X = 0$
(b) The reflection across the line $X = Y$
(c) The projection onto the line $Y = 0$

3.1.6. Suppose that $T : V \to \mathbf{R}^n$ and $U : V \to \mathbf{R}^m$ are linear. Show that the function $G : V \to \mathbf{R}^{n+m}$ defined by $G(v) = (T(v), U(v))$ is a linear transformation.

3.1.7. Suppose that $T : V \to W$ is linear and $T(v_1), T(v_2), \ldots, T(v_n)$ are linearly independent in W. Show that $v_1, v_2, \ldots, v_n$ are linearly independent in V.

3.1.8. (a) Consider the function $S : P^n \to P^{2n}$ defined by $S(f(X)) = (f(X))^2$. Is S linear?
(b) What about $G : P^n \to P^n$ given by $G(f(X)) = f(X + 1)$?

3.1.9. (a) Consider the function $T : C(\mathbf{R}, \mathbf{R}) \to \mathbf{R}$ defined by $T(f) = f(1) + f(2)$. Is T linear?
(b) What about $U : C(\mathbf{R}, \mathbf{R}) \to \mathbf{R}$ defined by $U(f) = f(1)f(2)$?

3.1.10. Recall that $M_{n \times n}(F)$ denotes the vector space of all $n \times n$ matrices over $\mathbf{F}$.
(a) Show that the function $S : M_{n \times n}(\mathbf{F}) \to M_{n \times n}(\mathbf{F})$ defined by $S(A) = A^t - A$ is a linear transformation.
(b) For $A = (a_{ij})$ we define the *trace* of A by

$$tr(A) = a_{11} + a_{22} + \cdots + a_{nn}$$

Show that $tr : M_{n \times n}(\mathbf{F}) \to \mathbf{F}$ is a linear transformation.
(c) Is the determinant function $det : M_{n \times n}(\mathbf{F}) \to \mathbf{F}$ a linear transformation?

3.1.11. Suppose that $T : V \to W$ is linear and U is a subspace of W. Show that $S = \{v \in V \mid T(v) \in U\}$ is a subspace of V.

3.1.12. Suppose that $T : V \to W$ is a linear transformation. Let U be a subspace of V. The *restriction* of T to U, $T|_U : U \to W$ is the function defined by $T|_U(u) = T(u)$ for all $u \in U$. Show that $T|_U$ is a linear transformation.

3.1.13. (a) Show that the set of linear transformations from $\mathbf{F}^n$ to $\mathbf{F}^m$, usually denoted by $Hom(\mathbf{F}^n, \mathbf{F}^m)$, is a vector space.
(b) Show that the function $Mat : Hom(\mathbf{F}^n, \mathbf{F}^m) \to M_{m \times n}(\mathbf{F})$ defined by $Mat(T) = [T]$ (the standard matrix of T) is a linear transformation.

3.2 Properties of Linear Transformations

Our goal in this section is to relate the important properties of linear transformations on $\mathbf{F}^n$ to the properties (for example rank) of their associated matrices. The result is a two-way street. Matrices can be used to study linear transformations. But as well, many (seemingly computational) matrix questions can be answered by thinking about the associated linear transformation from a functional point of view. We begin with two basic definitions.

Definition 3.2.1 Suppose $T : V \to W$ is a linear transformation.
(i) The *kernel* of T, usually denoted $ker(T)$, is defined by $ker(T) = \{v \in V \mid T(v) = \vec{0}\}$.
(ii) The *image* of T, usually denoted $im(T)$, is defined by $im(T) = \{T(v) \mid v \in V\}$.

Observe that $ker(T) \subseteq V$, while $im(T) \subseteq W$. It is important not to confuse where these subsets are located. Often, $ker(T)$ is called the *nullspace* of T, and $im(T)$ is called the *range space*, or simply the *range*, of T.

The kernel of $T : V \to W$ should be thought of as the set of solutions to the equation $T(v) = \vec{0}$. For instance, in Example 3.1.2 (vi) the kernel of $S : V \to V$ is precisely the set of solutions to the homogeneous linear differential equation $S(f) = 0$. The image of S also has a nice interpretation. By definition $im(S) = \{g \in V \mid \text{there exists } f \in V \text{ with } S(f) = g\}$. In other words, the image of S is precisely the set of functions $g \in V$ for which the differential equation $S(f) = g$ has a solution.

In the study of the kernel and image of a linear transformation, we need to apply general vector space theory. The next lemma shows that both the kernel and image of a linear transformation are vector spaces.

Lemma 3.2.2 *If* $T : V \to W$ *is a linear transformation, then* $ker(T)$ *is a subspace of* V *and* $im(T)$ *is a subspace of* W.

Proof: Suppose that $u, v \in ker(T)$ and k is a scalar. Then, since T is linear,
$$T(u + v) = T(u) + T(v) = \vec{0} + \vec{0} = \vec{0}$$
and
$$T(kv) = kT(v) = k\vec{0} = \vec{0}$$
Thus, $u + v$, $kv \in ker(T)$; this shows that $ker(T)$ is a subspace of V. Now choose $z, w \in im(T)$. By the definition of $im(T)$, there exist $u, v \in V$ with $T(u) = z$, $T(v) = w$. Then, since T is linear, $T(u + v) = z + w$ and $T(ku) = kz$. Hence, $z + w, kz \in im(T)$, so that $im(T)$ is a subspace of W. This proves the lemma. //

Now that we know that $im(T)$ and $ker(T)$ are vector spaces, we give special names to their dimensions.

Definition 3.2.3 Suppose that $T : V \to W$ is a linear transformation. The dimension of $ker(T)$ is called the *nullity* of T and is denoted by $null(T)$. The dimension of $im(T)$ is called the *rank* of T and is denoted $rk(T)$.

When studying a linear transformation between finite-dimensional vector spaces, the dimensions $rk(T)$ and $null(T)$ can be computed by using matrices. In the next theorem, we see that in the case of a linear transformation $T : \mathbf{F}^n \to \mathbf{F}^m$, the concepts of kernel, image, nullity, and rank coincide with the corresponding notions for the matrix representing T. Part (iv) of this theorem is the linear transformation version of the rank plus nullity theorem, which was given earlier in matrix form as Theorem 1.4.10.

Theorem 3.2.4 *Suppose that $T = T_A : \mathbf{F}^n \to \mathbf{F}^m$ is a linear transformation where A is an $m \times n$ matrix. Then*
(i)   *$ker(T)$ is the set of solutions to $AX = \vec{0}$.*
(ii) *$im(T)$ is the column space of A.*
(iii) *$rk(T) = rk(A)$.*
(iv) *(Rank Plus Nullity Theorem) $rk(T) + null(T) = n$.*

Proof: (i) By the definitions, $T_A(v) = \vec{0}$ if and only if $Av = \vec{0}$.

(ii) A vector $w \in \mathbf{F}^m$ lies in $im(T)$ if and only if there exists some $v \in \mathbf{F}^n$ such that $w = Av$. According to Theorem 1.1.4 (i) this occurs if and only if w is a linear combination of the columns of A, that is, if and only if w lies in the column space of A.

(iii) According to Theorem 1.4.6 (ii), $rk(A)$ is the dimension of the column space of A. Thus (iii) follows from (ii).

(iv) Finally, in view of Theorem 1.4.10, $n - rk(A) = dim\{v \in \mathbf{F}^n \mid Av = \vec{0}\}$, which by (i) is $dim(ker(T_A)) = null(T)$. This proves the theorem. //

Examples 3.2.5 (i) Consider the linear transformation $T_A : \mathbf{F}^2 \to \mathbf{F}^2$ where

$$A = \begin{pmatrix} 2 & 3 \\ 1 & 0 \end{pmatrix}$$

Evidently, $rk(A) = 2$ and consequently $im(T_A)$ is two-dimensional, that is, $im(T_A) = \mathbf{F}^2$. As $null(T_A) = 2 - 2 = 0$, we find that $ker(T_A) = \{\vec{0}\}$.

(ii) Consider the linear transformation $T_B : \mathbf{R}^3 \to \mathbf{R}^5$ where

$$B = \begin{pmatrix} 1 & -1 & 1 \\ 2 & 1 & 4 \\ 1 & 2 & 3 \\ 0 & 6 & 4 \\ -1 & -5 & -5 \end{pmatrix}$$

whose reduced row echelon form is

$$B = \begin{pmatrix} 1 & 0 & 5/3 \\ 0 & 1 & 2/3 \\ 0 & 0 & 0 \\ 0 & 0 & 0 \\ 0 & 0 & 0 \end{pmatrix}$$

By Theorem 3.2.4 (i) and (ii) we find that

$$ker(T_B) = span \left\{ \begin{pmatrix} -5/3 \\ -2/3 \\ 1 \end{pmatrix} \right\}$$

and

$$im(T_B) = span \left\{ \begin{pmatrix} 1 \\ 2 \\ 1 \\ 0 \\ -1 \end{pmatrix}, \begin{pmatrix} -1 \\ 1 \\ 2 \\ 6 \\ -5 \end{pmatrix} \right\}$$

So, $ker(T_B)$ is a one-dimensional subspace of $\mathbf{R}^3$ and $im(T_B)$ is a two-dimensional subspace of $\mathbf{R}^5$.

The rank plus nullity theorem given in part (iv) of Theorem 3.2.4 states that whenever $T : \mathbf{F}^n \to \mathbf{F}^m$ is a linear transformation, then one has an equation $n = dim(\mathbf{F}^n) = rk(T) + null(T)$. This result was a consequence of our analysis of matrix equations in Chap. 1, which in turn were dependent upon our study of systems of linear equations in Chap. 0. At that time we saw the number of free variables needed to describe all solutions to a homogeneous matrix equation $AX = \vec{0}$ was $n - rk(A)$, where n is the number of columns of A.

The next theorem generalizes the rank plus nullity theorem to the situation where $T : V \to W$ is a linear transformation and V is an arbitrary finite-dimensional vector space. This result also goes by the name rank plus nullity theorem. Observe that the value space W need not be finite-dimensional in this theorem. For this reason, we cannot hope to prove such a result using matrix methods. Instead we must use the general theory of vector spaces, linear independence, span, and so on. An important feature of this proof is that it gives an alternative viewpoint for both Theorem 3.2.4 (iv) and the earlier results from Chaps. 0 and 1.

Theorem 3.2.6 (Rank Plus Nullity Theorem) *Suppose that $T : V \to W$ is a linear transformation where V is a finite-dimensional vector space. Then $rk(T) + null(T) = dim(V)$.*

Proof: We assume that $dim(V) = n$. Choose $\{v_1, v_2, \ldots, v_s\}$ to be a basis for $ker(T)$. So $null(T) = s \leq n$. By the Basis-Extension Theorem 1.4.10 there exist $v_{s+1}, v_{s+2}, \ldots, v_n \in V$ such that $\{v_1, \ldots, v_s, v_{s+1}, \ldots, v_n\}$ is a basis for V. Set $w_{s+1} = T(v_{s+1})$, $w_{s+2} = T(v_{s+2}), \ldots,$ $w_n = T(v_n)$. In order to prove the theorem, it suffices to show that $\{w_{s+1}, w_{s+2}, \ldots, w_n\}$ is a basis for $im(T)$, since then $dim(im(T)) = rk(T) = n - s = dim(V) - null(T)$.

We first verify linear independence. Assume that $a_{s+1}, a_{s+2}, \ldots, a_n$ are scalars and $a_{s+1}w_{s+1} + \cdots + a_n w_n = \vec{0}$. It follows that $T(a_{s+1}v_{s+1} + \cdots + a_n v_n) = a_{s+1}w_{s+1} + \cdots + a_n w_n = \vec{0}$. Hence $a_{s+1}v_{s+1} + \cdots + a_n v_n \in ker(T)$. Since $\{v_1, v_2, \ldots, v_s\}$ is a basis for $ker(T)$, there exist scalars $a_1, a_2, \ldots, a_s$ so that $a_1 v_1 + a_2 v_2 + \cdots + a_s v_s = a_{s+1}v_{s+1} + \cdots + a_n v_n$. From this, we see that in V, $a_1 v_1 + a_2 v_2 + \cdots + a_s v_s - a_{s+1}v_{s+1} - \cdots - a_n v_n = \vec{0}$. Since $v_1, v_2, \ldots, v_n$ are linearly independent, it follows that $a_1 = 0$, $a_2 = 0, \ldots, a_n = 0$; in particular $a_{s+1} = 0$, $a_{s+2} = 0, \ldots,$ $a_n = 0$. This shows that $w_{s+1}, w_{s+2}, \ldots, w_n$ are linearly independent.

Next we show that $w_{s+1}, w_{s+2}, \ldots, w_n$ span $im(T)$. Let $w \in im(T)$. Since $v_1, v_2, \ldots, v_n$ span V, there exist scalars $a_1, a_2, \ldots, a_n$ so that $w = T(a_1 v_1 + a_2 v_2 + \cdots + a_n v_n)$. Thus $w = a_1 T(v_1) + \cdots + a_s T(v_s) + a_{s+1}T(v_{s+1}) + \cdots + a_n T(v_n) = \vec{0} + \cdots + \vec{0} + a_{s+1}T(v_{s+1}) + \cdots + a_n T(v_n) = a_{s+1}T(v_{s+1}) + \cdots + a_n T(v_n) = a_{s+1}w_{s+1} + \cdots + a_n w_n$. This shows that $w_{s+1}, \ldots, w_n$ span $im(T)$ and proves the theorem. //

One important consequence of Theorem 3.2.6 is that if $T : V \to W$ is a linear transformation where V is finite-dimensional, then $im(T)$ is finite-dimensional even though W may not be. It follows that we can view T as a linear transformation between finite-dimensional vector spaces, whereby we replace the value space W with $im(T)$. In other words, we think of T as a linear transformation $T : V \to im(T)$. As an example, we can consider the linear transformation $T : \mathbf{F}^n \to P$, where P is the vector space of all polynomials in the variable X with coefficients from $\mathbf{F}$ and where we define $T(a_1, \ldots, a_n) = a_n X^{n-1} + a_{n-1} X^{n-2} + \cdots + a_2 X + a_1$. P, of course, is an infinite-dimensional vector space. So to feel more comfortable, we could replace P by $im(T)$, which is the n-dimensional vector space P^{n-1} of all polynomials of degree less than n.

The proof of Theorem 3.2.6 can be refined to show that if $T : V \to W$ is a linear transformation, then V is finite-dimensional if and only if both $ker(T)$ and $im(T)$ are finite-dimensional. Thus the rank plus nullity theorem holds even when W is not originally assumed to be finite-dimensional. [Of course, if either $rk(T)$ or $null(T)$ is infinite, then V cannot be finite-dimensional.] In the next definition we collect some basic terminology from the theory of functions. We state these definitions for linear transformations, but they apply to general functions between sets as well.

Definition 3.2.7 A linear transformation $T : V \to W$ is called *one-one* if whenever $u, v \in V$ and $T(u) = T(v)$, then necessarily $u = v$. T is called *onto* if $im(T) = W$. T is called an *isomorphism* if T is both one-one and onto. Whenever there exists an isomorphism $T : V \to W$, we say that the vector spaces V and W are *isomorphic*. Often, one-one functions are called *injective*, onto functions are called *surjective*, and functions which are both one-one and onto are called *bijective*. An isomorphism $T : V \to V$ is often called a *nonsingular linear operator*.

To illustrate this terminology, we give a few examples. Let P denote the vector space of all polynomials with coefficients in the field $\mathbf{R}$. Consider the linear transformation $D : P \to P$ defined by differentiation; that is, $D(f(X)) = f'(X)$. Then D is not one-one, because $D(X^2) = D(X^2 + 1) = 2X$, while $X^2 \neq X^2 + 1$. On the other hand D is onto, because every polynomial is the derivative of some other polynomial. In other words, if $f(X) \in P$, then there is some $F(X) \in P$ with $F'(X) = f(X)$.

Now consider $S : P \to P$ defined by $S(f(X)) = f(X^2)$. Thus $S(X + 1) = X^2 + 1$, $S(X^2 - 2X) = X^4 - 2X^2$, and so forth. It is easy to check that S is a linear transformation. Clearly there is no polynomial $f(X) \in P$ with $S(f(X)) = X$, so S is not onto. However, S is one-one, because whenever $S(f(X)) = S(g(X))$, necessarily $f(X) = g(X)$. [Both $f(X)$ and $g(X)$ must be the polynomial obtained from the polynomial $S(f(X))$ by replacing each occurrence of X^{2n} by X^n.] Now consider $T : P \rightarrow P$ defined by $T(f(X)) = 2f(X)$. (We assume that $1 + 1 = 2 \neq 0 \in \mathbf{F}$.) Since $T((1/2)g(X)) = g(X)$ for all $g(X) \in P$, we see that T is onto. Also, if $T(f(X)) = T(g(X))$, then $2f(X) = 2g(X)$, so $f(X) = g(X)$ follows. Hence T is one-one as well. This shows that T is an isomorphism.

The concept of being onto often causes confusion. The reason for this confusion is that being onto is a relative notion and depends upon where the function is *viewed* as taking values, that is, upon the specified value space. For example, even if $T : V \rightarrow W$ is not onto, one can view (as explained above) T as a linear transformation $T : V \rightarrow im(T)$. In this latter context, T, as a function taking values in $im(T)$, is always onto. So whenever addressing the question of a transformation being onto, be sure that the value space is clearly understood.

If $f : X \rightarrow Y$ is a one-one and onto function between the sets X and Y, then each element of X corresponds, via f, to a unique element of Y and vice versa. In other words, if $x \in X$, then there is a unique $y \in Y$ for which $f(x) = y$, and for each $y \in Y$, there is a unique $x \in X$ with $f(x) = y$. Such functions $f(x)$ are often called *invertible* functions since they have a well-defined inverse function $f^{-1} : Y \rightarrow X$ given by $f^{-1}(y) = x$ whenever $f(x) = y$. Evidently $f^{-1}(f(x)) = x$ for all $x \in X$ and $f(f^{-1}(y)) = y$ for all $y \in Y$. Sometimes it is useful to weaken the concept of invertibility and consider right and left inverses. A *right inverse* of a function $f : X \rightarrow Y$ is a function $R : Y \rightarrow X$ which satisfies $f(R(y)) = y$ for all $y \in Y$, and a *left inverse* of f is a function $L : Y \rightarrow X$ which satisfies $L(f(x)) = x$ for all $x \in X$.

If $f : X \rightarrow Y$ is bijective (that is, both one-one and onto), the sets X and Y have the same number of elements. However, the function f does more than just count elements; it specifies an explicit correspondence between the two sets X and Y. In most applications of bijective functions, the particular correspondence is what is important. This is the case with isomorphisms. An isomorphism $T : V \rightarrow W$ between two vector spaces V and W is a bijective correspondence between V and W which also transfers the vector space operations of vector addition and scalar multiplication from V to W. This is because the linear transformation T has the property specified in Lemma 3.1.3.

In Theorem 1.5.5 we saw our first example of a vector space isomorphsim (although it was not called this at that time). It was shown there that if V is an n-dimensional vector space with (ordered) basis $\mathcal{B}$, then the function $U_\mathcal{B} : V \to \mathbf{F}^n$ defined by $U_\mathcal{B}(v) = (v)_\mathcal{B}$ is a linear transformation. Evidently, this T is both one-one and onto, and as remarked at that time, T provides a way of viewing V and $\mathbf{F}^n$ as the "same". This is true for general isomorphisms, but as we shall see shortly, isomorphisms have many other important uses as well.

In some cases it is important to view isomorphic vector spaces as different (this is why the terminology of "isomorphism" was invented). In other cases the isomorphism is a function from a vector space to itself, so we know the vector spaces are exactly the same. In this latter situation, the properties of the function are what is important. The situation is similar to what occurs in high school geometry, where a distinction is made between equal and congruent triangles. Two triangles are equal if they are exactly the same triangle drawn on a piece of paper (perhaps only named differently, for example, $\triangle ABC = \triangle BAC$). In contrast, two triangles are congruent if they have exactly the same-length sides (consequently same shape and area). While discussing isomorphisms, one needs to keep such distinctions in the back of one's mind and judge from the context what the proper interpretation should be.

The next theorem contains alternative characterizations for one-one and/or onto linear transformations. These results are used so often in the sequel that after a while reference to this theorem will be omitted. One consequence of this result is that if two vector spaces V and W are isomorphic, then $dim(V) = dim(W)$. This should be reasonable, for after all, vector spaces that "look the same" should have the same dimension.

Theorem 3.2.8 *Suppose that V and W are finite-dimensional vector spaces and $T : V \to W$ is a linear transformation. Then*
(i)   *T is one-one if and only if $ker(T) = \{\vec{0}\}$.*
(ii)   *T is one-one if and only if whenever $v_1, v_2, \ldots, v_n$ are linearly independent in V, $T(v_1), T(v_2), \ldots, T(v_n)$ are linearly independent in W.*
(iii)   *T is onto if and only if $rk(T) = dim(W)$.*
(iv)   *T is an isomorphism if and only if whenever $\{v_1, v_2, \ldots, v_n\}$ is a*

basis for V, $\{T(v_1), T(v_2), \ldots, T(v_n)\}$ *is a basis for* W.
(v)   *If* T *is an isomorphism, then* $dim(V) = dim(W)$.
(vi)   *If* T *is an isomorphism, then* T^{-1} *exists as a function and* T^{-1} :
$W \to V$ *is also an isomorphism.*

Proof: (i) Suppose that T is one-one. If $T(v) = \vec{0}$, then $T(v) = T(\vec{0}) = \vec{0}$. Since T is one-one, we must have $v = \vec{0}$. Hence $ker(T) = \{\vec{0}\}$. Conversely, assume $ker(T) = \{\vec{0}\}$ and $T(u) = T(v)$. Then $T(u-v) = \vec{0}$; that is, $u - v \in ker(T)$. Thus $u - v = \vec{0}$, and $u = v$ follows. This shows that T is one-one.

(ii)   Suppose that T is one-one. Assume $v_1, v_2, \ldots, v_n$ are linearly independent and $a_1 T(v_1) + a_2 T(v_2) + \cdots + a_n T(v_n) = \vec{0}$. Then $T(a_1 v_1 + a_2 v_2 + \cdots + a_n v_n) = a_1 T(v_1) + a_2 T(v_2) + \cdots + a_n T(v_n) = \vec{0}$. This shows that $a_1 v_1 + a_2 v_2 + \cdots + a_n v_n = \vec{0}$ since T is one-one. Using the linear independence of $v_1, v_2, \ldots, v_n$, we find that $a_1 = 0$, $a_2 = 0, \ldots, a_n = 0$. This shows that $T(v_1), T(v_2), \ldots, T(v_n)$ are linearly independent. Conversely, suppose that T is not one-one. Then by part (i), for some $v \neq \vec{0} \in V$, we have $T(v) = \vec{0}$. But v is linearly independent as $v \neq \vec{0}$, while $T(v) = \vec{0}$ is not linearly independent. This contradiction proves (ii).

(iii)   By definition, T is onto if and only if $im(T) = W$. However, $dim(im(T)) = rk(T)$. Thus, $im(T) = W$ if and only if $rk(T) = dim(W)$. This proves (iii).

(iv)   Suppose that T is an isomorphism and $\{v_1, v_2, \ldots, v_n\}$ is a basis for V. Since T is one-one, by (ii) $T(v_1), T(v_2), \ldots, T(v_n)$ are linearly independent. Let $w \in W$. Then as T is onto, there is some $v \in V$ with $T(v) = w$. Since $\{v_1, v_2, \ldots, v_n\}$ is a basis for V, we express $v = a_1 v_1 + a_2 v_2 + \cdots + a_n v_n$ for scalars $a_1, a_2, \ldots, a_n$. We find that $w = T(v) = a_1 T(v_1) + a_2 T(v_2) + \cdots + a_n T(v_n)$, so $T(v_1), T(v_2), \ldots, T(v_n)$ span W. Thus they form a basis for W.

Conversely, suppose that whenever $\{v_1, v_2, \ldots, v_n\}$ is a basis for V, $\{T(v_1), T(v_2), \ldots, T(v_n)\}$ is a basis for W. By (ii), T must be one-one. Let $\{v_1, v_2, \ldots, v_n\}$ be a basis for V. Then since $T(v_1), T(v_2), \ldots, T(v_n)$ span W and $span\{T(v_1), T(v_2), \ldots, T(v_n)\} \subseteq im(T)$, we see $im(T) = W$. Thus, T is an isomorphism.

(v)   Part (v) is an immediate consequence of (iv).

(vi)   Any one-one, onto function T has an inverse function T^{-1} which is both one-one and onto and is defined by $T^{-1}(w) = u$ for the unique u satisfying $T(u) = w$. So, we must only show that T^{-1} is linear. Suppose that $T(u) = z$ and $T(v) = w$. Then $T(u + v) = z + w$ and $T(kv) = kw$. Hence by the defintion of T^{-1}, $T^{-1}(z + w) = u + v = T^{-1}(z) + T^{-1}(w)$ and $T^{-1}(kw) = kz = kT^{-1}(z)$. This shows that T^{-1} is linear and proves the theorem. //

Discussion 3.2.9 Whenever $T : \mathbf{F}^n \to \mathbf{F}^m$ is the linear transformation associated with an $m \times n$ matrix, Theorem 3.2.8 can be interpreted in terms of the associated system of equations. As a result, we recapture some of the basic results proved in Chap. 0. This observation illustrates the connection between the general reasoning we are now using and the computational results obtained earlier.

(i) Suppose that B is an $m \times n$ matrix. Then by part (i) of Theorem 3.2.8, the linear transformation $T_B : \mathbf{F}^n \to \mathbf{F}^m$ is one-one if and only if $ker(T_B) = \{\vec{0}\}$. In other words, T_B is one-one if and only if the homogeneous system $BX = \vec{0}$ has only the trivial solution. Observe that T_B being one-one says that whenever $T_B(v) = w$ has a solution, then it has a unique solution. Interpreted in terms of the system of equations $BX = w$, we obtain exactly the result given in Theorem 0.5.7 (ii). This result says that there is a unique solution in this case if and only if $rk(B) = n$, which occurs if and only if $BX = \vec{0}$ has a unique solution.

(ii) Part (iii) of Theorem 3.2.8 shows that whenever B is $m \times n$, T_B is onto if and only if $rk(T_B) = dim(\mathbf{F}^m) = m$. But, according to Theorem 3.2.4 (iii), $rk(T_B) = rk(B)$. Thus, T_B is onto if and only if $rk(B) = m$. In terms of systems of equations this says that $BX = w$ has a solution for all $w \in \mathbf{F}^m$ if and only if $rk(B) = m$. This is an immediate consequence of Theorem 0.5.7 (i), since the rank of the augmented matrix for this system cannot exceed the number of rows of B.

(iii) Suppose that A is an $n \times n$ invertible matrix. Then the linear transformation $T_A : \mathbf{F}^n \to \mathbf{F}^n$ is an isomorphism. Theorem 3.2.8 (i) guarantees that T_A is one-one. By Theorem 3.2.6 $rk(T_A) = n = dim(\mathbf{F}^n)$, so T_A is onto by Theorem 3.2.8 (iii). In particular the invertibility of a matrix is equivalent to the invertibility (as a function) of the associated linear transformation. This justifies the usage of the term "invertible."

Consider the real matrices

$$A = \begin{pmatrix} 2 & 3 & 4 \\ 1 & 1 & 1 \end{pmatrix} \quad \text{and} \quad B = \begin{pmatrix} 1 & 0 & 0 \\ 1 & 2 & 0 \\ 0 & 1 & 1 \end{pmatrix}$$

Since $rk(A) = 2$, the above discussion shows that $T_A : \mathbf{F}^3 \to \mathbf{F}^2$ cannot be one-one, but it is onto. Consequently, the system of equations $AX = \vec{0}$ has non-zero solutions. Moreover, since T_A is onto, the system of equations $AX = v$ can always be solved for any $v \in \mathbf{F}^2$ and thus always has infinitely many solutions. The matrix B has rank 3, so we see that $T_B : \mathbf{F}^3 \to \mathbf{F}^3$ is an isomorphism and for any $w \in \mathbf{F}^3$ the system $BX = w$ always has a unique solution. In the next section we will show how to use these same matrix methods to analyze linear transformations $T : V \to W$ between arbitrary finite-dimensional vector spaces V and W.

PROBLEMS

3.2.1.   Find the standard matrix, rank, and nullity of each of the following linear transformations.
(a) $T : \mathbf{R}^3 \to \mathbf{R}^4$ defined by $T(x, y, z) = (x - z, x + z, 2x + 2z, z)$
(b) $R : \mathbf{R}^4 \to \mathbf{R}^2$ defined by $R(x, y, z, w) = (x + y - z - w, x + z)$
(c) $Z : \mathbf{R}^4 \to \mathbf{R}$ given by $Z(v) = 0$ for all $v \in \mathbf{R}^4$
(d) $W : \mathbf{R} \to \mathbf{R}$ where $W(r) = 3r$
(e) $N : \mathbf{R}^5 \to \mathbf{R}^5$ given by $N(v) = -v$ for all $v \in \mathbf{R}^5$

3.2.2. Find the rank and nullity and bases for the range and kernel of T_A whenever A is

(a) $\begin{pmatrix} 0 & 0 & 0 & 0 \\ 3 & 1 & 8 & 8 \\ 1 & 1 & 0 & 1 \end{pmatrix}$ (b) $\begin{pmatrix} 1 & -1 \\ 2 & 2 \\ 0 & 9 \end{pmatrix}$

(c) $\begin{pmatrix} 1 & 0 & 1 \\ 1 & 1 & 0 \\ 0 & 1 & 1 \end{pmatrix}$ (d) $\begin{pmatrix} 1 \\ 3 \\ 3 \\ 1 \\ 0 \end{pmatrix}$

(e) $(1 \quad 3 \quad 3 \quad 1 \quad 0)$ (f) $\begin{pmatrix} 2 & 4 \\ 3 & 6 \end{pmatrix}$

3.2.3. Find a linear transformation $T : \mathbf{R}^3 \to \mathbf{R}^3$ such that

(a) $im(T) = span\left\{ \begin{pmatrix} 1 \\ 1 \\ 0 \end{pmatrix}, \begin{pmatrix} 1 \\ 2 \\ 9 \end{pmatrix} \right\}$

(b) $ker(T) = span\left\{ \begin{pmatrix} 1 \\ 1 \\ 0 \end{pmatrix}, \begin{pmatrix} 1 \\ 2 \\ 9 \end{pmatrix} \right\}$

3.2.4. Find the nullspace and range of the linear transformation $T : P^3 \to P^3$ defined by $T(f(X)) = f(X+1)$.

3.2.5. Suppose that $T : \mathbf{R}^3 \to \mathbf{R}^3$ satisfies $T(1,0,0) = (2,1,2)$, $T(0,1,0) = (3,3,3)$ and $T(0,0,1) = (2,1,2)$. Is T invertible?

3.2.6. Suppose that $T : V \to W$ is a one-one linear transformation. Show that the vector spaces V and $im(T)$ are isomorphic.

3.2.7. Give an example of a linear transformation $T : \mathbf{R}^3 \to \mathbf{R}^3$ for which $T \neq 0$ but $T \circ T = 0$. What is the rank of your example?

3.2.8. Consider the matrices A given in Prob. 3.2.2. Which of the vector spaces $ker(T_A)$ and $im(T_A)$ are isomorphic to $\mathbf{R}^2$? Which are isomorphic to $\mathbf{R}^3$?

3.2.9. Suppose that $T : V \to V$ is a linear operator on a finite-dimensional vector space V and $rk(T \circ T) = rk(T)$. Show that the intersection $ker(T) \cap im(T) = \{\vec{0}\}$.

3.2.10. (a) Suppose that $T, U : V \to V$ are both one-one. Show that $T \circ U$ is also one-one.
(b) Suppose that $T, U : V \to V$ are both onto. Show that $T \circ U$ is also onto.

3.2.11. Suppose that $T : V \to V$ is a linear transformation, where V is finite-dimensional and $ker(T) = im(T)$. Show that V is even-dimensional.

3.2.12. Suppose that $T : V \to W$ is a linear transformation. A linear transformation $L : W \to V$ is called a *left inverse* for T if $L \circ T = I_V$, the identity transformation on V. A linear transformation $R : W \to V$ is called a *right inverse* for T if $T \circ R = I_W$.
(a) Show that if T has a right inverse, then T is onto.
(b) Show that if T has a left inverse, then T is one-one.

3.2.13. Suppose that $T : V \to W$ is one-one, $S : W \to U$ is onto, $dim(V) = m$, $dim(U) = n$, and $im(T) = ker(S)$. What is $dim(W)$?

3.2.14. Suppose that $R, S : V \to W$ are both linear transformations between finite-dimensional vector spaces. Define

$$(R + S)(v) = R(v) + S(v)$$

Show:
(a) $R + S : V \to W$ is a linear transformation.
(b) $rk(R) + rk(S) \geq rk(R + S)$.

3.2.15. Suppose that V is a vector space with basis $\{v_1, v_2, \ldots, v_n\}$. Assume that $T : V \to V$ is a linear transformation such that $T(v_1) = v_2$, $T(v_2) = v_3, \ldots$, $T(v_{n-1}) = v_n$, and $T(v_n) = \vec{0}$. Find the nullity of T^i for all i with $1 \leq i \leq n$.

3.2.16. Suppose that $T : V \to V$ is linear, V is finite-dimensional, and $T^n = 0$ for some n. Show that if $rk(T^i) > 0$, then $rk(T^i) > rk(T^{i+1})$.

3.2.17. Suppose that

$$B = \begin{pmatrix} -1 & 1 & 2 & 1 \\ 3 & -1 & 2 & -1 \\ -2 & -3 & 5 & -1 \\ -1 & 2 & -15 & 0 \end{pmatrix}$$

Show that $im(T_B)$ is the same as $\{(a, b, c, d) \in \mathbf{R}^4 \mid 3a + 2b + c + d = 0\}$.

3.2.18. Sharpen Theorem 3.2.6 and show that if $T : V \to W$ is a linear transformation, then V is finite-dimensional if and only if both $ker(T)$ and $im(T)$ are finite-dimensional.

3.2.19. Suppose that $T : V \to V$ is linear and $T = T \circ T$. Show that $ker(T) + im(T) = V$.

3.2.20. Suppose

$$A = \begin{pmatrix} 1 & -3 & 4 & 2 \\ -2 & 1 & -1 & -5 \\ 1 & 7 & -10 & 4 \end{pmatrix}$$

Find a basis for $ker(T_A)$.

3.2.21. Suppose that A is an $n \times n$ matrix such that $A^2 = 0$.
(a) Show that $im(T_A) \subseteq ker(T_A)$.
(b) Show $rk(A) \leq n/2$.

3.2.22. (a) Suppose that C is a $k \times p$ matrix and D is a $p \times r$ matrix. Show that $im(T_{CD})$ is a subspace of $im(T_C)$.

(b) Using (a), conclude that whenever C is a $k \times p$ matrix and D is a $p \times r$ matrix, $rk(CD) \le rk(C)$.

(c) Compare your method of proof in (b) to that used in solving Prob. 1.4.21.

3.2.23. If $L : \mathbf{R}^3 \to \mathbf{R}^4$ is defined by

$$L \begin{pmatrix} x \\ y \\ z \end{pmatrix} = \begin{pmatrix} 2y - x \\ z \\ x + z \\ y \end{pmatrix}$$

(a) Find A so that $L = T_A$.

(b) Find $rk(L)$ and $null(L)$.

(c) Is L one-one? Is L onto?

(d) Find a basis for $im(L)$.

3.2.24. Suppose that A and B are matrices and the product AB makes sense. Prove that $null(AB) \ge null(B)$ by showing that $ker(T_B) \subseteq ker(T_{AB})$.

3.2.25. If A is invertible, show that AB and B have the same nullspace and row space.

3.2.26. Let V be a finite-dimensional vector space and W be a subspace of V. Show that there is a linear transformation $T : V \to V$ with $ker(T) = W$. Also show that there is a linear transformation $S : V \to V$ with $im(S) = W$.

3.2.27. Suppose that $T : \mathbf{F}^4 \to \mathbf{F}^3$ is linear and

$$im(T) = \{(x, y, z) \mid x + y + z = 0\}$$

What is $null(T)$?

3.2.28. (a) Suppose that $S : U \to V$ and $T : V \to W$ are linear with finite ranks. Show that

$$rk(T \circ S) \le min\{rk(T), rk(S)\}$$

Furthermore, show that $rk(T \circ S) = rk(S)$ if and only if $S(U) \cap ker(T) = \{\vec{0}\}$ and that $rk(T \circ S) = rk(T)$ if and only if $S(U) + ker(T) = V$.

(b) Let $S : U \to V$ and $T : V \to W$ be linear with finite nullities. Show that

$$null(T \circ S) \le null(S) + null(T)$$

with equality if and only if $ker(T) \subseteq S(U)$.

3.3   Matrix Representations for Linear Transformations

In the previous sections we saw how to relate information about an $m \times n$ matrix A to the linear transformation $T_A : \mathbf{F}^n \to \mathbf{F}^n$. We now show how to represent arbitrary linear transformations between finite-dimensional vector spaces with matrices. Thus, even in the general situation, the techniques of matrix algebra may be applied to study linear transformations. The utility of this will become especially apparent in the sections which follow. We begin with a definition.

Definition 3.3.1   Suppose that $T : V \to W$ is a linear transformation, $\mathcal{B} = \{v_1, v_2, \ldots, v_n\}$ is an ordered basis for V, and $\mathcal{C} = \{w_1, w_2, \ldots, w_m\}$ is an ordered basis for W. We define the $m \times n$ matrix $[T]_{\mathcal{B},\mathcal{C}}$ by $[T]_{\mathcal{B},\mathcal{C}} = (a_{ij})$ where for each j the scalars a_{ij} are uniquely determined by the equations

$$T(v_j) = a_{1j} w_1 + a_{2j} w_2 + \cdots + a_{mj} w_m$$

In other words, $[T]_{\mathcal{B},\mathcal{C}}$ is the matrix whose jth column is the coordinates of $T(v_j)$ with respect to the basis $\mathcal{C}$ for W.

Note that before the matrix representation $[T]_{\mathcal{B},\mathcal{C}}$ could be defined, we had to *fix* the bases $\mathcal{B}$ and $\mathcal{C}$. This is crucial, because without specific ordered bases, there is no meaningful way to associate a matrix with a linear transformation. In fact, as we shall see later, a change in the choice of bases can wildly change the matrix which represents the linear transformation.

Definition 3.3.1 gives a recipe for computing the matrix $[T]_{\mathcal{B},\mathcal{C}}$. For example, suppose that $T : \mathbf{F}^3 \to \mathbf{F}^2$ is the linear transformation defined by $T(x, y, z) = (x - y, z + y)$. Let $\mathcal{B} = \{(1, 1, 0), (0, 1, 1), (0, 0, 1)\}$ be an ordered basis for $\mathbf{F}^3$ and $\mathcal{C} = \{(1, 1), (1, -1)\}$ be an ordered basis for $\mathbf{F}^2$. To compute $[T]_{\mathcal{B},\mathcal{C}}$, we must compute the $\mathcal{C}$ coordinates of the images of the basis elements in $\mathcal{B}$ and list these coordinates as the columns of the matrix $[T]_{\mathcal{B},\mathcal{C}}$. We find $T(1, 1, 0) = (0, 1), T(0, 1, 1) = (-1, 2)$, and $T(0, 0, 1) = (0, 1)$. We next compute that $(T(1, 1, 0))_{\mathcal{C}} = (1/2, -1/2), (T(0, 1, 1))_{\mathcal{C}} = (1/2, -3/2)$, and $(T(0, 0, 1))_{\mathcal{C}} = (1/2, -1/2)$. (Do not forget this second step of computing the C coordinates!) From this we obtain

$$[T]_{B,C} = \begin{pmatrix} 1/2 & 1/2 & 1/2 \\ -1/2 & -3/2 & -1/2 \end{pmatrix}$$

Consider the special case where $T : \mathbf{F}^n \to \mathbf{F}^m$ and B, C are the standard bases for $\mathbf{F}^n, \mathbf{F}^m$ respectively. Then the matrix $[T]_{B,C}$ just defined is precisely the standard matrix of T; that is, $[T]_{B,C} = A$ where $T = T_A$. This occurs because the recipe just described gives the standard matrix for T whenever the standard bases are used.

Recall that if B is an ordered basis for V and $v \in V$, $(v)_B$ denotes the column matrix of the coordinates of v with respect to B. The next result shows that the matrix $[T]_{B,C}$, describes the linear transformation T by matrix multiplication, except that one represents vectors by their B and C coordinates.

Theorem 3.3.2 *Let $T : V \to W$ be a linear transformation between finite-dimensional vector spaces, where B is an ordered basis for V and C is an ordered basis for W. Then for all $v \in V$*

$$[T]_{B,C}(v)_B = (T(v))_C$$

Moreover, this condition uniquely determines the matrix $[T]_{B,C}$.

Proof: We assume that $B = \{v_1, v_2, \ldots, v_n\}$ is an ordered basis for V and $C = \{w_1, w_2, \ldots, w_m\}$ is an ordered basis for W. Note that for any $v_i \in B$, since $v_i = 0v_1 + \cdots + 0v_{i-1} + 1v_i + 0v_{i+1} + \cdots + 0v_n$,

$$(v_i)_B = \begin{pmatrix} 0 \\ \vdots \\ 0 \\ 1 \\ 0 \\ \vdots \\ 0 \end{pmatrix}$$

where the single 1 is in the ith row. Since the ith column of $[T]_{B,C}$ is the coordinates of $T(v_i)$ with respect to C, we see that $[T]_{B,C}(v_i)_B = (T(v_i))_C$. This also shows that any matrix A satisfying the equation $A(v)_B = (T(v))_C$ is necessarily $[T]_{B,C}$, because the ith column is completely determined by the value $T(v_i)$. Thus the uniqueness assertion of the theorem follows once it is shown that $[T]_{B,C}$ has the stated properties.

Suppose $v = a_1 v_1 + a_2 v_2 + \cdots + a_n v_n \in V$. Then

$$(v)_B = \begin{pmatrix} a_1 \\ \vdots \\ a_n \end{pmatrix} = a_1 (v_1)_B + \cdots + a_m (v_m)_B$$

where the addition is that of $n \times 1$ matrices. Applying this, together with the previous paragraph, gives

$$\begin{aligned} [T]_{B,C}(v)_B &= [T]_{B,C}(a_1 (v_1)_B + \cdots + a_n (v_n)_B) \\ &= a_1 [T]_{B,C}(v_1)_B + \cdots + a_n [T]_{B,C}(v_n)_B \\ &= a_1 (T(v_1))_C + \cdots + a_n (T(v_n))_C = (T(v))_C \end{aligned}$$

where the last equality uses Corollary 1.5.5. This proves the theorem.
//

The idea behind the proof of the preceding theorem can be illustrated in the following diagram:

$$\begin{array}{ccc} & T & \\ V & \longrightarrow & W \\ U_B \downarrow & & \downarrow U_C \\ \mathbf{F}^n & \longrightarrow & \mathbf{F}^m \\ & T_A & \end{array}$$

where A is the $m \times n$ matrix $[T]_{B,C}$ and U_B, U_C are the isomorphisms given by coordinate maps. [So, $U_B(v) = (v)_B$ and $U_C(w) = (w)_C$ for $v \in V$ and $w \in W$ described in Theorem 1.5.5.] In this diagram, there is an equality of composition of functions: $T \circ U_C = U_B \circ T_A$. The point is that the matrix $A = [T]_{B,C}$ is nothing other than the standard matrix of the linear transformation from $\mathbf{F}^n$ to $\mathbf{F}^m$ that arises from T when V and W are identified with $\mathbf{F}^n$ and $\mathbf{F}^m$ via the coordinate maps associated with the bases B and C.

Example 3.3.3 Let V be the two-dimensional subspace of $\mathbf{R}^3$ with ordered basis

$$B = \left\{ \begin{pmatrix} 1 \\ 1 \\ 0 \end{pmatrix}, \begin{pmatrix} 0 \\ 1 \\ 1 \end{pmatrix} \right\}$$

and let W be the three dimensional subspace of $\mathbf{R}^5$ with ordered basis

$$C = \left\{ \begin{pmatrix} 1 \\ 1 \\ 1 \\ 0 \\ 0 \end{pmatrix}, \begin{pmatrix} 0 \\ 0 \\ 1 \\ 1 \\ 0 \end{pmatrix}, \begin{pmatrix} 0 \\ 0 \\ 0 \\ 1 \\ 1 \end{pmatrix} \right\}$$

Define $T : V \to W$ by

$$T \begin{pmatrix} x \\ y \\ z \end{pmatrix} = \begin{pmatrix} x + y + z \\ x + y + z \\ x - y + z \\ -2y \\ 0 \end{pmatrix} \quad \text{for} \quad \begin{pmatrix} x \\ y \\ z \end{pmatrix} \in V$$

Then,

$$T\,(1 \quad 1 \quad 0) = \begin{pmatrix} 2 \\ 2 \\ 0 \\ -2 \\ 0 \end{pmatrix} = 2\begin{pmatrix} 1 \\ 1 \\ 1 \\ 0 \\ 0 \end{pmatrix} + (-2)\begin{pmatrix} 0 \\ 0 \\ 1 \\ 1 \\ 0 \end{pmatrix} + 0\begin{pmatrix} 0 \\ 0 \\ 0 \\ 1 \\ 1 \end{pmatrix}$$

and

$$T\,(0 \quad 1 \quad 1) = \begin{pmatrix} 2 \\ 2 \\ 0 \\ -2 \\ 0 \end{pmatrix} = 2\begin{pmatrix} 1 \\ 1 \\ 1 \\ 0 \\ 0 \end{pmatrix} + (-2)\begin{pmatrix} 0 \\ 0 \\ 1 \\ 1 \\ 0 \end{pmatrix} + 0\begin{pmatrix} 0 \\ 0 \\ 0 \\ 1 \\ 1 \end{pmatrix}$$

This shows

$$\left(T \begin{pmatrix} 1 \\ 1 \\ 0 \end{pmatrix} \right)_C = \begin{pmatrix} 2 \\ -2 \\ 0 \end{pmatrix} \quad \text{and} \quad \left(T \begin{pmatrix} 0 \\ 1 \\ 1 \end{pmatrix} \right)_C = \begin{pmatrix} 2 \\ -2 \\ 0 \end{pmatrix}$$

which shows that

$$[T]_{B,C} = \begin{pmatrix} 2 & 2 \\ -2 & -2 \\ 0 & 0 \end{pmatrix}$$

Using this matrix, we can see that $im(T)$ consists precisely of all scalar multiples of the vector in W whose C coordinates are $(2, -2, 0)$. It follows that $rk(T) = 1$, and since $dim(V) = 2$, $null(T) = 1$. The same is true for the matrix $[T]_{B,C}$. (We shall see this is true more generally in Corollary 3.3.8.) $[T]_{B,C}$ also can be used to compute values of T. For example,

$$\begin{pmatrix} 3 \\ 4 \\ 1 \end{pmatrix}_B = \begin{pmatrix} 3 \\ 1 \end{pmatrix}$$

Hence,

$$\left(T \begin{pmatrix} 3 \\ 4 \\ 1 \end{pmatrix} \right)_C = \begin{pmatrix} 2 & 2 \\ -2 & -2 \\ 0 & 0 \end{pmatrix} \begin{pmatrix} 3 \\ 1 \end{pmatrix} = \begin{pmatrix} 8 \\ -8 \\ 0 \end{pmatrix}$$

From this we obtain

$$T \begin{pmatrix} 3 \\ 4 \\ 1 \end{pmatrix} = 8 \begin{pmatrix} 1 \\ 1 \\ 1 \\ 0 \\ 0 \end{pmatrix} + -8 \begin{pmatrix} 0 \\ 0 \\ 1 \\ 1 \\ 0 \end{pmatrix} + 0 \begin{pmatrix} 0 \\ 0 \\ 0 \\ 1 \\ 1 \end{pmatrix} = \begin{pmatrix} 8 \\ 8 \\ 0 \\ -8 \\ 0 \end{pmatrix}$$

It is important to learn to work with different bases when studying linear transformations. Suppose that a linear transformation $T : V \to W$ is specified by a matrix $[T]_{B_1,C_1}$ for specific ordered bases B_1 and C_1. An important problem is finding the matrix for this transformation with respect to a new pair of bases B_2 and C_2. The next three theorems solve this problem. The first of these generalizes a result that was given earlier in Theorem 1.5.4.

Theorem 3.3.4 *Let B_1 and B_2 be two bases for a n-dimensional vector space V. Then there is an invertible $n \times n$ matrix P such that for every $v \in V$*

$$P(v)_{B_1} = (v)_{B_2} \quad \text{and} \quad P^{-1}(v)_{B_2} = (v)_{B_1}$$

Proof: Suppose that $B_1 = \{v_1, v_2, \ldots, v_n\}$. We define P by taking the ith column of P to be $(v_i)_{B_2}$. Then in fact, P is the matrix $[I]_{B_1,B_2}$ where $I : V \to V$ is the identity function. (This is because the ith column of P is the coordinates of v_i itself in the new basis $\mathcal{B}_2$. Thus P is the matrix of the linear transformation that takes each element of $\mathcal{B}_1$ to itself. By Theorem 3.1.3 this transformation is necessarily I.) It now follows from Theorem 3.3.2 that $P(v)_{\mathcal{B}_1} = [I]_{\mathcal{B}_1,\mathcal{B}_2}(v)_{\mathcal{B}_1} = (I(v))_{\mathcal{B}_2} = (v)_{\mathcal{B}_2}$, as desired. The matrix P is invertible since its columns are linearly independent. Thus, $P^{-1}(v)_{\mathcal{B}_2} = P^{-1}P(v)_{\mathcal{B}_1} = (v)_{\mathcal{B}_1}$, as required. The theorem is proved. //

It is readily checked that $P^{-1} = [I]_{\mathcal{B}_2,\mathcal{B}_1}$. $[I]_{\mathcal{B}_2,\mathcal{B}_1}(v)_{\mathcal{B}_2} = (v)_{\mathcal{B}_1}$ and therefore $[I]_{\mathcal{B}_2,\mathcal{B}_1}P(v)_{\mathcal{B}_1} = (v)_{\mathcal{B}_1}$. Since $[I]_{\mathcal{B}_2,\mathcal{B}_1}PN = N$ for every $n \times 1$ matrix N (N can be taken to be the coordinate matrix of arbitrary vectors in V), we see that $[I]_{\mathcal{B}_2,\mathcal{B}_1}P = I_n$.

Definition 3.3.5 The matrix $P = [I]_{\mathcal{B}_1,\mathcal{B}_2}$ characterized in Theorem 3.3.4 will be called the *transition* or *coordinate-change matrix* from the basis $\mathcal{B}_1$ to the basis $\mathcal{B}_2$.

For example, consider the two ordered bases $\mathcal{B}_1 = \{(1,2),(1,-1)\}$ and $\mathcal{B}_2 = \{(0,1),(1,0)\}$ for $\mathbf{F}^2$. To compute the transition matrix between these bases, we must compute the matrix $[I]_{\mathcal{B}_1,\mathcal{B}_2}$. For this, we see that $(I(1,2))_{\mathcal{B}_2} = (1,2)_{\mathcal{B}_2} = (2,1)$ and likewise $(I(1,-1))_{\mathcal{B}_2} = (-1,1)$. (Note that coordinates in the $\mathcal{B}_2$ basis are the standard coordinates in reverse order.) This shows that the transition matrix is

$$P = [I]_{\mathcal{B}_1,\mathcal{B}_2} = \begin{pmatrix} 2 & -1 \\ 1 & 1 \end{pmatrix}$$

To see how this matrix is used, consider the vector $(4,5) \in \mathbf{F}^2$. We can compute that

$$\begin{pmatrix} 4 \\ 5 \end{pmatrix}_{\mathcal{B}_1} = \begin{pmatrix} 3 \\ 1 \end{pmatrix}$$

Multiplying these coordinates by P, we get

$$\begin{pmatrix} 2 & -1 \\ 1 & 1 \end{pmatrix}\begin{pmatrix} 3 \\ 1 \end{pmatrix} = \begin{pmatrix} 5 \\ 4 \end{pmatrix}$$

which gives the $\mathcal{B}_2$ coordinates of the original vector $(4,5)$.

We saw earlier, in Theorem 3.1.7, that functional composition of linear transformations on $\mathbf{F}^m$ corresponded to matrix multiplication. The next result gives this in the general setting.

Theorem 3.3.6 *Suppose that $S : V \to W$ and $T : U \to V$ are linear transformations (so $S \circ T : U \to W$ is linear). Suppose that B, C, D are ordered bases for U, V, W respectively. Then*

$$[S \circ T]_{B,D} = [S]_{C,D}[T]_{B,C}$$

Proof: Using the defining properties of $[T]_{B,C}$ and $[S]_{C,D}$, we compute for any $u \in U$

$$[S]_{C,D}[T]_{B,C}(u)_B = [S]_{C,D}(T(u))_C = (S(T(u)))_D = ((S \circ T)(u))_D$$

This shows that the product $[S]_{C,D}[T]_{B,C}$ satisfies the defining equation for $[S \circ T]_{B,D}$. The uniqueness assertion of Theorem 3.3.2 shows these matrices are equal, and this proves the theorem. //

Using the same ideas as in the proof of the previous theorem, we obtain the next result which shows how the representing matrix changes when coordinate systems are changed. The special case of this result where $T : V \to V$ will be exploited greatly in the next section, as well as in Chap. 6.

Theorem 3.3.7 *Suppose that $T : V \to W$ is a linear transformation, B_1, B_2 are two bases of V, and C_1, C_2 are two bases of W. Let P denote the transition matrix from B_1 to B_2, and let Q denote the transition matrix from C_1 to C_2. Then*

$$[T]_{B_2,C_2} = Q[T]_{B_1,C_1}P^{-1}$$

In the special case where $V = W$, $B_1 = C_1$, and $B_2 = C_2$, we obtain

$$[T]_{B_2,B_2} = P[T]_{B_1,B_1}P^{-1}$$

Proof: According to Theorems 3.3.2 and 3.3.4,

$$Q[T]_{B_1,C_1}P^{-1}(v)_{B_2} = Q[T]_{B_1,C_1}(v)_{B_1} = Q(T(v))_{C_1} = (T(v))_{C_2}$$

This shows that the matrix $Q[T]_{B_1,C_1}P^{-1}$ satisfies the defining property of $[T]_{B_2,C_2}$. The theorem follows from the uniqueness assertion of Theorem 3.3.2. The special case also follows, because then $P = Q$. //

The special case of Theorem 3.3.7 shows that if two $n \times n$ matrices A_1 and A_2 represent the same linear transformation $T : \mathbf{F}^n \to \mathbf{F}^n$

except with respect to different bases, then $A_1 = PA_2P^{-1}$ for some invertible $n \times n$ matrix P. This motivates the following definition.

Definition 3.3.8 Suppose that A and B are $n \times n$ matrices. If there exists an invertible matrix P such that $A = PBP^{-1}$, we say that A and B are *similar*.

Observe that if $A = PBP^{-1}$, then $B = P^{-1}AP$, so the definition of similarity is symmetric. This next corollary gives the converse of this special case of Theorem 3.3.7. Although slightly peculiar on a first reading, this corollary proves to be a convenient result.

Corollary 3.3.9 *Suppose that $T : V \to V$ is a linear transformation, where V is an n-dimensional vector space. Assume that $[T]_{B_1, B_1} = A$ for some ordered basis B_1 of V. Let P be an invertible $n \times n$ matrix. Then there is a basis B_2 of V for which $[T]_{B_2, B_2} = PAP^{-1}$.*

Proof: Suppose that $B_1 = \{v_1, v_2, \ldots, v_n\}$. Let w_i be the unique vector in V with B_1 coordinates in the ith column of P^{-1}. (In other words $P^{-1} = ((w_1)_{B_1} \quad \cdots \quad (w_n)_{B_1})$. If we set $B_2 = \{w_1, w_2, \ldots, w_n\}$, then it follows that for each basis element w_i of B_2, $P^{-1}(w_i)_{B_2} = (w_i)_{B_1}$. This shows that P^{-1} is the transition matrix from B_2 to B_1, and consequently, P is the transition matrix from B_1 to B_2. The corollary is now a consequence of Theorem 3.3.7. //

Theorem 3.2.4 shows that the rank of a linear transformation $T_A : \mathbf{F}^n \to \mathbf{F}^m$ is the same as the rank of the $m \times n$ matrix A which represents T_A. This is an extremely powerful computational tool, since it means that matrix algebra techniques (in particular gaussian elimination) can be applied in studying T_A. This remains true when representing a linear transformation by a matrix with arbitrary bases and is the subject of the next theorem.

Theorem 3.3.10 *Suppose that $T : V \to W$ is a linear transformation between finite-dimensional vector spaces. For* any *ordered basis B of V and* any *ordered basis C of W, $rk(T) = rk([T]_{B,C})$.*

Proof: If the rank of $[T]_{B,C}$ is r, then there are r columns of $[T]_{B,C}$ which are linearly independent and span $col([T]_{B,C})$ (the column space) by Theorem 1.4.6 (ii). By the definition of $[T]_{B,C}$ these columns are the C coordinates of r vectors $T(v_1),\ldots,T(v_r) \in W$. Moreover, by Theorem 1.5.5, as the coordinates of $T(v_1),\ldots,T(v_r)$ span the column space of $[T]_{B,C}$, it follows that $T(v_1),\ldots,T(v_r)$ must span $im(T)$. For the same reason, $T(v_1),\ldots,T(v_r)$ must be linearly independent. This shows that $dim(im(T)) = r$, that is, the rank of T is r. This proves the theorem. //

The following corollary is a useful application of Theorem 3.3.10.

Corollary 3.3.11 *If A and B are similar matrices, then $rk(A) = rk(B)$.*

Proof: Suppose that $B = PAP^{-1}$. Let $B = \{Pe_1, Pe_2, \ldots, Pe_n\}$ be the basis for $\mathbf{F}^n$ given by the columns of P. The proof of Theorem 3.3.4 shows that P is the transition matrix from the standard basis S for $\mathbf{F}^n$ to the basis B for $\mathbf{F}^n$. Applying Theorem 3.3.7, we find

$$B = PAP^{-1} = P[T_A]_{S,S}P^{-1} = [T_A]_{B,B}$$

Theorem 3.3.10 gives that $rk(A) = rk(T_A) = rk([T_A]_{B,B}) = rk(B)$, as required. //

Example 3.3.12 Consider $T : P^2 \to P^2$ defined by $T(a+bX+cX^2) = (a + b + c) + 2(b + c)X + 3cX^2$. Then, as $T(1) = 1$, $T(X) = 1 + 2X$, $T(X^2) = 1+2X+3X^2$, we find that with respect to the standard basis $S = \{1, X, X^2\}$

$$[T]_{S,S} = \begin{pmatrix} 1 & 1 & 1 \\ 0 & 2 & 2 \\ 0 & 0 & 3 \end{pmatrix}$$

Now consider the basis $C = \{1, 1 + X, 3 + 4X + 2X^2\}$ for P^2. The transition matrix from C to S is the matrix whose columns are the standard coordinates of the elements of C; that is, it is

$$P^{-1} = \begin{pmatrix} 1 & 1 & 3 \\ 0 & 1 & 4 \\ 0 & 0 & 2 \end{pmatrix}$$

The transition matrix from S to C is given by

$$P = \begin{pmatrix} 1 & -1 & 1/2 \\ 0 & 1 & -2 \\ 0 & 0 & 1/2 \end{pmatrix}$$

The special case of Theorem 3.3.7 now gives that

$$
\begin{aligned}
[T]_{C,C} &= P[T]_{S,S}\,P^{-1} \\
&= \begin{pmatrix} 1 & -1 & 1/2 \\ 0 & 1 & -2 \\ 0 & 0 & 1/2 \end{pmatrix} \begin{pmatrix} 1 & 1 & 1 \\ 0 & 2 & 2 \\ 0 & 0 & 3 \end{pmatrix} \begin{pmatrix} 1 & 1 & 3 \\ 0 & 1 & 4 \\ 0 & 0 & 2 \end{pmatrix} \\
&= \begin{pmatrix} 1 & 0 & 0 \\ 0 & 2 & 0 \\ 0 & 0 & 3 \end{pmatrix}
\end{aligned}
$$

Example 3.3.13 Let $T_A : \mathbf{R}^3 \to \mathbf{R}^4$, where (in the standard basis)

$$
A = \begin{pmatrix} 1 & 4 & 7 \\ 3 & 2 & 6 \\ 4 & 6 & 9 \\ 1 & 1 & 1 \end{pmatrix}
$$

Consider

$$
\mathcal{B} = \left\{ \begin{pmatrix} -1 \\ -1 \\ 1 \end{pmatrix}, \begin{pmatrix} 0 \\ 1 \\ 2 \end{pmatrix}, \begin{pmatrix} 1 \\ 1 \\ 1 \end{pmatrix} \right\}
$$

a basis for $\mathbf{R}^3$ and

$$
\mathcal{C} = \left\{ \begin{pmatrix} -1 \\ 0 \\ 1 \\ 0 \end{pmatrix}, \begin{pmatrix} -5 \\ 1 \\ 3 \\ 0 \end{pmatrix}, \begin{pmatrix} 7 \\ -1 \\ -4 \\ 0 \end{pmatrix}, \begin{pmatrix} 0 \\ 0 \\ 0 \\ 1 \end{pmatrix} \right\}
$$

a basis for $\mathbf{R}^4$. Then

$$
P^{-1} = \begin{pmatrix} -1 & 0 & 1 \\ -1 & 1 & 1 \\ 1 & 2 & 1 \end{pmatrix}
$$

is the transition matrix from $\mathcal{B}$ to the standard basis, and

$$
Q^{-1} = \begin{pmatrix} -1 & -5 & 7 & 0 \\ 0 & 1 & -1 & 0 \\ 1 & 3 & -4 & 0 \\ 0 & 0 & 0 & 1 \end{pmatrix}
$$

is the transition matrix from C to the standard basis. It follows from Theorem 3.3.7 that $[T_A]_{B,C} = QAP^{-1}$. Inverting Q^{-1}, one finds

$$Q = \begin{pmatrix} 1 & -1 & 2 & 0 \\ 1 & 3 & 1 & 0 \\ 1 & 2 & 1 & 0 \\ 0 & 0 & 0 & 1 \end{pmatrix}$$

Hence,

$$[T_A]_{B,C} = \begin{pmatrix} 1 & -1 & 2 & 0 \\ 1 & 3 & 1 & 0 \\ 1 & 2 & 1 & 0 \\ 0 & 0 & 0 & 1 \end{pmatrix} \begin{pmatrix} 1 & 4 & 7 \\ 3 & 2 & 6 \\ 4 & 6 & 9 \\ 1 & 1 & 1 \end{pmatrix} \begin{pmatrix} -1 & 0 & 1 \\ -1 & 1 & 1 \\ 1 & 2 & 1 \end{pmatrix}$$

$$= \begin{pmatrix} -1 & 52 & 39 \\ 4 & 84 & 64 \\ 3 & 70 & 53 \\ -1 & 3 & 3 \end{pmatrix}$$

PROBLEMS

3.3.1. For each of the following pairs of ordered bases, find the transition matrix from the basis B to the basis C.

(a) $B = \{(1,2),(2,1)\}$ and $C = \{(1,0),(0,1)\}$ for $\mathbf{R}^2$.

(b) $B = \{(1,0),(0,1)\}$ and $C = \{(2,5),(-1,-3)\}$ for $\mathbf{R}^2$.

(c) $B = \{(1,0,0),(0,1,0),(0,0,1)\}$ and $C = \{(0,-1,2),(1,0,3),(0,0,1)\}$ for $\mathbf{R}^3$.

(d) $B = \{(-3,-2,5),(5,5,-2),(-3,-1,2)\}$ and $C = \{(0,-1,1),(1,0,1),(-1,1,0)\}$ for $\mathbf{R}^3$.

3.3.2. For each of the following pairs of ordered bases, find the transition matrix from the basis B to the basis C.

(a) $B = \{(1,0),(0,1)\}$ and $C = \{(r,s),(t,u)\}$ for $\mathbf{R}^2$.

(b) $B = \{1, X, X^2\}$ and $C = \{1 - X, X^2, 1 + X\}$ for P^2.

(c) $B = \{1, X, X^2, X^3\}$ and $C = \{1, 1+X, 1+X+X^2, 1+X+X^2+X^3\}$ for P^3.

(d) $B = \{X^3, X^2, X, 1\}$ and $C = \{1, 1+X, 1+X+X^2, 1+X+X^2+X^3\}$ for P^3.

3.3.3. Show that $[T]_{B,B}$ is the zero matrix for every ordered basis B of V if and only if $T(v) = \vec{0}$ for all $v \in V$.

3.3.4. Let
$$B = \left\{ \begin{pmatrix} 0 \\ 1 \\ 1 \end{pmatrix}, \begin{pmatrix} 0 \\ 0 \\ 1 \end{pmatrix}, \begin{pmatrix} 1 \\ 0 \\ 1 \end{pmatrix} \right\}$$

be an ordered basis of $\mathbf{R}^3$, and let

$$C = \left\{ \begin{pmatrix} 1 \\ 1 \end{pmatrix}, \begin{pmatrix} 1 \\ -1 \end{pmatrix} \right\}$$

be an ordered basis for $\mathbf{R}^2$. Suppose that $T : \mathbf{R}^3 \to \mathbf{R}^2$ is given by

$$T \begin{pmatrix} x \\ y \\ z \end{pmatrix} = \begin{pmatrix} x - 3z \\ y \end{pmatrix}$$

Find:
(a) The standard matrix for T, that is, $[T]$
(b) The matrix $[T]_{B,C}$
(c) Invertible matrices P, Q such that $[T]_{B,C} = Q[T]P^{-1}$
Suppose that $U : \mathbf{R}^2 \to \mathbf{R}^3$ is given by

$$U \begin{pmatrix} x \\ y \end{pmatrix} = \begin{pmatrix} x \\ x - y \\ y \end{pmatrix}$$

Find:
(d) The standard matrix for U, that is, $[U]$
(e) The matrix $[U]_{C,B}$
(f) Invertible matrices P, Q such that $[U]_{C,B} = Q[U]P^{-1}$

3.3.5. Find the matrix representation, with respect to the standard bases, of the linear transformation $Int : P^3 \to P^4$ defined by $Int(f)(x) = \int_0^x f(t)\,dt$.

3.3.6. Find the matrix representation, with respect to your favorite basis, of the linear operator given by the transpose $Tps : M_{2\times2}(\mathbf{F}) \to M_{2\times2}(\mathbf{F})$.

3.3.7. Suppose that $B = \{v_1, v_2, v_3\}$ is an ordered basis of the vector space V. Assume that $T : V \to V$ is a linear transformation for which $T(v_1) = v_2 + v_3$, $T(v_2) = v_3$, and $T(v_3) = v_1 - v_2$.
(a) Find $[T]_{B,B}$.
(b) If B' is the ordered basis $B' = \{v_2, v_3 + v_1, v_1 - v_2\}$, find $[T]_{B',B'}$.

3.3.8. Suppose that V is a n-dimensional $\mathbf{F}$ vector space and W is an m-dimensional $\mathbf{F}$ vector space. Show that the vector space of linear transformations from V to W, $Hom(V, W)$, is isomorphic to $M_{n \times m}(\mathbf{F})$.

3.3.9.   View $\mathbf{C}$ as a real vector space with ordered basis $\{1, i\}$. Find the matrix for the linear operator given by complex conjugation with respect to this basis.

3.3.10.   In each case below, find a linear transformation T with the specified properties (in which case give its standard representing matrix) or else explain *why* no such T can exist.
(a) $T : \mathbf{F}^3 \rightarrow \mathbf{F}^2$ such that $T(1, -1, 0) = (1, 0)$, $T(0, 1, -1) = (0, 1)$, $T(1, 1, 1) = (0, 0)$
(b) $T : \mathbf{F}^4 \rightarrow \mathbf{F}^3$ with $im(T) = span\{(1, 0, 0), (1, 1, 0)\}$ and $ker(T) = span\{(1, 1, 0, 0)\}$
(c) $T : \mathbf{F}^3 \rightarrow \mathbf{F}$ with $T(1, -1, 0) = 1$, $T(0, 1, -1) = -1$, $T(-1, 0, 1) = 0$
(d) $T : \mathbf{F}^3 \rightarrow \mathbf{F}$ with $T(0, 1, -2) = 1$, $T(0, 1, -1) = -1$, $T(-1, 0, 1) = 0$
(e) $T : \mathbf{F}^3 \rightarrow \mathbf{F}^3$ with $im(T) = span\{(1, -1, 0), (0, 1, -1)\}$ and $ker(T) = span\{(-1, 0, 1)\}$
(f) $T : \mathbf{F}^3 \rightarrow \mathbf{F}^3$ with $T(1, 0, 0) = (-1, 0, 0)$ and $im(T) \subseteq ker(T)$
(g) $T : \mathbf{F}^3 \rightarrow \mathbf{F}^3$ with $T(1, 0, 0) = (0, 1, 0)$ and $im(T) \subseteq ker(T)$

3.3.11.   (a) If the standard matrix of $T : \mathbf{R}^3 \rightarrow \mathbf{R}^2$ is the matrix

$$A = \begin{pmatrix} 1 & 0 & -2 \\ 0 & 1 & 3 \end{pmatrix}$$

find the matrix for T with respect to the ordered bases

$$\mathcal{B}_1 = \left\{ \begin{pmatrix} 1 \\ 1 \\ 1 \end{pmatrix}, \begin{pmatrix} 0 \\ 1 \\ 1 \end{pmatrix}, \begin{pmatrix} 0 \\ 0 \\ 1 \end{pmatrix} \right\} \quad \text{and} \quad \mathcal{B}_2 = \left\{ \begin{pmatrix} 1 \\ 1 \end{pmatrix}, \begin{pmatrix} 0 \\ 1 \end{pmatrix} \right\}$$

(b) Now assume that the matrix of T with respect to $\mathcal{B}_1$ and $\mathcal{B}_2$ is given by A, and find the standard matrix for T.

3.3.12. Let $T : P^2 \rightarrow P^2$ be defined by $T(a + bX + cX^2) = (a + c) + bX^2$. If S is the standard basis of P^2 and $\mathcal{B} = \{1 + X, X, 1 + X^2\}$ is another ordered basis,
(a) Find $[T]_{S,S}$.
(b) Find $[T]_{\mathcal{B},\mathcal{B}}$.
(c) Find an invertible matrix C so that $C[T]_{\mathcal{B},\mathcal{B}} C^{-1} = [T]_{S,S}$.

3.3.13. Suppose that $T : V \to W$ is a linear transformation where $dim(V) = dim(W) = n$. Show that there exist ordered bases $\mathcal{B}$ of V and $\mathcal{C}$ of W such that $[T]_{\mathcal{B},\mathcal{C}}$ is a diagonal matrix.

3.3.14. Suppose that $\mathcal{B} = \{v_1, v_2, \ldots, v_n\}$ is an ordered basis of V.
(a) Define $T : V \to V$ by $T(v_1) = v_2$, $T(v_2) = v_3, \ldots, T(v_{n-1}) = v_n$, and $T(v_n) = \vec{0}$. Find $[T]_{\mathcal{B},\mathcal{B}}$.
(b) Define $T : V \to V$ by $T(v_1) = v_2$, $T(v_2) = v_3, \ldots, T(v_{n-1}) = v_n$, and $T(v_n) = a_1v_1 + a_2v_2 + \cdots + a_nv_n$. Find $[T]_{\mathcal{B},\mathcal{B}}$. When is this T one-one? When is this T onto?

3.4 Eigenvalues and Eigenvectors

Throughout the next two sections we will consider only linear transformations $T : V \to V$ where V is a vector space. Such linear transformations, which are functions from a vector space to itself, are called *linear operators*. Often, we shall simply call them *operators*. If V is finite-dimensional, for any ordered basis $\mathcal{B}$ of V we denote the matrix $[T]_{\mathcal{B},\mathcal{B}}$ by $[T]_{\mathcal{B}}$.

Linear operators are a particularly important type of linear transformation. They arise naturally in many situations. For example, the derivative $D : C^\infty(\mathbf{R}, \mathbf{R}) \to C^\infty(\mathbf{R}, \mathbf{R})$ defined by $D(f(x)) = f'(x)$ is a linear operator. Whenever $F : \mathbf{R}^n \to \mathbf{R}^n$ is a differentiable function, at any point $v \in \mathbf{R}^n$, the jacobian matrix $F'(v)$ defines a linear operator on $\mathbf{R}^n$ (see Digression 0.3.10 for the definition of F'). It turns out that the properties of this linear operator are closely linked to the behavior of F near the point v.

In this section we study eigenvalues and eigenvectors for linear operators. These are defined next. They play a crucial role in the determination of the behavior of linear operators.

Definition 3.4.1 Suppose that $T : V \to V$ is a linear operator. If $v \in V$ is a nonzero vector and there exists a scalar k such that $T(v) = kv$, we call v an *eigenvector* of T. The scalar k is called the *eigenvalue* of T associated with the eigenvector v. Often we find the terminology characteristic values and characteristic vectors for eigenvalues and eigenvectors respectively. If A is an $n \times n$ matrix, then by an eigenvalue (respectively eigenvector) of A we mean an eigenvalue (respectively eigenvector) of T_A.

As an example consider the matrix

$$A = \begin{pmatrix} 1 & 2 \\ 2 & 1 \end{pmatrix}$$

Since

$$T_A \begin{pmatrix} 1 \\ 1 \end{pmatrix} = \begin{pmatrix} 1 & 2 \\ 2 & 1 \end{pmatrix} \begin{pmatrix} 1 \\ 1 \end{pmatrix} = \begin{pmatrix} 3 \\ 3 \end{pmatrix}$$

and

$$T_A \begin{pmatrix} 1 \\ -1 \end{pmatrix} = \begin{pmatrix} 1 & 2 \\ 2 & 1 \end{pmatrix} \begin{pmatrix} 1 \\ -1 \end{pmatrix} = \begin{pmatrix} -1 \\ 1 \end{pmatrix}$$

we see that the linear operator $T_A : F^2 \to F^2$ has eigenvectors

$$\begin{pmatrix} 1 \\ 1 \end{pmatrix} \quad \text{and} \quad \begin{pmatrix} 1 \\ -1 \end{pmatrix}$$

with associated eigenvalues 3 and -1 respectively. If we consider the 3×3 matrix

$$B = \begin{pmatrix} 6 & -1 & 4 \\ 5 & 1 & -11/2 \\ 2 & 0 & -1 \end{pmatrix}$$

then we can check that

$$v = \begin{pmatrix} 3 \\ 4 \\ 2 \end{pmatrix} \in F^3$$

satisfies $T_B(v) = 2v$. Thus, v is an eigenvector for T_B with associated eigenvalue 2.

Elementary calculus also provides nice examples of eigenvectors with associated eigenvalues $k \in \mathbf{R}$. The function $e^{kx} \in C^\infty(\mathbf{R}, \mathbf{R})$ has derivative $(e^{kx})' = ke^{kx}$. This shows that e^{kx} is an eigenvector for the differentiation linear operator D with associated eigenvalue k. If P denotes the vector space of all real polynomials, it is readily checked that $T : P \to P$ defined by $T(p(X)) = Xp'(X)$ is a linear operator. Since $(X^n)' = nX^{n-1}$, it follows that $T(X^n) = nX^n$, so that X^n is an eigenvector of T with eigenvalue n.

We shall give shortly, in Lemma 3.4.5, a procedure for computing all the eigenvalues of a linear operator on a finite-dimensional vector space using an $n \times n$ matrix which represents the operator. In order to do this, we need to define addition and scalar multiplication for operators. Whenever T_1, $T_2 : V \to V$ are linear operators and a is a scalar, we can define new linear operators $T_1 + T_2$ and aT_1 by setting

$$(T_1 + T_2)(v) = T_1(v) + T_2(v) \quad \text{for all} \ \ v \in V$$

and

$$(aT_1)(v) = a[T_1(v)] \quad \text{for all} \ \ v \in V$$

We omit the easy verification that $T_1 + T_2$ and aT_1 are linear operators.

In the context of linear operators, we shall use I to denote the identity linear operator $I : V \to V$. [Here, $I(v) = v$ for all $v \in V$.] With this notation we have the following lemma, which gives an alternative characterization of eigenvectors.

Lemma 3.4.2 *Suppose that $T : V \to V$ is a linear operator. Then*
(i) v is an eigenvector of T with associated eigenvalue r if and only if $\vec{0} \neq v \in ker(rI - T)$.
(ii) If r is an eigenvalue of T, then the set of eigenvectors of T with associated eigenvalue r together with the vector $\vec{0}$ is a subspace of V.

Proof: (i) Observe that $(rI - T)(v) = \vec{0}$ if and only if $rI(v) - T(v) = rv - T(v) = \vec{0}$. This is the same as $T(v) = rv$. Hence, (i) follows.

(ii) This is a consequence of (i), since the kernel of any linear transformation is always a subspace. This proves the lemma. $//$

The fact that the collection of eigenvectors with a specified eigenvalue together with $\vec{0}$ is a subspace is important. For example, consider the linear operator $D^2 : \mathcal{C}^\infty(\mathbf{R}, \mathbf{R}) \to \mathcal{C}^\infty(\mathbf{R}, \mathbf{R})$ defined by $D^2 = D \circ D$, that is, $D^2(f(x)) = f''(x)$. One easily checks that $D^2(\sin kx) = -k^2 \sin kx$ and $D^2(\cos kx) = -k^2 \cos kx$. It now follows from Lemma 3.4.2 that every function of the form $A\sin kx + B\cos kx$, with $A, B \in \mathbf{R}$ (not both zero), is an eigenvector of D^2 with eigenvalue $-k^2$. Another way of viewing this is to see that each of these functions is a solution to the homogeneous linear differential equation $f''(x) + k^2 f(x) = 0$. [In fact, the proof of Lemma 3.4.2 would, in this case, say that these eigenvectors lie in the kernel of the linear operator which takes a function $f(x) \in \mathcal{C}^\infty(\mathbf{R}, \mathbf{R})$ to the function $f''(x) + k^2 f(x)$.] The ability to choose A and B arbitrarily is important in the theory of differential equations, as it means that this differential equation $f''(x) + f(x) = 0$ has solutions $f(x)$ which satisfy any pair of initial conditions of the form $f(p) = r$ and $f'(p) = s$ where $p, r, s \in \mathbf{R}$.

As a consequence of Lemma 3.4.2 (ii) it makes sense to give the following definition.

Definition 3.4.3 Let r be an eigenvalue of $T : V \to V$. The subspace of V of all eigenvectors with associated eigenvalue r together with the vector $\vec{0}$ is called the *eigenspace* associated with r. The eigenspace associated with r is denoted by E_r.

It must be emphasized that the sum of two eigenvectors with different eigenvalues is *not* an eigenvector. The sum of two eigenvectors is another eigenvector only in the case where they have the same eigenvalue. For example, consider the matrix

$$A = \begin{pmatrix} 1 & 0 & 0 \\ 1 & 2 & 0 \\ 0 & 0 & 1 \end{pmatrix}$$

One can see that the vectors

$$v_1 = \begin{pmatrix} 1 \\ -1 \\ 0 \end{pmatrix}, \quad v_2 = \begin{pmatrix} 0 \\ 0 \\ 0 \end{pmatrix}, \quad \text{and} \quad v_3 = \begin{pmatrix} 0 \\ 1 \\ 0 \end{pmatrix}$$

are eigenvectors for T_A. v_1 and v_2 have associated eigenvalue 1 and v_3 has associated eigenvalue 2. Note that the vector $v_1 + v_3$ is *not* an eigenvector for T_A since

$$T_A(v_1 + v_3) = \begin{pmatrix} 1 & 0 & 0 \\ 1 & 2 & 0 \\ 0 & 0 & 1 \end{pmatrix} \begin{pmatrix} 1 \\ 0 \\ 0 \end{pmatrix} = \begin{pmatrix} 1 \\ 1 \\ 0 \end{pmatrix}$$

which is not a scalar multiple of $v_1 + v_3$. Lemma 3.4.2 shows that the eigenspace E_1 contains the two-dimensional subspace $span\{v_1, v_2\}$ and the eigenspace E_2 contains the one-dimensional subspace $span\{v_3\}$. In fact, E_1 must be two-dimensional; if it were larger, it would be all of $\mathbf{F}^3$. This is impossible since not every vector is an eigenvector for T_A. In addition, E_2 can only be one-dimensional; if it were two-dimensional, it would have to have a nonzero common vector with E_1. This is impossible since an eigenvector cannot have two different eigenvalues. Thus $E_1 = span\{v_1, v_2\}$ and $E_2 = span\{v_3\}$.

In order to study more examples we must show how to determine the eigenvalues of a linear operator from a matrix representation (in the finite-dimensional case). We saw, in Lemma 3.4.2 (i) that a vector v was an eigenvector for T with eigenvalue r precisely if $v \in ker(rI - T)$. This shows that in the case where $T = T_A$ for some matrix A, the eigenvectors v for T_A satisfy $(rI_n - A)v = \vec{0}$. In order to find the possible values of r for which such a v can exist, we are motivated to give the next definition. In this definition we consider a matrix with polynomial entries and utilize its determinant. Strictly speaking, we defined determinants only for matrices with scalar entries. However, a polynomial $f(X)$ can be viewed as an element of the field $\mathbf{F}(X)$, which by definition consists of quotients $g(X)/h(X)$ of polynomials where $h(X) \neq 0$. Hence, polynomials can be considered as "scalars," where by "scalars" we now mean elements of the field $\mathbf{F}(X)$. Consequently, the theory of determinants developed in Chap. 2 can be applied to matrices with polynomial entries.

Definition 3.4.4 For any $n \times n$ matrix A we define the *characteristic polynomial* of A, $C_A(X)$, to be the polynomial (in the variable X) which is the determinant of the matrix $(XI_n - A)$; that is,

$$C_A(X) = det(XI_n - A)$$

For example, suppose

$$A = \begin{pmatrix} 1 & 3 \\ 2 & 4 \end{pmatrix}$$

Then the definition gives that

$$C_A(X) = det\left(X \begin{pmatrix} 1 & 0 \\ 0 & 1 \end{pmatrix} - \begin{pmatrix} 1 & 3 \\ 2 & 4 \end{pmatrix} \right) = det \begin{pmatrix} X - 1 & -3 \\ -2 & X - 4 \end{pmatrix}$$
$$= (X - 1)(X - 4) - 6 = X^2 - 5X - 2$$

Note that $C_A(X)$ is a polynomial of degree 2.

In general, if A is an $n \times n$ matrix, then the polynomial $C_A(X)$ is a polynomial of degree n. The reason for this is that the determinant of $(XI_n - A)$ is a sum of (signed) products of entries of this matrix, where each summand consists of a product of entries representing exactly one row and exactly one column. The only entries of $(XI_n - A)$ which contain the variable X occur on the diagonal (that is, are the iith entries). Thus, exactly one summand in the determinant expansion, namely the product of the diagonal entries, is a polynomial of degree n. All other summands are polynomials of degree less than $n-1$ since they must omit at least two diagonal entries. This shows that $C_A(X)$ is a polynomial of degree n. As the iith entry of $(XI_n - A)$ is $X - a_{ii}$, this also shows that the coefficient of X^n in $C_A(X)$ is 1. Thus, whenever A is an $n \times n$ matrix, $C_A(X)$ is a monic polynomial of degree n.

The characteristic polynomial is probably the most important algebraic invariant of a matrix. The next lemma shows that the roots of the characteristic polynomial of A are the eigenvalues of T_A.

Lemma 3.4.5 *Let V be an n-dimensional vector space. Suppose that $A = [T]_\mathcal{B}$, where $\mathcal{B}$ is an ordered basis for V and $T : V \to V$ is a linear operator.*
(i) *A scalar r is an eigenvalue of T if and only if $C_A(r) = 0$.*
(ii) *If r is an eigenvalue of T, the dimension of the eigenspace E_r of eigenvalue r is $null(rI_n - A) = n - rk(rI_n - A)$.*

Proof: (i) Suppose that r is an eigenvalue of T. Then there exists some nonzero vector $v \in V$ such that $T(v) = rv$. From this we see that $(T(v))_\mathcal{B} = r(v)_\mathcal{B}$. In particular, $\vec{0} = r(v)_\mathcal{B} - A(v)_\mathcal{B} = rI_n(v)_\mathcal{B} - A(v)_\mathcal{B} = (rI_n - A)(v)_\mathcal{B}$. However, $v \neq \vec{0}$, so $(v)_\mathcal{B} \neq \vec{0}$, and as $(rI_n - A)(v)_\mathcal{B} = \vec{0}$, we find that $rk(rI_n - A) < n$. Hence, by Theorem 0.6.9 (iii) $(rI_n - A)$ is not invertible. It follows from Theorem 2.1.12 that $0 = det(rI_n - A) = C_A(r)$. Conversely, in case $C_A(r) = 0$, we find that $det(rI_n - A) = 0$. Hence $(rI_n - A)$ is not invertible; that is, $rk(rI_n - A) < n$. From this, by Corollary 0.5.8 (i) there exists some nonzero solution w to the equation $(rI_n - A)w = \vec{0}$. Such a solution w satisfies $Aw = rw$; that is, w is the coordinates of a (nonzero) eigenvector of T_A with associated eigenvalue r. This proves (i).

(ii) We note that by the observations in the proof of part (i) $v \in E_r$ if and only if $(rI_n - A)(v)_\mathcal{B} = \vec{0}$. It now follows from Theorem 1.5.5 that any basis for $ker(rI_n - A)$ will be the coordinates of the vectors in a basis for E_r. Consequently, $dim(E_r) = dim(ker(rI_n - A)) = null(rI_n - A)$, as required. //

We easily obtain the following consequence of Lemma 3.4.5.

Corollary 3.4.6 *If V is an n-dimensional vector space, then $T : V \to V$ has at most n distinct eigenvalues.*

Proof: Note that $A = [T]_B$ is an $n \times n$ matrix for any ordered basis B of V. As observed above, the degree of the polynomial $C_A(X)$ is n. The result now follows from Lemma 3.4.5, since $C_A(X)$ can have at most n distinct roots. //

Examples 3.4.7 (i) Consider

$$A = \begin{pmatrix} 2 & 3 \\ 3 & 2 \end{pmatrix}$$

We compute that

$$C_A(X) = \det \begin{pmatrix} X - 2 & -3 \\ -3 & X - 2 \end{pmatrix} = (X - 2)(X - 2) - 9$$
$$= X^2 - 4X - 5 = (X - 5)(X + 1)$$

which has the two roots -1 and 5. Hence, $T_A : \mathbf{R}^2 \to \mathbf{R}^2$ has the two eigenvalues -1 and 5. To find the associated eigenspaces, we must compute the nullspaces of $-I - A$ and $5I - A$. The eigenspace associated with -1 is

$$E_{-1} = \ker(-I - A) = \ker \begin{pmatrix} -3 & -3 \\ -3 & -3 \end{pmatrix} = span \left\{ \begin{pmatrix} -1 \\ 1 \end{pmatrix} \right\}$$

and the eigenspace associated with 5 is

$$E_5 = \ker(5I - A) = \ker \begin{pmatrix} 3 & -3 \\ -2 & 2 \end{pmatrix} = span \left\{ \begin{pmatrix} 1 \\ 1 \end{pmatrix} \right\}$$

(ii) Suppose

$$B = \begin{pmatrix} 3 & 2 & 2 \\ 1 & 2 & 2 \\ -1 & -1 & 0 \end{pmatrix}$$

Then the characteristic polynomial

$$C_B(X) = \det \begin{pmatrix} X - 3 & -2 & -2 \\ -1 & X - 2 & -2 \\ 1 & 1 & X \end{pmatrix}$$
$$= (X-3)[(X-2)(X) + 2] + 2[(-X) + 2] - 2[-1 - (X-2)]$$
$$= X^3 - 5X^2 - 8X + 4$$
$$= (X - 1)(X - 2)^2$$

It follows that the eigenvalues of T_B are 1 and 2. The eigenspace associated with 1 is

$$
ker \begin{pmatrix} -2 & -2 & -2 \\ -1 & -1 & -2 \\ 1 & 1 & 1 \end{pmatrix} = ker \begin{pmatrix} 1 & 1 & 0 \\ 0 & 0 & 1 \\ 0 & 0 & 0 \end{pmatrix} = span \left\{ \begin{pmatrix} -1 \\ 1 \\ 0 \end{pmatrix} \right\}
$$

and the eigenspace associated with 2 is

$$
ker \begin{pmatrix} -1 & -2 & -2 \\ -1 & 0 & -2 \\ 1 & 1 & 2 \end{pmatrix} = ker \begin{pmatrix} 1 & 0 & 2 \\ 0 & 1 & 0 \\ 0 & 0 & 0 \end{pmatrix} = span \left\{ \begin{pmatrix} 2 \\ 0 \\ -1 \end{pmatrix} \right\}
$$

(iii) It is occasionally easier to use row operations when computing the determinant which gives the characteristic polynomial. Consider

$$
D = \begin{pmatrix} 2 & 0 & 1 \\ 1 & 1 & 0 \\ 1 & 0 & 1 \end{pmatrix}
$$

Then we find that

$$
\begin{aligned}
C_D(X) &= det \begin{pmatrix} X-2 & 0 & -1 \\ -1 & X-1 & 0 \\ -1 & 0 & X-1 \end{pmatrix} \\
&= \frac{1}{X-1} det \begin{pmatrix} (X-2)(X-1) & 0 & -(X-1) \\ -1 & X-1 & 0 \\ -1 & 0 & X-1 \end{pmatrix} \\
&= \frac{1}{X-1} det \begin{pmatrix} (X-2)(X-1)-1 & 0 & 0 \\ -1 & X-1 & 0 \\ -1 & 0 & X-1 \end{pmatrix} \\
&= \frac{1}{X-1}[(X-2)(X-1)-1](X-1)^2 \\
&= [(X-2)(X-1)-1](X-1) \\
&= (X^2-3X+1)(X-1)
\end{aligned}
$$

Since the roots of X^2-3X+1 are $(1/2)(3\pm\sqrt{9-4})$, it follows that the matrix D has the three eigenvalues 1, $(1/2)(3+\sqrt5)$, and $(1/2)(3-\sqrt5)$.

The preceding lemma shows that if $T : V \to V$ is a linear operator and B is an ordered basis for V, then the roots of the characteristic polynomial of $[T]_\mathcal{B}$ are exactly the eigenvalues of T. Thus, no matter which ordered basis $\mathcal{B}$ is chosen, the roots of the characteristic polynomial of $[T]_\mathcal{B}$ remain unchanged. In fact, more is true. The characteristic polynomial itself remains unchanged. In other words, if $\mathcal{B}$ and $\mathcal{C}$ are two ordered bases for a finite-dimensional vector space V and $T : V \to V$ is a linear operator, then the characteristic polynomials of $[T]_\mathcal{B}$ and $[T]_\mathcal{C}$ are the same. This is proved next. A consequence of this result is that the characteristic polynomials make sense for linear operators on any finite-dimensional vector space. One can define the characteristic polynomial of a linear operator to be the characteristic polynomial of any matrix that represents the operator.

Lemma 3.4.8 *Suppose that V is a finite-dimensional vector space and $T : V \to V$ is a linear operator. Let $\mathcal{B}_1$ and $\mathcal{B}_2$ be ordered bases for V. If $A_1 = [T]_{\mathcal{B}_1}$ and $A_2 = [T]_{\mathcal{B}_2}$, then $C_{A_1}(X) = C_{A_2}(X)$.*

Proof: As noted earlier, the results of Chap. 2 apply to determinants of $n \times n$ matrices where entries are polynomials. We know from Theorem 3.3.7 that there exists an invertible matrix P for which $PA_2P^{-1} = A_1$. We thus can compute that

$$
\begin{aligned}
C_{A_1}(X) &= det(XI - A_1) = det(XPIP^{-1} - PA_2P^{-1}) \\
&= det(PXIP^{-1} - PA_2P^{-1}) = det(P(XIP^{-1} - A_2P^{-1})) \\
&= det(P(XI - A_2)P^{-1}) = det(P)det(XI - A_2)det(P^{-1}) \\
&= det(XI - A_2) = C_{A_2}(X)
\end{aligned}
$$

This proves the lemma. $//$

In view of the lemma we now can define the characteristic polynomial of a linear operator.

Definition 3.4.9 If $T : V \to V$ is a linear operator on a finite-dimensional vector space V, then $C_T(X)$, the *characteristic polynomial of T*, is defined to be $C_A(X)$ where $A = [T]_\mathcal{B}$ and $\mathcal{B}$ is some ordered basis for V.

Examples 3.4.10 (i) Let P^n be the real $(n+1)$-dimensional vector space of all polynomials of degree at most n. We consider $D : P^n \to P^n$, the differentiation operator. We let $\mathcal{S} = \{1, X, X^2, \ldots, X^n\}$ be the standard basis of P^n. It is easy to see [using the fact that $(X^n)' = nX^{n-1}$] that

$$[D]_S = \begin{pmatrix} 0 & 1 & 0 & 0 & \cdots & 0 \\ 0 & 0 & 2 & 0 & \cdots & 0 \\ 0 & 0 & 0 & 3 & \cdots & 0 \\ \vdots & \vdots & \vdots & \vdots & \ddots & \vdots \\ 0 & 0 & 0 & 0 & \cdots & n \\ 0 & 0 & 0 & 0 & \cdots & 0 \end{pmatrix}$$

Consequently, the characteristic polynomial of D is X^{n+1}. Observe that the rank of $D - 0I_{n+1} = D$ is n. Thus, according to Lemma 3.4.5 (ii), the dimension of E_0 is 1, so in particular $E_0 = span\{1\}$. We obtain (via eigenspace theory) the well-known fact that the derivative of a polynomial is 0 if and only if the polynomial is a constant.

(ii) Recall the linear operator $T : P^n \rightarrow P^n$ considered earlier in this section and defined by $T(p(X)) = Xp'(X)$. We have that

$$[T]_S = \begin{pmatrix} 0 & 0 & 0 & \cdots & 0 & 0 \\ 0 & 1 & 0 & \cdots & 0 & 0 \\ 0 & 0 & 2 & \cdots & 0 & 0 \\ \vdots & \vdots & \vdots & \ddots & \vdots & \vdots \\ 0 & 0 & 0 & \cdots & n-1 & 0 \\ 0 & 0 & 0 & \cdots & 0 & n \end{pmatrix}$$

Consequently we obtain that $C_T(X) = X(X-1)(X-2)\cdots(X-n)$. This confirms what we observed earlier when we saw that each of the natural numbers $0, 1, 2, \ldots, n$ was an eigenvalue for T.

PROBLEMS

3.4.1. Find the characteristic polynomial, the eigenvalues, and the eigenspaces of each of the following matrices.

(a) $\begin{pmatrix} 4 & 3 \\ 4 & 0 \end{pmatrix}$ (b) $\begin{pmatrix} 0 & 1 \\ 1 & 0 \end{pmatrix}$ (c) $\begin{pmatrix} 0 & 1 \\ -1 & 0 \end{pmatrix}$ (d) $\begin{pmatrix} 0 & 1 & 0 \\ 0 & 0 & 1 \\ 2 & 1 & -2 \end{pmatrix}$

3.4.2. Find the characteristic polynomial, the eigenvalues, and the eigenspaces of each of the following matrices.

(a) $\begin{pmatrix} 0 & 8 & 1 \\ 1 & 0 & 0 \\ 0 & 4 & -1 \end{pmatrix}$ (b) $\begin{pmatrix} 0 & 0 & 0 & 1 \\ 1 & 0 & 0 & 0 \\ 0 & 1 & 0 & 0 \\ 0 & 0 & 1 & 0 \end{pmatrix}$ (c) $\begin{pmatrix} 0 & 0 & -2 \\ 1 & 0 & 3 \\ 0 & 1 & 0 \end{pmatrix}$

(d) $\begin{pmatrix} 0 & 0 & 1 \\ 1 & 0 & -3 \\ 0 & 1 & 3 \end{pmatrix}$ (e) $\begin{pmatrix} 1 & 0 & 0 & 0 \\ 0 & 0 & 0 & 1 \\ 0 & 1 & 0 & -3 \\ 0 & 0 & 1 & 3 \end{pmatrix}$

3.4.3. Find a 2×2 real matrix with eigenvalues 2 and 1 and with

$$E_2 = span \left\{ \begin{pmatrix} 1 \\ 1 \end{pmatrix} \right\} \quad \text{and} \quad E_1 = span \left\{ \begin{pmatrix} 2 \\ 1 \end{pmatrix} \right\}$$

3.4.4. Prove that an $n \times n$ matrix is singular if and only if 0 is an eigenvalue. Deduce that an $n \times n$ matrix A is nonsingular if and only if $C_A(X)$ has a nonzero constant term.

3.4.5. Recall that a square matrix N is called *nilpotent* if N^k is the zero matrix for some positive integer k. Show that if N is nilpotent, then 0 is the only eigenvalue of N.

3.4.6. (a) Show that if k is an eigenvalue of the matrix A, then k^2 is an eigenvalue of the matrix A^2.
(b) Show that if k is an eigenvalue of an invertible matrix A, then k^{-1} is an eigenvalue of A^{-1}.

3.4.7. Suppose that A is an $n \times n$ matrix with n distinct eigenvalues $c_1, c_2, \ldots, c_n$.
(a) Show that $det(A) = c_1 c_2 \cdots c_n$.
(b) The *trace* of A, denoted by $tr(A)$, is defined to be the sum of the diagonal entries of A. Show that $tr(A) = c_1 + c_2 + \cdots + c_n$.

3.4.8. Suppose that A is a real 3×3 matrix. Show that there is some real number c with $rk(A - cI) \leq 2$.

3.4.9. Show that for any square matrix A, A and A^t have the same characteristic polynomial.

3.4.10. (a) Let $T : P^2 \to P^2$ be defined by $T(f(X)) = f(X) + f'(X)$ for all real polynomials. Find all eigenvalues of T and all their eigenspaces.
(b) Let $T : P^2 \to P^2$ be defined by $T(f(X)) = f(X) + Xf'(X)$ for all real polnomials. Find all eigenvalues of T and all their eigenspaces.

3.4.11. Suppose that A is an $n \times n$ matrix and the sum of the entries of each row is the same number r. Show that r is an eigenvalue of A.

3.4.12. Let $Tps : M_{n \times n}(\mathbf{F}) \to M_{n \times n}(\mathbf{F})$ be the linear operator defined by transposition; that is, $Tps(A) = A^t$ for all $A \in M_{n \times n}(\mathbf{F})$.

(a) Show that ± 1 are the only eigenvalues of Tps.

(b) Determine the eigenspaces of Tps.

3.4.13. Suppose that A is an $n \times n$ matrix with characteristic polyomial $X^n + a_{n-1}X^{n-1} + \cdots + a_1 X + a_0$. Show that $det(A) = (-1)^n a_0$. Deduce that for any linear operator $T : V \rightarrow V$ on a finite-dimensional vector space V, T is invertible if and only if $C_T(0) \neq 0$.

3.4.14. Suppose that A is an $n \times n$ matrix with eigenvalue c. Show that for any polynomial $f(X)$, $f(c)$ is an eigenvalue of $f(A)$.

3.5   Eigenspaces and Diagonalizability

Diagonal matrices have many nice properties. For example, if the diagonal entries of a diagonal matrix D are $d_1, d_2, \ldots, d_n$, then the characteristic polynomial of D is $(X - d_1)(X - d_2) \cdots (X - d_n)$. In particular, the eigenvalues of T_D are the $d_1, d_2, \ldots, d_n$. Moreover, an eigenvector with eigenvalue d_i is the ith basis element e_i. If a diagonal matrix $D = [T]_B$ is the matrix of a linear operator T in some ordered basis B, then these same observations apply, except that now an eigenvector for d_i is the ith basis vector in B. In order to take advantage of these observations and apply them in operator theory, we give the following definition.

Definition 3.5.1 Suppose that V is a vector space over the scalar field $\mathbf{F}$. A linear operator $T : V \rightarrow V$ is called *diagonalizable* over $\mathbf{F}$ if there exists an ordered basis B of V for which $[T]_B$ is a diagonal matrix. A matrix A will be called diagonalizable over $\mathbf{F}$ if A is similar (over $\mathbf{F}$) to a diagonal matrix. Whenever the field $\mathbf{F}$ is understood from context, we shall delete the phrase "over $\mathbf{F}$."

It is very important to note that in the above definition the concept of diagonalizability depends upon the field $\mathbf{F}$. For example the matrix

$$G = \begin{pmatrix} 0 & 1 \\ -1 & 0 \end{pmatrix}$$

is *not* diagonalizable over the field of real numbers, since the characteristic polynomial of G is $C_G(X) = X^2 + 1$, which does not have any real roots. In other words, G is not diagonalizable over the reals because G does not have any real eigenvalues. However, G is diagonalizable when viewed as a complex matrix, because over the complex numbers $\mathbf{C}$ we have $C_G(X) = (X + i)(X - i)$. We see that

$$G\begin{pmatrix} 1 \\ i \end{pmatrix} = \begin{pmatrix} i \\ -1 \end{pmatrix} = i \begin{pmatrix} 1 \\ i \end{pmatrix} \quad \text{and} \quad G\begin{pmatrix} 1 \\ -i \end{pmatrix} = \begin{pmatrix} -i \\ -1 \end{pmatrix} = -i \begin{pmatrix} 1 \\ -i \end{pmatrix}$$

This shows that in the ordered basis

$$\mathcal{B} = \left\{ \begin{pmatrix} 1 \\ i \end{pmatrix}, \begin{pmatrix} 1 \\ -i \end{pmatrix} \right\}$$

we have

$$[T_G]_\mathcal{B} = \begin{pmatrix} i & 0 \\ 0 & -i \end{pmatrix}$$

As another example consider the matrix

$$D = \begin{pmatrix} 1 & 0 & 0 \\ 1 & 0 & -2 \\ -1 & 1 & -3 \end{pmatrix}$$

Consider the basis $\mathcal{B} = \{v_1, v_2, v_3\}$ for $\mathbf{F}^3$ where

$$v_1 = \begin{pmatrix} 1 \\ 1 \\ 0 \end{pmatrix}, \quad v_2 = \begin{pmatrix} 0 \\ 2 \\ 1 \end{pmatrix} \quad \text{and} \quad v_3 = \begin{pmatrix} 0 \\ 1 \\ 1 \end{pmatrix}$$

Then we readily compute that $Dv_1 = v_1$, $Dv_2 = -v_2$ and $Dv_3 = -2v_3$. This shows that T_D is diagonalizable. In fact applying Theorem 3.3.7, we obtain that

$$\begin{aligned}
[T_D]_\mathcal{B} &= \begin{pmatrix} 1 & 0 & 0 \\ -1 & 1 & -1 \\ 1 & -1 & 2 \end{pmatrix} \begin{pmatrix} 1 & 0 & 0 \\ 1 & 0 & -2 \\ -1 & 1 & -3 \end{pmatrix} \begin{pmatrix} 1 & 0 & 0 \\ 1 & 2 & 1 \\ 0 & 1 & 1 \end{pmatrix} \\
&= \begin{pmatrix} 1 & 0 & 0 \\ 0 & -1 & 0 \\ 0 & 0 & -2 \end{pmatrix}
\end{aligned}$$

Our next task is to investigate the different eigenspaces of a linear operator and see how they are related. This is crucial to the study of when a linear operator is diagonalizable. If r and s are different eigenvalues for an operator T, the intersection of the two eigenspaces $E_r \cap E_s = \{\vec{0}\}$, because any nonzero eigenvector can have only one associated eigenvalue. However, more is true. In this next result we show that the eigenspaces of a linear operator are independent.

Lemma 3.5.2 *Suppose that $v_1, v_2, \ldots, v_n$ are eigenvectors of a linear operator $T : V \to V$ each corresponding to different eigenvalues. Then $v_1, v_2, \ldots, v_n$ are linearly independent.*

Proof: Assume the contrary. Denote by k_i the eigenvalue associated with v_i. Suppose that

$$a_1 v_1 + a_2 v_2 + \cdots + a_n v_n = \vec{0} \qquad (*)$$

where not all $a_i = 0$. Assume further that this expression is chosen to have a *minimal* number of nonzero coefficients a_i among the collection of all such expressions. Since each v_i is a nonzero vector, we see that at least two a_i must be nonzero. In particular, since the eigenvalues $k_1, k_2, \ldots, k_n$ are distinct, there is some i with $a_i \neq 0$ and $k_i \neq a0$. Without loss of generality, we can assume that $i = 1$, that is, both $a_1 \neq 0$ and $k_1 \neq 0$. We find that $\vec{0} = T(\vec{0}) = T(a_1 v_1 + a_2 v_2 + \cdots + a_n v_n) = a_1 T(v_1) + a_2 T(v_2) + \cdots + a_n T(v_n) = a_1 k_1 v_1 + a_2 k_2 v_2 + \cdots + a_n k_n v_n$. Since $k_1 \neq 0$, this gives

$$a_1 v_1 + a_2 (k_2/k_1) v_2 + \cdots + a_n (k_n/k_1) v_n = \vec{0} \qquad (**)$$

Subtracting $(**)$ from our original expression $(*)$, we find that

$$0 v_1 + a_2 (1 - k_2/k_1) v_2 + \cdots + a_n (1 - k_n/k_1) v_n = \vec{0} \qquad (***)$$

Note that as $k_1, k_2, \ldots, k_n$ are all distinct, each of $1 - k_2/k_1$, $1 - k_3/k_1, \ldots, 1 - k_n/k_1$ are nonzero. Hence $(***)$ is a nontrivial linear combination of $v_1, v_2, \ldots, v_n$ that is $\vec{0}$. But $(***)$ has fewer nonzero coefficients than does $(*)$, since v_1 has 0 as a coefficient. This contradicts our original choice of $(*)$. It follows that $v_1, v_2, \ldots, v_n$ must be linearly independent, and this proves the lemma. $//$

We next give another proof of Lemma 3.5.2. This proof is included because the ideas it contains are generalized and exploited in Chap. 6. This proof also illustrates the change in point of view alluded to in the beginning of this chapter. In this proof we study different operators, rather than computing with specific vectors.

Proof: We again denote by k_i the eigenvalue of v_i. For each j with $1 \leq j \leq n$ we consider the operator

$$U_j = (T - k_1 I) \circ (T - k_2 I) \circ \cdots \circ (T - \widehat{k_j I}) \circ \cdots \circ (T - k_n I)$$

where the "hat" means delete. Observe that as $(T - k_s I)(v_i) = k_i v_i - k_s v_i = (k_i - k_s)v_i$, each v_i is an eigenvector of the operator $T - k_s I$ with eigenvalue $(k_i - k_s)$. Since each U_j is a composition of $n - 1$ of these operators, we see that $U_j(v_i) = \vec{0}$ whenever $i \neq j$ and

$$U_j(v_j) = \left(\prod_{j=1,\ldots,n; j \neq i} (k_i - k_j) \right) v_j$$

In particular, $U_j(v_j) \neq \vec{0}$ as the eigenvalues $k_1, k_2, \ldots, k_n$ are all distinct.

Suppose that $a_1 v_1 + \cdots + a_n v_n = \vec{0}$, where $a_1, a_2, \ldots, a_n \in \mathbf{F}$. We apply the operator U_j to this expression, and by what we saw above,

$$\vec{0} = U_j(a_1 v_1 + \cdots + a_n v_n) = a_j U_j(v_j)$$

As $U_j(v_j) \neq \vec{0}$, we conclude that $a_j = 0$. This proves the linear independence of $v_1, v_2, \ldots, v_n$. //

As an example, consider $e^x, e^{2x}, e^{3x}, \ldots, e^{nx} \in \mathcal{C}^\infty(\mathbf{R}, \mathbf{R})$. Each of these vectors is an eigenvector for the differentiation operator $D : \mathcal{C}^\infty(\mathbf{R}, \mathbf{R}) \rightarrow \mathcal{C}^\infty(\mathbf{R}, \mathbf{R})$. Moreover, their eigenvalues are all distinct. Consequently, Lemma 3.5.2 shows that $e^x, e^{2x}, e^{3x}, \ldots, e^{nx}$ are linearly independent.

In the following theorem we show how to interpret diagonalizability in terms of eigenvectors. It shows that for a matrix A, the operator T_A is diagonalizable if and only if there is a basis $\mathcal{B}$ for $\mathbf{F}^n$ consisting of eigenvectors for T_A. Such a basis $\mathcal{B}$ is often called an *eigenbasis* for T_A.

Theorem 3.5.3 *Suppose $T : V \rightarrow V$ is a linear operator where V is finite-dimensional. Then the following are equivalent.*
(i)   *T is diagonalizable.*
(ii)   *There is an ordered basis $\mathcal{B}$ for V which consists entirely of eigenvectors of T.*
(iii) *For any ordered basis $\mathcal{C}$ of V, $[T]_\mathcal{C}$ is diagonalizable.*

Proof: We first show for any ordered basis $\mathcal{B}$ of V that $[T]_\mathcal{B}$ is diagonal if and only if $\mathcal{B}$ is a basis of eigenvectors for T. This will establish the equivalence of (i) and (ii). Suppose $\mathcal{B} = \{v_1, v_2, \ldots, v_n\}$ and $T(v_i) =$

$k_i v_i$ for each $i = 1, 2, \ldots, n$. Since $T(v_i) = 0v_1 + \cdots + k_i v_i + \cdots + 0v_n$ we find that

$$(T(v_i))_\mathcal{B} = \begin{pmatrix} 0 \\ \vdots \\ 0 \\ k_i \\ 0 \\ \vdots \\ 0 \end{pmatrix}$$

with at most one nonzero entry. As the above column is the ith column of the matrix $[T]_\mathcal{B}$, this matrix must be diagonal. Conversely, whenever $[T]_\mathcal{B}$ is diagonal, since the ith column of $[T]_\mathcal{B}$ has the above form, we see that $T(v_i) = k_i v_i$ for some scalar k_i. This establishes the equivalence of (i) and (ii).

Suppose that $\mathcal{B}$ is an ordered basis of eigenvectors of T and $\mathcal{C}$ is an ordered basis of V. Let P denote the transition matrix from the basis $\mathcal{C}$ to the basis $\mathcal{B}$. Then according to Theorem 3.3.7, we have that

$$P[T]_\mathcal{C} P^{-1} = [T]_\mathcal{B}$$

According to the above, $[T]_\mathcal{B}$ is diagonal. We have shown that condition (iii) is a consequence of (ii). Conversely, assuming condition (iii), if $\mathcal{C}$ is an ordered basis of V we know that there is an invertible matrix P for which $P[T]_\mathcal{C} P^{-1}$ is diagonal. By Corollary 3.3.9 there is a basis $\mathcal{B}$ for which $[T]_\mathcal{B}$ is this diagonal matrix. This shows that T is diagonalizable. The theorem is proved. //

Examples 3.5.4 (i) The matrix

$$A = \begin{pmatrix} 17 & -30 \\ 9 & -16 \end{pmatrix}$$

is diagonalizable. We compute that A has characteristic polynomial $C_A(X) = (X - 2)(X + 1)$ and eigenvectors

$$\begin{pmatrix} 2 \\ 1 \end{pmatrix} \quad \text{and} \quad \begin{pmatrix} 5 \\ 3 \end{pmatrix}$$

of eigenvalue 2 and -1 respectively. It follows that

$$P^{-1}AP = \begin{pmatrix} 2 & 0 \\ 0 & -1 \end{pmatrix} \quad \text{where} \quad P = \begin{pmatrix} 2 & 5 \\ 1 & 3 \end{pmatrix}$$

We use this to compute A^4. For any square matrix A and any invertible matrix P,

$$(PAP^{-1})^2 \; = \; (PAP^{-1})(PAP^{-1}) \; = \; PA(P^{-1}P)AP^{-1}$$
$$= \; PAAP^{-1} \; = \; PA^2P^{-1}$$

The same calculation shows that $(PAP^{-1})^n = PA^nP^{-1}$ for all $n \geq 1$. We find for the above A

$$A^4 \; = \; \left(P \begin{pmatrix} 2 & 0 \\ 0 & -1 \end{pmatrix} P^{-1} \right)^4 = P \begin{pmatrix} 2 & 0 \\ 0 & -1 \end{pmatrix}^4 P^{-1}$$

$$= \; \begin{pmatrix} 2 & 5 \\ 1 & 3 \end{pmatrix} \begin{pmatrix} 16 & 0 \\ 0 & 1 \end{pmatrix} \begin{pmatrix} 3 & -5 \\ -1 & 2 \end{pmatrix}$$

$$= \; \begin{pmatrix} 32 & 5 \\ 16 & 3 \end{pmatrix} \begin{pmatrix} 3 & -5 \\ -1 & 2 \end{pmatrix} = \begin{pmatrix} 91 & -150 \\ 45 & -74 \end{pmatrix}$$

This process of computing matrix powers is quite useful in the study of systems of linear differential equations.

(ii) Consider the matrix

$$M = \begin{pmatrix} 1 & 1 \\ 0 & 1 \end{pmatrix}$$

Then $C_M(X) = (X - 1)^2$. Therefore the eigenspace

$$E_1 = span \left\{ \begin{pmatrix} 1 \\ 0 \end{pmatrix} \right\}$$

is one-dimensional. Consequently, there is no ordered basis of eigenvectors for M. This shows that M is not diagonalizable over any field.

(iii) The matrix

$$A = \begin{pmatrix} 2 & 3 \\ 3 & 2 \end{pmatrix}$$

of Example 3.4.7 is diagonalizable by Theorem 3.5.3. A basis of eigenvectors of T_A is

$$B = \left\{ \begin{pmatrix} -1 \\ 1 \end{pmatrix}, \begin{pmatrix} 1 \\ 1 \end{pmatrix} \right\}$$

It follows that if

$$P = \begin{pmatrix} -1 & 1 \\ 1 & 1 \end{pmatrix}$$

then since P^{-1} is the basis change matrix from the standard basis to the $\mathcal{B}$ basis,

$$P^{-1}AP = \begin{pmatrix} -1/2 & 1/2 \\ 1/2 & 1/2 \end{pmatrix}\begin{pmatrix} 2 & 3 \\ 3 & 2 \end{pmatrix}\begin{pmatrix} -1 & 1 \\ 1 & 1 \end{pmatrix} = \begin{pmatrix} -1 & 0 \\ 0 & 5 \end{pmatrix}$$

(iv) Consider the 3×3 matrix

$$H = \begin{pmatrix} 1 & 3 & 0 \\ 0 & -2 & 0 \\ 0 & 6 & 1 \end{pmatrix}$$

H has characteristic polynomial $C_H(X) = (X-1)^2(X+2)$. The eigenspaces E_1 and E_{-2} have bases

$$\mathcal{B}_1 = \left\{\begin{pmatrix} 1 \\ 0 \\ 0 \end{pmatrix}, \begin{pmatrix} 0 \\ 0 \\ 1 \end{pmatrix}\right\} \quad \text{and} \quad \mathcal{B}_{-2} = \left\{\begin{pmatrix} 1 \\ -1 \\ 2 \end{pmatrix}\right\}$$

respectively. This shows that $\mathcal{B} = \mathcal{B}_1 \cup \mathcal{B}_{-2}$ is a basis for $\mathbf{F}^3$ of eigenvectors of T_H. It follows that H is diagonalizable. The matrix $[T_H]_{\mathcal{B}} = P^{-1}HP$ will be diagonal where P is the transition matrix from the basis $\mathcal{B}$ to the standard basis. We have

$$P^{-1}HP = \begin{pmatrix} 1 & 1 & 0 \\ 0 & 2 & 1 \\ 0 & -1 & 0 \end{pmatrix}\begin{pmatrix} 1 & 3 & 0 \\ 0 & -2 & 0 \\ 0 & 6 & 1 \end{pmatrix}\begin{pmatrix} 1 & 0 & 1 \\ 0 & 0 & -1 \\ 0 & 1 & 2 \end{pmatrix}$$

$$= \begin{pmatrix} 1 & 0 & 0 \\ 0 & 1 & 0 \\ 0 & 0 & -2 \end{pmatrix}$$

(v) The matrix B of Example 3.4.7 (ii) is not diagonalizable, since the calculations there show that there cannot be a basis for $\mathbf{R}^3$ of eigenvectors of T_B.

(vi) The matrix D of Example 3.4.7 (iii) is diagonalizable over $\mathbf{R}$, since a basis for $\mathbf{R}^3$ of eigenvectors for T_D was exhibited. Note that D is not diagonalizable over the rational numbers $\mathbf{Q}$, since not all eigenvalues of D are rational.

The preceding examples show that a matrix may or may not be diagonalizable. We also saw that diagonalizability can depend upon the field of scalars. In any case, we can consider the subspace spanned by all the eigenvectors.

Definition 3.5.5 Let $T : V \to V$ be a linear operator with eigenspaces $E_1, E_2, \ldots, E_s$. The subspace $W = E_1 + E_2 + \cdots + E_s$ spanned by the E_i inside V is called the *eigenspace sum* of T.

According to Theorem 3.5.3 we see that the eigenspace sum of a linear operator $T : V \to V$ is all of V precisely when T is diagonalizable. Of course, this is not always the case. For example, over the reals, if

$$G = \begin{pmatrix} 0 & 1 \\ -1 & 0 \end{pmatrix}$$

is the matrix considered earlier, then T_G has as a eigenspace sum $\{\vec{0}\}$ since it has no eigenvalues. Over $\mathbf{C}$, however, T_G has as eigenspace sum $\mathbf{C}^2$.

Lemma 3.5.2 enables us readily to compute the dimension of the eigenspace sum W. The result is the following.

Lemma 3.5.6 *Suppose that $T : V \to V$ is a linear operator, with eigenspaces $E_1, E_2, \ldots, E_s$. If $\mathcal{B}_i$ is a basis of the eigenspace E_i, then the union $\mathcal{B} = \mathcal{B}_1 \cup \mathcal{B}_2 \cup \cdots \cup \mathcal{B}_s$ of these bases is a basis of the eigenspace sum W of T. Consequently, if the dimension of E_i is m_i, then the dimension of W is $m_1 + m_2 + \cdots + m_s$.*

Proof: Let $\mathcal{B}_i = \{v_{i1}, v_{i2}, \ldots, v_{im_i}\}$ be a basis for E_i. We claim that the collection $\mathcal{B}$ of all the v_{ij} is a basis for W. For this it suffices to verify that the v_{ij} are linearly independent, since by definition W is spanned by the E_i. Suppose that

$$\vec{0} = a_{11}v_{11} + a_{12}v_{12} + \cdots + a_{sm_s}v_{sm_s}$$

for scalars a_{ij}. Then we can write

$$\begin{aligned} \vec{0} = \ & (a_{11}v_{11} + \cdots + a_{1m}v_{1m}) + (a_{21}v_{21} + \cdots + a_{2m}v_{2m}) \\ & + \cdots + (a_{s1}v_{s1} + \cdots + a_{sm}v_{sm}) \end{aligned}$$

where the sum in the ith set of parentheses lies inside E_i.

By the definition of the v_{ij}, we see that each sum inside the ith pair of parentheses is an eigenvector with eigenvalue k_i. Applying Lemma

3.3.1, since these are eigenvectors with distinct eigenvalues, we see each must be $\vec{0}$. So we have

$$
\begin{array}{rcl}
(a_{11}v_{11} + \cdots + a_{1m_1}v_{1m_1}) &=& \vec{0} \\
(a_{21}v_{21} + \cdots + a_{2m_2}v_{2m_2}) &=& \vec{0} \\
\vdots \qquad \ddots \qquad \vdots & & \vdots \\
(a_{s1}v_{s1} + \cdots + a_{sm_s}v_{sm_s}) &=& \vec{0}
\end{array}
$$

Since each of the collections $v_{i1}, v_{i2}, \ldots, v_{im_i}$ are linearly independent, we see that each $a_{ij} = 0$. This shows that all the v_{ij} are linearly independent. Hence $\mathcal{B}$ is a basis for W and the lemma is proved. //

We next study diagonalizability by analyzing the dimension of the eigenspace sum of a linear operator. The key is the relationship between the dimension of the eigenspace associated with some eigenvalue k and the highest power of $(X - k)$ which divides the characteristic polynomial. We give the necessary terminology in the following definition.

Definition 3.5.7 Suppose that A is an $n \times n$ matrix and k is an eigenvalue of T_A. The highest power of $(X - k)$ which divides $C_A(X)$ is called the *algebraic multiplicity* of the eigenvalue k. The dimension of the eigenspace E_k of T_A associated with k is called the *geometric multiplicity* of the eigenvalue k.

We shall see later (in Chap. 6), that the geometric multiplicity of an eigenvalue never exceeds its algebraic multiplicity. However, it often occurs that the geometric multiplicity is less that the algebraic multiplicity. For example, consider the matrix

$$
J = \begin{pmatrix} 2 & 1 & 0 \\ 0 & 2 & 1 \\ 0 & 0 & 2 \end{pmatrix}
$$

Evidently, $C_J(X) = (X - 2)^3$, so 2 is an eigenvalue of T_J with algebraic multiplicity 3. Since $rk(2I_3 - J) = 2$, we see that E_2 is a one-dimensional subspace, that is, the eigenvalue 2 has geometric multiplicity 1.

It is not hard to see how to generalize the preceding example (see Examples 3.5.10) to construct a matrix with an eigenvalue having algebraic multiplicity n and geometric multiplicity any integer m with

$1 \leq m \leq n$. In terms of diagonalizability, the case where $m = n$ is what is important, as is illustrated in the following theorem.

Theorem 3.5.8 *Let A be an $n \times n$ matrix and suppose that $C_A(X)$ factors into linear (degree 1) factors.*
(i)   *T_A is diagonalizable if and only if the sum of the geometric multiplicities of its eigenvalues is n.*
(ii)   *If the geometric multiplicity of each eigenvalue of T_A is the same as the algebraic multiplicity, then T_A is diagonalizable.*
(iii)   *If all the factors of C_A are distinct (that is, each algebraic multiplicity is 1), then T_A is diagonalizable.*

Proof: (i) Suppose that the distinct eigenvalues of T_A are $k_1, k_2, \ldots, k_s$. Denote by E_i the eigenspace associated to k_i, and suppose that the dimension of E_i is m_i. We must show that $\mathbf{F}^n$ has a basis of eigenvectors for T_A if and only if $m_1 + m_2 + \cdots + m_s = n$.

For this, let $W = E_1 + E_2 + \cdots + E_s$ be the eigenspace sum of T_A. By Lemma 3.5.6, we know that $dim(W) = m_1 + m_2 + \cdots + m_s \leq n$. In case $m_1 + m_2 + \cdots + m_s = n$, we see that $W = \mathbf{F}^n$, and thus the collection of bases for the E_i give a basis for $\mathbf{F}^n$ of eigenvectors for T_A. Conversely, if $m_1 + m_2 + \cdots + m_s < n$, we see that it is impossible for $\mathbf{F}^n$ to have a spanning set of eigenvectors for T_A as all eigenvectors for T_A are contained in $W \neq \mathbf{F}^n$. From this, (i) follows.

(ii) This follows from (i), as the sum of the algebraic multiplicities in this case must be n.

(iii) This is a special case of (ii), as the dimension of any eigenspace associated with any eigenvalue must be at least 1. Consequently, the sum of the geometric multiplicities must be at least n. This proves the theorem. //

Examples 3.5.9 (i)   The following matrices each have the same characteristic polynomial $(X - 3)^4$, and thus they each have the single eigenvalue 3.

$$\begin{pmatrix} 3 & 0 & 0 & 0 \\ 0 & 3 & 0 & 0 \\ 0 & 0 & 3 & 0 \\ 0 & 0 & 0 & 3 \end{pmatrix}, \begin{pmatrix} 3 & 1 & 0 & 0 \\ 0 & 3 & 0 & 0 \\ 0 & 0 & 3 & 0 \\ 0 & 0 & 0 & 3 \end{pmatrix}, \begin{pmatrix} 3 & 1 & 0 & 0 \\ 0 & 3 & 1 & 0 \\ 0 & 0 & 3 & 0 \\ 0 & 0 & 0 & 3 \end{pmatrix}, \begin{pmatrix} 3 & 1 & 0 & 0 \\ 0 & 3 & 1 & 0 \\ 0 & 0 & 3 & 1 \\ 0 & 0 & 0 & 3 \end{pmatrix}$$

Subtracting $3I_4$ from each matrix gives matrices of rank 0, 1, 2, 3 respectively. This shows that the geometric multiplicities of the eigenvalue 3 are 4, 3, 2, 1 respectively. Since the algebraic multiplicity of the eigenvalue 3 is 4 in each case, only the first matrix is diagonalizable (and is diagonalized already!).

(ii) Consider the matrices

$$\begin{pmatrix} 4 & 1 & 0 & 0 \\ 0 & 3 & 1 & 0 \\ 0 & 0 & 2 & 1 \\ 0 & 0 & 0 & 1 \end{pmatrix} \quad \text{and} \quad \begin{pmatrix} 1 & 2 & 5 & 9 \\ 0 & 9 & 5 & 3 \\ 0 & 0 & 8 & 3 \\ 0 & 0 & 0 & 5 \end{pmatrix}$$

In each case the characteristic polynomial has distinct linear factors. Thus both matrices are diagonalizable.

Examples 3.5.10 (i) Let $V_1 = span\{\sin x, \cos x, e^x\} \subset C(\mathbf{R}, \mathbf{R})$. Let $D : V_1 \rightarrow V_1$ denote the differentiation operator. In terms of the ordered basis $\mathcal{B} = \{\sin x, \cos x, e^x\}$ we have

$$[D]_\mathcal{B} = \begin{pmatrix} 0 & -1 & 0 \\ 1 & 0 & 0 \\ 0 & 0 & 1 \end{pmatrix}$$

and consequently, $C_D = (X^2+1)(X-1)$ (on V_1). The operator D is not diagonalizable over $\mathbf{R}$ since its eigenspace sum is the one-dimensional vector space $span\{e^x\}$. However, if we view $D : V_2 \rightarrow V_2$ where $V_2 = span\{\sin x, \cos x, e^x\} \subset C(\mathbf{C}, \mathbf{C})$, then D becomes diagonalizable, since its characteristic polynomial factors as $C_D(X) = (X+i)(X-i)(X-1)$.

(ii) We now consider $V_3 = span\{e^x, xe^x, x^2e^x\} \subset C(\mathbf{R}, \mathbf{R})$. The differentiation operator $D : V_3 \rightarrow V_3$ is not diagonalizable. In this case the matrix for D in the basis $\mathcal{B} = \{e^x, xe^x, x^2e^x\}$ is

$$[D]_\mathcal{B} = \begin{pmatrix} 1 & 1 & 0 \\ 0 & 1 & 2 \\ 0 & 0 & 1 \end{pmatrix}$$

and consequently, $C_D = (X - 1)^3$. The eigenspace E_1 for D in this case is $span\{e^x\}$. Consequently, D is not diagonalizable. Note that D also fails to be diagonalizable when V_3 is "complexified" to $V_4 = span\{e^x, xe^x, x^2e^x\} \subset C(\mathbf{C}, \mathbf{C})$. This is because the eigenspace E_1 inside V_4 is still one-dimensional.

PROBLEMS

3.5.1. For each of the following matrices, determine all real eigenvalues, all eigenspaces, and if the matrix is diagonalizable. If the matrix is diagonalizable, show how to change bases to make it diagonal.

(a) $\begin{pmatrix} 0 & 1 & 1 \\ 1 & 0 & 1 \\ 1 & 1 & 0 \end{pmatrix}$ (b) $\begin{pmatrix} 0 & 0 & 1 \\ 1 & 0 & -3 \\ 0 & 1 & 3 \end{pmatrix}$ (c) $\begin{pmatrix} 0 & -1 \\ 1 & -1 \end{pmatrix}$

(d) $\begin{pmatrix} 1 & 0 & 0 & 0 \\ 0 & 1 & 1 & 0 \\ 0 & 0 & 1 & 0 \\ 0 & 0 & 0 & 1 \end{pmatrix}$ (e) $\begin{pmatrix} 1 & 0 & 0 & 0 \\ 0 & 2 & 1 & 0 \\ 0 & 0 & 1 & 0 \\ 0 & 0 & 0 & 1 \end{pmatrix}$

3.5.2. For each of the following matrices, determine all real eigenvalues, all eigenspaces, and if the matrix is diagonalizable. If the matrix is diagonalizable, show how to change bases to make it diagonal.

(a) $\begin{pmatrix} 3 & 1 & 0 \\ 1 & 2 & 0 \\ 1 & 1 & 0 \end{pmatrix}$ (b) $\begin{pmatrix} 1 & 0 & 2 \\ 1 & 1 & -1 \\ -1 & 0 & 2 \end{pmatrix}$ (c) $\begin{pmatrix} 0 & r & 0 & 0 \\ 0 & 0 & s & 0 \\ 0 & 0 & 0 & t \\ 0 & 0 & 0 & 0 \end{pmatrix}$

(d) $\begin{pmatrix} 1 & 2 & 3 & 4 \\ 0 & 2 & 3 & 4 \\ 0 & 0 & 3 & 4 \\ 0 & 0 & 0 & 4 \end{pmatrix}$ (e) $\begin{pmatrix} 2 & 0 & 0 & 0 \\ 1 & 1 & 0 & 0 \\ 3 & 2 & 1 & 0 \\ 0 & 0 & 0 & 1 \end{pmatrix}$

3.5.3. Suppose that A is a nonzero 2×2 matrix such that $T_A \circ T_A = 0$. Show that there exists an invertible matrix P such that

$$PAP^{-1} = \begin{pmatrix} 0 & 0 \\ 1 & 0 \end{pmatrix}$$

3.5.4. (a) Suppose that $dim(V) = n$ and $T : V \to V$ is a diagonalizable linear operator with only one eigenvalue. For each basis $\mathcal{B}$ of V show that $[T]_{\mathcal{B}} = kI_n$ for some scalar k.
(b) Write down five 5×5 matrices, each with the single eigenvalue 2, with all possible geometric multiplicities.

3.5.5. Show that any triangular matrix with distinct diagonal entries is diagonalizable. Find a triangular matrix that is not diagonalizable.

3.5.6. Find all real values of z for which $(zI - A)$ fails to be invertible if

$$A = \begin{pmatrix} 0 & -1 & 3 \\ 1 & 0 & 2 \\ -3 & -2 & 0 \end{pmatrix}$$

3.5.7. Suppose that A is a $n \times n$ real matrix with n distinct eigenvalues, B is another $n \times n$ matrix, and $AB = BA$. Show that B is diagonalizable.

3.5.8. Diagonalize the following real matrices. Display the basis-change matrix in each case.

(a) $\begin{pmatrix} 2 & 4 \\ 5 & 3 \end{pmatrix}$ (b) $\begin{pmatrix} 3 & 2 \\ -2 & 3 \end{pmatrix}$ (c) $\begin{pmatrix} 1 & 0 \\ 2 & 1 \end{pmatrix}$

3.5.9. Determine if the following pairs of matrices are similar.

(a) $\begin{pmatrix} 1 & 1 \\ 0 & 1 \end{pmatrix}$ and $\begin{pmatrix} 1 & 0 \\ 0 & 1 \end{pmatrix}$

(b) $\begin{pmatrix} 1 & 1 \\ 0 & 1 \end{pmatrix}$ and $\begin{pmatrix} 0 & 0 \\ 0 & 1 \end{pmatrix}$

(c) $\begin{pmatrix} 1 & 2 \\ 3 & 4 \end{pmatrix}$ and $\begin{pmatrix} 4 & 3 \\ 2 & 1 \end{pmatrix}$

3.5.10. Suppose that A and B are diagonalizable matrices with precisely the same eigenspaces (but not necessarily the same eigenvalues). Prove that $AB = BA$. (Hint: Diagonalize A and B in the same basis.)

3.5.11. Let $T : V \rightarrow V$ be a linear operator on a finite-dimensional vector space V. Prove that the eigenspace sum W of T is *invariant* under T; that is, prove that $T(w) \in W$ whenever $w \in W$.

3.5.12. Using the idea from Example 3.5.5 (i), find a formula for A^n where

$$A = \begin{pmatrix} 1 & 2 \\ 5 & 4 \end{pmatrix}$$

3.5.13. Suppose that $T : V \rightarrow V$ is a linear operator where V is n-dimensional. If $ker(T)$ is $(n - 1)$-dimensional and T has a nonzero eigenvalue, show that T is diagonalizable.

3.5.14. Prove that every matrix over $\mathbf{C}$ is similar to an upper-triangular matrix using the following outline as a guide: Proceed by induction on n where A is an $n \times n$ complex matrix. If A has the eigenvalue c, show that A is similar to the partitioned matrix

$$
\left(
\begin{array}{c|ccc}
c & b_2 & \cdots & b_n \\
\hline
0 & & & \\
\vdots & & B & \\
0 & & &
\end{array}
\right)
$$

where B is an $(n-1) \times (n-1)$ complex matrix. Apply the induction hypothesis to B and derive the result.

3.5.15. Suppose that $T : V \to V$ is diagonalizable. Show that for any n, $T^n : V \to V$ is also diagonalizable.

3.6 Applications of Eigenvalues: Markov Chains, Difference Equations, and Gerschgorin's Theorem

In this section we indicate a few applications of the earlier results from this chapter. The purpose is not to give a thorough treatment of these subjects, but rather to give the reader some indication why linear operators and eigenvalues are useful. Our first topic is that of Markov chains.

Markov Chains

Markov chains are useful tools in certain kinds of probabilistic models. They make use of matrix algebra and the theory of eigenvalues in a powerful way. The basic idea is the following: Suppose that you are observing some collection of objects that are changing through time. Assume that the total number of objects is not changing, but rather their "states" (position, color, disposition, etc.) are changing. Further, assume that the proportion of state-A objects changing to state B is constant and these changes occur at discrete stages, one after the next. Then you are in a good position to model the changes by a Markov chain.

As an example we consider the three-story aviary at a local zoo which houses 300 small birds. The aviary has three levels, and the birds spend their day flying around from one favorite perch to the next. Thus at any given time the birds seem to be randomly distributed throughout the three levels, except at feeding time when they all fly to the bottom level. Our problem is to determine what the probability is of a given bird being at a given level of the aviary at a given time. Of course, the birds are always flying from one level to another, so the bird population on each level is constantly fluctuating. We shall use a Markov chain to model this situation.

Consider a 3×1 matrix

$$p = \begin{pmatrix} p_1 \\ p_2 \\ p_3 \end{pmatrix}$$

where p_1 is the percentage of total birds on the first level, p_2 is the percentage on the second level, and p_3 is the percentage on the third level. Note that $p_1 + p_2 + p_3 = 1 = 100$ percent. After 5 min we have a new matrix

$$p' = \begin{pmatrix} p_1' \\ p_2' \\ p_3' \end{pmatrix}$$

giving the new distribution of the birds. We shall assume that the change from the p matrix to the p' matrix is given by a linear operator on $\mathbf{R}^3$. In other words there is a 3×3 matrix T, known as the *transition matrix* for the Markov chain (and not to be confused with the coordinate-change matrix), for which $Tp = p'$. After another 5 min we have another distribution $p'' = Tp'$ (using the same matrix T), and so forth. This type of model is known as a *Markov chain*. The *same* matrix T is used since we are assuming that the probability of a bird moving to another level is independent of time. (The model ignores the fact that at some point the birds will again be hungry and return to the bottom level for their next snack.)

In order to develop the mathematics behind Markov chains, we begin with a few definitions. We shall return to the aviary shortly. All vectors and matrices in this discussion will have real entries.

Definition 3.6.1 A vector $(a_1, a_2, \ldots, a_n) \in \mathbf{R}^n$ is called a *probability vector* (sometimes a *distribution vector*) if each $a_i \geq 0$ and $a_1 + a_2 + \cdots + a_n = 1$. An $n \times n$ matrix T is called a *stochastic matrix* if each column is a probability vector.

The idea behind a Markov chain is to measure each state of the system by a probability vector. Then each change in state, that is each transition, will be given by multiplying the probability vector by a stochastic matrix (the transition matrix). Before carrying this out for our example, we give a lemma that ensures that whenever this is done, we obtain a new probability vector to describe the system.

Lemma 3.6.2 (i) *A matrix T is a stochastic matrix if and only if all entries are nonnegative and whenever p is a probability vector, Tp is also a probability vector.*
(ii) *If T_1 and T_2 are stochastic matrices, then T_1T_2 is a stochastic matrix.*

Proof: (i) Assume that for all probability vectors p, Tp is also a probability vector. Since the standard basis vector e_i is a probability vector, it follows that Te_i, the ith column of T, is a probability vector. This shows that T is stochastic. Conversely, suppose that $T = (t_{ij})$ is stochastic and $p = (p_i)$ is a probability vector. Then the sum of the entries of Tp is

$$\sum_{i=1}^{n}(\sum_{j=1}^{n} t_{ij}p_j) = \sum_{j=1}^{n}(\sum_{i=1}^{n} t_{ij})p_j = \sum_{j=1}^{n} 1 \cdot p_j = 1$$

so that Tp is also a probability vector.

(ii) This is an immediate consequence of (i), because whenever p is a probability vector, $(T_1T_2)p = T_1(T_2p)$ must also be a probability vector. Furthermore, since all entries of T_1 and T_2 are nonnegative, the same is true for T_1T_2. This proves the lemma. //

According to the lemma, whenever T is a stochastic matrix and p is a probability vector, then Tp, T^2p, $T^3p, \ldots$ are all probability vectors. This shows that stochastic matrices are precisely the right type of matrix for carrying out the iterative Markov chain.

We now show how a stochastic matrix can be found for our aviary example. Assume that whenever a bird is on any level of the aviary, the probability of that bird being on the same level 5 min later is $1/2$. We shall assume that if the bird is on the first level, the probability of moving to the second level in 5 min is $1/3$ and of moving to the third in the same 5 min is $1/6$. For a bird on the second level, we shall assume that the probability of moving to either the first or the third level is $1/4$. Finally we assume for a bird on the third level, the probability of moving to the second level is $1/3$ and of moving to the first is $1/6$. Note that in each case, the sum of the possible probabilities of where a bird can go is 1.

Using these assumptions, we can construct the following stochastic matrix

$$T = \begin{pmatrix} 1/2 & 1/4 & 1/6 \\ 1/3 & 1/2 & 1/3 \\ 1/6 & 1/4 & 1/2 \end{pmatrix}$$

where the ijth entry of T is the probability that a bird on the jth level will move to the ith level in 5 min. Note of course, that the columns of T sum to 1, so that T is in fact stochastic.

Using T, we now can compute what happens to the bird distribution at 5-min intervals. Suppose that immediately after breakfast all the birds are in the dining area on the first level. Where are they in 5 min? The probability matrix at time 0 is

$$\begin{pmatrix} 1 \\ 0 \\ 0 \end{pmatrix}$$

According to our Markov chain the bird distribution after 5 min is

$$T \begin{pmatrix} 1 \\ 0 \\ 0 \end{pmatrix} = \begin{pmatrix} 1/2 \\ 1/3 \\ 1/6 \end{pmatrix}$$

After another 5 min the bird distribution becomes

$$T \begin{pmatrix} 1/2 \\ 1/3 \\ 1/6 \end{pmatrix} = \begin{pmatrix} 13/36 \\ 7/18 \\ 1/4 \end{pmatrix}$$

and so forth. What happens after one hour? To simplify the computations involved, we use some eigenvalue theory.

Definition 3.6.3 A stochastic matrix T is *regular* if for some n, T^n has no zero entries. A probability vector is called a *stable* vector for T if it is an eigenvector for T with eigenvalue 1.

If a stochastic matrix is regular, then eventually (after enough iterations) there is a positive probability that an object can change from its original state into any other state. Markov chains that model natural phenemena usually have associated stochastic matrices that are regular. Note that there can be zeros in a regular stochastic matrix T; the point is that all entries are nonzero in a sufficiently high power of T.

Suppose that T is an $n \times n$ stochastic matrix. Consider the matrix $(I_n - T)$, whose rows we denote by $R_1, R_2, \ldots, R_n$. Since T is stochastic, it follows that the sum of the entries of any column of $(I_n - T)$ is 0. Consequently, $R_1 + R_2 + \cdots + R_n = \vec{0}$. This shows that the rows of $(I_n - T)$ are linearly dependent, that is, $rk(I_n - T) < n$. It follows that 1 is an eigenvalue of T. In fact much more is true. It turns out that for a regular stochastic matrix, there is always a unique stable vector s for T. Furthermore, for any initial probability vector, whenever powers $T^n v$ are computed, (that is, the Markov chain is iterated), the result gets closer and closer to the stable vector s. We spell out this result in the next theorem. (See the book by Jones in the Bibliography for a proof.)

Theorem 3.6.4 *Let T be a regular stochastic matrix. Then*
(i) *1 is an eigenvalue of T and T has a unique stable vector v.*
(ii) *For any probability vector v the limit $\lim_{n \to \infty} T^n v$ exists and is the unique stable vector s.*

We now apply Theorem 3.6.4 to our example. We compute that the eigenspace of eigenvalue 1 for our matrix T is

$$ker \begin{pmatrix} -1/2 & 1/4 & 1/6 \\ 1/3 & -1/2 & 1/3 \\ 1/6 & 1/4 & -1/2 \end{pmatrix} = span \left\{ \begin{pmatrix} 1 \\ 4/3 \\ 1 \end{pmatrix} \right\}$$

Consequently, our stochastic matrix T has the unique stable vector

$$\begin{pmatrix} 3/10 \\ 4/10 \\ 3/10 \end{pmatrix}$$

According to our discussion, Theorem 3.6.4 (ii) shows that if we visit the birds sufficiently long after a feeding, they will not be as randomly distributed as we might guess. Forty percent of the birds will be located on the middle level, while 30 percent of the birds will be located on each of the first and third levels.

Difference Equations

In Markov chains one uses the very special stochastic matrices to describe the change from one state of a system to the next. The reason such matrices occurred was that the parameters being measured in the system under study were not growing. Often this is not the case, and one still wants to model the system by a sequence of discrete linear changes. This leads to the concept of a difference equation.

An elementary and familiar example of iterative linear changes is that of compounding interest. If 6 percent anual interest is compounded monthly, then the montly interest rate can be taken as 0.5 percent. Hence, if you have $\$D$ invested in such an account, at the end of 1 mon you have $\$(1.005)D$. At the end of 2 mon you have $\$(1.005)(1.005)D = \$(1.005)^2D$, and so forth.

More interesting examples arise when you assume that new values for the system depend upon more that the single previous value. A famous example is the Fibonacci sequence of natural numbers. This sequence, which starts as $1, 1, 2, 3, 5, 8, \ldots$, has its nth term x_n defined by the *difference equation* $x_n = x_{n-2} + x_{n-1}$. In other words the nth Fibonacci number is the sum of the two preceding Fibonacci numbers. The process of finding the nth term can be represented by multiplication by a 2×2 matrix as follows: The pair (x_{n-1}, x_n) is transformed (by the difference equation) to the pair (x_n, x_{n+1}) by the matrix multiplication

$$\begin{pmatrix} 0 & 1 \\ 1 & 1 \end{pmatrix} \begin{pmatrix} x_{n-1} \\ x_n \end{pmatrix} = \begin{pmatrix} x_n \\ x_{n+1} \end{pmatrix}$$

Such a matrix expression is quite useful. For example, suppose that we are interested in the ratio x_n/x_{n-1} for $n = 500$. To compute x_{500} by the definition would be a horrendous task. In order to understand the powers of

$$\Phi = \begin{pmatrix} 0 & 1 \\ 1 & 1 \end{pmatrix}$$

we first compute its eigenvalues and diagonalize. The characteristic polynomial of Φ is $X^2 - X - 1$, whose roots are $(1 \pm \sqrt{5})/2$. To abbreviate, we let $(1 + \sqrt{5})/2 = \gamma_1$ and $(1 - \sqrt{5})/2 = \gamma_2$. Then we check that $(1, \gamma_1)$ is an eigenvector for Φ with eigenvalue γ_1 and $(1, \gamma_2)$ is an eigenvector of Φ with eigenvalue γ_2. Consequently, if we set

$$P = \begin{pmatrix} 1 & 1 \\ \gamma_1 & \gamma_2 \end{pmatrix}, \quad \text{then} \quad P \begin{pmatrix} \gamma_1 & 0 \\ 0 & \gamma_2 \end{pmatrix} P^{-1} = \Phi$$

Since the pair
$$\begin{pmatrix} x_n \\ x_{n+1} \end{pmatrix} \quad \text{is given by} \quad \Phi^n \begin{pmatrix} 0 \\ 1 \end{pmatrix}$$

we find that

$$\begin{pmatrix} x_n \\ x_{n+1} \end{pmatrix} = \left(P \begin{pmatrix} \gamma_1 & 0 \\ 0 & \gamma_2 \end{pmatrix} P^{-1} \right)^n \begin{pmatrix} 1 \\ 0 \end{pmatrix}$$

$$= P \begin{pmatrix} \gamma_1^n & 0 \\ 0 & \gamma_2^n \end{pmatrix} P^{-1} \begin{pmatrix} 1 \\ 0 \end{pmatrix} = \begin{pmatrix} \gamma_1^n - \gamma_2^n \\ \gamma_1^{n+1} - \gamma_2^{n+1} \end{pmatrix}$$

using the fact that

$$P^{-1} = (\gamma_1 - \gamma_2)^{-1} \begin{pmatrix} -\gamma_2 & 1 \\ \gamma_1 & 1 \end{pmatrix}$$

We find that the nth Fibonacci number is

$$x_n = \frac{1}{\sqrt{5}} \left[\left(\frac{1 + \sqrt{5}}{2} \right)^n - \left(\frac{1 - \sqrt{5}}{2} \right)^n \right]$$

Using this expression, we see that the limit

$$lim_{n \to \infty} \frac{x_{n+1}}{x_n} = lim_{n \to \infty} \frac{[(1 + \sqrt{5})/2]^{n+1} - [(1 - \sqrt{5})/2]^{n+1}}{[(1 + \sqrt{5})/2]^n - [(1 - \sqrt{5})/2]^n}$$

$$= \frac{1 + \sqrt{5}}{2}$$

the famous "golden ratio."

The difference equation describing compound interest can be written as $X_n = (1 + s)X_{n-1}$, where X_0 represents an initial deposit, X_n represents the value of the deposit after n compoundings of the interest, and s is the yearly interest rate divided by the number of compoundings per year. Since X_n depends only upon the value X_{n-1}, this type of difference equation is called a *first-order* difference equation. Observe that the values of all the X_n are completely determined by the initial value X_0. Analogously, the difference equation $X_n = X_{n-1} + X_{n-2}$, which describes the Fibonacci sequence, is called a *second-order* difference equation since each X_n is determined by the preceding two values. Consequently the equation determines the values of all the X_n once the first two values are specified. The next definition generalizes these two examples.

Definition 3.6.5 A *kth-order linear difference equation* is an equation of the form

$$a_0 X_n + a_1 X_{n-1} + \cdots + a_k X_{n-k} = c_k$$

where $a_0, a_1, \ldots, a_k$ are fixed scalars and $c_0, c_1, c_2, \ldots$ is a sequence of scalars. A *solution* to such an equation is a sequence $x_0, x_1, x_2, \ldots$ of scalars that solves the equation for each $n \geq k$.

Strictly speaking, a kth-order linear difference equation is really an infinite sequence of equations. However, since the constants $a_0, a_1, \ldots,$ a_{k-1} are the same for each equation, it is common to refer to these equations as a single equation. In both examples above, the sequence $c_0, c_1, c_2, \ldots$ was identically zero. Such difference equations are called *homogeneous*, and the difference equation obtained from a given difference equation by replacing the sequence of c_i's by zeros is called the *associated homogeneous difference equation*. The solutions to a difference equation are infinite sequences of scalars, which are elements of the vector space $\mathbf{F}^\infty$ of all such sequences. The next theorem describes the nature of these solutions.

Theorem 3.6.6 *Given any kth-order linear difference equation*

$$a_0 X_n + a_1 X_{n-1} + \cdots + a_k X_{n-k} = c_k$$

and any $x_0, x_1, \ldots, x_{k-1} \in \mathbf{F}$, there is a unique solution to this equation with these initial values. The set of solutions to a kth-order homogeneous linear difference equation is a k-dimensional subspace of $\mathbf{F}^\infty$. All solutions to a general kth-order difference equation can be obtained from a single solution, by adding to this solution the solutions of the associated homogeneous difference equation.

The proof of Theorem 3.6.6 is not difficult and is left as an exercise. The reader may note the similarity of the nature of solutions to kth-order linear difference equations to the nature of solutions to kth-order linear differential equations. This is not merely a coincidence. Difference equations are discrete versions of differential equations.

For example, if the value of an account accruing interest is viewed as a function of time, then the compounding equation $X_n = (1+s)X_{n-1}$ is a difference equation governing the solution. In other words, suppose there are N yearly compoundings at a given interest rate r. If $F(t)$

represents the value of the account after t years, then $F(t) = x_{tN}$ where $x_0, x_1, x_2, \ldots$ is a solution to the difference equation with $F(0) = x_0$. The values of t can be chosen as any rational numbers M/N. The difference equation, in terms of F, says that $F(t + 1/N) = (1 + s)F(t)$. The ratio s in this expression is $s = r/N$ (which is the yearly interest rate divided by the number of yearly compoundings).

The rate of growth in such an account at time t by definition is

$$R(t) = \frac{1}{1/N} \left(F\left(t + \frac{1}{N}\right) - F(t) \right)$$

Applying our difference equation, we see that

$$\begin{aligned} R(t) &= N(F(t + 1/N) - F(t)) = N((1 + s)F(t) - F(t)) \\ &= NsF(t) = rF(t) \end{aligned}$$

We obtain (the expected) result that the rate of growth of such an account is the product of the interest rate and the current balance. The value $R(t)$ is really a discrete version of the derivative $F'(t)$, so we can interpret our difference equation as a discrete analogue of the differential equation $F'(t) = rF(t)$. The continuous solution $F(t) = Pe^{rt}$, [where $P = F(0)$] of this differential equation is the limit function obtained by compounding interest over smaller and smaller time intervals (so-called continuous compounding).

The relationship between difference and differential equations, together with the fact that matrix algebra proved useful in solving the second-order difference equation, suggests that matrix techniques may be useful in studying linear differential equations as well. This is indeed the case, and the way they are used is described in Sec. 7.2. The idea of diagonalizing the matrix describing the difference equation (as we did for the Fibonacci sequence) will also be exploited when we study systems of linear differential equations.

Gerschgorin's Theorem

It is often important to know the approximate location of the eigenvalues of a matrix, without necessarily determining their exact values. Gerschgorin's theorem is extremely useful in this connection. Suppose that A is an $n \times n$ complex matrix (so the result can be applied to real matrices as well). Gerschgorin's theorem says that all the eigenvalues (real or complex) of A lie inside a collection of specific disks in the complex plane.

Theorem 3.6.7 (Gerschgorin's Disk Theorem) *Suppose that $A = (a_{ij})$ is an $n \times n$ complex matrix and $r_i = \sum_{j=1, j \neq i}^{n} |a_{ij}|$. Let D_i denote the disk in $\mathbf{C}$ with center a_{ii} and radius r_i. Then every eigenvalue of A lies inside some D_i.*

Proof: Suppose that c is an eigenvalue of A with a corresponding eigenvector $v \in \mathbf{C}^n$. Express $v = (b_1, b_2, \ldots, b_n)$, and suppose that b_i has the largest absolute value among the coordinates of v. (Since $v \neq \vec{0}$, $|b_i| > 0$.) Applying the definition of matrix multiplication, the ith component of $cv = Av$ is $cb_i = \sum_{j=1}^{n} a_{ij}b_j$. Thus, $cb_i - a_{ii}b_i = \sum_{j=1, j \neq i}^{n} a_{ij}b_j$. Recall the triangle inequality for complex numbers: $|r + s| \leq |r| + |s|$ for all $r, s \in \mathbf{C}$. We obtain that

$$|cb_i - a_{ii}b_i| \leq \sum_{j=1, j \neq i}^{n} |a_{ij}b_j|$$

Consequently, the maximality of $|b_i|$ gives

$$|b_i| \cdot |c - a_{ii}| \leq |b_i| \left(\sum_{j=1, j \neq i}^{n} |a_{ij}| \right)$$

From this we obtain $|c - a_{ii}| \leq r_i$ and $c \in D_i$ follows. This proves the theorem. //

In applying Gerschgorin's theorem, one can instead use the radii r_i defined by $r_i = \sum_{j=i, j \neq i}^{n} |a_{ji}|$ (that is, one can sum along the columns instead of the rows). This is because the eigenvalues of a matrix are the same as its transpose. This observation can often be useful.

Example 3.6.8 Consider the matrix

$$A = \begin{pmatrix} 2 & 1 & 0 \\ 1 & 4 & 1 \\ 1 & 0 & 3 \end{pmatrix}$$

In the notation of Gerschgorin's theorem we have that $r_1 = 1 + 0 = 1$, $r_2 = 1 + 1 = 2$, and $r_3 = 1 + 0 = 1$. The Gerschgorin disks for A are the disk of radius 1 around 2, the disk of radius 2 around 4, and the disk of radius 1 around 3.

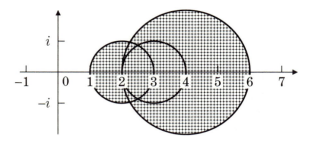

Observe that 0 does not lie in any of these disks. (A matrix for which 0 does not lie in any Gerschgorin disk is called *diagonally dominant.*) Consequently, 0 is not an eigenvalue of A, and so in particular we find that A is invertible. The matrix A will necessarily have at least one real eigenvalue (its characteristic polynomial has odd degree), so we learn that A has a positive real eigenvalue. This eigenvalue c must satisfy $1 \leq c \leq 6$. To learn more about the eigenvalues of A, we must compute the characteristic polynomial or use other methods.

Example 3.6.9 Consider

$$B = \begin{pmatrix} 7 & 1 & 1 \\ 1 & 15 & 1 \\ 2 & 1 & 25 \end{pmatrix}$$

The three Gerschgorin disks D_1, D_2, D_3 have radii 2 around the real numbers 7, 15, 25 (taking into account both row and column radii). B is diagonally dominant (as defined in the example above) and consequently we know that B is invertible. But more can be deduced. Consider the matrix

$$B_t = \begin{pmatrix} 7 & t & t \\ t & 15 & t \\ 2t & t & 25 \end{pmatrix}$$

Then $B = B_1$ and B_0 is a diagonal matrix with characteristic polynomial $C_{B_0}(X) = (X - 7)(X - 15)(X - 25)$. Consider the characteristic polynomials $C_{B_t}(X)$ for $0 \leq t \leq 1$. The roots of these polynomials (in $\mathbf{C}$) are continuous functions of t and by Gerschgorin's theorem must lie in the disjoint disks D_1, D_2, D_3 since the Gerschgorin disks for the

B_t $(0 \le t \le 1)$ lie inside these circles. As the roots $(7, 15,$ and $25)$ of $C_{B_0}(X)$ lie in distinct disks, continuity guarantees that each of the roots of $C_{B_t}(X)$ for $0 \le t \le 1$ lies in a distinct disk. In particular, $C_{B_1}(X) = C_B(X)$ has three distinct roots, one in each Gerschgorin disk.

Next note that each of the three roots of $C_B(X)$ must be real. $C_B(X)$ is a real polynomial, and therefore any nonreal roots must occur in conjugate pairs (that is, of the form $r \pm si$ where $r, s \in \mathbf{R}$). Since all the Gerschgorin disks for B are centered on the real line, any conjugate pair of complex numbers must lie in the same disk. Thus, the fact that the roots of $C_B(X)$ lie in distinct disks shows they are necessarily real. We have shown that B is diagonalizable over $\mathbf{R}$.

PROBLEMS

3.6.1. Which of the following stochastic matrices are regular?

(a) $\begin{pmatrix} 1/2 & 0 \\ 1/2 & 1 \end{pmatrix}$ (b) $\begin{pmatrix} 0 & 1/2 & 1/2 \\ 1/2 & 0 & 1/2 \\ 1/2 & 1/2 & 0 \end{pmatrix}$ (c) $\begin{pmatrix} 2/3 & 1/3 & 1/3 \\ 1/3 & 1/3 & 1/3 \\ 0 & 1/3 & 1/3 \end{pmatrix}$

3.6.2. Apply Theorem 3.6.4 (ii) to show that if T is a regular stochastic matrix, then the matrix limit $lim_{n \to \infty} T^n$ exists and consists of a matrix with identical columns.

3.6.3. What is the $lim_{n \to \infty} P^n$ if

$$P = \begin{pmatrix} 3/4 & 1/4 \\ 1/4 & 3/4 \end{pmatrix}$$

3.6.4. If r is an eigenvalue of a stochastic matrix, prove that $|r| \le 1$.

3.6.5. Suppose that the stochastic matrix describing our aviary problem in this section was the matrix

$$\begin{pmatrix} 3/4 & 1/8 & 0 \\ 1/4 & 3/4 & 1/4 \\ 0 & 1/8 & 3/4 \end{pmatrix}$$

(These are lazy birds.) What will the limiting distribution of the birds be in this case?

3.6.6. Compute the nth "Fibonacci" number in the sequence that is defined exactly the same way as the Fibonacci sequence, except starts with $a_1 = 1$ and $a_2 = 2$. What is the limit of the ratios a_n/a_{n+1} as $n \to \infty$ for this sequence?

3.6.7. What information does Gerschgorin's theorem give for the following matrices:

(a) $\begin{pmatrix} 1 & 0 & 0 \\ 2 & 6 & 1 \\ 1 & 2 & 4 \end{pmatrix}$ (b) $\begin{pmatrix} 1 & 0 & 1 \\ -1 & 2 & 3 \\ -1 & -1 & -1 \end{pmatrix}$ (c) $\begin{pmatrix} 0 & 0 & 0 & -1 \\ 1 & 0 & 0 & 0 \\ 0 & 1 & 0 & -2 \\ 0 & 0 & 1 & 0 \end{pmatrix}$

3.6.8. What information does Gerschgorin's theorem give for the following matrices. Compare this information with what you can easily deduce from inspection.

(a) $\begin{pmatrix} 0 & 1 & 0 \\ 1 & 0 & 1 \\ 0 & 1 & 0 \end{pmatrix}$ (b) $\begin{pmatrix} 2 & 2 & 2 \\ 2 & 2 & 2 \\ 2 & 2 & 2 \end{pmatrix}$ (c) $\begin{pmatrix} 1 & 0 & 1 \\ 0 & 1 & 0 \\ 0 & 0 & 10 \end{pmatrix}$

3.6.9. Suppose that A is the matrix

$$\frac{1}{10} \begin{pmatrix} 4 & 1 & 5 \\ 2 & 7 & 1 \\ 2 & 2 & 6 \end{pmatrix}$$

Find $lim_{n \to \infty} A^n$.

3.6.10. Prove Theorem 3.6.6.

3.6.11. Find two 2×2 real matrices A, B so that the limits $lim_{n \to \infty} A^n$, $lim_{n \to \infty} B^n$, and $lim_{n \to \infty}(AB)^n$ exist, but $(lim_{n \to \infty} A^n)(lim_{n \to \infty} B^n) \neq lim_{n \to \infty}(AB)^n$.

3.6.12. Generalize the idea of Example 3.6.9. State and prove a general theorem which gives sufficient conditions for a real matrix to be diagonalizable over $\mathbf{R}$.

Chapter 4

Norms and Inner Products

4.1 The Definitions and Basic Properties

In the previous chapters, our work with vectors and vector spaces was more algebraic than geometric. In particular, the concept of distance did not play any role. There are many reasons for this, the main reason being that we studied vector spaces over *arbitrary* fields, and in general we do not have a distance concept for an arbitrary field. This chapter studies real and complex vector spaces exclusively and begins by addressing the two key issues of distance and orthogonality.

In this section we introduce norms for real and complex vector spaces. Norms are (essentially) distance functions. We also introduce inner products, which generalize the concept of the dot product on $\mathbf{R}^n$. As we shall show, inner products are closely related to norms. Consequently, the ideas we deal with in this chapter will be more geometric. However, we shall continue to give algebraic proofs of the main results (although we will be guided by our geometric intuition). One reason for this is that we can then apply these results to infinite-dimensional vector spaces and obtain important applications.

We begin this section with the definition of a norm on an arbitrary real or complex vector space. Recall that for an arbitrary complex number $a + bi \in \mathbf{C}$, with $a, b \in \mathbf{R}$, the absolute value $|a + bi|$ is the nonnegative real number $\sqrt{a^2 + b^2}$.

Definition 4.1.1 A *norm* on a real or complex vector space V is a real-valued function, usually denoted by $\| \ \|$, which satisfies

(i) $\| v \| \geq 0$ for all $v \in V$, and $\| v \| = 0$ only if $v = \vec{0}$.

(ii) $\| kv \| = | k | \cdot \| v \|$ for all scalars k and $v \in V$.

(iii) $\| u + v \| \leq \| u \| + \| v \|$ for all $u, v \in V$.

If $\| \ \|$ is a norm on a vector space V and $v, w \in V$, then we define the distance between v and w (relative to $\| \ \|$) by $d(v, w) = \| v - w \|$.

Property (i) is known as the positive-definite property of $\| \ \|$, and property (iii) is known as the triangle inequality. Observe that property (ii) guarantees that $d(v, w) = \| v - w \| = | -1 | \cdot \| w - v \| = \| w - v \|$ $= d(w, v)$, so the distance function is symmetric (as it ought to be).

Examples 4.1.2 (i) On $\mathbf{R}^n$ define

$$\| (a_1, a_2, \ldots, a_n) \| = \sqrt{a_1^2 + a_2^2 + \cdots + a_n^2}$$

Then, $\| v \|$ is the distance from $(0, 0, \ldots, 0)$ to v in $\mathbf{R}^n$. (For $n = 1, 2, 3$ this should be familiar from high school geometry.) That $\| \ \|$ satisfies (i), (ii), (iii) above is well known. We shall refer to this norm $\| \ \|$ on $\mathbf{R}^n$ as the usual *euclidean* norm.

(ii) For $(a_1, \ldots, a_n) \in \mathbf{R}^n$, we define

$$\| (a_1, a_2, \ldots, a_n) \|_t = | a_1 | + | a_2 | + \cdots + | a_n |$$

That properties (i) and (ii) hold for $\| \ \|_t$ is clear. We check property (iii): For any pair of real numbers $a, b, | a + b | \leq | a | + | b |$. Hence $\| (a_1 + b_1, \ldots, a_n + b_n) \|_t = | a_1 + b_1 | + \cdots + | a_n + b_n | \leq | a_1 | + | b_1 |$ $+ \cdots + | a_n | + | b_n | = \| (a_1, \ldots, a_n) \|_t + \| b_1, \ldots, b_n) \|_t$. Thus $\| \ \|_t$ defines a norm on $\mathbf{R}^n$, which is sometimes called the taxicab norm.

(iii) Again on $\mathbf{R}^n$, define

$$\| (a_1, \ldots, a_n) \|_{max} = max\{| a_1 |, \ldots, | a_n |\}$$

It is easy to check that $\| \ \|_{max}$ is a norm on $\mathbf{R}^n$.

(iv) Let $\mathcal{C}([0, 1], \mathbf{R})$ denote the vector space of all continuous real-valued functions defined on $[0, 1]$. For $f(x) \in \mathcal{C}([0, 1], \mathbf{R})$ we define

$$\| f(x) \|_1 = \int_0^1 | f(x) | \ dx$$

It follows from calculus that $\| \ \|_1$ is a norm on $\mathcal{C}([0,1], \mathbf{R})$. (Note: To prove that $\| \ \|_1$ is positive definite, we need that $f(x)$ is continuous.)

(v) On $\mathbf{C}^n$ we have the usual norm given by

$$\| (c_1, c_2, \ldots, c_n) \| = \sqrt{|c_1|^2 + |c_2|^2 + \cdots + |c_n|^2}$$

If one views $\mathbf{C}^n$ as $\mathbf{R}^{2n}$ (using the identification of $\mathbf{C}$ with $\mathbf{R}^2$ as usual), this norm gives the usual distance function on $\mathbf{R}^{2n}$.

(vi) If we denote by $\mathcal{C}([0,1], \mathbf{C})$ the complex vector space of all continuous complex functions on the unit interval, then the formula given in (iv) above gives a norm on this vector space. Note that the integrand is that of a positive real-valued function, and so the argument that it gives a norm will be the same.

Closely related to the concept of a norm is that of an inner product. Indeed, we shall prove that every inner product gives rise to a norm in a natural way (although the converse is *not* true). It is through inner products that we will derive the basic properties of the norms we are interested in. We next give the definition of a real inner product, deferring the complex case to the end of the section. The real and complex cases are quite similar, and in fact the proofs of all our theorems could have been given simultaneously. However, when first studying this subject, it is probably best to examine the familiar real case first and then see how the complex case generalizes the real case. In Secs. 4.2 and 4.3, all the proofs given will apply to both cases.

Definition 4.1.3 Let V be a real vector space. A *real inner product* on V is a real valued function on $V \times V$, usually denoted by $< \ , \ >$, which satisfies:

(i) $<u, v> = <v, u>$ for all $u, v \in V$.
(ii) $<ku, v> = k <u, v>$ for all $k \in \mathbf{R}$ and all $u, v \in V$.
(iii) $<v_1 + v_2, u> = <v_1, u> + <v_2, u>$ for all $u, v_1, v_2 \in V$.
(iv) $<v, v> \geq 0$ for all $v \in V$, and $<v, v> = 0$ only if $v = \vec{0}$.

It is customary to refer to the vector space V, together with some inner product $< \ , \ >$, as a *real inner product space*. A finite-dimensional real inner product space is often called a *euclidean space*.

Property (i) of a real inner product is known as symmetry, and (iv) is referred to as the positive-definite property. Note that because of (i), properties (ii) and (iii) hold as well on the opposite sides of the inner product bracket. Specifically, we have that

(ii') $<u, kv> = k <u, v>$ for all $k \in \mathbf{R}$ and all $u, v \in V$.

(iii') $<u, v_1 + v_2> = <u, v_1> + <u, v_2>$ for all $u, v_1, v_2 \in V$.

Properties (ii), (ii'), (iii), and (iii') taken together are called *bilinearity*. We emphasize that property (i) changes when considering complex inner product spaces (see Definition 4.1.3').

Examples 4.1.4 (i) The dot product on $\mathbf{R}^n$ is an inner product. Recall that for $(a_1, \ldots, a_n), (b_1, \ldots, b_n) \in \mathbf{R}^n$ their dot product is defined by

$$(a_1, \ldots, a_n) \cdot (b_1, \ldots, b_n) = a_1 b_1 + \cdots + a_n b_n$$

That properties (i), (ii), (iii), and (iv) hold for the dot product is well known and easily checked.

(ii) The function $< , >: \mathbf{R}^3 \times \mathbf{R}^3 \to \mathbf{R}$ defined by

$$<(x_1, y_1, z_1), (x_2, y_2, z_2)> = x_1 x_2 - 2(x_1 y_2 + y_1 x_2) + 5 y_1 y_2 + 3 z_1 z_2$$

is an inner product on $\mathbf{R}^3$. The positive-definite property follows because $<(x, y, z), (x, y, z)> = (x - 2y)^2 + y^2 + 3z^2$ is positive whenever $(x, y, z) \neq (0, 0, 0)$.

(iii) Let $V = \mathcal{C}([0, 1], \mathbf{R})$ be the vector space of continuous real-valued functions on the interval $[0, 1]$. For $f(x), g(x) \in V$ we can define an inner product by

$$<f(x), g(x)> = \int_0^1 f(x) g(x) \, dx$$

That properties (i), (ii), (iii), and (iv) hold follows from elementary properties of the integral.

We turn to the geometry of inner product spaces. Whenever $u, v \in V$ and $< , >$ is an inner product on V, we say that u and v are *orthogonal* (which is often denoted by $u \perp v$) if $<u, v> = 0$. In the case of the dot product on $\mathbf{R}^2$ and $\mathbf{R}^3$, two vectors are orthogonal if and only if they are perpendicular (geometrically). This follows from the formula (proved in most calculus courses)

$$u \cdot v = \| u \| \| v \| \cos \theta$$

where θ is the angle between the two vectors u and v. Using the dot product, this formula enables us to extend the geometric notions of angle and perpendicularity from $\mathbf{R}^2$ and $\mathbf{R}^3$ to $\mathbf{R}^n$. Mutually perpendicular nonzero vectors in $\mathbf{R}^n$ are linearly independent, although *not* conversely. This is actually a general fact about inner products, as we now prove.

Theorem 4.1.5 *Suppose that $v_1, v_2, \ldots, v_n$ are all mutually orthogonal nonzero vectors in a vector space V with an inner product $< , >$. Then $v_1, v_2, \ldots, v_n$ are linearly independent.*

Proof: Suppose that $a_1 v_1 + a_2 v_2 + \cdots + a_n v_n = \vec{0}$. It follows that for each i,

$$
\begin{aligned}
0 &= <v_i, \vec{0}> \\
&= <v_i, a_1 v_1 + a_2 v_2 + \cdots + a_n v_n> \\
&= <v_i, a_1 v_1> + \cdots + <v_i, a_n v_n> \\
&= a_1 <v_i, v_1> + \cdots + a_i <v_i, v_i> + \cdots + a_n <v_i, v_n> \\
&= a_1 0 + \cdots + a_i <v_i, v_i> + \cdots + a_n 0 \\
&= a_i <v_i, v_i>
\end{aligned}
$$

Since $v_i \neq \vec{0}$, we find by the positive definite property $<v_i, v_i> \neq 0$. Thus the equation $a_i <v_i, v_i> = 0$ gives $a_i = 0$. This shows that $v_1, v_2, \ldots, v_n$ are linearly independent, as desired. //

Consider two vectors $u, v \in \mathbf{R}^2$. If θ is the angle between these two vectors, then by elementry trigonometry $\| u \| \cos \theta$ is the length of the projection of the vector u onto the vector v. (See the figure on the next page.) Consequently, the projection of u onto v [denoted $proj_v(u)$] is the vector

$$
\| u \| \cos \theta \left(\frac{v}{\| v \|} \right) = \frac{\| u \| \cos \theta}{\| v \|} v
$$

That $\| v \| = (v \cdot v)^{1/2}$ and $u \cdot v = \| u \| \| v \| \cos \theta$, shows

$$
\begin{aligned}
proj_v(u) &= \frac{\| u \| \| v \| \cos\theta}{\| v \| \| v \|} v \\
&= \frac{u \cdot v}{\| v \|^2} v = \frac{u \cdot v}{v \cdot v} v
\end{aligned}
$$

The projection of a vector u onto a vector v is pictured below.

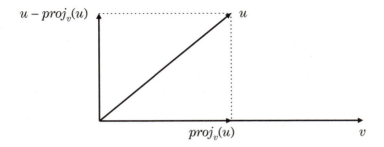

This motivates the following definition.

Definition 4.1.6 If $< \, , \, >$ is an inner product on V and $u, v \neq \vec{0} \in V$, we define the *orthogonal projection* of u onto v by

$$proj_v(u) = \frac{<u, v>}{<v, v>} v$$

Observe in the picture above that the vector $w = u - proj_v(u)$ is perpendicular to the vector v. This is a general fact about inner product spaces, whose (purely algebraic) proof is given next.

Lemma 4.1.7 *Whenever* $u, v \in V$ *and* $< \, , \, >$ *is a real inner product on* V, *the vectors* v *and* $(u - proj_v(u))$ *are orthogonal.*

Proof: We compute, using the properties of a real inner product:

$$
\begin{aligned}
<v, u - proj_v(u)> \quad &= \quad <v, u - \frac{<u, v>}{<v, v>} v> \\
&= \quad <v, u> - <v, \frac{<u, v>}{<v, v>} v> \\
&= \quad <v, u> - \frac{<u, v>}{<v, v>} <v, v> \\
&= \quad <v, u> - <u, v> \\
&= \quad 0 \qquad\qquad\qquad //
\end{aligned}
$$

In euclidean space, $\mathbf{R}^n$, we have that $v \cdot v = \| v \|^2$. We use this same idea and define the norm associated with an arbitrary inner product.

Definition 4.1.8 If V is a vector space, $v \in V$, and $< , >$ is an inner product on V, we define $\| v \| = <v, v>^{1/2}$. This function $\| \ \|: V \to \mathbf{R}$ is called the *norm* associated with $< , >$.

The norm associated with an inner product is fundamentally related to the inner product. One fact we shall use often is that whenever v and w are orthogonal, $\| v + w \|^2 = \| v \|^2 + \| w \|^2$. This fact is easily checked: $\| v + w \|^2 = <v + w, v + w> = <v, v> + <v, w> + <w, v> + <w, w> = <v, v> + <w, w> = \| v \|^2 + \| w \|^2$. We show in Theorem 4.1.10 that $\| \ \|$ satisfies all the conditions of Definition 4.1.1 for being a norm on V. In order to prove Theorem 4.1.10, we need the following theorem, which is a useful result in its own right.

Theorem 4.1.9 (Schwarz' Inequality) *Suppose that V is a real inner product space with inner product $< , >$. For $u, v \in V$,*

$$<u, v> \ \leq \ \| u \| \cdot \| v \|$$

Equality holds if and only if $u = kv$ for some $k \geq 0$.

Proof: If $v = \vec{0}$, the result is easy. Thus we may assume that $v \neq \vec{0}$. As $proj_v(u)$ is a scalar multiple of v, Lemma 4.1.7 shows that $<proj_v(u), u - proj_v(u)> = 0$. Using this, the positive definite property of $< , >$, together with the linearity of inner products, we have:

$$
\begin{aligned}
0 \ &\leq \ <u - proj_v(u), u - proj_v(u)> \\
&= \ <u, u - proj_v(u)> \ - \ <proj_v(u), u - proj_v(u)> \\
&= \ [<u, u> \ - \ <u, proj_v(u)>] - 0 \\
&= \ <u, u> \ - \ <u, \frac{<u, v>}{<v, v>} v> \\
&= \ <u, u> \ - \ \frac{<u, v>}{<v, v>} <u, v> \\
&= \ <u, u> \ - \ \frac{(<u, v>)^2}{<v, v>}
\end{aligned}
$$

Hence,

$$\frac{(<u, v>)^2}{<v, v>} \ \leq \ <u, u>$$

Multiplying this inequality by the positive real number $<v, v>$ shows that $<u, u> <v, v> \geq <u, v>^2$. The inequality of the theorem follows taking by square roots.

In case $u = kv$ where $k \geq 0$, then $\| u \| = k \| v \|$ and $<u, v> = <kv, v> = k \| v \|^2 = \| u \| \| v \|$. In case $<u, v> = \| u \| \cdot \| v \|$, then the chain of equalities above shows that $<u - proj_v(u), u - proj_v(u)> = 0$. By the positive definite property of the inner product we have $u - proj_v(u) = \vec{0}$, so $u = proj_v(u)$. But then u is a scalar multiple of v; that is, $u = kv$ for $k \in \mathbf{R}$. If $k < 0$, then $<u, v> < 0 < \| u \| \cdot \| v \|$. Hence $k \geq 0$ and the theorem is proved. //

We now show that the norm associated with a real inner product satisfies all the conditions given in Definition 4.1.1. This result is well known for the standard inner product on $\mathbf{R}^n$. Many of the important applications of this result are to the infinite-dimensional case. We give a glimpse of such uses in Secs. 7.3 and 7.4.

Theorem 4.1.10 *If $< , >$ is a real inner product on V, then $\| \|$: $V \to \mathbf{R}$ defined by $\| v \| = <v, v>^{1/2}$ is a norm on V.*

Proof: (i) That $\| \|$ is positive definite follows immediately from the fact that $< , >$ is positive definite.

(ii) For $v \in V$, $k \in \mathbf{R}$, we find that

$$
\begin{aligned}
\| kv \| &= <kv, kv>^{1/2} \\
&= (k <v, kv>)^{1/2} \\
&= (k^2 <v, v>)^{1/2} \\
&= | k | <v, v>^{1/2} \\
&= | k | \cdot \| v \|
\end{aligned}
$$

(iii) Let $u, v \in V$. Then

$$
\begin{aligned}
\| u + v \|^2 &= <u + v, u + v> \\
&= <u, u> + <u, v> + <v, u> + <v, v> \\
&= <u, u> + 2 <u, v> + <v, v> \\
&\leq \| u \|^2 + 2 \| u \| \| v \| + \| v \|^2 \\
&= (\| u \| + \| v \|)^2
\end{aligned}
$$

(where the inequality follows by Definition 4.1.8 and Theorem 4.1.9). Taking square roots gives the triangle inequality, and this proves the theorem. //

Example 4.1.11 The inner product defined on $V = \mathcal{C}([0, 1], \mathbf{R})$ by

$$<f(x), g(x)> = \int_0^1 f(x)g(x)\, dx$$

must satisfy Schwarz' inequality. Written out explicitly, we obtain that for all $f(x)$, $g(x) \in V$:

$$\int_0^1 f(x)g(x)\, dx \leq \left(\int_0^1 f(x)^2\, dx \right)^{1/2} \left(\int_0^1 g(x)^2\, dx \right)^{1/2}$$

In order to see how useful the techniques of linear algebra are, we invite the reader to try to prove this result using only the tools of calculus. The norm associated with this inner product is the norm described in Example 4.1.2 (iv).

We now turn to the case of inner products on complex vector spaces. The situation is remarkably similar to the case of real inner products. In fact, in the remaining sections of this chapter, whenever possible these two cases will be treated simultaneously. After giving the definition of complex inner products, we will repeat the proofs of Lemma 4.1.7 and Theorems 4.1.9 and 4.1.10. We do this so that the reader can spot precisely where the differences are.

If $a + bi \in \mathbf{C}$ where $a, b \in \mathbf{R}$, recall that the *complex conjugate* of $a + bi$ is $\overline{a + bi} = a - bi$. Two important properties of complex conjugation we need are that for all $r, s \in \mathbf{C}$, $\overline{r} + \overline{s} = \overline{r + s}$ and $\overline{r} \cdot \overline{s} = \overline{r \cdot s}$.

Definition 4.1.3′ Let V be a complex vector space. A *complex* (or *hermitian*) *inner product* on V is a complex-valued function on $V \times V$, usually denoted by $< , >$, which satisfies:

(i) $<u, v> = \overline{<v, u>}$ for all $u, v \in V$.

(ii) $<ku, v> = k <u, v>$ for all $k \in \mathbf{C}$ and all $u, v \in V$.

(iii) $<v_1 + v_2, u> = <v_1, u> + <v_2, u>$ for all $u, v_1, v_2 \in V$.

(iv) $<v, v> \in \mathbf{R}$ and $<v, v> \geq 0$ for all $v \in V$. Moreover, $<v, v> = 0$ only if $v = \vec{0}$.

It is customary to refer to the vector space V, together with some complex inner product $< , >$, as a *hermitian inner product space*. A finite-dimensional complex inner product space is called a *unitary space*.

Property (i) of a complex inner product is known as *conjugate symmetry*, in contrast to the corresponding symmetric property from the real case. Note that because of (i), properties (ii) and (iii) have analogues on the opposite sides of the inner product bracket, but we do *not* have bilinearity in the complex case. Specifically, we have that

(ii') $<u, kv> = \bar{k} <u, v>$ for all $k \in \mathbf{C}$ and all $u, v \in V$.

(iii') $<u, v_1 + v_2> = <u, v_1> + <u, v_2>$ for all $u, v_1, v_2 \in V$.

Do not forget to conjugate scalars pulled out from the right side of a complex inner product. Taken together, properties (ii), (ii'), (iii), and (iii') are called *sesquilinearity* (or *hermitian bilinearity*). We shall see shortly how these changes affect the proofs of the main theorems given already in this section. First a few examples.

Examples 4.1.4' (i) The dot product on $\mathbf{R}^n$ can be generalized to give a hermitian inner product on $\mathbf{C}^n$. For $(a_1, \ldots, a_n)$, $(b_1, \ldots, b_n) \in \mathbf{C}^n$ we define the inner product by

$$<(a_1, \ldots, a_n), (b_1, \ldots, b_n)> = a_1\bar{b_1} + \cdots + a_n\bar{b_n}$$

With this inner product, $\mathbf{C}^n$ is often called *unitary n space*. That properties (i), (ii), (iii), and (iv) hold for this inner product is easily checked by using the previously mentioned properties of complex conjugation.

(ii) Let $V = \mathcal{C}([0, 1], \mathbf{C})$ be the vector space of continuous complex-valued functions on the interval $[0, 1]$. For $f(x), g(x) \in V$ we can define an inner product by

$$<f(x), g(x)> = \int_0^1 f(x)\overline{g(x)} \, dx$$

That properties (i), (ii), (iii), and (iv) hold follows from elementary properties of the integral.

The definitions of orthogonality, projections, and associated norms are precisely the same for complex inner products as they are in the real case. The proof of Theorem 4.1.5, giving the linear independence of mutually orthogonal vectors, applies without modification to the complex case. We will not repeat these definitions and results. The first change in our earlier arguments comes in Lemma 4.1.7. Here is this lemma, with the appropriately modified proof.

Lemma 4.1.7′   *Whenever $u, v \in V$ and $< , >$ is a complex inner product on V, the vectors v and $(u - proj_v(u))$ are orthogonal.*

Proof: We compute, using the properties of a complex inner product:

$$
\begin{aligned}
<v, u - proj_v(u)> &= <v, u - \frac{<u, v>}{<v, v>}v> \\
&= <v, u> - <v, \frac{<u, v>}{<v, v>}v> \\
&= <v, u> - \frac{\overline{<u, v>}}{<v, v>}<v, v> \\
&= <v, u> - \overline{<u, v>} \\
&= <v, u> - <v, u> \\
&= 0 \qquad\qquad //
\end{aligned}
$$

Note that the preceding proof was copied vebatim from the proof in the real case, except for the two occurences of complex conjugation, which cancel each other out. (Also needed was the fact that $<v, v> \in \mathbf{R}$ so that $\overline{<v, v>} = <v, v>$.) Next comes the complex version of Schwarz' inequality.

Theorem 4.1.9′   (Schwarz' Inequality)   *Suppose that V is a complex inner product space with inner product $< , >$. For $u, v \in V$,*

$$| <u, v> |^2 \leq (\| u \| \cdot \| v \|)^2$$

Consequently, $| <u, v> | \leq \| u \| \cdot \| v \|$.

Proof: Since $proj_v(u)$ is a scalar multiple of v, Lemma 4.1.7′ shows that $<proj_v(u), u - proj_v(u)> = 0$. Using this, as $< , >$ is positive definite and $r\bar{r} = | r |^2$ for all $r \in \mathbf{C}$, we have:

$$
\begin{aligned}
0 &\leq <u - proj_v(u), u - proj_v(u)> \\
&= <u, u - proj_v(u)> - <proj_v(u), u - proj_v(u)> \\
&= [<u, u> - <u, proj_v(u)>] - 0 \\
&= <u, u> - <u, \frac{<u, v>}{<v, v>}v> \\
&= <u, u> - \frac{\overline{<u, v>}}{<v, v>}<u, v> \\
&= <u, u> - \frac{| <u, v> |^2}{<v, v>}
\end{aligned}
$$

Multiplying this inequality by the positive real number $<v, v>$ shows that $<u, u><v, v> \geq |<u, v>|^2$. The result follows. //

We conclude this section with the complex version of Theorem 4.1.10.

Theorem 4.1.10′ *If $< , >$ is a complex inner product on V, then $\| \ \| : V \rightarrow \mathbf{R}$, defined by $\| v \| = <v, v>^{1/2}$, is a norm on V.*

Proof: (i) That $\| \ \|$ is positive definite follows immediately from the fact that $<,>$ is positive definite.

(ii) For $v \in V$ and $k \in \mathbf{C}$ we find that $\| kv \| = (<kv, kv>)^{1/2} = (k<v, kv>)^{1/2} = (k\bar{k} <v, v>)^{1/2} = (|k|^2 <v, v>)^{1/2} = |k| \cdot \| v \|$.

(iii) Let $u, v \in V$. For any complex number $a + bi$, with $a, b \in \mathbf{R}$, we denote by $re(a + bi) = a$, the real part of $a + bi$. It is easy to see that for any complex number $s \in \mathbf{C}$, $s + \bar{s} = 2re(s)$ and $re(s) \leq |s|$. We now have

$$
\begin{aligned}
\| u + v \|^2 &= <u + v, u + v> \\
&= <u, u> + <u, v> + <v, u> + <v, v> \\
&= <u, u> + <u, v> + \overline{<u, v>} + <v, v> \\
&= <u, u> + 2re(<u, v>) + <v, v> \\
&\leq <u, u> + 2|<u, v>| + <v, v> \\
&\leq \| u \|^2 + 2 \| u \| \cdot \| v \| + \| v \|^2 \\
&= (\| u \| + \| v \|)^2
\end{aligned}
$$

(The latter inequality follows applying Definition 4.1.8 and Theorem 4.1.9′.) Taking square roots gives the triangle inequality, and this proves the theorem. //

PROBLEMS

4.1.1. Let $<,>$ be an inner product on a vector space V.
(a) Show that $<\vec{0}, v> = 0$ for all $v \in V$.
(b) If $T : V \rightarrow V$ is a linear transformation and $<T(v), w> = 0$ for all $w \in V$, show that $T = 0$.

4.1.2. (a) If $S \subseteq V$ is a subset of an inner product space V, show that $S^{\perp} = \{v \in V \mid <s, v> = 0 \text{ for all } s \in S\}$ is a subspace of V.

(b) Find $S^\perp \subseteq \mathbf{R}^3$ if $S = \{(1,1,1),(2,1,0)\}$ (dot product).

(c) Find $S^\perp \subseteq \mathbf{R}^3$ if $S = \{(1,1,1),(2,1,0),(1,0,-1)\}$ (dot product).

4.1.3. (a) Show that $<(x,y),(z,w)> = 2xz + 3yw$ is an inner product on $\mathbf{R}^2$.

(b) Compute $<(1,1),(-1,1)>$ and $\|(2,7)\|$ with respect to this inner product.

(c) Find $\{v \in \mathbf{R}^2 \mid <(1,1),v> = 0\}$ for this inner product.

4.1.4. Prove the parallelogram law for an arbitrary inner product

$$\| v + w \|^2 + \| v - w \|^2 = 2 \| v \|^2 + 2 \| w \|^2 \quad \text{for all} \quad v, w \in V$$

4.1.5. Suppose that $<\ ,\ >$ is an inner product on $\mathbf{R}^n$ and A is an invertible $n \times n$ matrix. Show that the function defined by

$$<v, w>' = <Av, Aw>$$

for all $v, w \in \mathbf{R}^n$ is an inner product on $\mathbf{R}^n$.

4.1.6. Prove that for two vectors v, w in a real inner product space $\| v + w \|^2 = \| v \|^2 + \| w \|^2$ if and only if v and w are orthogonal.

4.1.7. Suppose that $\{v_1, v_2, \ldots, v_n\}$ is a basis for a vector space V with inner product $<,>$. Show for every sequence of real numbers, $r_1, r_2, \ldots, r_n$ there is a unique vector $w \in V$ such that $<v_i, w> = r_i$ for all $1, 2, 3, \ldots, n$.

4.1.8. If V is a vector space with norm $\| \ \|$, show:

(a) $d(v, w) \geq 0$.

(b) $d(v, w) = 0$ if and only if $v = w$.

(c) $d(v, w) = d(w, v)$.

(d) $d(u, v) \leq d(u, w) + d(v, w)$.

4.1.9. Show that the function $<A, B> = tr(AB^t)$ defines an inner product on the vector space of real $n \times n$ matrices $M_{n \times n}(\mathbf{R})$. (Here, tr denotes the trace, that is, the sum of the diagonal entries of the matrix.)

4.1.10. If V is an inner product space show that

$$\| u - v \|^2 = \| u \|^2 - 2re <u, v> + \| v \|^2$$

for all $u, v \in V$.

4.1.11. Consider the vector space $Hom(\mathbf{R}^n, \mathbf{R}^m)$ of all linear transformations from $\mathbf{R}^n$ to $\mathbf{R}^m$. If $\| \ \|$ denotes the standard norms on $\mathbf{R}^n$ and

$\mathbf{R}^m$, prove that the function $\| T \| = max\{ \| T(v) \| / \| v \|$ where $v \in \mathbf{R}^n$ and $\| v \| = 1\}$ is a norm on $Hom(\mathbf{R}^n, \mathbf{R}^m)$.

4.1.12.   Suppose that $a_1, a_2, \ldots, a_n$ are positive real numbers. Show that $\| (x_1, \ldots, x_n) \| = a_1 \, | \, x_1 \, | + a_2 \, | \, x_2 \, | + \cdots + a_n \, | \, x_n \, |$ is a norm on $\mathbf{R}^n$. Show also that $\| (x_1, \ldots, x_n) \| = max\{ a_1 \, | \, x_1 \, |, \ldots, a_n \, | \, x_n \, | \}$ is a norm on $\mathbf{R}^n$.

4.1.13.   Is $\| (x, y) \| = (x^2 + 2xy + 4y^2)^{1/2}$ a norm on $\mathbf{R}^2$? How about $\| (x, y) \| = (x^2 + 3xy + 2y^2)^{1/2}$?   What conditions must $a, b, c \in \mathbf{R}$ satisfy in order that $\| (x, y) \| = (ax^2 + bxy + cy^2)^{1/2}$ is a norm on $\mathbf{R}^2$?

4.1.14.   Let $K(t)$ be any continuous real valued function on $[a, b]$ with $K(t) > 0$ for all $t \in (a, b)$. Verify that

$$<f(x), g(x)> = \int_a^b f(t)g(t)K(t) \, dt$$

defines an inner product on $\mathcal{C}([a, b], \mathbf{R})$, the vector space of continuous real-valued functions on $[a, b]$.

4.1.15.   Suppose that V is a finite-dimensional inner product space and $T : V \to V$ is a linear operator. If $\| T(v) \| = \| v \|$ for all $v \in V$, show that T is an isomorphism.

4.1.16.   (a) If V is a real inner product space, prove the *polar identity*

$$<u, v> = (1/4) \| u + v \|^2 - (1/4) \| u - v \|^2$$

for all $u, v \in V$.
(b) If V is a complex inner product space, prove the *polar identity*

$$<u, v> = (1/4) \left(\| u + v \|^2 + i \| u + iv \|^2 \right.$$
$$\left. - \| u - v \|^2 - i \| u - iv \|^2 \right)$$

for all $u, v \in V$.

4.2   Orthogonal Bases

When working with vectors in $\mathbf{R}^n$, the situation is so familiar that it is easy to take for granted some of the special features of cartesian coordinates. For example, the standard basis vectors e_i are mutually orthogonal, and each has length 1. Consequently, for any vector $w =$

$(a_1, \ldots, a_n) \in \mathbf{R}^n$, the projection $proj_{e_i}(w) = a_i e_i$. In general, it is extremely useful to have bases for an inner product space with these properties. Not only are such bases useful for computations, they help us understand the behavior of the inner product as well.

Throughout the next two sections the phrase "$<\,,\,>$ is an inner product on V" will mean that $<\,,\,>$ is either a real or a complex innner product on the real or complex vector space V. All definitions and calculations will apply equally well to either case. Thus there will be no need for separate proofs as we gave in the last section. We begin with the definition of orthogonal and orthonormal bases for arbitrary inner product spaces.

Definition 4.2.1 Suppose that $<\,,\,>$ is an inner product on a vector space V. We say that a basis $\{u_1, u_2, \ldots, u_n\}$ is an *orthogonal* basis for V if the u_i are mutually orthogonal; that is, $<u_i, u_j> = 0$ whenever $i \neq j$. In addition, if each $\| u_i \| = 1$ (that is, each u_i is a *unit vector*), we say that the basis is *orthonormal*.

The standard bases of $\mathbf{R}^n$ and of $\mathbf{C}^n$ are orthonormal. However, there are many other possible orthonormal bases for these and other inner product spaces. Orthonormal bases behave extremely well when it comes to coordinatization. In fact, given an orthonormal basis for a finite-dimensional inner product space, the innner product computes the coordinates for you. This is the content of the next theorem. As noted above, the next result follows by definition in the special case of the standard basis for $\mathbf{R}^n$.

Theorem 4.2.2 *Suppose that $u_1, u_2, \ldots, u_n$ is an orthonormal basis of a vector space V with inner product $<\,,\,>$. Then for any $v \in V$,*

$$v = a_1 u_1 + a_2 u_2 + \cdots + a_n u_n$$

where $a_i = <v, u_i>$.

Proof: Since $u_1, u_2, \ldots, u_n$ is a basis of V, we know that there is a unique expression of the form $v = a_1 u_1 + a_2 u_2 + \cdots + a_n u_n$. We must show that $a_i = <v, u_i>$. We compute:

$$
\begin{aligned}
<v, u_i> \;&=\; <a_1 u_1 + a_2 u_2 + \cdots + a_n u_n, u_i> \\
&=\; <a_1 u_1, u_i> + \cdots + <a_i u_i, u_i> + \cdots + <a_n u_n, u_i>
\end{aligned}
$$

$$
\begin{aligned}
&= \; a_1 <u_1, u_i> + \cdots + a_i <u_i, u_i> + \cdots + a_n <u_n, u_i> \\
&= \; a_1 0 + \cdots + a_i 1 + \cdots + a_n 0 \\
&= \; a_i
\end{aligned}
$$

This proves the theorem. //

Often one needs to find an orthogonal (orthonormal) basis for an inner product space. Suppose a basis for the vector space is given. There is an algorithm for producing an orthonormal basis from a given basis, known as the *Gram-Schmidt procedure*. The key component of this algorithm is given in the following theorem.

Theorem 4.2.3 (Gram-Schmidt Theorem) *Suppose that* $w_1, \ldots, w_m$ *are mutually orthogonal and nonzero vectors in an inner product space* V. *Let* $y \in V$ *and set*

$$
\begin{aligned}
w_{m+1} &= \; y - \sum_{j=1}^{m} \frac{<y, w_j>}{<w_j, w_j>} w_j \\
&= \; y - \sum_{j=1}^{m} proj_{w_j}(y)
\end{aligned}
$$

Then the vectors $w_1, w_2, \ldots, w_m, w_{m+1}$ *are mutually orthogonal and* $span\{w_1, \ldots, w_m, y\} = span\{w_1, \ldots, w_m, w_{m+1}\}$. *Further,* $w_{m+1} = \vec{0}$ *if and only if* $y \in span\{w_1, \ldots, w_m\}$.

Proof: By the definition of w_{m+1}, one sees immediately that $y \in span\{w_1, \ldots, w_m, w_{m+1}\}$ and $w_{m+1} \in span\{w_1, \ldots, w_m, y\}$. From this, $span\{w_1, \ldots, w_m, w_{m+1}\} = span\{w_1, \ldots, w_m, y\}$ follows. In order to prove the orthogonality of $w_1, \ldots, w_m, w_{m+1}$, as $w_1, \ldots, w_m$ are mutually orthogonal, it remains to show that $<w_{m+1}, w_k> = 0$ for each $k = 1, 2, \ldots, m$. For this, we compute, using the definitions:

$$
\begin{aligned}
<w_{m+1}, w_k> &= \; <y - \sum_{j=1}^{m} \frac{<y, w_j>}{<w_j, w_j>} w_j, w_k> \\
&= \; <y, w_k> - \sum_{j=1}^{m} \frac{<y, w_j>}{<w_j, w_j>} <w_j, w_k> \\
&= \; <y, w_k> - \frac{<y, w_k>}{<w_k, w_k>} <w_k, w_k> \\
&= \; <y, w_k> - <y, w_k> \\
&= \; 0
\end{aligned}
$$

For the final statement, if $w_{m+1} = \vec{0}$, then $y \in span\{w_1, \ldots, w_m\}$. Conversely, suppose $y \in span\{w_1, \ldots, w_m\}$ and set $u_i = (1/\| w_i \|)w_i$. Then $\{u_1, \ldots, u_m\}$ is an orthonormal basis for $span\{w_1, \ldots, w_m\}$, and Theorem 4.2.2 shows that

$$y = \sum_{i=1}^{m} <y, u_i> u_i = \sum_{i=1}^{m} \frac{1}{\| w_i \|^2} <y, w_i> w_i$$

From this $w_{m+1} = \vec{0}$ follows and the theorem is proved. //

The next theorem follows by induction from the Gram-Schmidt theorem stated above. The proof is the algorithm known as the Gram-Schmidt procedure.

Theorem 4.2.4 *Any finite-dimensional inner product space V has an orthogonal basis (and hence an orthonormal basis).*

Proof: Suppose that $\{v_1, v_2, \ldots, v_n\}$ is a basis for V. We construct an orthogonal basis $\{w_1, w_2, \ldots, w_n\}$ for V using $v_1, v_2, \ldots, v_n$. First set $w_1 = v_1$ ($\neq \vec{0}$). Of course, $span\{w_1\} = span\{v_1\}$. For k with $1 < k \leq n$ we define w_k inductively. We set

$$w_k = v_k - \sum_{j=1}^{k-1} \frac{<v_k, w_j>}{<w_j, w_j>} w_j$$

(the Gram-Schmidt orthogonalization). Theorem 4.2.3 guarantees that the w_1, w_2, $\ldots$, w_k are mutually orthogonal, $span\{w_1, \ldots, w_k\} = span\{w_1, \ldots, w_{k-1}, v_k\} = span\{v_1, \ldots, v_k\}$. Applying the linear independence of the y_1, y_2, $\ldots$, y_k also gives that w_k is nonzero (and thus the definition of w_{k+1} will make sense). After n steps, we have a collection of vectors $w_1, w_2, \ldots, w_n$ which are mutually orthogonal and nonzero and for which $span\{w_1, w_2, \ldots, w_n\} = span\{v_1, v_2, \ldots, v_n\} = V$. By Theorem 4.1.5, the $w_1, w_2, \ldots, w_n$ are linearly independent and hence form an orthogonal basis for V.

If an orthonormal basis is desired, one sets $u_i = (1/ \| w_i \|)w_i$ so that $\| u_i \| = 1$. Clearly, $u_1, u_2, \ldots, u_n$ are mutually orthogonal, since $w_1, w_2, \ldots, w_n$ were. This proves the theorem. //

Remark 4.2.5 The definition of w_k in the preceding proof is the key step in the Gram-Schmidt process. Often, it is convenient to normalize

w_k to u_k immediately after its computation. When doing so, one can use the u_j for $1 \leq j < k$ in defining w_k. This has the advantage that $<u_j, u_j> = 1$ in this calculation. However, in some applications it is not convenient to normalize until the end.

Example 4.2.6 Consider the subspace V of $\mathbf{R}^4$ with basis the three vectors $v_1 = (2, 1, 0, 2)$, $v_2 = (0, 1, 1, 4)$, $v_3 = (3, 3, 3, 0)$. The vector space V has an inner product given by the usual dot product on $\mathbf{R}^4$. We apply the Gram-Schmidt process, and we will normalize our w_k to u_k as we go along. Since $\| (2, 1, 0, 2) \| = 3$, we obtain that $u_1 = (2/3, 1/3, 0, 2/3)$. Next, $w_2 = v_2 - proj_{u_1}(v_1) = (0, 1, 1, 4) - (9/3)(2/3, 1/3, 0, 2/3) = (-2, 0, 1, 2)$. Since $\| w_2 \| = 3$, we obtain that $u_2 = (-2/3, 0, 1/3, 2/3)$. Finally, we set $w_3 = v_3 - proj_{u_1}(v_3) - proj_{u_2}(v_3) = (3, 3, 3, 0) - (3)(2/3, 1/3, 0, 2/3) - (-1)(-2/3, 0, 1/3, 2/3) = (1/3, 2, 10/3, -4/3)$. Then $\| w_2 \| = \sqrt{153/9} = \sqrt{17}$. We set $u_3 = \sqrt{17}w_3 = \sqrt{17}(1/51, 2/17, 10/51, -4/51)$. $\{u_1, u_2, u_3\}$ is the desired orthonormal basis of V.

Example 4.2.7 Consider the n^2-dimensional complex vector space $M_{n,n}(\mathbf{C})$ of all complex $n \times n$ matrices. For any $A = (a_{ij}) \in M_{n,n}(\mathbf{C})$ we define the *trace* of A by $tr(A) = a_{11} + a_{22} + \cdots + a_{nn}$. It is easy to check that $tr : M_{n \times n}(\mathbf{C}) \to \mathbf{C}$ is a linear transformation. (This is true over any field.) Using the trace, we can define a hermitian inner product on $M_{n \times n}(\mathbf{C})$ by setting $< A, B > = tr(A\overline{B}^t)$, where $\overline{B}^t$ is the transpose of the matrix obtained from B by replacing each entry of B by its complex conjugate. Properties (i), (ii), and (iii) of a complex inner product are easily verified by using the properties of matrix multiplication. To check property (iv), we observe that the iith diagonal entry of $A\overline{A}^t$ is $a_{i1}\overline{a}_{i1} + a_{i2}\overline{a}_{i2} + \cdots + a_{in}\overline{a}_{in}$ which is a positive real number. This inner product space has a natural orthonormal basis given by the elements $E_{ij} \in M_{n \times n}(\mathbf{C})$ where E_{ij} is defined to be the $n \times n$ matrix which has 0 as every entry except the ijth entry which is 1.

Example 4.2.8 Consider the vector space P^2 of all real polynomials of degree at most 2. P^2 has an inner product given by $<f(X), g(X)> = \int_0^1 f(X)g(X)\,dX$. The standard basis $\{1, X, X^2\}$ for P^2 is not orthogonal for this inner product. We apply the Gram-Schmidt procedure to obtain an orthogonal basis: We take $w_1 = 1$. Easily, we see $<1,1>=1$ and $<X,1>=1/2$. Consequently, we have $w_2 = X-(1/2)1 = X-1/2$. To compute w_3, we calculate that $<w_2, w_2> = \int_0^1 (X-1/2)^2\,dX = 1/3 - 1/2 + 1/4 = 1/12$, $<X^2, 1> = 1/3$, and $<X_2, X - 1/2> = 1/4 - 1/6 = 1/12$. Consequently, $w_3 = X^2 - (1/3)1 - (1/12)/(1/12)(X - 1/2) = X^2 - X + 1/6$. This gives an orthogonal basis for P^2.

Suppose that $v_1, v_2, \ldots, v_m$ are linearly independent vectors in $\mathbf{R}^n$ and $A = (v_1 v_2 \ldots v_m)$ is the $n \times m$ matrix whose columns are the v_i. The (normalized version of the) Gram-Schmidt process, applied to the v_i, gives an orthonormal set of vectors $q_1, q_2, \ldots, q_m$ with the same span as the v_i. Suppose that Q is the matrix $(q_1\ q_2\ \cdots\ q_m)$. The relationship between the matrix A and the matrix Q is important. Since the Gram-Schmidt process was used to find the q_i, the vector q_1 is a multiple of v_1, the vector q_2 is a linear combination of v_1 and v_2, the vector q_3 is a linear combination of v_1, v_2, v_3, and so forth. We express $q_1 = r_{11}v_1$, $q_2 = r_{12}v_1 + r_{22}v_2$, and in general $q_j = r_{1j}v_1 + r_{2j}v_2 + \cdots + r_{jj}v_j$ for each j. If we define $r_{ij} = 0$ whenever $m \geq i > j$ and set T to be the $m \times m$ matrix (r_{ij}), the definition of matrix multiplication gives $Q = AT$. The matrix T is upper triangular, and since each entry r_{jj} is necessarily nonzero, T is invertible. It follows that $A = QR$, where $R = T^{-1}$ is also upper triangular.

Whenever A is a real $n \times m$ matrix of rank m, an equation of the form $A = QR$, where R is an upper triangular invertible matrix and Q has orthonormal columns, is called a *Q-R factorization* of A. The Gram-Schmidt algorithm guarantees any such A has a Q-R factorization.

Example 4.2.9 Consider

$$A = \begin{pmatrix} 1 & 1 \\ 0 & 1 \\ 1 & 1 \end{pmatrix}$$

To find a Q-R factorization of A, we apply the Gram-Schmidt procedure to the columns of A. We obtain the orthonormal vectors

$$u_1 = \frac{1}{\sqrt{2}} \begin{pmatrix} 1 \\ 0 \\ 1 \end{pmatrix} \quad \text{and} \quad u_2 = \begin{pmatrix} 0 \\ -1 \\ 0 \end{pmatrix} = \begin{pmatrix} 1 \\ 0 \\ 1 \end{pmatrix} - \begin{pmatrix} 1 \\ 1 \\ 1 \end{pmatrix}$$

This gives the matrix equation

$$\begin{pmatrix} 1/\sqrt{2} & 2 \\ 0 & -1 \\ 1/\sqrt{2} & 0 \end{pmatrix} = \begin{pmatrix} 1 & 1 \\ 0 & 1 \\ 1 & 1 \end{pmatrix} \begin{pmatrix} 1/\sqrt{2} & 1 \\ 0 & -1 \end{pmatrix}$$

From which we obtain the Q-R factorization

$$\begin{pmatrix} 1 & 1 \\ 0 & 1 \\ 1 & 1 \end{pmatrix} = \begin{pmatrix} 1/\sqrt{2} & 2 \\ 0 & -1 \\ 1/\sqrt{2} & 0 \end{pmatrix} \begin{pmatrix} \sqrt{2} & \sqrt{2} \\ 0 & -1 \end{pmatrix}$$

PROBLEMS

4.2.1. Apply the Gram-Schmidt proceedure to the vectors $(0, 3, 4)$, $(2, 0, 0)$, $(0, 1, 1)$ in $\mathbf{R}^3$ to find an orthonormal basis (usual dot product).

4.2.2. (a) Find an orthogonal basis for the subspace of $\mathbf{R}^4$ spanned by the vectors $v_1 = (1, 1, 0, 0)$, $v_2 = (1, 0, 1, 0)$, $v_3 = (1, 0, 0, 1)$ (usual dot product).
(b) Find a basis for the subspace $W = \{w \in \mathbf{R}^4 \,|\, <w, v_i> = 0, i = 1, 2, 3\}$.

4.2.3. Find an orthogonal basis for $\mathbf{R}^2$ with the inner product given by

$$<(x_1, y_1), (x_2, y_2)> = x_1 x_2 + (1/2)(x_1 y_2 + x_2 y_1) + y_1 y_2$$

4.2.4. Find an orthogonal basis for $\mathbf{R}^3$ with the inner product given by

$$<(x_1, y_1, z_1), (x_2, y_2, z_2)> = x_1 y_1 + (1/2)x_1 y_3 + (1/2)x_3 y_1 + 2x_2 y_2 + x_3 y_3$$

4.2.5. Let P^2 be the vector space of all polynomial functions $f : \mathbf{R} \to \mathbf{R}$ of degree at most 2. Find an orthonormal basis for V, using the inner product given by

$$<f(x), g(x)> = \int_{-1}^{1} f(x)g(x)\, dx$$

4.2.6. Let (a_1, b_1, c_1), $(a_2, b_2, c_2) \in \mathbf{R}^3$. Recall that the cross product is defined by $(a_1, b_1, c_1) \times (a_2, b_2, c_2) = (b_1 c_2 - c_1 b_2, c_1 a_2 - a_1 c_2, a_1 b_2 - a_2 b_1)$.

Show that $(a_1, b_1, c_1) \times (a_2, b_2, c_2)$ is orthogonal to both (a_1, b_1, c_1) and (a_2, b_2, c_2).

4.2.7.   Prove or disprove: If $u, v, w \in V$ where V is an inner product space, $u \perp v$, and $v \perp w$, then $u \perp w$.

4.2.8.   If $\{u_1, u_2, \ldots, u_n\}$ is an orthonormal basis of an inner product space V, show that for any $v \in V$, $\| v \|^2 = <v, u_1>^2 + <v, u_2>^2 + \cdots + <v, u_n>^2$.

4.2.9.   Find Q-R factorizations of the following matrices.

(a) $\begin{pmatrix} 1 & 1 \\ 0 & 1 \end{pmatrix}$ (b) $\begin{pmatrix} 1 & 0 \\ 1 & 1 \\ 0 & 1 \end{pmatrix}$ (c) $\begin{pmatrix} 1 & -1 & 1 \\ 1 & 0 & 0 \\ 1 & -2 & 0 \\ 1 & 1 & 1 \end{pmatrix}$

4.2.10.   (a) Let $\mathcal{L}$ be the line spanned by a nonzero vector $v \in \mathbf{R}^n$. Let $w \in \mathbf{R}^n$ (viewed as a point, as well as a vector). Show that the distance between w and $\mathcal{L}$ is $\| w - proj_v(w) \|$.

(b) Generalize (a) above and describe how, using the Gram- Schmidt theorem, to find the distance between a point $w \in \mathbf{R}^n$ and the subspace $span\{v_1, v_2, \ldots, v_s\}$.

(c) Find the distance between the point $(1, 1, 1, 0)$ and the subspace $V = span\{(2, 0, 0, 1), (1, 1, 0, 0)\}$ in $\mathbf{R}^4$.

4.3   Orthogonal Projections and Direct Sums

Our first task in this section is to generalize the notion of an orthogonal projection onto a vector (Definition 4.1.6) to that of an orthogonal projection onto a subspace. These ideas lead naturally to the concept of a direct sum, which plays a big role in linear algebra. We shall study direct sums more extensively in Chap. 6, when we return to the study of linear operators.

   An orthogonal projection is like shadows cast at midday on a flat plane, the shadows being the images of three-dimensional objects under the projection. (At times other than midday, the projection is not orthogonal.) To give a precise algebraic description of the orthogonal projection onto a subspace, one can use projections onto various vectors. In this description one first projects onto an orthonormal basis for the subspace, and then the sum of the results is the desired projection.

   Unfortunately, this construction is technically inconvenient, because it depends upon the choice of a (fixed) orthonormal basis of the subspace in question. Thus we must check that different orthonormal bases do not result in different projection functions (that is, we must show that our description is well-defined). In Definition 4.3.1 we adopt a different approach and define the orthogonal projection to be a linear operator on V which satisfies some special conditions. Theorem 4.3.3, which follows the definition, establishes the existence and uniqueness of such projection operators onto finite-dimensional subspaces by checking that the above description has the desired properties. An alternative (matrix-oriented) computation of projection operators on $\mathbf{R}^n$ is given in Theorem 4.4.6.

Definition 4.3.1 Suppose that V is an inner product space and S is a subset of V. The *orthogonal complement* of S in V is defined to be the subspace $S^\perp$ (read "S perp") defined by $S^\perp = \{v \in V \mid <v, w> = 0 \text{ for all } w \in S\}$. If W is a finite-dimensional subspace of V, the *orthogonal projection* $Proj_W : V \to V$ is the (unique) linear operator for which $Proj_W(w) = w$ whenever $w \in W$, and $Proj_W(v) = \vec{0}$ whenever $v \in W^\perp$.

For example, consider the subspace W of $\mathbf{R}^4$ spanned by the vectors $v_1 = (2, 1, 0, 0)$, $v_2 = (1, 1, 1, 1)$, and $v_3 = (0, 0, 1, 2)$. These three vectors are linearly independent, so W is three-dimensional. It is easy to see that a vector w' lies in the orthogonal complement of W if and only if w' is orthogonal to each of these three vectors which span W. Thus, $W^\perp$ is the subspace $\{w' \in \mathbf{R}^4 \mid w' \cdot v_1 = w' \cdot v_2 = w' \cdot v_3 = 0\}$. Carrying out the dot product calculation for $w' = (w_1, w_2, w_3, w_4)$ gives the system of equations $2w_1 + w_2 + 0w_3 + 0w_4 = 0$, $w_1 + w_2 + w_3 + w_4 = 0$, and $0w_1 + 0w_2 + w_3 + 2w_4 = 0$. The solution set to this system is the subspace $W^\perp = \{k(-1, 2, -2, 1) \mid k \in \mathbf{R}\}$.

This first lemma is an important consequence of the Gram-Schmidt Theorem 4.2.3. It says that a finite-dimensional vector space is always spanned by a subspace together with its orthogonal complement. We point out that the existence assertion in this lemma is strengthened to an existence and uniqueness assertion in Corollary 4.3.7.

Lemma 4.3.2 *Let W be a finite-dimensional subspace of a vector space V with inner product $< , >$. Then every $y \in V$ can be expressed $y = w + w'$ with $w \in W$ and $w' \in W^\perp$; that is, $V = W + W^\perp$.*

Proof: Let $y \in V$. Since W is finite-dimensional, it has an orthonormal basis $\{u_1, u_2, \ldots, u_s\}$. In case $y \in span\{u_1, u_2, \ldots, u_s\} = W$, then $y = y + \vec{0}$ is the desired expression. In case $y \notin span\{u_1, u_2, \ldots, u_s\}$, then by the Gram-Schmidt Theorem 4.2.3 we find that $w' = y - \sum_{j=1}^{s} <y, u_j> u_j$ is orthogonal to each of $u_1, u_2, \ldots, u_s$. Since $u_1, u_2, \ldots, u_s$ span W, any element $z \in W$ is a linear combination of $u_1, u_2, \ldots, u_s$ and consequently, $<w', z> = 0$ for all $z \in W$. This shows that $w' \in W^{\perp}$. Setting $w = \sum_{j=1}^{s} <y, u_j> u_j$, we find that $y = w + w'$ with $w \in W$ and $w' \in W^{\perp}$. This proves the lemma. //

As defined, the orthogonal projection $Proj_W : V \rightarrow W$ is a linear operator satisfying certain properties. We must show that such an operator exists and the specified conditions determine the operator uniquely. The latter justifies the terminology of "the" orthogonal projection.

Theorem 4.3.3 *Suppose that $<\ ,\ >$ is an inner product on V and W is a finite-dimensional subspace of V. Then there is a unique orthogonal projection $Proj_W : V \rightarrow V$. Let $u_1, u_2, \ldots, u_s$ be an orthonormal basis of W. The orthogonal projection $Proj_W$ can be computed by the formula*

$$Proj_W(v) = <v, u_1> u_1 + <v, u_2> u_2 + \cdots + <v, u_s> u_s$$

Proof: Consider the function $Proj_W : V \rightarrow V$ defined by $Proj_W(v) = <v, u_1> u_1 + <v, u_2> u_2 + \cdots + <v, u_s> u_s$. In view of the linearity properties (ii) and (iii) of $<\ ,\ >$, $Proj_W$ is a linear transformation. By Theorem 4.2.2 we find that whenever $w \in W$, $Proj_W(w) = w$. If $v \in W^{\perp}$, then $<v, u_i> = 0$ for each $i = 1, 2, \ldots, s$, so $Proj_W(v) = \vec{0}$. Thus $Proj_W$ meets the criteria of Definition 4.3.1 for being an orthogonal projection.

We must show that the projection $Proj_W$ is unique. For this, let P_W be some linear transformation which satisfies the two criteria of Definition 4.3.1. We must show that $P_W = Proj_W$ as linear operators on V. For $y \in V$, we know by Lemma 4.3.2 (and its proof) that there is an expression $y = w + w'$ where $w' \in W^{\perp}$ and $w = \sum_{j=1}^{s} <y, u_j> u_j = Proj_W(y) \in W$. Since P_W is an orthogonal projection, $P_W(w') = \vec{0}$, $P_W(w) = w$, and consequently $P_W(y) = P_W(w) + P_W(w') = w + \vec{0} = w = Proj_W(y)$. This proves the theorem. //

In order to illustrate Theorem 4.3.3, we recall the three-dimensional subspace $W = span\{(2, 1, 0, 0), (0, 0, 1, 2), (1, 1, 1, 1)\} \subseteq \mathbf{R}^4$, considered earlier in this section. We observed that $W^{\perp} = span\{(1, -2, 2, -1)\}$. One can check that

$$\{(1/\sqrt{5})(2, 1, 0, 0), (1/\sqrt{10})(1, -2, -2, 1), (1/\sqrt{5})(0, 0, 1, 2)\}$$

is an orthonormal basis for W and $\{(1/\sqrt{10})(1, -2, 2, -1)\}$ is an orthonormal basis for $W^{\perp}$. The projection formula in Theorem 4.3.3 shows that whenever $(x, y, z, w) \in \mathbf{R}^4$,

$$\begin{aligned} Proj_W(x, y, z, w) &= (1/5)(2x + y)(2, 1, 0, 0) \\ &\quad + (1/10)(x - 2y - 2z + w)(1, -2, -2, 1) \\ &\quad + (1/5)(z + 2w)(0, 0, 1, 2) \end{aligned}$$

and

$$Proj_{W^{\perp}}(x, y, z, w) = (1/10)(x - 2y + 2z - w)(1, -2, 2, -1)$$

It follows that the matrices corresponding to these projection operators are

$$[Proj_W] = \frac{1}{10} \begin{pmatrix} 9 & 2 & -2 & 1 \\ 2 & 6 & 4 & -2 \\ -2 & 4 & 6 & 2 \\ 1 & -2 & 2 & 9 \end{pmatrix}$$

and

$$[Proj_{W^{\perp}}] = \frac{1}{10} \begin{pmatrix} 1 & -2 & -2 & -1 \\ -2 & 4 & 4 & 2 \\ 2 & -4 & 4 & -2 \\ -1 & 2 & -2 & 1 \end{pmatrix}$$

The reader should notice that $[Proj_W] + [Proj_{W^{\perp}}] = I_4$, and consequently the linear operator $Proj_W + Proj_{W^{\perp}}$ is the identity. This is an important feature of orthogonal projections and is proved more generally in Corollary 4.3.7 (ii).

If W is a subspace of a finite-dimensional inner product space V, then the relationship between W and $W^{\perp}$ is extremely important. Note that if $w \in W \cap W^{\perp}$, then as $Proj_W(w) = w$ and $Proj_{W^{\perp}}(w) = \vec{0}$, necessarily $w = \vec{0}$. In particular, $W + W^{\perp} = V$ and $W \cap W^{\perp} = \{\vec{0}\}$. We isolate these important conditions in the following definition of the direct sum.

Definition 4.3.4 Suppose that W_1 and W_2 are subspaces of a vector space V. Then $V = W_1 \oplus W_2$ if

(i) $W_1 \cap W_2 = \{\vec{0}\}$.
(ii) $V = W_1 + W_2$, that is, every $v \in V$ can be expressed as $v = w_1 + w_2$ where $w_1 \in W_1$ and $w_2 \in W_2$.
When this occurs we say that W is the *direct sum* of W_1 and W_2.

Note that the definition of the direct sum has nothing to do with inner products. The direct sums that arise when considering orthogonal complements of a subspace are only a very special type. The general situation will be investigated in detail in Chap. 6 (see Definition 6.3.1). The condition that $W_1 \cap W_2 = \{0\}$ is the same as saying that whenever $w_1 \in W_1$ and $w_2 \in W_2$ are nonzero vectors, then w_1 and w_2 are linearly independent. (This is because two nonzero vectors are linearly independent if and only if they are not scalar multiples of one another.) Consequently, whenever two subspaces W_1 and W_2 satisfy $W_1 \cap W_2 = \{\vec{0}\}$, it is customary to say that W_1 and W_2 are *independent*.

Examples 4.3.5 (i)   As pointed out above, if $<\,,\,>$ is an inner product on a finite-dimensional vector space V and W is a subspace of V, then $V = W \oplus W^\perp$.
(ii) Suppose that $\{v_1, \ldots, v_s, v_{s+1}, \ldots, v_n\}$ is a basis of a vector space V. Set $W_1 = span\{v_1, \ldots, v_s\}$ and $W_2 = span\{v_{s+1}, \ldots, v_n\}$. Then $V = W_1 \oplus W_2$, because if $v = a_1 v_1 + a_2 v_2 + \cdots + a_n v_n \in V$, then $v = (a_1 v_1 + \cdots + a_s v_s) + (a_{s+1} v_{s+1} + \cdots + a_n v_n) \in W_1 + W_2$. Now suppose that $v \in W_1 \cap W_2$. Then, $v = a_1 v_1 + a_2 v_2 + \cdots + a_s v_s = a_{s+1} v_{s+1} + \cdots + a_n v_n$; this gives $a_1 v_1 + a_2 v_2 + \cdots + a_s v_s - a_{s+1} v_{s+1} - \cdots - a_n v_n = \vec{0}$. The linear independence of $v_1, \ldots, v_n$ shows that $a_1 = 0, a_2 = 0, \ldots, a_n = 0$, and thus $v = \vec{0}$.

This next lemma gives two key properties of direct-sum decompositions of vector spaces.

Lemma 4.3.6 *Suppose that $V = W_1 \oplus W_2$. Then*
(i)   *Every $v \in V$ can be expressed* uniquely *as $v = w_1 + w_2$ where $w_1 \in W_1$ and $w_2 \in W_2$.*
(ii)   *If $\{v_1, \ldots, v_r\}$ is a basis for W_1 and $\{u_1, \ldots, u_s\}$ is a basis for W_2, then $\{w_1, \ldots, w_r, u_1, \ldots, u_s\}$ is a basis for V. In particular, $dim(V) = dim(W_1) + dim(W_2)$.*

Proof: (i) By definition such an expression for v exists. We must show uniqueness. Suppose that $v = w_1 + w_2 = w_1' + w_2'$ where $w_1, w_1' \in W_1$ and $w_2, w_2' \in W_2$. Then $(w_1 - w_1') = (w_2' - w_2) \in W_1 \cap W_2$. It follows that $w_1 - w_1' = \vec{0} = w_2' - w_2$. Hence $w_1 = w_1'$ and $w_2 = w_2'$; this gives uniqueness.

(ii) Assume that $\{v_1, \ldots, v_r\}$ is a basis for W_1 and $\{u_1, \ldots, u_s\}$ is a basis for W_2. Let $v \in V$. Expressing $v = w_1 + w_2$ with $w_1 \in W_1$ and $w_2 \in W_2$ and then writing $w_1 = a_1 v_1 + \cdots + a_r v_r$ and $w_2 = b_1 u_1 + \cdots + b_s u_s$ gives that $v = w_1 + w_2 = a_1 v_1 + \cdots + a_r v_r + b_1 u_1 + \cdots + b_s u_s$. This shows that $v_1, \ldots, v_r, u_r, \ldots, u_s$ span V.

Next suppose that $\vec{0} = a_1 v_1 + \cdots + a_r v_r + b_1 u_1 + \cdots + b_s u_s$. If $w_1 = a_1 v_1 + \cdots + a_r v_r$ and $w_2 = b_1 u_1 + \cdots + b_s u_s$, then the expression $\vec{0} = \vec{0} + \vec{0} = w_1 + w_2$ together with the uniqueness assertion of part (i) gives that $\vec{0} = w_1$ and $\vec{0} = w_2$. The linear independence of $v_1, \ldots, v_r$ gives that $a_1 = 0, \ldots, a_r = 0$, and the linear independence of $u_1, \ldots, u_s$ gives that $b_1 = 0, b_2 = 0, \ldots, b_s = 0$. This shows that $v_1, \ldots, v_r, u_1, \ldots, u_s$ are linearly independent, and this proves the lemma. //

Applying Lemma 4.3.6 to the case of inner product spaces immediately gives the following.

Corollary 4.3.7 *Suppose V is a finite-dimensional inner product space and W is a subspace of V.*
(i) *If $dim(V) = n$ and $dim(W) = s$, then $dim(W^\perp) = n - s$.*
(ii) *For every $v \in V$, there exist* unique *$w_1 \in W$ and $w_2 \in W^\perp$ such that $v = w_1 + w_2$. In this situation $w_1 = Proj_W(v)$ and $w_2 = Proj_{W^\perp}(v)$, which shows that $Proj_W + Proj_{W^\perp} = I$.*

Proof: (i) This is immediate in view of Lemma 4.3.6 (ii).

(ii) The first statement of (ii) follows from Lemma 4.3.6 (i). To prove the second statement of (ii), let $\{v_1, \ldots, v_r\}$ be an orthonormal basis of W and $\{u_1, \ldots, u_t\}$ be an orthonormal basis of $W^\perp$. Then $\{v_1, \ldots, v_s, u_1, \ldots, u_t\}$ is an orthonormal basis of V. If $v \in V$, then

$$
\begin{aligned}
v &= <v, v_1> v_1 + \cdots + <v, v_r> v_r + <v, u_1> u_1 + \cdots + <v, u_s> u_s \\
&= (<v, v_1> v_1 + \cdots + <v, v_r> v_r) + (<v, u_1> u_1 + \cdots + <v, u_s> u_s) \\
&= Proj_W(v) + Proj_{W^\perp}(v)
\end{aligned}
$$

by Theorem 4.3.3. It now follows from the uniqueness assertion of Lemma 4.3.6 (i) that $w_1 = Proj_W(v)$ and $w_2 = Proj_{W^\perp}(v)$. This gives the corollary. //

Projection operators and direct sums are intimately related. The properties given next can be used to give a definition of general projection operators, outside of the context of an inner product space. See Prob. 4.3.10 for details.

Lemma 4.3.8 *Suppose that V is a finite-dimensional inner product space and W is a subspace of V. Then:*
(i)   *$im(Proj_W) = W$ and $ker(Proj_W) = W^\perp$ and consequently $V = im(Proj_W) \oplus ker(Proj_W)$.*
(ii)   *$Proj_W \circ Proj_W = Proj_W$.*
(iii) *If $W' \subseteq W^\perp$, then $Proj_W \circ Proj_{W'} = 0$.*

Proof: (i) For any subspace W of V, the projection operator $Proj_W : V \to V$ is defined by the two conditions that $Proj_W(w) = w$ for all $w \in W$ and $Proj_W(v) = \vec{0}$ for all $v \in W^\perp$. It immediately follows from this characterization that $im(Proj_W) = W$. Moreover as $W^\perp \subseteq ker(Proj_W)$ and $null(Proj_W) = dim(V) - rk(Proj_W) = dim(V) - dim(W) = dim(W^\perp)$, we have $ker(Proj_W) = W^\perp$. As $V = W \oplus W^\perp$, (i) follows.

(ii) The two defining conditions of $Proj_W$ also show that the composition $Proj_W(Proj_W(v)) = Proj_W(v)$ for every $v \in W$ or $v \in W^\perp$. But, W and $W^\perp$ span V, so this is true for every $v \in V$; that is, $Proj_W \circ Proj_W = Proj_W^2 = Proj_W$ as an operator. This gives (ii).

(iii) Finally, whenever $w' \in W' \subseteq W^\perp$, necessarily $Proj_{W'}(w') = w' \in W^\perp$, and consequently $Proj_W(Proj_{W'}(w')) = Proj_W(w') = \vec{0}$. This establishes (iii) and proves the lemma. $//$

We conclude this section with a simple result which proves to be quite useful when working with the inner product space $\mathbf{R}^n$. Recall that for a matrix A, $ker(A)$ denotes the nullspace of A, $row(A)$ denotes the row space of A, and $col(A)$ denotes the column space of A.

Lemma 4.3.9 *Let A be any $m \times n$ real matrix. Then $ker(A) = (row(A))^\perp \subseteq \mathbf{R}^n$.*

Proof: By the definition of matrix multiplication, we see that $v \in ker(A)$ (that is, $Av = \vec{0}$) if and only if $R_i v = 0$ for each row R_i of A. But, the matrix product $R_i v$ is precisely the dot product of R_i and v (when viewed as vectors) in $\mathbf{R}^n$. Since $row(A)$ is the subspace of $\mathbf{R}^n$

spanned by the rows of A, we have that $v \in (row(A))^{\perp}$ if and only if $v \in ker(A)$. This proves the lemma. //

Recall the subspace $W = span\{(2, 1, 0, 0, 0), (1, 1, 1, 1), (0, 0, 1, 2)\} \subset \mathbf{R}^4$, considered earlier in this section. Lemma 4.3.9 says that $W^{\perp} = ker(A)$ where A is the matrix

$$A = \begin{pmatrix} 2 & 1 & 0 & 0 \\ 1 & 1 & 1 & 1 \\ 0 & 0 & 1 & 2 \end{pmatrix}$$

In computing $ker(A)$, one is required to solve the system of equations $2w_1 + w_2 = 0$, $w_1 + w_2 + w_3 + w_4 = 0$, $w_3 + 2w_4 = 0$. This is precisely the same system of equations considered earlier when it was shown that $W^{\perp} = span\{(-1, 2, -2, 1)\}$.

Observe that if A is an $m \times n$ matrix of rank s then the nullity $null(A) = dim(ker(A)) = n - s$. By Lemma 4.3.7 we find that $dim(row(A)^{\perp}) = n - s$ as well. Applying Corollary 4.3.6 shows that $dim(row(A)) = n - (n - s) = s$. Of course, we knew this already, for $s = rk(A)$, but this shows how the dimensionality relations work. The case where $rk(A) = n$ is nice, as is illustrated by the following corollary.

Corollary 4.3.10 *Suppose that A is an $m \times n$ real matrix of rank n. Then $A^t A$ is an invertible $n \times n$ matrix.*

Proof: It suffices to show that the system of equations $(A^t A)\vec{X} = \vec{0}$ has a unique solution. If $v \in \mathbf{R}^n$ is a solution, then as $A^t(Av) = \vec{0}$, we see that $Av \in ker(A^t)$. However, we also know that $Av \in col(A) = row(A^t)$. As $row(A^t)^{\perp} = ker(A^t)$, we conclude that $Av \in row(A^t) \cap row(A^t)^{\perp} = \{\vec{0}\}$; that is, $Av = \vec{0}$. Since $v \in \mathbf{R}^n$ and $rk(A) = n$, the linear transformation T_A is injective. Consequently, $v = \vec{0}$. This proves the corollary. //

PROBLEMS

4.3.1. Find the matrix of the orthogonal projection operator $P : \mathbf{R}^3 \to \mathbf{R}^3$ onto the following subspaces of $\mathbf{R}^3$:
(a) $span\{(1, 2, 2), (0, 1, -1)\}$
(b) $\{(x, y, z) \mid x - 2y + 3z = 0\}$
(c) $span\{(1, 2, -4)\}$

4.3.2. Suppose that

$$W = span \left\{ \begin{pmatrix} 1 \\ 1 \\ 0 \end{pmatrix}, \begin{pmatrix} 0 \\ 1 \\ 1 \end{pmatrix} \right\} \subseteq \mathbf{R}^3$$

(a) Find a basis for $W^\perp$ (with respect to the standard dot product).
(b) Find the standard matrix of the orthogonal projection $Proj_W$: $\mathbf{R}^3 \to \mathbf{R}^3$.
(c) What is the standard matrix of $P_{W^\perp} : \mathbf{R}^3 \to W^\perp$?

4.3.3. Find the orthogonal complement of each of the subspaces of $\mathbf{R}^3$ given in Prob. 4.3.1.

4.3.4. Consider the subspace W_1 of $\mathbf{R}^3$ spanned by $(2, 1, 0)$ and $(2, 3, 1)$. Find two different subspaces W_2 and W_2' of $\mathbf{R}^3$ for which $\mathbf{R}^3 = W_1 \oplus W_2$ and $\mathbf{R}^3 = W_1 \oplus W_2'$.

4.3.5. Suppose that V is an inner product space and U and W are subspaces of V. Prove the following:
(a) $W \subseteq (W^\perp)^\perp$, with equality if V is finite-dimensional.
(b) If $U \subseteq W$, then $W^\perp \subseteq U^\perp$.
(c) $(U + W)^\perp = U^\perp \cap W^\perp$.
(d) $U^\perp + W^\perp \subseteq (U \cap W)^\perp$, with equality if V is finite-dimensional.

4.3.6. Suppose that $Proj_W : V \to V$ is the orthogonal projection onto the subspace W of a finite-dimensional inner product space V. Denote the identity function by $I : V \to V$. Show that the linear operator $I - Proj_W : V \to V$ is the orthogonal projection onto $W^\perp$.

4.3.7. Consider the inner product on $\mathcal{C}([-1, 1], \mathbf{R})$ defined by

$$<f(x), g(x)> = \int_{-1}^{1} f(x)g(x)\, dx$$

A function $f(x)$ is called *odd* if $f(-x) = -f(x)$ for all $x \in [-1, 1]$. Let O_{dd} denote the subspace of $\mathcal{C}([-1, 1], \mathbf{R})$ consisting of all the odd functions. Find $O_{dd}^\perp$.

4.3.8. Give an alternative proof of Corollary 4.3.10 by showing that $ker(A^t A) = ker(A)$, so that $rk(A^t A) = rk(A)$. (Hint: Compute $x^t A^t A x$.)

4.3.9. Suppose that $V = V_1 \oplus V_2$, and V_1, V_2, W are finite-dimensional $\mathbf{F}$ vector spaces. Recall that $Hom(V, W)$ denotes the $\mathbf{F}$ vector space of all linear transformations from V to W. Show that $Hom(V,W)$ is isomorphic to $Hom(V_1,W) \oplus Hom(V_2,W)$.

4.3.10. (a) For an arbitrary vector space V any operator $P : V \to V$ which satisfies $P^2 = P$ is called a *projection operator*. Suppose that P_1 and P_2 are projection operators on V which satisfy $P_1 + P_2 = I$ (the identity operator) and $P_1 \circ P_2 = P_2 \circ P_1 = 0$ (the zero operator). Prove that $V = im(P_1) \oplus im(P_2)$.
(b) State and prove the converse to (a) above.

4.3.11. (a) Suppose that $T : V \to V$ is an orthogonal projection onto some subspace of V. Show that $\| T(v) \| \leq \| v \|$ for all $v \in V$.
(b) Suppose that $T : V \to V$ is a projection operator (see Prob.4.3.10) on a finite-dimensional inner product space V. If $\| T(v) \| \leq \| v \|$ for all $v \in V$, show that T is an orthogonal projection.

4.3.12. If $\{v_1, v_2, \ldots, v_n\}$ is an orthonormal basis of an inner product space V, prove *Parseval's identity*: For all $u, w \in V$

$$<u, w> = \sum_{i=1}^{n} <u, v_i> \overline{<w, v_i>}$$

4.3.13. Prove *Bessel's Inequality* that whenever $v_1, v_2, \ldots, v_n \in V$ are othogonal and nonzero and $v \in V$, then

$$\sum_{i=1}^{n} \frac{| <v, v_i> |^2}{<v_i, v_i>} \leq <v, v>$$

(Hint: Write $v = w_1 + w_2$ with $w_1 \in span\{v_1, \ldots, v_n\}$ and $w_2 \in span\{v_1, \ldots, v_n\}^{\perp}$.)

4.4 Least-Squares Approximations

In some practical uses of linear algebra it is necessary to "solve" an inconsistent system of linear equations. Yes, strange as it sounds, this is often the case. Why does this happen and what can be done? These questions are discussed in this section.

We begin by stating the problem. Suppose A is an $m \times n$ matrix and the system of equations $A\vec{X} = \vec{b}$ is inconsistent. Such a situation often arises when studying experimental or statistical data. In such circumstances, we might repeat an experiment many times (to gather a large amount of data), and consequently the matrix A will have (many) more rows than columns. However, although the system is inconsistent, the inconsistencies should be small and we expect to find values for $\vec{X}$ which come close to solving the system $A\vec{X} = \vec{b}$. The problem is to find a vector $\vec{c} \in \mathbf{R}^n$ which is the "closest to a solution" in that the overall collection of errors is small.

The phrase "closest to a solution" can be interpreted in many ways. In linear algebra the easiest notion of "closest" is the following.

Definition 4.4.1 Let A be a real $m \times n$ matrix. A vector $\vec{c} \in \mathbf{R}^n$ is called a *least-squares solution* to the (possibly inconsistent) system $A\vec{X} = \vec{b}$ if the distance

$$\| A\vec{c} - \vec{b} \| = \sqrt{<A\vec{c} - \vec{b}, A\vec{c} - \vec{b}>}$$

in $\mathbf{R}^n$ is a minimum (among all possible choices for $\vec{c}$).

The reason for the terminology "least squares" is explained in the following.

Discussion 4.4.2 Suppose that a collection of n points $p_1 = (r_1, s_1)$, $\dots, p_n = (r_n, s_n) \in \mathbf{R}^2$ are given, and for sake of argument, we assume that they all lie reasonably close to a line in the plane $\mathbf{R}^2$. Such points might arise, for example, after plotting experimental data. The problem is to find the equation of a line which is a best fit for these points.

We assume that the equation of a possible line is expressed as $Y = mX + b$, where $m, b \in \mathbf{R}$. Our goal is to best approximate the s_i values by substituting r_i into this equation. Thus we desire $s_i \sim mr_i + b$. For a best fit, we shall require that the sum of the squares of the errors of this linear approximation be minimized. In other words, we require that $\sum_{i=1}^{n} (s_i - mr_i - b)^2$ be minimal among all possible choices of m and b.

This criterion coincides with the least-squares solution as defined above. To see this, we reinterpret the question as an n-dimensional problem. Our "best" equation $Y = mX + b$ represents an attempt to solve the (presumably) inconsistent system for m and b:

$$\begin{pmatrix} r_1 & 1 \\ r_2 & 1 \\ \vdots & \vdots \\ r_n & 1 \end{pmatrix} \begin{pmatrix} m \\ b \end{pmatrix} = \begin{pmatrix} s_1 \\ s_2 \\ \vdots \\ s_n \end{pmatrix} \qquad (*)$$

We denote the column of the r_i by $\vec{r}$, the column of 1s by $\vec{1}$, and the column of the s_i by $\vec{s}$. We can then represent this equation by

$$(\vec{r} \ \ \vec{1}) \begin{pmatrix} m \\ b \end{pmatrix} = \vec{s}$$

According to Definition 4.4.1, the least-squares solution is one for which the distance (in $\mathbf{R}^n$)

$$\| (\vec{r} \ \ \vec{1}) \begin{pmatrix} m \\ b \end{pmatrix} - \vec{s} \|$$

is minimal. However the square of this distance is precisely $(mr_1 + b - s_1)^2 + \cdots + (mr_n + b - s_n)^2$, which is the original sum we sought to minimize. This shows that a least-squares solution to $(*)$ will give the desired m and b. For this reason, the best linear approximation $Y = mX + b$ is called the *least-squares line* for the data $p_1, \ldots, p_n$.

We next solve for m and b. The vector

$$\vec{w} = (\vec{r} \ \ \vec{1}) \begin{pmatrix} m \\ b \end{pmatrix} \in \mathbf{R}^n$$

is a linear combination of the two vectors $\vec{r}, \vec{1}$; that is, $\vec{w} \in span\{\vec{r}, \vec{1}\} = W$. In order that $\| \vec{w} - \vec{s} \|$ be minimal, we must have $\vec{w} = Proj_W(\vec{s})$. (This is because the shortest distance from a point to a plane is along a line perpendicular to the plane.) This means that the first step is to compute $\vec{w} = Proj_W(\vec{s})$. The second step is to compute m and b by solving the equation

$$(\vec{r} \ \ \vec{1}) \begin{pmatrix} m \\ b \end{pmatrix} = \vec{w}$$

Assuming $r_1 \neq r_2$, m and b will be the (unique) solutions to $r_1 m + b = w_1$ and $r_2 m + b = w_2$.

An alternative approach to minimizing $\sum_{i=1}^{n}(s_i - mr_i - b)^2$ is to minimize this function of the two real variables m and b by computing the two partial derivatives and finding a critical point. These partial derivatives are linear equations in m and b and thus have a unique solution. (A minimum must exist, as otherwise this sum of squares would be a constant function of m and b.) This process gives the same answer. We shall, however, use the methods of linear algebra rather than calculus.

This next example explicitly computes a least-squares line.

Example 4.4.3 Find the least-squares line which approximates the points $(2, 8)$, $(3, 10)$, $(-5, -3)$. For this we set $W = span\{v_1, v_2\}$ and consider $\vec{s} \in \mathbf{R}^3$ where

$$v_1 = \begin{pmatrix} 2 \\ 3 \\ -5 \end{pmatrix} \quad , \quad v_2 = \begin{pmatrix} 1 \\ 1 \\ 1 \end{pmatrix} \quad \text{and} \quad \vec{s} = \begin{pmatrix} 8 \\ 10 \\ -3 \end{pmatrix}$$

We must compute the projection operator $Proj_W : \mathbf{R}^3 \to \mathbf{R}^3$. Since the vectors v_1 and v_2 are orthogonal, we find that $Proj_W = proj_{\vec{r}} + proj_{\vec{1}}$. This shows that

$$
\begin{aligned}
Proj_W(\vec{s}) &= proj_{\vec{r}}(\vec{s}) + proj_{\vec{1}}(\vec{s}) \\
&= \frac{51}{38}\begin{pmatrix} 2 \\ 3 \\ -5 \end{pmatrix} + \frac{15}{3}\begin{pmatrix} 1 \\ 1 \\ 1 \end{pmatrix} = \frac{1}{38}\begin{pmatrix} 292 \\ 343 \\ -65 \end{pmatrix}
\end{aligned}
$$

To solve for m and b, we must solve the system $2m + b = 292/38$ and $3m + b = 343/38$. We find that $m = 51/38$ and $b = 190/38$.

The preceding discussion showed that there are really two parts in solving a least-squares problem. First, there is the reduction of the problem to solving a consistent linear system. Second, there is the problem of solving this new linear system. This second problem was easy in the example above, because it reduced to solving a 2×2 system. For the remainder of this section we address the first problem of reducing the equations to a consistent system.

This next lemma explicitly proves that any least-squares solution is a solution to the original equation modified by appropriately projecting the value matrix. This formalizes the geometric idea that was utilized in Discussion 4.4.2.

Lemma 4.4.4 *Consider the (possibly inconsistent) system $A\vec{X} = \vec{b}$ and let W be the column space of A. Then $A\vec{X} = \vec{b}$ has a least-squares solution and the least-squares solutions are precisely the solutions to the consistent system $A\vec{X} = Proj_W(\vec{b})$. Consequently, if $\vec{c}$ is a least-squares solution to $A\vec{X} = \vec{b}$, then $A\vec{c} - \vec{b}$ is orthogonal to the column space of A.*

Proof: Observe as $Proj_W(\vec{b}) \in col(A)$, the system $A\vec{X} = Proj_W(\vec{b})$ is consistent. Thus, the existence of a least-squares solution to $A\vec{X} = \vec{b}$ follows once we establish that the set of least-squares solutions to $A\vec{X} = \vec{b}$ is precisely the same as the set of solutions to $A\vec{X} = Proj_W(\vec{b})$.

Set $w = Proj_W(\vec{b})$ and let $v \in W$ be arbitrary. Since $(v - w) \in W$ and $(\vec{b} - w) = \vec{b} - Proj_W(\vec{b}) \in W^{\perp}$, we compute that $\| \vec{b} - v \|^2 = <\vec{b} - v, \vec{b} - v> = <(\vec{b} - w) + (w - v), (\vec{b} - w) + (w - v)> = <\vec{b} - w, \vec{b} - w> + <w - v, w - v> = \| \vec{b} - w \|^2 + \| w - v \|^2$. This shows that

$$\| \vec{b} - w \|^2 \leq \| \vec{b} - w \|^2 + \| w - v \|^2 = \| \vec{b} - v \|^2$$

and consequently $\| \vec{b} - w \| \leq \| \vec{b} - v \|$ for any $v \in W$. Hence, $w = Proj_W(\vec{b})$ is an element of the column space of A with $\| \vec{b} - w \|$ minimal. Consequently any solution to $A\vec{X} = w$ will be a least squares solution to $A\vec{X} = \vec{b}$. Note further that whenever $v \in V$ and $v \neq w$, we have $\| \vec{b} - w \| < \| \vec{b} - v \|$. This shows for any least squares solution $\vec{c}$ of $A\vec{X} = \vec{b}$, since $\| A\vec{c} - w \|$ is minimal, we necessarily have $A\vec{c} = w$, and hence $\vec{c}$ is a solution to $A\vec{X} = Proj_W(\vec{b})$. This proves the lemma. //

For example, consider the least-squares problem

$$\begin{pmatrix} 1 & 2 & 3 \\ 1 & 2 & 3 \\ 1 & 2 & 3 \end{pmatrix} \begin{pmatrix} X \\ Y \\ Z \end{pmatrix} = \begin{pmatrix} 7 \\ 8 \\ 9 \end{pmatrix}$$

The column space of the coefficient matrix is the one-dimensional subspace $W = span\{(1, 1, 1)\} \subset \mathbf{R}^3$. One computes that $Proj_W(7, 8, 9) = (8, 8, 8)$. From this we see that the least-squares solutions are precisely the solutions to the system

$$\begin{pmatrix} 1 & 2 & 3 \\ 1 & 2 & 3 \\ 1 & 2 & 3 \end{pmatrix} \begin{pmatrix} X \\ Y \\ Z \end{pmatrix} = \begin{pmatrix} 8 \\ 8 \\ 8 \end{pmatrix}$$

that is, the solutions are the plane in $\mathbf{R}^3$ described by $X + 2Y + 3Z = 8$.

This next theorem gives an alternative consistent system which has as solutions the least squares solutions to the original system. This formula has the advantage that one need not directly compute the projection operator.

Theorem 4.4.5 *If A is an $m \times n$ matrix, then the least squares solutions to the (presumably inconsistent) system $A\vec{X} = \vec{b}$ are precisely the solutions to the (consistent) system $(A^t A)\vec{X} = A^t \vec{b}$.*

Proof: We set $W = col(A)$, the column space of A, and we note by Lemma 4.3.9 that $W^\perp = ker(A^t)$ (the nullspace of A^t). If $\vec{c}$ is a least-squares solution to $A\vec{X} = \vec{b}$, then by the previous lemma, $(A\vec{c} - \vec{b}) \in W^\perp$. Hence, $A^t(A\vec{c} - \vec{b}) = \vec{0}$; that is, $A^t A\vec{c} = A^t \vec{b}$. Conversely, if $A^t A\vec{c} = A^t \vec{b}$, then $A\vec{c} - \vec{b} \in ker(A^t) = W^\perp$. Since $A\vec{c} \in W$ and $\vec{b} = (A\vec{c}) - (A\vec{c} - \vec{b})$, necessarily $Proj_W(\vec{b}) = A\vec{c} + 0 = A\vec{c}$ by Corollary 4.3.7 (ii). Thus $\vec{c}$ is a least-squares solution to $A\vec{X} = \vec{b}$. The consistency of $(A^t A)\vec{X} = A^t \vec{b}$ now follows from the existence of a least-squares solution to $A\vec{X} = \vec{b}$. This proves the theorem. //

Continuing with the preceding example, Theorem 4.4.5 shows that the solutions to the least squares problem are the solutions to the system of equations

$$\begin{pmatrix} 1 & 1 & 1 \\ 2 & 2 & 2 \\ 3 & 3 & 3 \end{pmatrix} \begin{pmatrix} 1 & 2 & 2 \\ 1 & 2 & 3 \\ 1 & 2 & 3 \end{pmatrix} \begin{pmatrix} X \\ Y \\ Z \end{pmatrix} = \begin{pmatrix} 1 & 1 & 1 \\ 2 & 2 & 2 \\ 3 & 3 & 3 \end{pmatrix} \begin{pmatrix} 7 \\ 8 \\ 9 \end{pmatrix}$$

that is, the least squares solutions are the solutions to

$$\begin{pmatrix} 3 & 6 & 9 \\ 6 & 12 & 18 \\ 9 & 18 & 27 \end{pmatrix} \begin{pmatrix} X \\ Y \\ Z \end{pmatrix} = \begin{pmatrix} 24 \\ 48 \\ 72 \end{pmatrix}$$

We obtain the same answer as before, namely the plane described by $X + 2Y + 3Z = 8$.

The equation $(A^t A)\vec{X} = A^t \vec{b}$ is known as the *normal equation* associated with the least squares problem $A\vec{X} = \vec{b}$. Its solutions can always be found by gaussian elimination if necessary. When the rank of the $m \times n$ matrix A is n, then by Corollary 4.3.10 the matrix $A^t A$ is invertible and there is a unique least squares solution. We close this section with an explicit description of the unique solution in this (full) rank n case.

While nice in theory, the results of Sec. 4.3 describing projection operators are not usually practical for computational purposes. This next theorem gives an alternative method for computing the matrices which give projection operators.

Theorem 4.4.6 *Let A be an $m \times n$ real matrix of rank n. Set W to be the column space of A [so $W \subseteq \mathbf{R}^m$ and $dim(W) = n$]. Then the standard matrix $[Proj_W] = A(A^t A)^{-1} A^t$.*

Proof: Observe that as $rk(A) = n$ and A is $m \times n$, $A^t A$ is invertible by Corollary 4.3.10. Thus the expression $A(A^t A)^{-1} A^t$ makes sense, and for convenience we denote it by $\tilde{A}$. By definition W is spanned by the columns of A, and according to Lemma 4.3.9 the vector space $W^\perp$ is precisely the nullspace of A^t. Recall that a linear transformation is determined by its action on a spanning set (Lemma 3.1.4). Thus, in order to show $[Proj_W] = \tilde{A}$ it suffices to show that $\tilde{A}\vec{v} = \vec{0}$ whenever $\vec{v} \in null(A^t)$, and that $\tilde{A}C = C$ for any column C of A.

If $\vec{v} \in null(A^t)$, then $\tilde{A}\vec{v} = A(A^t A)^{-1}(A^t \vec{v}) = A(A^t A)^{-1}\vec{0} = \vec{0}$, as required. If C is the ith column of A, then we know that $\tilde{A}C$ is the ith column of $\tilde{A}A$. Observe that $\tilde{A}A = A(A^t A)^{-1}(A^t A) = A$. From this, $\tilde{A}C = C$ follows. This proves the theorem. //

Observe that given any basis $\{w_1, \ldots, w_n\}$ for a subspace W of $\mathbf{R}^m$, one can form the $m \times n$ matrix whose columns are the vectors $w_1, \ldots, w_n$. Then $col(A) = W$, and the linear independence of the $w_1, \ldots, w_n$ guarantees that $rank(A) = n$. This matrix A can be used to compute the matrix $[Proj_W]$ according to Theorem 4.4.6. From this point of view, the assumption that $rk(A)$ is maximal in Theorem 4.4.6 is not as restrictive as it might have seemed at first.

The following corollary is an immediate consequence of the description of the projection given in Theorem 4.4.6, together with Lemma 4.4.5. Note that the difference between this result and Theorem 4.4.5 is that the coefficient matrix for the new equation is our original matrix A. We remark that the formula it gives for the least squares solution is by no means the most efficient. In most specific cases it is more efficient to solve the normal equation directly.

Corollary 4.4.7 *If A is an $m \times n$ rank n real matrix, then the (unique) least squares solution to the (presumably inconsistent) system $A\vec{X} = \vec{b}$ is the solution to the system $A\vec{X} = A(A^tA)^{-1}A^t\vec{b}$. In particular, the least squares solution is $\vec{X} = (A^tA)^{-1}A^t\vec{b}$.*

Example 4.4.8 Suppose five points in $\mathbf{R}^2$ are given: $(-2, s)$, $(-1, t)$, $(0, u)$, $(1, v)$, and $(2, w)$. The problem is to find a quadratic polynomial $Y = aX^2 + bX + c$ which best approximates these five points (called the *least-squares quadratic*). Translating this into a least squares problem, it turns out that what we are after is the least squares solution to

$$
\begin{pmatrix}
4 & -2 & 1 \\
1 & -1 & 1 \\
0 & 0 & 1 \\
1 & 1 & 1 \\
4 & 2 & 1
\end{pmatrix}
\begin{pmatrix}
a \\
b \\
c
\end{pmatrix}
=
\begin{pmatrix}
s \\
t \\
u \\
v \\
w
\end{pmatrix}
$$

The coefficient matrix arose by substituting the X values $-2, -1, 0, 1, 2$ into the monomials X^2, X and 1 into the polynomial $aX^2 + bX + c$. The matrix A has rank 3, so we can solve the least-squares problem by applying Corollary 4.4.7. We compute that

$$
A^tA =
\begin{pmatrix}
34 & 0 & 10 \\
0 & 10 & 0 \\
10 & 0 & 5
\end{pmatrix}
\quad \text{and} \quad
(A^tA)^{-1} =
\begin{pmatrix}
1/14 & 0 & -1/7 \\
0 & 1/10 & 0 \\
-1/7 & 0 & 17/35
\end{pmatrix}
$$

In particular,

$$
(A^tA)^{-1}A^t =
\begin{pmatrix}
1/7 & -1/14 & -1/7 & -1/14 & 1/7 \\
-1/5 & -1/10 & 0 & 1/10 & 1/5 \\
-3/35 & 12/35 & 17/35 & 12/35 & -3/35
\end{pmatrix}
$$

so we find that the solution (in terms of the unspecified s, t, u, v, w) is given by

$$
\begin{aligned}
a &= (1/14)(2s - t - 2u - v + 2w) \\
b &= (1/10)(-2s - t + v + 2w) \\
c &= (1/35)(-3s + 12t + 17u + 12v - 3w)
\end{aligned}
$$

The conclusion of Corollary 4.4.7 simplifies in case we have determined the Q-R factorization of the matrix A (see Sec. 4.2). If we have a Q-R factorization $A = QR$, then since the matrix Q has orthonormal columns, we have $Q^tQ = I_m$, and consequently $A^tA = (QR)^tQR = R^t(Q^tQ)R = R^tR$. We find that in this case the least square solution to $A\vec{X} = \vec{b}$ is given by $\vec{X} = (A^tA)^{-1}A^t\vec{b} = (R^tR)^{-1}(QR)^t\vec{b} = R^{-1}(R^t)^{-1}R^tQ^t\vec{b} = R^{-1}Q^t\vec{b}$.

PROBLEMS

4.4.1. Find the normal equation and use it to find the least-squares solution of the following.

(a)
$$\begin{aligned} X + Y &= 1 \\ X + 2Y &= 3 \\ X + 4Y &= 6 \end{aligned}$$

(b)
$$\begin{aligned} X + Y + Z &= 3 \\ X + + 3Z &= 4 \\ 2X + 4Y &= 4 \\ Y + Z &= 3 \end{aligned}$$

(c)
$$\begin{aligned} 2X - Y &= 4 \\ 2X - 3Y &= -2 \\ X + Y &= 6 \end{aligned}$$

4.4.2. If A is a rank n, $m \times n$ matrix and $\tilde{A} = A(A^tA)^{-1}A^t$, show that $\tilde{A}$ is symmetric and $\tilde{A}^2 = \tilde{A}$.

4.4.3. Find the least-squares solution for the following systems of equations, by computing the projection operators.

(a)
$$\begin{aligned} X + Y &= 5 \\ X - Y &= 3 \\ 2X + 3Y &= 10 \end{aligned}$$

(b)
$$\begin{aligned} X + 0Y &= 6 \\ 2X - 3Y &= 9 \\ X - Y &= 5 \\ 0X + Y &= 2 \end{aligned}$$

4.4.4. Find the least-squares solution for the following systems of equations.

(a)
$$\begin{aligned} X + Y &= 4 \\ X - Y &= 0 \\ 2X + 3Y &= 11 \end{aligned}$$

(b)
$$\begin{aligned} X + Y &= 6 \\ 2X - Y &= 7 \\ X - Y &= 2 \\ X + Y &= 5 \end{aligned}$$

4.4.5. Find the least-squares line for the the points $(2, 4), (4, 6), (6, 10)$, and $(8, 14)$.

4.4.6. (a) Find the least-squares line for the the points $(3, 5)$, $(5, 11)$, $(7, 14)$, and $(9, 19)$.
(b) Find the least-squares quadratic for the points given in (a).

4.4.7. Using Theorem 4.4.6 compute the projection operator $Proj_W$:
$\mathbf{R}^4 \to \mathbf{R}^4$ for
(a) $W = span\{(1, 1, 0, 0), (0, 0, 1, 1)\}$.
(b) $W = span\{(2, 4, 3, 1)\}$.
(c) $W = span\{(1, 1, 0, 0), (0, 1, 0, 1), (0, 0, 1, 1)\}$.

4.4.8.   Using the Q-R factorizations of A computed in Prob. 4.2.9,
solve the least-squares problem $A\vec{X} = \vec{b}$ for arbitrary $\vec{b}$, where A is as
below.

(a) $\begin{pmatrix} 1 & 1 \\ 0 & 1 \end{pmatrix}$ (b) $\begin{pmatrix} 1 & 0 \\ 1 & 1 \\ 0 & 1 \end{pmatrix}$ (c) $\begin{pmatrix} 1 & -1 & 1 \\ 1 & 0 & 0 \\ 1 & -2 & 0 \\ 1 & 1 & 1 \end{pmatrix}$

4.5   Isometries, Symmetric Operators, and Spectral Theory

Real symmetric matrices have especially nice properties. They are always diagonalizable, and in fact there is always an orthogonal basis of
eigenvectors. This result is particularly important in advanced calculus. It is crucial for the study of the matrix of second-order derivatives
of a twice-differentiable function on $\mathbf{R}^n$ (which is symmetric as the
order of mixed partials does not matter). In order to prove this result,
the key is to use the theory of inner products.
    We begin with the definitions of orthogonal and unitary matrices
and with the concept of isometries of inner product spaces.

Definition 4.5.1 An $n \times n$ real matrix U is called *orthogonal* if $UU^t = I_n$; that is, U is invertible and $U^{-1} = U^t$. If A is an $n \times n$ complex
matrix, we denote by $A^* = (\overline{A})^t$ the transpose of the matrix obtained
by conjugating all entries. An $n \times n$ complex matrix U is called *unitary*
if $UU^* = I_n$; that is, U is invertible and $U^{-1} = U^*$.

    Orthogonal and unitary matrices have the following characteriza-
tion in terms of the usual inner products on $\mathbf{R}^n$ and $\mathbf{C}^n$.

Lemma 4.5.2   *A real matrix U is orthogonal if and only if its rows
(or columns) are orthonormal with respect to the dot product on $\mathbf{R}^n$.
A complex matrix U is unitary if and only if its rows (or columns) are
orthonormal in unitary space $\mathbf{C}^n$.*

Proof: Suppose that U is an $n \times n$ real orthogonal matrix. Let R_i denote the ith row of U. Then R_j is the jth column of U^t. It follows from the definition of matrix multiplication that the ijth entry of UU^t is the dot product $R_i \cdot R_j$. Hence, $UU^t = I_n$ if and only if $R_i \cdot R_j = 1$ for $i = j$ and $R_i \cdot R_j = 0$ for $i \neq j$. This proves the result in the orthogonal case. The proof in the unitary case is exactly the same, except that the ijth entry of UU^* is, because of the conjugation, the unitary inner product $< R_i, R_j >$. The lemma follows. //

As examples the reader can check that the following matrices are orthogonal:

$$\begin{pmatrix} 0 & -1 \\ 1 & 0 \end{pmatrix}, \quad \begin{pmatrix} 4/5 & 3/5 \\ 3/5 & -4/5 \end{pmatrix}, \quad \begin{pmatrix} -2/3 & 1/3 & 2/3 \\ 2/3 & 2/3 & 1/3 \\ 1/3 & -2/3 & 2/3 \end{pmatrix}$$

and the following matrices are unitary:

$$\begin{pmatrix} 0 & i \\ i & 0 \end{pmatrix}, \quad \frac{1}{2}\begin{pmatrix} 1+i & 1+i \\ 1-i & -1+i \end{pmatrix}, \quad \frac{1}{2}\begin{pmatrix} 1+i & 0 & 1+i \\ 1 & \sqrt{2} & -1 \\ 1 & -\sqrt{2} & -1 \end{pmatrix}$$

Observe that real orthogonal matrices are always unitary as well, since complex conjugation does not change real numbers.

The linear transformations defined by orthogonal and unitary matrices are important because they "respect" the behavior of the inner products. To make this precise, we need the next definition, which is given for arbitrary inner product spaces.

Definition 4.5.3 Let $< , >_i \colon V_i \times V_i \to \mathbf{F}$ be inner products, where $i = 1, 2$. An *isometry* of inner product spaces $T \colon V_1 \to V_2$ is an isomorphism for which $< v, w >_1 = < T(v), T(w) >_2$ for all $v, w \in V_1$. If V_1 and V_2 are inner product spaces and there is an isometry $T \colon V_1 \to V_2$, then we say that V_1 and V_2 are *isometric*.

Recall that the real vector space P^2 of all polynomials of degree at most 2 is an inner product space with inner product given by $< f(X), g(X) > = \int_0^1 f(X)g(X)\, dX$. P^2 is isometric to $\mathbf{R}^3$. To see this, we define $T \colon \mathbf{R}^3 \to P^2$ by $T(a, b, c) = (a - (1/2)b + (1/6)c) + (b - c)X + cX^2$. It is messy (but not difficult) to verify that T is an isometry by directly checking the criteria required by Definition 4.5.3.

Using Lemma 4.5.4, which is given next, we shall immediately see that T is an isometry.

An isometry $T : V \to W$ between inner product spaces preserves the distance and orthogonality relations that are associated with the inner products. The next lemma exploits these facts and gives an alternative characterization of isometries. The lemma again illustrates why orthonormal bases are useful.

Lemma 4.5.4 *Suppose that V and W are finite-dimensional inner product spaces. Let $T : V \to W$ be a linear transformation. The following are equivalent:*
(i) T is an isometry.
(ii) For some orthonormal basis $\{u_1, \ldots, u_n\}$ for V, $\{T(u_1), \ldots, T(u_n)\}$ is an orthonormal basis for W.
(iii) Whenever $\{u_1, \ldots, u_n\}$ is an orthonormal basis for V, the image $\{T(u_1), \ldots, T(u_n)\}$ is an orthonormal basis for W.

Proof: (i) $\implies$ (iii). If T is an isometry and $\{u_1, \ldots, u_n\}$ is an orthonormal basis for V, then $<T(u_i), T(u_j)> = <u_i, u_j> = 1$ whenever $i = j$ and is 0 otherwise. Hence $T(u_1), \ldots, T(u_n)$ are orthonormal in W and necessarily form a basis for W since T is an isomorphism.

(iii) $\implies$ (ii) is trivial.

(ii) $\implies$ (i). Suppose that $\{u_1, \ldots, u_n\}$ and $\{T(u_1), \ldots, T(u_n)\}$ are orthonormal bases. We see immediately that T is an isomorphism, since the image of a basis for V under T is a basis for W. To see that T is an isometry, we assume that $v = \sum a_i u_i$, $w = \sum b_i u_i \in V$ and compute directly (using the properties (ii) and (iii) of $< , >$ repeatedly):

$$
\begin{aligned}
<v, w> &= \sum_{i=1}^{n} a_i < u_i, w > \\
&= \sum_{i=1}^{n} \sum_{j=1}^{n} a_i \overline{b_j} < u_i, u_j > \\
&= \sum_{i=1}^{n} \sum_{j=1}^{n} a_i \overline{b_j} < T(u_i), T(u_j) > \\
&= \sum_{i=1}^{n} a_i < T(u_i), T(w) > \\
&= < T(v), T(w) >
\end{aligned}
$$

(Note that in the case of a real inner product, $\overline{b_j} = b_j$ in the preceding formulas.) This shows that T is an isometry and proves the lemma. //

To see that $T : \mathbf{R}^3 \to P^2$ defined above is an isometry, we apply part (ii) of Lemma 4.5.4 together with the observation of Example 4.2.8 that $\{1, X - 1/2, X^2 - X - 1/6\}$ is an orthonormal basis of P^2. Evidently, $T(1,0,0) = 1, T(0,1,0) = X-1/2$, and $T(0,0,1) = X^2-X+1/6$, which shows that part (ii) of Lemma 4.5.4 applies to T. Another application of Lemma 4.5.4 is given in the following lemma. It shows that isometries from $\mathbf{R}^n$ to $\mathbf{R}^n$ are described by orthogonal matrices and isometries from $\mathbf{C}^n$ to $\mathbf{C}^n$ are described by unitary matrices.

Lemma 4.5.5 (i) *A linear operator $T : \mathbf{R}^n \to \mathbf{R}^n$ is an isometry (with respect to the dot product) if and only if the standard matrix $[T]$ is orthogonal. Consequently, if U is an $n \times n$ orthogonal matrix, $\| Uv \| = \| v \|$ for all $v \in \mathbf{R}^n$.*
(i') *A linear operator $T : \mathbf{C}^n \to \mathbf{C}^n$ is an isometry of unitary space if and only if the standard matrix $[T]$ is unitary. Consequently, if U is an $n \times n$ unitary matrix, $\| Uv \| = \| v \|$ for all $v \in \mathbf{C}^n$.*
(ii) *Two finite-dimensional real inner product spaces are isometric if and only if they have the same dimension.*
(ii') *Two finite-dimensional complex inner product spaces are isometric if and only if they have the same dimension.*

Proof: (i) and (i') Assume first that $T : \mathbf{F}^n \to \mathbf{F}^n$ is an isometry where $\mathbf{F}$ is either $\mathbf{R}$ or $\mathbf{C}$. Then since the standard basis $\{e_1, e_2, \ldots, e_n\}$ is an orthonormal basis for $\mathbf{F}^n$, Lemma 4.5.4 shows that the vectors $T(e_1), T(e_2), \ldots, T(e_n)$ also form an orthonormal basis for $\mathbf{F}^n$. Since these vectors form the columns of $[T]$, we see by Lemma 4.5.2 that $[T]$ is orthogonal in the real case and is unitary in the complex case.

Conversely, suppose $[T]$ is orthogonal or unitary. Then since $T(e_1)$, $T(e_2)$, $\ldots$, $T(e_n)$ are the columns of $[T]$, they form an orthonormal basis for $\mathbf{F}^n$ by Lemma 4.5.2. Hence T is an isometry by Lemma 4.5.4. This gives (i) and (i').

(ii) and (ii') An isometry is an isomorphism, so dimensions agree. Conversely, if V_1 and V_2 are inner product spaces with $dim(V_1) = dim(V_2) = n$, we can apply Theorem 4.2.6 to find orthonormal bases $\{v_1, \ldots, v_n\}$ for V_1 and $\{u_1, \ldots, u_n\}$ for V_2. According to Lemma 3.1.4, there is a linear transformation $T : V_1 \to V_2$ for which $T(v_i) = u_i$. By Lemma 4.5.4, such a linear transformation is an isometry. This proves (ii) and (ii'). //

We next give the operator analogue of symmetric matrices.

Definition 4.5.6 Suppose that $T : V \to V$ is a linear operator on an inner product space V. We say that T is *self-adjoint* if for all $v, w \in V$, $<T(v), w> = <v, T(w)>$. In the real case, such operators are also called *symmetric*, and in the complex case *hermitian*. A real matrix A is called *self-adjoint* or *symmetric* if $A^t = A$, and a complex matrix is called *self-adjoint* or *hermitian* if $A^* = A$.

The result which follows is exactly what one would expect, given the terminology. Note that if $v, w \in \mathbf{C}^n$ are column vectors, then the hermitian inner product $<v, w> = v^t \overline{w} \in \mathbf{C}$ where $\overline{w}$ is the vector whose entries are the conjugates of the entries of w.

Lemma 4.5.7 (i) *If $A \in M_{n \times n}(\mathbf{C})$, then for all $v, w \in \mathbf{C}^n$ $<Av, w> = <v, A^* w>$.*
(ii) *Let $\mathbf{F} = \mathbf{R}$ or $\mathbf{C}$. If $T : \mathbf{F}^n \to \mathbf{F}^n$ is a linear operator, then T is self-adjoint if and only if the standard matrix $[T]$ is self-adjoint.*

Proof: (i) For all $v, w \in \mathbf{C}^n$,

$$
\begin{aligned}
<Av, w> &= (Av)^t \overline{w} = v^t (A^t \overline{w}) \\
&= v^t (\overline{A^* w}) = v^t (\overline{A^* w}) \\
&= <v, A^* w>
\end{aligned}
$$

This gives (i).
(ii) Suppose that $[T] = A = (a_{ij})$. If A is self-adjoint, then by (i)

$$
\begin{aligned}
<T(v), w> &= <Av, w> \\
&= <v, A^* w> \\
&= <v, Aw> \\
&= <v, T(w)>
\end{aligned}
$$

so T is self-adjoint. Conversely, suppose T is a self-adjoint operator. By Theorem 4.2.2, $< T(e_j), e_i >$ is the ith coordinate of the vector

$T(e_j)$. This means that $a_{ij} = < T(e_j), e_i >$. But, T is self-adjoint, so $a_{ij} = < T(e_j), e_i > = < e_j, T(e_i) > = \overline{< T(e_i), e_j >} = \overline{a_{ji}}$. Hence, A is self-adjoint. //

The next theorem provides the induction step for the analysis of self-adjoint matrices and self-adjoint operators. The diagonalizability of real symmetric operators follows in Theorem 4.5.9.

Theorem 4.5.8 *Every self-adjoint operator $T : V \to V$ on a finite-dimensional inner product space V has a* real *eigenvalue. Moreover, all eigenvalues are real.*

Proof: Since every finite-dimensional inner product space V is isometric to $\mathbf{R}^n$ or $\mathbf{C}^n$ for some n, we shall assume that $V = \mathbf{F}^n$ where $\mathbf{F} = \mathbf{R}$ or $\mathbf{C}$. In this case, $T = T_S$ for some self-adjoint matrix S. In the real case, the matrix S is symmetric. But S can be viewed as a complex matrix, in which case it is also self-adjoint (because a real symmetric matrix is its own conjugate transpose). Thus the complex operator $T_S : \mathbf{C}^n \to \mathbf{C}^n$ is self adjoint. If we know that complex self-adjoint operators only have real eigenvalues, then since S has some (at first possibly complex) eigenvalue, S will have some (and *only*) real eigenvalues. Hence, in order to prove the theorem, it suffices to establish that all eigenvalues of a complex self-adjoint matrix are real.

Consider a root $h \in \mathbf{C}$ of the characteristic polynomial $C_T(X) = C_S(X)$. There is some nonzero vector $v \in \mathbf{C}^n$ with $Sv = hv$. Since T_S is self-adjoint, we find that

$$
\begin{aligned}
h < v, v > \quad &= \quad < hv, v > = < Sv, v > \\
&= \quad < v, Sv > = < v, hv > = \overline{h} < v, v >
\end{aligned}
$$

Since $< v, v > \neq 0$, we conclude that $h = \overline{h} \in \mathbf{R}$. This proves the theorem. //

We can now prove the key result about self-adjoint operators. The theorem says that all self-adjoint operators are diagonalizable. Moreover, there is an orthonormal basis of eigenvectors. Geometrically, this result means that after an isometric change of basis, self adjoint operators stretch or shrink $\mathbf{R}^n$ (or $\mathbf{C}^n$) along the coordinate axes.

Theorem 4.5.9 (Principal Axis Theorem) *Suppose $T : V \to V$ is a self-adjoint operator, where V is a finite-dimensional inner product space. Then V has an orthonormal basis of eigenvectors for T with only real eigenvalues. Consequently, if S is an $n \times n$ real symmetric matrix, then there is a real orthogonal matrix U such that $U^{-1}SU = U^t SU$ is diagonal.*

Proof: The proof proceeds by induction on $n = dim(V)$. If $n = 1$, the result is trivial, so we assume the result for dimension $n - 1$. We know by Theorem 4.5.8 that T has an eigenvector u_1 with real eigenvalue k_1. We can assume that $\| u_1 \| = 1$. Let $W = span\{u_1\}$, and note that $dim(W^\perp) = n - 1$. Let $v \in W^\perp$. We find that $< u_1, T(v) > = < T(u_1), v > = < k_1 u_1, v > = k_1 < u_1, v > = 0$. Hence $T(v) \in W^\perp$. This shows that restricting T to $W^\perp$ gives a linear operator $T|_{W^\perp} : W^\perp \to W^\perp$, which is necessarily also a self-adjoint operator. By induction, $W^\perp$ has an orthonormal basis of eigenvectors for $T|_{W^\perp}$. This basis, together with u_1, provides the desired basis for V.

The second statement is an immediate consequence of the first, since (by Lemma 4.5.2) the basis change to the orthonormal basis of eigenvectors for $\mathbf{R}^n$ comes from an orthogonal matrix. This proves the theorem. //

Example 4.5.10 Consider

$$S = \begin{pmatrix} 2 & -2 & -4 \\ -2 & 5 & -2 \\ -4 & -2 & 2 \end{pmatrix}$$

Then it is easily checked that $C_S(X) = (X - 6)^2(X + 3)$. Computing the eigenspaces of S we get

$$E_6 = ker \begin{pmatrix} -4 & -2 & -4 \\ -2 & -1 & -1 \\ -4 & -2 & -4 \end{pmatrix} = ker \begin{pmatrix} 2 & 1 & 2 \\ 0 & 0 & 0 \\ 0 & 0 & 0 \end{pmatrix}$$

$$= span \left\{ \begin{pmatrix} 0 \\ -2 \\ 1 \end{pmatrix}, \begin{pmatrix} -1 \\ 2 \\ 0 \end{pmatrix} \right\}$$

and

$$E_{-3} = ker \begin{pmatrix} 5 & -2 & -4 \\ -2 & 8 & -2 \\ -4 & -2 & 5 \end{pmatrix} = ker \begin{pmatrix} 1 & -4 & 1 \\ 0 & 18 & -9 \\ 0 & 0 & 0 \end{pmatrix}$$

$$= \text{ker} \begin{pmatrix} 1 & 0 & -1 \\ 0 & 1 & -1/2 \\ 0 & 0 & 0 \end{pmatrix} = \text{span} \left\{ \begin{pmatrix} 2 \\ 1 \\ 2 \end{pmatrix} \right\}$$

We apply the Gram-Schmidt process to this basis of E_6 to obtain an orthonormal basis for E_6:

$$u_1 = \frac{1}{\sqrt{5}} \begin{pmatrix} 0 \\ -2 \\ 1 \end{pmatrix} , \quad w_2 = \begin{pmatrix} -1 \\ 2 \\ 0 \end{pmatrix} - \frac{-4}{5} \begin{pmatrix} 0 \\ -2 \\ 1 \end{pmatrix} = \begin{pmatrix} -1 \\ 2/5 \\ 4/5 \end{pmatrix}$$

so that

$$u_2 = \frac{\sqrt{5}}{3} \begin{pmatrix} -1 \\ 2/5 \\ 4/5 \end{pmatrix}$$

Normalizing

$$\begin{pmatrix} 2 \\ 1 \\ 2 \end{pmatrix} \quad \text{to} \quad \begin{pmatrix} 2/3 \\ 1/3 \\ 2/3 \end{pmatrix}$$

gives u_3. The basis $\{u_1, u_2, u_3\}$ is an orthonormal basis of eigenvectors for T_S. Setting

$$U = \begin{pmatrix} 0 & -\sqrt{5}/3 & 2/3 \\ -2/\sqrt{5} & 2\sqrt{5}/15 & 1/3 \\ 1/\sqrt{5} & 4\sqrt{5}/15 & 2/3 \end{pmatrix}$$

gives an orthogonal matrix for which

$$U^{-1}SU = \begin{pmatrix} 6 & 0 & 0 \\ 0 & 6 & 0 \\ 0 & 0 & -3 \end{pmatrix}$$

We remark the matrix U is by no means unique, because there are many orthonormal bases of eigenvectors for T_S. For example, the reader can check that if

$$U' = \begin{pmatrix} -2/3 & 1/3 & 2/3 \\ 2/3 & 2/3 & 1/3 \\ 1/3 & -2/3 & 2/3 \end{pmatrix}$$

then we also have that $(U')^t S U' = U^t S U$ is diagonal.

Remark 4.5.11 Over general fields, it is *not* true that symmetric matrices can be diagonalized. Note that the proof of Theorem 4.5.8 depends heavily upon the fact that the complex numbers **C** are algebraically closed (that is, every polynomial over **C** factors into linear factors).

The final result of this section is a reformulation of Theorem 4.5.9 in terms of orthogonal projections. This particular formulation will be useful in the next section.

Theorem 4.5.12 (Spectral Theorem) *Suppose that $T : V \to V$ is a self-adjoint operator on a finite-dimensional inner product space V. Then there exist mutually orthogonal subspaces $W_1, W_2, \ldots, W_s$ of V, together with distinct real numbers $r_1, r_2, \ldots, r_s$ such that*

$$T = r_1 Proj_{W_1} + r_2 Proj_{W_2} + \cdots + r_s Proj_{W_s}$$

and

$$I_V = Proj_{W_1} + Proj_{W_2} + \cdots + Proj_{W_s}$$

where $I_V : V \to V$ is the identity operator.

The decomposition of T given by Theorem 4.5.12 is called a *spectral resolution* of T.

Proof: By Theorem 4.5.9 T has an orthonormal basis of eigenvectors. We list such an orthonormal eigenbasis for T as $u_{11}, u_{12}, \ldots, u_{1j_1}, \ldots,$ $u_{s1}, u_{s2}, \ldots, u_{sj_s}$ where the associated eigenvalue of u_{ij} is r_i. (So, we have grouped together the eigenvectors with common eigenvalue.) Denote by $W_i = span\{u_{i1}, u_{i2}, \ldots, u_{ij_i}\}$. Then W_i is the j_i-dimensional eigenspace associated to r_i. By construction, the W_i are mutually orthogonal. Also $Proj_{W_k}(u_{ij}) = \vec{0}$ if $i \neq k$ while $Proj_{W_i}(u_{ij}) = u_{ij}$. Since $T(u_{ij}) = r_i u_{ij}$, we see that T and $r_1 Proj_{W_1} + r_2 Proj_{W_2} + \cdots + r_s Proj_{W_s}$ agree on the basis $\{u_{ij}\}$. This proves the theorem. //

Recall from Lemma 4.3.8 that for any subspace W of $\mathbf{R}^n$, the projection operator $Proj_W : \mathbf{R}^n \to \mathbf{R}^n$ satisfies $Proj_W \circ Proj_W = Proj_W$, and whenever $W' \subseteq W^\perp$, $Proj_W \circ Proj_{W'} = 0$. Applying these rules of composition to a spectral decomposition of a self-adjoint operator T gives the spectral decompositon of T^2, T^3, and so forth. For example, if $T = r_1 Proj_{W_1} + \cdots + r_s Proj_{W_s}$, then $T^2 = r_1^2 Proj_{W_1} + \cdots + r_s^2 Proj_{W_s}$.

Remark 4.5.13 A *normal matrix* is a complex (or real) matrix for which $AA^* = A^*A$. Evidently, every self-adjoint matrix is normal. It can be shown that the class of normal matrices is precisely the class of matrices A for which the linear operator $T_A : \mathbf{F}^n \to \mathbf{F}^n$ has a spectral decomposition (as given in Theorem 4.5.12 for self-adjoint operators). In particular, the conclusions of Theorem 4.5.12 hold for the operator T_A whenever A is normal (that is, $\mathbf{F}^n$ has an orthogonal basis of eigenvectors for T_A).

Example 4.5.14 We continue Example 4.5.10. The symmetric operator $T_S : \mathbf{R}^3 \to \mathbf{R}^3$ has two eigenspaces

$$E_6 = span\left\{\begin{pmatrix} 0 \\ -2 \\ 1 \end{pmatrix}, \begin{pmatrix} -1 \\ 2 \\ 0 \end{pmatrix}\right\} \quad \text{and} \quad E_{-3} = span\left\{\begin{pmatrix} 2 \\ 1 \\ 2 \end{pmatrix}\right\}$$

We can apply Theorem 4.4.6 to calculate the projection operators onto these subspaces: For E_6, we set

$$A = \begin{pmatrix} 0 & -1 \\ -2 & 2 \\ 1 & 0 \end{pmatrix}$$

so that

$$A^t A = \begin{pmatrix} 5 & -4 \\ 1 & 0 \end{pmatrix}$$

Hence,

$$(A^t A)^{-1} = \begin{pmatrix} 5/9 & 4/9 \\ 4/9 & 5/9 \end{pmatrix}$$

We conclude that

$$\tilde{A} = A(A^t A)^{-1} A^t = \begin{pmatrix} 0 & -1 \\ -2 & 2 \\ 1 & 0 \end{pmatrix} \begin{pmatrix} 5/9 & 4/9 \\ 4/9 & 5/9 \end{pmatrix} \begin{pmatrix} 0 & -2 & 1 \\ -1 & 2 & 0 \end{pmatrix}$$

$$= \frac{1}{9} \begin{pmatrix} 5 & -2 & -4 \\ -2 & 8 & -2 \\ -4 & -2 & 5 \end{pmatrix} = Proj_{E_6}$$

Similarly, for E_{-3}, we compute that for

$$B = \begin{pmatrix} 2 \\ 1 \\ 2 \end{pmatrix}, \quad B^t B = (9)$$

so that $(B^t B)^{-1} = (1/9)$. Hence,

$$\tilde{B} = \begin{pmatrix} 2 \\ 1 \\ 2 \end{pmatrix} \frac{1}{9} (2 \quad 1 \quad 2) = \frac{1}{9} \begin{pmatrix} 4 & 2 & 4 \\ 2 & 1 & 2 \\ 4 & 2 & 4 \end{pmatrix} = Proj_{E_{-3}}$$

The spectral decomposition $T_S = 6Proj_{E_6} + (-3)Proj_{E_{-3}}$ gives the following decomposition of our original matrix S:

$$S = \begin{pmatrix} 2 & -2 & -4 \\ -2 & 5 & -2 \\ -4 & -2 & 2 \end{pmatrix}$$

$$= 6\left(\frac{1}{9}\right) \begin{pmatrix} 5 & -2 & -4 \\ -2 & 8 & -2 \\ -4 & -2 & 5 \end{pmatrix} + -3\left(\frac{1}{9}\right) \begin{pmatrix} 4 & 2 & 4 \\ 2 & 1 & 2 \\ 4 & 2 & 4 \end{pmatrix}$$

PROBLEMS

4.5.1. Which of the following matrices are orthogonal (unitary)?

(a) $\begin{pmatrix} 0 & 1 & 0 \\ 1 & 0 & 0 \\ 0 & 0 & 1 \end{pmatrix}$ (b) $\begin{pmatrix} 1/2 & 1/2 & 0 \\ 1/2 & -1/2 & 0 \\ 0 & 0 & 1 \end{pmatrix}$ (c) $\begin{pmatrix} \sin x & \cos x \\ \cos x & \sin x \end{pmatrix}$

(d) $\begin{pmatrix} i & 0 & 0 \\ 0 & i & 0 \\ 0 & 0 & i \end{pmatrix}$ (e) $\begin{pmatrix} 0 & e^i & 0 \\ i & 0 & 0 \\ 0 & 0 & -1 \end{pmatrix}$

4.5.2. Show that if U is unitary, $| det(U) | = 1$.

4.5.3. Show that if A is a real orthogonal matrix, then all (real) eigenvalues of A are ± 1.

4.5.4. Suppose that A and B are self-adjoint. Show that AB is self-adjoint if and only if $AB = BA$.

4.5.5. If S is a skew-symmetric real matrix (that is, $S^t = -S$), show that $I + S$ is nonsingular and $(I - S)(I + S)^{-1}$ is orthogonal.

4.5.6. Show for all invertible $A \in M_{n,n}(\mathbf{C})$, $(A^*)^{-1} = (A^{-1})^*$.

4.5.7. Show that the characteristic polynomial of any self-adjoint linear operator has only real coefficients.

4.5.8. Find a unitary matrix which diagonalizes:

(a) $\begin{pmatrix} 7 & 2+i \\ 2-i & 3 \end{pmatrix}$ (b) $\begin{pmatrix} 0 & e^{i\theta} \\ e^{-i\theta} & 0 \end{pmatrix}$

4.5.9.   Find the spectral resolutions of the following real symmetric matrices:

(a) $\begin{pmatrix} 2 & 3 \\ 3 & 2 \end{pmatrix}$ (b) $\begin{pmatrix} 2 & 0 & -1 \\ 0 & 2 & 0 \\ -1 & 0 & 2 \end{pmatrix}$

4.5.10.   (a) Show that hermitian and skew-hermitian matrices are normal. (A is *skew-hermitian* if $A = -\overline{A}^t$.)
(b) If H is hermitian, show that iH is normal.
(c) Show that the eigenvalues of a skew-hermitian matrix are pure imaginary. (Hint: Apply Lemma 4.5.7 (i).)

4.5.11. Show that if two hermitian matrices have the same characteristic polynomial, then they are similar.

4.5.12. Suppose that $T : V \to V$ is a symmetric operator on the real inner product space V. Show
(a) $T(V) \subseteq (ker(T))^{\perp}$ with equality in case V is finite-dimensional.
(b) $ker(T) \subseteq (T(V))^{\perp}$ with equality in case V is finite-dimensional.

4.5.13. Suppose that A is an $n \times n$ real matrix and there is some orthogonal matrix U with $U^{-1}AU$ diagonal. Prove that A is symmetric.

4.5.14. Let A be a complex matrix.
(a) Show that $A + A^*$ and AA^* are self-adjoint.
(b) Show that $null(AA^*) = null(A)$.
(c) Show that $rk(AA^*) = rk(A) = rk(A^*)$.

4.5.15. Show that orthogonal projections are self-adjoint.

4.5.16. Suppose that $T : V \to W$ is a linear transformation between finite-dimensional inner product spaces.
(a) Show that there exists a unique linear transformation $T^* : W \to V$ characterized by the property that $< T(u), v > = < u, T(v) >$ for all $u, v \in V$.
(b) If $dim(V) = dim(W)$ and $\mathcal{B}$ and $\mathcal{C}$ are orthonormal bases for V and W respectively, show that $[T^*]_{\mathcal{C},\mathcal{B}} = [T]^*_{\mathcal{B},\mathcal{C}}$.

4.5.17. Suppose that T is a self-adjoint operator on a hermitian inner product space V.
(a) Show that $\| T(v) \pm iv \|^2 = \| T(v) \|^2 + \| v \|^2$ for all $v \in V$.
(b) Show that $(T + iI)$ is invertible with $((T + iI)^{-1})^* = (T - iI)^{-1}$.
(c) Show that $(T + iI)(T - iI)^{-1}$ is a unitary operator on V.

4.5.18. Let V be a finite-dimensional inner product space. Let $W \subseteq V$ be a subspace, and recall that $V = W \oplus W^\perp$. Define $T : V \to V$ by $T(w_1 + w_2) = w_1 + aw_2$ where $w_1 \in W$ and $w_2 \in W^\perp$. For which real values of a is T orthogonal? If V is hermitian, for which complex values of a is T unitary?

4.6 The Singular-Value Decomposition, the Pseudoinverse, and the Polar Decompositon

In Sec. 4.4 we found that the least-squares problem $A\vec{X} = \vec{b}$ has a unique solution given by $\vec{X} = (A^t A)^{-1} A^t \vec{b}$, whenever A has full column rank (that is, $rk(A) = n$ if A is $m \times n$). What happens if A does not have full column rank? The study of this situation is the main project of this section. It turns out that the least-squares problem, as stated in Sec. 4.4, does not have a unique solution when A fails to have full column rank. However, there is a uniquely determined solution which is characterized by having minimal norm among all least-squares solutions. In this section we produce a special matrix, $A^\dagger$, known as the pseudoinverse of A, for which $A^\dagger \vec{b}$ is this desired special least-squares solution.

The route to defining (and computing) the pseudoinverse is through a matrix factorization of A known as the singular value decomposition. To begin, we need some results about symmetric matrices of the form $A^t A$. The results from Sec. 4.5 will play a crucial role in this investigation.

Definition 4.6.1 A symmetric operator $S : V \to V$ on a real inner product space V (analogously a real symmetric matrix) is called *positive-semidefinite* if $<v, S(v)> \geq 0$ for all $v \in V$.

Our first lemma shows that the operator associated with a real symmetric matrix of the form $A^t A$ is positive semidefinite. We therefore observe, using the spectral theorem, that positive-semidefinite operators always have square roots. This lemma is true as well for complex matrices, using instead the matrix $A^* A$. We omit the details.

Lemma 4.6.2 (i)   *For any $m \times n$ real matrix A, if $S = A^t A$, then T_S is a positive-semidefinite symmetric operator on $\mathbf{R}^n$.*

(ii)   *A symmetric operator S on a real inner product space V is positive-semidefinite if and only if all its eigenvalues are nonnegative (so 0 is a possible eigenvalue).*

(iii) *If S is a positive-semidefinite real symmetric matrix, then there is a positive semidefinite real symmetric matrix U with $U^2 = S$.*

Proof: (i) We compute that $< v, T_S(v) > = < v, A^t A v > = v^t A^t A v = (Av)^t A v = < Av, Av > \geq 0$, as required.

(ii) Suppose that S is positive semidefinite and let $v \in V$ be an eigenvector for S with eigenvalue c. Then as $c < v, v > = < v, cv > = < v, S(v) > \geq 0$, we conclude $c \geq 0$. Conversely, if S is a symmetric operator with nonnegative eigenvalues, we apply the Spectral Theorem 4.5.12 and express

$$S = r_1 Proj_{W_1} + r_2 Proj_{W_2} + \cdots + r_s Proj_{W_s}$$

where each $Proj_{W_i}$ is the projection operator onto mutually orthogonal subspaces W_i. Each r_i is an eigenvalue of S, so each $r_i \geq 0$. Recall that

$$I_V = Proj_{W_1} + Proj_{W_2} + \cdots + Proj_{W_s}$$

Thus for any $v \in V$, $v = Proj_{W_1}(v) + Proj_{W_2}(v) + \cdots + Proj_{W_s}(v)$ and

$$
\begin{aligned}
< v, S(v) > \ &= \ < P_{W_1}(v) + \cdots + P_{W_s}(v), r_1 P_{W_1}(v) + \cdots + r_s P_{W_s}(v) > \\
&= \ < P_{W_1}(v), r_1 P_{W_1}(v) > + \cdots + < P_{W_s}(v), r_s P_{W_s}(v) > \\
&= \ r_1 < P_{W_1}(v), P_{W_1}(v) > + \cdots + r_s < P_{W_s}(v), P_{W_s}(v) > \\
&\geq \ 0
\end{aligned}
$$

This shows that S is positive semidefinite.

(iii) Suppose S is a positive-semidefinite symmetric matrix, so by the Spectral Theorem 4.5.12 $T_S = r_1 P_{W_1} + r_2 P_{W_2} + \cdots + r_s P_{W_s}$ where each $r_i \geq 0$ and the W_i are subspaces of $\mathbf{R}^n$.   We define $T_U = \sqrt{r_1} P_{W_1} + \sqrt{r_2} P_{W_2} + \cdots + \sqrt{r_s} P_{W_s}$ (where U is the standard matrix of this operator). By Lemma 4.3.8 we know the composition $P_{W_i} \circ P_{W_j} = 0$ if $i \neq j$ and $P_{W_i} \circ P_{W_i} = P_{W_i}$. It follows that $T_U \circ T_U = T_{U^2} = T_S$; that is, $U^2 = S$. This proves the lemma. //

The next result shows how the eigenvectors of the symmetric $n \times n$ matrix $S = A^t A$ are related to A. We recall that Theorem 4.5.9 guarantees that $\mathbf{R}^n$ has an orthornormal basis of eigenvectors for S.

Lemma 4.6.3 *Suppose that A is an $m \times n$ real matrix and set $S = A^t A$. Let $u_1, \ldots, u_n$ be an orthonormal basis of eigenvectors for T_S where the eigenvalue of u_i is r_i^2 $(r_i \geq 0)$. Then the vectors $Au_1, \ldots, Au_n$ are orthogonal in $\mathbf{R}^m$ and $\| Au_i \| = r_i$.*

Proof: We can compute that $< Au_i, Au_j >= (Au_i)^t Au_j = (u_i)^t A^t Au_j$ $= (u_i)^t S u_j = < u_i, S u_j >=< u_i, r_j^2 u_j >= r_j^2 < u_i, u_j >$. If $i \neq j$, then we know that $< u_i, u_j >= 0$; this shows that $< Au_i, Au_j >= 0$. Hence the vectors $Au_1, \ldots, Au_n$ are orthogonal. If $i = j$, then we have that $< Au_i, Au_i >= r_i^2 < u_i, u_i >= r_i^2$. Hence $\| Au_i \| = r_i$ follows, proving the lemma. //

Note that some of the vectors $Au_1, Au_2, \ldots, Au_n$ given by Lemma 4.6.3 may be zero. (In fact, if $rk(A) < n$, some of them will have to be zero, since the column space of A cannot have n linearly independent vectors.) We now define pseudodiagonal matrices and the singular-value decomposition.

Definition 4.6.4 (i) An $m \times n$ matrix $D = (d_{ij})$ is called a *pseudodiagonal matrix* if $d_{ij} = 0$ whenever $i \neq j$.
(ii) Let A be an $m \times n$ matrix. We say that a product $A = U_1 D U_2^t$ is a *singular-value decomposition* for A if U_1 is an $m \times m$ orthogonal matrix, D is an $m \times n$ pseudodiagonal matrix all of whose entries are nonnegative, and U_2 is an $n \times n$ orthogonal matrix. The diagonal entries of D are called the *singular values* of A.

Note that a pseudodiagonal matrix need not be square. All that is required is that the matrix be zero everywhere off the main diagonal starting at the upper left corner. If S is a real symmetric matrix, we saw in the last section that S can be orthogonally diagonalized; that is, there exists an orthogonal matrix U and a diagonal matrix D such that $S = U D U^t$. Such a decomposition is a singular-value decomposition (but of a very special type). This decomposition shows that real symmetric operators are characterized geometrically as appropriate stretchings (or shrinkings) along some orthogonal basis. The idea behind the singular value decompositon is to give similar geometric information. The orthogonal matrices represent basis changes via isometries, and the pseudodiagonal matrix maps one orthogonal basis to another, stretched accordingly.

Our next task is to show that singular-value decompositions exist, give some indication of how to find them, and deduce the uniqueness of the singular values.

Theorem 4.6.5 *Let A be an $m \times n$ real matrix. Then A has a singular-value decomposition. Moreover, the singular values of A are uniquely determined (with multiplicity).*

Proof: We assume that $n \leq m$; the case of $n > m$ follows from this case by considering the matrix A^t. Set $S = A^t A$ and apply Lemma 4.6.3 to obtain an orthonormal basis of $\mathbf{R}^n$, $\{u_1, \ldots, u_n\}$, for which the vectors $Au_1, \ldots, Au_n$ are orthogonal in $\mathbf{R}^m$. We set $r_i = \| Au_i \|$, and we know by Lemma 4.6.3 that u_i is an eigenvector of T_S with eigenvalue r_i^2. We normalize each of the nonzero Au_i by setting $w_i = (1/r_i)Au_i$, and we extend these orthonormal w_i to an orthonormal basis $\{w_1, \ldots, w_n, \ldots, w_m\}$ of $\mathbf{R}^m$.

We set U_2 to be the $n \times n$ orthogonal matrix with columns $u_1, \ldots, u_n$, and we set U_1 to be the $m \times m$ orthogonal matrix with columns $w_1, \ldots, w_m$. Let $D = (d_{ij})$ be the pseudodiagonal $m \times n$ matrix with $d_{ii} = r_i$ for $i = 1, 2, \ldots, n$ and all other $d_{ij} = 0$. We denote by $\{e_1, \ldots, e_n\}$ the standard basis for $\mathbf{R}^n$ and by $\{e_1', \ldots, e_m'\}$ the standard basis for $\mathbf{R}^m$. The orthogonal matrices U_1, U_2 are basis-change matrices, and consequently $U_2^t u_i = e_i$ and $U_1 e_j' = w_j$. From this we see for $i = 1, 2, \ldots, n$,

$$(U_1 D U_2^t)(u_i) = (U_1 D)(e_i) = (U_1)(r_i e_i') = r_i w_i = Au_i$$

This shows that $U_1 D U_2^t = A$, since the two linear transformations they define agree on the basis $\{u_1, \ldots, u_n\}$ of $\mathbf{R}^n$.

With the existence of the singular value decomposition established, it remains to show that the matrix D is uniquely determined (up to the order of the diagonal entries). The entries d_{ii} on the diagonal of D in the above construction were precisely the nonnegative square roots of the eigenvalues of the symmetric matrix $A^t A$ with multiplicity the dimension of the associated eigenspace. As these eigenvalues, with multiplicity, are uniquely determined by the matrix $A^t A$, it suffices to show that whenever $A = U_1 D U_2^t$ is a singular value decomposition of A, the diagonal entries of D are necessarily the eigenvalues of $A^t A$. Computing (remember $U_i^t = U_i^{-1}$) $A^t A = (U_1 D U_2)^t U_1 D U_2^t = U_2 D^t D U_2^t$, we see that the matrices $A^t A$ and $D^t D$ are similar. In particular,

$A^t A$ and $D^t D$ have the same eigenvalues (with multiplicity). But $D^t D$ is an $n \times n$ diagonal matrix with diagonal entries d_{ii}^2, which are its eigenvalues. This proves the theorem. //

For example, consider the matrix

$$A = \begin{pmatrix} 1 & 2 \\ 1 & 2 \\ 1 & 2 \end{pmatrix}$$

A has rank 1. We follow the proceedure given in the proof of Theorem 4.6.5 to determine its singular-value decomposition. We form the symmetric matrix

$$S = A^t A = \begin{pmatrix} 3 & 6 \\ 6 & 12 \end{pmatrix}$$

The characteristic polynomial of S is $C_S(X) = X(X-15)$, and it is easily checked that $(1, 2)$ is an eigenvector with eigenvalue 15, while $(2, -1)$ is an eigenvector with eigenvalue 0. Consequently, $\{(1/\sqrt{5}, 2/\sqrt{5}), (2/\sqrt{5}, -1/\sqrt{5})\}$ is an orthonormal basis for $\mathbf{R}^2$ of eigenvectors of S. (This basis gives the vectors denoted by $u_1, \ldots, u_n$ in the proof of Theorem 4.6.5.)

Following the proof of Theorem 4.6.5, we compute

$$A \begin{pmatrix} 1/\sqrt{5} \\ 2/\sqrt{5} \end{pmatrix} = \begin{pmatrix} \sqrt{5} \\ \sqrt{5} \\ \sqrt{5} \end{pmatrix}$$

which has norm $\sqrt{15}$, the square root of the eigenvalue of the eigenvector $(1/\sqrt{5}, 2/\sqrt{5})$ of S. This vector is normalized to the unit vector $(1/\sqrt{3}, 1/\sqrt{3}, 1/\sqrt{3})$, which we extend to an orthonormal basis of $\mathbf{R}^3$, say

$$\{(1/\sqrt{3}, 1/\sqrt{3}, 1/\sqrt{3}), (1/\sqrt{2}, 0, -1/\sqrt{2}), (1/\sqrt{6}, -2/\sqrt{6}, 1/\sqrt{6})\}$$

This gives the following singular-value decomposition of A:

$$\begin{pmatrix} 1 & 2 \\ 1 & 2 \\ 1 & 2 \end{pmatrix} = \begin{pmatrix} 1/\sqrt{3} & 1/\sqrt{2} & 1/\sqrt{6} \\ 1/\sqrt{3} & 0 & -2/\sqrt{6} \\ 1/\sqrt{3} & -1/\sqrt{2} & 1/\sqrt{6} \end{pmatrix} \begin{pmatrix} \sqrt{15} & 0 \\ 0 & 0 \\ 0 & 0 \end{pmatrix} \begin{pmatrix} 1/\sqrt{5} & 2/\sqrt{5} \\ 2/\sqrt{5} & -1/\sqrt{5} \end{pmatrix}$$

We now return to the least-squares problem. Suppose that a least-squares problem $A\vec{X} = \vec{b}$ does not have a unique solution. Since all solutions to the least-squares problem are solutions to the normal equation $A^t A \vec{X} = A^t \vec{b}$, this is the same as saying that $A^t A$ is not invertible; that is, $rk(A) < n$ (where A is $m \times n$). Among all possible solutions to the normal equation, we seek a special solution known as the *optimal solution*. This is defined next.

Definition 4.6.6 The *optimal solution* to a least-squares problem $A\vec{X} = \vec{b}$ is the least squares-solution $\vec{c}$ with $\| \vec{c} \|$ minimal among all possible least-square solutions.

It is not immediately clear from the definition that there is always a unique optimal solution to a least-squares problem. That this is true is given in the next lemma.

Lemma 4.6.7 *Any least-squares problem $A\vec{X} = \vec{b}$ has a unique optimal solution $\vec{c}$, which is characterized by $Proj_{W'}(\vec{c}) = \vec{0}$, where W' is the subspace $ker(A^t A)$.*

Proof: Let $\vec{d}$ be a solution to the normal equation $A^t A \vec{X} = A^t \vec{b}$ of the least-squares problem $A\vec{X} = \vec{b}$. Then we know that all the solutions to this equation have the form $\vec{d} + \vec{w}$ where $\vec{w} \in ker(A^t A) = W'$. As $Proj_{W'}(\vec{d} + \vec{w}) = Proj_{W'}(\vec{d}) + \vec{w}$, it follows that there is a unique solution $\vec{c}$ with $Proj_{W'}(\vec{c}) = \vec{0}$ [that is, take $w = -Proj_{W'}(\vec{d})$]. All solutions to the least-squares problem now can be written as $\vec{c} + \vec{w}$, $\vec{w} \in W'$. Moreover, as $\vec{c}$ is orthogonal to W', we find that $\| \vec{c} + \vec{w} \|^2 = \| \vec{c} \|^2 + \| \vec{w} \|^2$. From this we see that the chosen $\vec{c}$ is the unique vector with minimal norm among all the solutions $\vec{c} + \vec{w}$. This proves the lemma. //

We now describe the pseudoinverse, from which the optimal solution to any least-squares problem can be obtained. We saw in Corollary 4.4.7 that if the $m \times n$ matrix A has rank n, then the unique (and hence optimal) least squares solution to $A\vec{X} = \vec{b}$ was given by $(A^t A)^{-1} A^t \vec{b}$. Thus in this case it is reasonable to take the matrix $(A^t A)^{-1} A^t$ to be the pseudoinverse of A. Observe in fact, $(A^t A)^{-1} A^t A = I_m$, so that $(A^t A)^{-1} A^t$ is indeed a left inverse of A.

The definition of the pseudoinverse is awkward because one must use the singular-value decomposition. The reason for this is that with matrices of low rank one cannot expect to find right or left inverses. Consequently, one must construct the pseudoinverse of a matrix A to be a matrix whose product with A is a pseudodiagonal matrix with only nonzero entries 1 along the main diagonal. All of this is described next. Note: The well-definition of the pseudoinverse is proved in the theorem which follows.

Definition 4.6.8 (i) An $n \times n$ pseudodiagonal matrix E is called a *pseudoidentity matrix*, if the nonzero diagonal entries are 1.
(ii) If $D = (d_{ij})$ is an $m \times n$ pseudodiagonal matrix, the *pseudoinverse* of D, denoted $D^\dagger$, is defined to be the $n \times m$ matrix given by $D^\dagger = (f_{kl})$ where $f_{ii} = d_{ii}^{-1}$ if $d_{ii} \neq 0$ and $f_{kl} = 0$ otherwise.
(iii) If A is an $m \times n$ matrix with singular-value decomposition $A = U_1 D U_2^t$, we define the *pseudoinverse* of A, $A^\dagger$, to be the $n \times m$ matrix $A^\dagger = U_2 D^\dagger U_1^t$. The pseudoinverse is also known as the *Moore-Penrose generalized inverse*.

For example we have the singular-value decomposition

$$M = \begin{pmatrix} 7/3 & 8 \\ -10/3 & 10 \\ 34/3 & -4 \end{pmatrix}$$

$$= \begin{pmatrix} 2/3 & 1/3 & -2/3 \\ 1/3 & 2/3 & 2/3 \\ 2/3 & -2/3 & 1/2 \end{pmatrix} \begin{pmatrix} 10 & 0 \\ 0 & 15 \\ 0 & 0 \end{pmatrix} \begin{pmatrix} 4/5 & 3/5 \\ -3/5 & 4/5 \end{pmatrix}$$

Thus we compute the pseudoinverse of M, $M^\dagger$ by

$$M^\dagger = \begin{pmatrix} 4/5 & -3/5 \\ 3/5 & 4/5 \end{pmatrix} \begin{pmatrix} 1/10 & 0 & 0 \\ 0 & 1/15 & 0 \end{pmatrix} \begin{pmatrix} 2/3 & 1/3 & 2/3 \\ 1/3 & 2/3 & -2/3 \\ -2/3 & 2/3 & 1/3 \end{pmatrix}$$

$$= \frac{1}{450} \begin{pmatrix} 18 & 0 & 36 \\ 26 & 25 & 2 \end{pmatrix}$$

One can check readily that $M^\dagger M = I_2$; that is, $M^\dagger$ is a left inverse for M.

There are many possible rank s, $n \times n$ pseudoidentity matrices. Also observe that in case D is a pseudodiagonal $m \times n$ matrix, then $DD^\dagger$ is an $m \times m$ pseudoidentity matrix. Moreover, $rk(D) = rk(D^\dagger) = rk(DD^\dagger)$ since each has the same number of nonzero diagonal entries. We generalize this fact to arbitrary pseudoinverses in the following.

Theorem 4.6.9 *Suppose that A is an $m \times n$ matrix and $A = U_1 D U_2^t$ is a singular-value decomposition of A. We let $A^\dagger = U_2 D^\dagger U_1^t$. Then:*
(i) $rk(A) = rk(A^\dagger)$.
(ii)   *The optimal solution to the least-squares problem $A\vec{X} = \vec{b}$ is given by $\vec{X} = A^\dagger \vec{b}$. Consequently, the pseudoinverse of A, $A^\dagger$, as given in Definition 4.6.8 is well-defined.*
(iii) *If $rk(A) = n$, then $A^\dagger = (A^t A)^{-1} A^t$.*

Proof: (i) Expressing $A = U_1 D U_2^t$, as both U_1 and U_2 are invertible, we conclude that $rk(A) = rk(D)$. [Recall that if U is invertible, then for any matrix B, B and UB are row equivalent since U can be expressed as a product of elementary matrices. Consequently, $rk(B) = rk(UB)$. Our assertion follows from this.] But $rk(D) = rk(D^\dagger)$, and the same reasoning shows $rk(D^\dagger) = rk(A^\dagger)$. Hence $rk(A) = rk(A^\dagger)$.

(ii) We begin with the case where $A = D = (d_{ij})$ is pseudodiagonal. First we check that $D^\dagger \vec{b}$ is a least-squares solution to $D\vec{X} = \vec{b}$. Let $W_D = col(D)$ and observe that $W_D = span\{e_i \in \mathbf{R}^m \mid d_{ii} \neq 0\}$. Evidently, $DD^\dagger$ is the $m \times m$ pseudoidentity matrix with column space W_D, and consequently $DD^\dagger = [Proj_{W_D}]$. Thus $Proj_{W_D}(\vec{b}) = (DD^\dagger)\vec{b} = D(D^\dagger \vec{b})$, which shows by Lemma 4.4.4 that $D^\dagger \vec{b}$ is a least-squares solution to $D\vec{X} = \vec{b}$. Next observe that $W_D' = ker(D^t D) = span\{e_j \in \mathbf{R}^n \mid d_{jj} = 0\}$ and $D^\dagger \vec{b} \in span\{e_j \in \mathbf{R}^n \mid d_{jj} \neq 0\} = W_D'^\perp$. This shows that $Proj_{W_D'}(D^\dagger \vec{b}) = 0$, so by Lemma 4.6.7, $D^\dagger \vec{b}$ is the optimal least-squares solution to $D\vec{X} = \vec{b}$.

Now consider the general case where A has the singular-value decomposition $A = U_1 D U_2^t$. Recall for any orthogonal $m \times m$ matrix U and any $v \in \mathbf{R}^m$, that $\| v \| = \| Uv \|$ (because $T_U : \mathbf{R}^m \to \mathbf{R}^m$ is an isometry by Lemma 4.5.5 (i)). Using this fact, we compute the norm:

$$
\begin{aligned}
\| A\vec{X} - \vec{b} \| &= \| U_1 D U_2^t \vec{X} - \vec{b} \| \\
&= \| U_1^t(U_1 D U_2^t \vec{X} - \vec{b}) \| \\
&= \| D U_2^t \vec{X} - U_1^t \vec{b} \|
\end{aligned}
$$

Consider any $\vec{c} \in \mathbf{R}^n$ which is a least squares solution to $A\vec{X} = \vec{b}$. Since $\| \vec{c} \| = \| U_2^t \vec{c} \|$, it follows that $\vec{c}$ is the optimal solution to $A\vec{X} = \vec{b}$ if and only if $U_2^t \vec{c}$ is the optimal solution (for Y) to $AU_2 Y = \vec{b}$. Multiplying on the left by U_1^t, we obtain $U_1^t A U_2 Y = U_1^t \vec{b}$. Using the fact that

$A = U_1 D U_2^t$, this equation becomes $D\vec{Y} = U_1^t \vec{b}$. By the pseudodiagonal case, we know that $D^\dagger(U_1^t\vec{b})$ is the optimal least-squares solution to $D\vec{Y} = U_1^t\vec{b}$. Multiplying this solution by $(U_2^t)^{-1} = U_2$, we obtain that $U_2 D^\dagger U_1^t \vec{b} = A^\dagger \vec{b}$ is the optimal least-squares solution to the original system $A\vec{X} = \vec{b}$.

The well-definition of $A^\dagger$ now follows. Let $\vec{b} \in \mathbf{R}^n$ be arbitrary. Then $A^\dagger$ is the unique matrix for which $A^\dagger\vec{b}$ is the unique optimal solution to the least-squares problem $A\vec{X} = \vec{b}$. This proves (ii).

(iii) This follows for the same reason as (ii). Whenever $rk(A) = m$, we know that $(A^t A)^{-1} A^t \vec{b}$ is always the unique (and hence optimal) solution to the least-squares problem $A\vec{X} = \vec{b}$. This proves the theorem. //

Example 4.6.10 We earlier obtained the singular-value decomposition of the matrix

$$A = \begin{pmatrix} 1 & 2 \\ 1 & 2 \\ 1 & 2 \end{pmatrix}$$

$$= \begin{pmatrix} 1/\sqrt{3} & 1/\sqrt{2} & 1/\sqrt{6} \\ 1/\sqrt{3} & 0 & -2/\sqrt{6} \\ 1/\sqrt{3} & -1/\sqrt{2} & 1/\sqrt{6} \end{pmatrix} \begin{pmatrix} \sqrt{15} & 0 \\ 0 & 0 \\ 0 & 0 \end{pmatrix} \begin{pmatrix} 1/\sqrt{5} & 2/\sqrt{5} \\ 2/\sqrt{5} & -1/\sqrt{5} \end{pmatrix}$$

We now can compute the pseudoinverse of **A**. The pseudoinverse of

$$\begin{pmatrix} \sqrt{15} & 0 \\ 0 & 0 \\ 0 & 0 \end{pmatrix} \quad \text{is} \quad \begin{pmatrix} 1/\sqrt{15} & 0 & 0 \\ 0 & 0 & 0 \end{pmatrix}$$

so we find that

$$A^\dagger = \begin{pmatrix} 1/\sqrt{5} & 2/\sqrt{5} \\ 2/\sqrt{5} & -1/\sqrt{5} \end{pmatrix} \begin{pmatrix} 1/\sqrt{15} & 0 & 0 \\ 0 & 0 & 0 \end{pmatrix} \begin{pmatrix} 1/\sqrt{3} & 1/\sqrt{3} & 1/\sqrt{3} \\ 1/\sqrt{2} & 0 & -1/\sqrt{2} \\ 1/\sqrt{6} & -2/\sqrt{6} & 1/\sqrt{6} \end{pmatrix}$$

$$= \begin{pmatrix} 1/15 & 1/15 & 1/15 \\ 2/15 & 2/15 & 2/15 \end{pmatrix}$$

From this it follows that the optimal least-squares solution to the the system

$$A\vec{X} = \begin{pmatrix} r \\ s \\ t \end{pmatrix}$$

is given by

$$\vec{X} = A^\dagger \begin{pmatrix} r \\ s \\ t \end{pmatrix} = \frac{1}{15} \begin{pmatrix} r+s+t \\ 2(r+s+t) \end{pmatrix}$$

We note that this system has infinitely many different (nonoptimal) least-squares solutions, because the normal equation associated with

$$A\vec{X} = \begin{pmatrix} r \\ s \\ t \end{pmatrix} \quad \text{is} \quad A^t A\vec{X} = A^t \begin{pmatrix} r \\ s \\ t \end{pmatrix}$$

that is,

$$\begin{pmatrix} 3 & 6 \\ 6 & 12 \end{pmatrix} \vec{X} = (r+s+t) \begin{pmatrix} 1 \\ 2 \end{pmatrix}$$

This shows that all the solutions to the normal equation have the form $(X, (1/6)(r + s + t - 3X))$ and the optimal solution is the particular least squares solution with $X = (1/15)(r + s + t)$.

The final result of this section gives a geometric interpretation of a general linear operator $T : \mathbf{R}^n \to \mathbf{R}^n$ and is a nice application of the existence of the singular-value decomposition. What it says is roughly this: Every linear operator on $\mathbf{R}^n$ can be decomposed into a composition of two operators. The first is a stretching (or shrinking) operator along perpendicular subspaces. It is a symmetric operator. The second is a rotation (plus a possible reflection). It is an isometry. Such an expression for a linear operator on $\mathbf{R}^n$ is known as a *polar decomposition*. This same result is true for operators on unitary space $\mathbf{C}^n$. In that case one replaces "symmetric" by "self-adjoint" and "orthogonal" by "unitary".

Theorem 4.6.11 (Polar Decomposition) *If $T : \mathbf{R}^n \to \mathbf{R}^n$ is a linear operator, then there is a positive-semidefinite symmetric operator T_S and an isometry T_U such that $T = T_U \circ T_S$. In other words, every real $n \times n$ matrix A can be written as $A = US$ where S is a positive-semidefinite symmetric matrix, and U is orthogonal.*

Proof: We set $A = [T]$ in the standard basis. The matrix A has a singular-value decomposition $A = U_1 D U_2^t$ where U_1, U_2 are orthogonal $n \times n$ matrices and D is a diagonal $n \times n$ matrix. Set $U = U_1 U_2^t$ and $S = U_2 D U_2^t$. Then U is orthogonal and S is symmetric. Moreover, $US = (U_1 U_2^t)(U_2 D U_2^t) = U_1 D U_2^t = A$. This proves the theorem. //

Example 4.6.12 We consider the 3×3 real matrix

$$A = \begin{pmatrix} -2 & 1 & 0 \\ 14 & 2 & 12 \\ 13 & -2 & 12 \end{pmatrix}$$

Computing, we get

$$A^t A = \begin{pmatrix} 369 & 0 & 324 \\ 0 & 9 & 0 \\ 324 & 0 & 288 \end{pmatrix} = 9 \begin{pmatrix} 41 & 0 & 36 \\ 0 & 1 & 0 \\ 36 & 0 & 24 \end{pmatrix} = 9 \begin{pmatrix} 5 & 0 & 4 \\ 0 & 1 & 0 \\ 4 & 0 & 4 \end{pmatrix}^2$$

Denoting

$$S = \begin{pmatrix} 15 & 0 & 12 \\ 0 & 3 & 0 \\ 12 & 0 & 12 \end{pmatrix}$$

we see the construction of the singular-value decomposition shows that the polar decomposition of A will be of the form $A = US$, for some orthogonal matrix U. Since S is invertible, we can compute $U = AS^{-1}$. This shows that

$$U = \frac{1}{3} \begin{pmatrix} -2 & 1 & 2 \\ 2 & 2 & 1 \\ 1 & -2 & -2 \end{pmatrix}$$

which is readily checked to be orthogonal. This gives the polar decompositon of A.

PROBLEMS

4.6.1. Find the singular-value decompostions and pseudoinverses of

(a) $\begin{pmatrix} 1 & 1 \\ 0 & 0 \end{pmatrix}$ (b) $\begin{pmatrix} 1 & 0 \\ 1 & 1 \\ 0 & 1 \end{pmatrix}$ (c) $\begin{pmatrix} 1 & 1 & 0 \\ 0 & 1 & 0 \\ 0 & 0 & 0 \end{pmatrix}$

4.6.2. If P and Q are orthogonal, show that A and PAQ have the same singular values.

4.6.3. Show that for any matrix A, the matrices $A^\dagger$ and A^t have the same column space. Then show that $A^\dagger A$ gives the orthogonal projection onto the column space of A^t.

4.6.4. Suppose that P is an orthogonal matrix diagonalizing $A^t A$, in other words $P^t(A^t A)P$ is diagonal. Let v_i be the ith column of P, so that the v_i form an orthonormal basis of eigenvectors for $A^t A$. Suppose that the eigenvalue of v_i is s_i [for which, by Lemma 4.6.2 (ii), we know $s_i \geq 0$]. Assume that $s_1, \ldots, s_r > 0$ and $s_{r+1} = 0 = \cdots = s_n$. [So, $rank(A^t A) = r$.]

(a) Show that

$$\begin{pmatrix} x_1 \\ x_2 \\ \vdots \\ x_n \end{pmatrix}$$

is a solution to $A^t A \vec{X} = A^t \vec{b}$ if and only if

$$\begin{pmatrix} x_1 \\ x_2 \\ \vdots \\ x_r \\ x_{r+1} \\ \vdots \\ x_n \end{pmatrix} = P \begin{pmatrix} <v_1, b> \\ <v_2, b> \\ \vdots \\ <v_r, b> \\ u_1 \\ \vdots \\ u_{n-r} \end{pmatrix}$$

where $<\,,\,>$ is the dot product and $u_1, \ldots, u_{n-r}$ are arbitrary real numbers.

(b) Using the fact that the solutions to $A^t A \vec{X} = A^t \vec{b}$ are the least-squares solutions to $A\vec{X} = \vec{b}$, together with (a), what is the unique such least-squares solution of minimal length, in terms of $P, v_1, \ldots, v_r$?

4.6.5. By computing an appropriate pseudoinverse, find the optimal solution to the following least-squares problems:

(a) $\begin{pmatrix} 1 & 1 \\ 1 & 1 \\ 1 & 1 \end{pmatrix} \begin{pmatrix} x \\ y \end{pmatrix} = \begin{pmatrix} 3/2 \\ 2 \\ 1 \end{pmatrix}$ (b) $\begin{pmatrix} 1 & 1 & 1 \\ 1 & 1 & 2 \\ 2 & 2 & 3 \\ 1 & 1 & 1 \end{pmatrix} \begin{pmatrix} x \\ y \\ z \end{pmatrix} = \begin{pmatrix} 5 \\ 4 \\ 9 \\ 6 \end{pmatrix}$

4.6.6. Two matrices A and B are called *equivalent* if there exist invertible matrices P and Q such that $A = PBQ$. Show that if A is a rank r, $m \times n$ matrix, then A is equivalent to the rank r, $m \times n$ pseudoidentity matrix.

4.6.7. Compute a polar decomposition for each of the following real matrices

(a) $\begin{pmatrix} -1 & -9 \\ 7 & 13 \end{pmatrix}$ (b) $\begin{pmatrix} 1 & 0 & 2 \\ 2 & 3 & 0 \\ 4 & 2 & 1 \end{pmatrix}$

Part II: Advanced Topics

Chapter 5

Bilinear Forms, Symmetric Matrices, and Quadratic Forms

5.1   Matrices and Bilinear Forms

Real symmetric matrices were studied in Chap. 4. In this chapter we will study symmetric matrices over arbitrary fields, but with a different objective in mind. They will be used in the study of symmetric bilinear forms, instead of being used to study special types of linear operators. At the same time we will develop the closely related theory of quadratic forms over arbitrary fields.

Throughout this chapter we assume that the characteristic of our scalar field is different from 2; this means that $1 + 1 \neq 0$. (This is certainly true for $\mathbf{R}$ and $\mathbf{C}$, but not for the field $\mathbf{Z}/2\mathbf{Z}$ used in Sec. 1.6.) Our first task is to establish an equivalence between the three different classes of objects — symmetric matrices, symmetric bilinear forms, and quadratic forms. This equivalence is powerful, in that questions which may be obscure in one case become simple when considered for an equivalent class. This first section treats the relationship between matrices and bilinear forms.

The reader will note that the definition of a bilinear form (Definition 5.1.1) is a generalization of the concept of a real inner product. In fact, many of the results which follow are direct generalizations of results proved earlier for real inner products.

Definition 5.1.1 Suppose that V is a vector space over the scalar field $\mathbf{F}$. A *bilinear form* is a function $(\ , \ ) : V \times V \to \mathbf{F}$ which satisfies the following:

(i) $(u + v, w) = (u, w) + (v, w)$, $(u, v + w) = (u, v) + (u, w)$, and $(kv, w) = k(v, w) = (v, kw)$ for all $u, v, w \in V$ and all $k \in \mathbf{F}$.

$(\ , \ )$ is called a *symmetric bilinear form* if, in addition, it satisfies:

(ii) $(v, w) = (w, v)$ for all $v, w \in V$.

Remark In Chapter 8 we will study n-linear forms, of which bilinear forms are the special case where $n = 2$. It is also important to study bilinear functions on $V \times W$ where V and W are (possibly) distinct $\mathbf{F}$ vector spaces. (Usually the terminology "bilinear pairing" is used in this case.) This more general situation is also treated in Chap. 8 (see Prob. 5.1.4 as well). We chose to restrict Definition 5.1.1 as above in order to emphasize the connection between symmetric bilinear forms, symmetric matrices, and quadratic forms.

Examples 5.1.2 (i) Any inner product on a real vector space (see Chap. 4) is a symmetric bilinear form. The only additional condition required for a real symmetric bilinear form to be an inner product is the positive-definite conditition. Note that a hermitian inner product is *not* bilinear because it fails to be linear in the second variable (instead it is conjugate bilinear).

(ii) This example introduces some notation we need for the rest of this chapter. Suppose that A is an $n \times n$ matrix. For $u, v \in \mathbf{F}^n$ (which we denote by column vectors) we define

$$(u, v)_A = u^t A v \in \mathbf{F}$$

We can quickly check that $(\ , \ )_A$ is a bilinear form: $(u + v, w)_A = (u + v)^t A w = (u^t + v^t) A w = u^t A w + v^t A w = (u, w)_A + (v, w)_A$ and $(kv, w)_A = (kv)^t A w = k(v^t A w) = k(v, w)_A$. Similar reasoning establishes linearity on the right.

(iii) Suppose that S is a symmetric $n \times n$ matrix. Then the preceding bilinear form $(\ , \ )_S$ is symmetric as $(u, v)_S = (u, v)_S^t = (u^t S v)^t = v^t S^t u = v^t S u = (v, u)_S$.

We will show that every (symmetric) bilinear form on $\mathbf{F}^n$ arises from a (symmetric) matrix exactly as in Example 5.1.2 (ii). (This is why we gave a short list of initial examples.) In fact, we shall prove a more general result, which applies to arbitrary bases of a **F** vector space V (see Theorem 5.1.4 below). First we consider an easy lemma which justifies the use of the terminology "bilinear."

Lemma 5.1.3 *Suppose that* $(,) : V \times V \to \mathbf{F}$ *is a bilinear form. For any* $v \in V$ *the functions* R_v *and* $L_v : V \to \mathbf{F}$ *defined by* $R_v(u) = (v, u)$ *and* $L_v(u) = (u, v)$ *are linear transformations.*

Proof: $L_v(u + w) = (u + w, v) = (u, v) + (w, v) = L_v(u) + L_v(w)$, and $L_v(ku) = (ku, v) = k(u, v) = kL_v(u)$. This shows that L_v is a linear transformation. That R_v is linear is proved similarly. //

The next theorem gives the above mentioned fact that all (symmetric) bilinear forms on $\mathbf{F}^n$ arise from (symmetric) matrices. This result should not come as a surprise, since we know that linear transformations are given by matrices and, by the preceding lemma, the functions R_v and L_v are linear. Observe that the proof uses exactly the same ideas as the proof of Theorem 3.1.5, which showed that all linear transformations from $\mathbf{F}^n$ to $\mathbf{F}^m$ arise from matrix multiplication.

Theorem 5.1.4 *Suppose that* $(,) : V \times V \to \mathbf{F}$ *is a bilinear form.*
(i) *Let* $\mathcal{B} = \{u_1, u_2, \ldots, u_n\}$ *be an ordered basis for* V. *Then there is a unique* $n \times n$ *matrix* $A_{\mathcal{B}} = (a_{ij})$ *such that for all* $v, w \in V$, $(v, w) = (v)_{\mathcal{B}}^t A_{\mathcal{B}} (w)_{\mathcal{B}}$. $A_{\mathcal{B}}$ *is given by* $a_{ij} = (u_i, u_j)$.
(ii) $(,) : V \times V \to \mathbf{F}$ *is symmetric if and only if the matrix* $A_{\mathcal{B}}$ *is symmetric.*
(iii) *If* $V = \mathbf{F}^n$, *there is a unique* $n \times n$ *matrix* A *such that* $(,) = (,)_A$. *Moreover,* $(,)$ *is symmetric if and only if* A *is symmetric.*

Proof: (i) and (ii) Let $A = (a_{ij})$ be any matrix which satisfies $(v, w) = (v)_{\mathcal{B}}^t A(w)_{\mathcal{B}}$ for all $v, w \in V$. Then, $(u_i, u_j) = (u_i)_{\mathcal{B}}^t A(u_j)_{\mathcal{B}} = e_i^t A e_j = a_{ij} \in \mathbf{F}$, so A must be the matrix $A_{\mathcal{B}}$ given in (i). Thus the uniqueness statement of (i) will follow once we establish that $A_{\mathcal{B}}$ has the stated property. In case $(,)$ is a symmetric bilinear form, we have that $(u_i, u_j) = (u_j, u_i)$, so $a_{ij} = a_{ji}$. Consequently, $A_{\mathcal{B}}$ is a symmetric matrix. Conversely, if $A_{\mathcal{B}}$ is a symmetric matrix, then $(v, w) = (v)_{\mathcal{B}}^t A_{\mathcal{B}}(w)_{\mathcal{B}} = ((v)_{\mathcal{B}}^t A_{\mathcal{B}}(w)_{\mathcal{B}})^t = (w)_{\mathcal{B}}^t A_{\mathcal{B}}^t(v)_{\mathcal{B}} = (w)_{\mathcal{B}}^t A_{\mathcal{B}}(v)_{\mathcal{B}} = (w, v)$. Hence $(,)$ is symmetric.

To complete the proof, we must show that $(u, w) = (u)_{\mathcal{B}}^t A_{\mathcal{B}} (w)_{\mathcal{B}}$ holds for all $u, w \in V$. For any $v \in V$ we apply Lemma 5.1.3 and consider the two linear transformations R_v and R'_v defined by $R_v(u) = (v, u)$ and $R'_v(u) = (v)_{\mathcal{B}}^t A_{\mathcal{B}} (u)_{\mathcal{B}}$. In order to establish the equation $(u, w) = (u)_{\mathcal{B}}^t A_{\mathcal{B}} (w)_{\mathcal{B}}$, it suffices to establish the equality of functions $R_v = R'_v$.

We assume that

$$(v)_{\mathcal{B}} = \begin{pmatrix} b_1 \\ b_2 \\ \vdots \\ b_n \end{pmatrix}$$

so that $v = b_1 u_1 + b_2 u_2 + \cdots + b_n u_n$. Using the multilinearity properties of (,), we evaluate

$$
\begin{aligned}
R_v(u_k) &= (v, u_k) = (b_1 u_1 + b_2 u_2 + \cdots + b_n u_n, u_k) \\
&= b_1(u_1, u_k) + b_2(u_2, u_k) + \cdots + b_n(u_n, u_k) \\
&= b_1 a_{1k} + b_2 a_{2k} + \cdots + b_n a_{nk} \\
&= (b_1 \ b_2 \ \cdots \ b_n) \begin{pmatrix} a_{1k} \\ \vdots \\ a_{nk} \end{pmatrix} \\
&= (b_1 \ b_2 \ \cdots \ b_n)(A_{\mathcal{B}} e_k) \\
&= (v)_{\mathcal{B}}^t A_{\mathcal{B}} (u_k)_{\mathcal{B}} \\
&= R'_v(u_k)
\end{aligned}
$$

From this we see that R_v and R'_v agree on a basis for $\mathbf{F}^n$. According to Lemma 3.1.4, these two linear transformations must be the same; that is, $R_v = R'_v$. Consequently, $A_{\mathcal{B}}$ has the desired properties. This gives (i) and (ii).

(iii) This follows from (i) and (ii) by taking $\mathcal{B}$ to be the standard basis of $\mathbf{F}^n$. This proves the theorem. $/\!/$

The representing matrix $A = (a_{ij})$, where $a_{ij} = (u_i, u_j)$ is sometimes called the *Gram matrix* of the form (,) relative to the basis $\{u_1, u_2, \ldots, u_n\}$. As an example, consider the function $(\ ,\ )_W : \mathbf{F}^3 \times \mathbf{F}^3 \to \mathbf{F}$ defined by $((a, b, c), (r, s, t))_W = a(s + t) - 2cr$. It is easy to check that $(\ ,\ )_W$ is a bilinear form on $\mathbf{F}^3$. Computing in the standard basis, we find that $(e_1, e_2)_W = 1$, $(e_1, e_3)_W = 1$, $(e_3, e_1)_W = -2$, and all other $(e_i, e_j)_W = 0$. Consequently the standard (Gram) matrix for this bilinear form is

$$\begin{pmatrix} 0 & 1 & 1 \\ 0 & 0 & 0 \\ -2 & 0 & 0 \end{pmatrix}$$

We can check this by carrying out the matrix multiplication

$$(a \quad b \quad c) \begin{pmatrix} 0 & 1 & 1 \\ 0 & 0 & 0 \\ -2 & 0 & 0 \end{pmatrix} \begin{pmatrix} r \\ s \\ t \end{pmatrix} = (a \quad b \quad c) \begin{pmatrix} s + t \\ 0 \\ -2r \end{pmatrix}$$

$$= a(s + t) - 2cr$$

Suppose that one desires to find the matrix representation of the bilinear form $(,)_W$ above with respect to some second basis of $\mathbf{F}^n$. One could use the proceedure given in part (i) of Theorem 5.1.4. Alternatively, one can use the next theorem. This result is the bilinear-form analogue of the Basis-Change Theorem 3.3.7 for linear transformations.

Theorem 5.1.5 *Let $\mathcal{B} = \{u_1, u_2, \ldots, u_n\}$ be an ordered basis for $\mathbf{F}^n$. Let P be the transition matrix from $\mathcal{B}$ coordinates to standard coordinates. For the bilinear form $(,)_A : \mathbf{F}^n \times \mathbf{F}^n \to \mathbf{F}$, we have $A_{\mathcal{B}} = P^t A P$.*

Note Be sure to observe the occurrence of the transpose P^t in the change-of-basis equation. This is *not* the same as in the case of linear transformations, where one has the inverse P^{-1}. The matrices $P^t A P$ and A are, in general, not similar. They are *congruent*, a relationship discussed in detail in the next section.

Proof: We recall that for $v \in \mathbf{F}^n$, $P(v)_{\mathcal{B}} = v$. From this we see that $(v)_{\mathcal{B}}^t (P^t A P)(w)_{\mathcal{B}} = ((v)_{\mathcal{B}}^t P^t) A(P(w)_{\mathcal{B}}) = (P(v)_{\mathcal{B}})^t A(P(w)_{\mathcal{B}}) = (v)^t A(w) = (v, w)_S$. Hence, $A_{\mathcal{B}} = P^t A P$ by the uniqueness assertion of Theorem 5.1.4 (i). //

Continuing with our example $(,)_W$ on $\mathbf{F}^3$, we consider the ordered basis $\mathcal{B} = \{(0, 1, 1), (1, 0, 0), (0, 0, -1)\}$ of $\mathbf{F}^3$. The transition matrix from $\mathcal{B}$ coordinates to standard coordinates is the matrix

$$P = \begin{pmatrix} 0 & 1 & 0 \\ 1 & 0 & 0 \\ 1 & 0 & -1 \end{pmatrix}$$

Consequently, the bilinear form $(\ , \ )_W$ is represented by the matrix

$$\begin{pmatrix} 0 & 1 & 1 \\ 1 & 0 & 0 \\ 0 & 0 & -1 \end{pmatrix} \begin{pmatrix} 0 & 1 & 1 \\ 0 & 0 & 0 \\ -2 & 0 & 0 \end{pmatrix} \begin{pmatrix} 0 & 1 & 0 \\ 1 & 0 & 0 \\ 1 & 0 & -1 \end{pmatrix} = \begin{pmatrix} 0 & -2 & 0 \\ 2 & 0 & -1 \\ 0 & 2 & 0 \end{pmatrix}$$

The following notion proves to be convenient when studying bilinear forms.

Definition 5.1.6 Suppose that $(\ , \ ) : V \times V \to \mathbf{F}$ is a bilinear form on a finite-dimensional vector space V. The *rank* of $(\ , \ )$ is defined to be the rank of any matrix A which represents $(\ , \ )$ with respect to some ordered basis of V.

Observe that whenever P is invertible, the rank of A is the same as the rank of $P^t A P$. Thus Theorem 5.1.5 shows that the rank of a bilinear form on a finite-dimensional vector space is well-defined. Evidently, the rank of our example $(\ , \ )_W$ above is 2.

PROBLEMS

5.1.1. Find the $n \times n$ matrix A for which $(\ , \ ) = (\ , \ )_A$ for each of the following bilinear forms:
(a) $(\ , \ ) : \mathbf{F}^2 \times \mathbf{F}^2 \to \mathbf{F}$ defined by $((a, b), (c, d)) = ad - 2bc + bd$
(b) $(\ , \ ) : \mathbf{F}^3 \times \mathbf{F}^3 \to \mathbf{F}$ defined by $((a, b, c), (d, e, f)) = ad - be + af - ce$
(c) $(\ , \ ) : \mathbf{F}^3 \times \mathbf{F}^3 \to \mathbf{F}$ defined by $((a, b, c), (d, e, f)) = ad + 2bf + cf + 2ce - ae + be - db$

5.1.2. Consider the bilinear form $((a, b), (c, d)) = ac - bd$ on $\mathbf{F}^2$. Find the symmetric matrix which represents $(\ , \ )$ in each of the following bases:
(a) $\mathcal{B}_1 = \{(1, 0), (0, 1)\}$.
(b) $\mathcal{B}_2 = \{(1, 1), (1, -1)\}$.

5.1.3. Consider the bilinear form $(\ , \ )_A$ on $\mathbf{F}^3$ given by the matrix

$$A = \begin{pmatrix} 0 & 1 & 0 \\ 1 & 0 & 2 \\ 0 & 2 & 1 \end{pmatrix}$$

Apply Theorem 5.1.5 to find the symmetric matrix which represents $(\ , \ )_A$ in each of the following bases:

(a) $\mathcal{B}_1 = \{(1,0,0),(0,1,1),(1,0,1)\}$.
(b) $\mathcal{B}_2 = \{(1,1,1),(1,-1,0),(0,0,1)\}$.

5.1.4.   The theory of bilinear forms developed in this section can be generalized to the following situation. Suppose that V and W are both $\mathbf{F}$ vector spaces. A bilinear form $(\,,\,) : V \times W \to \mathbf{F}$ is a function which is linear in each variable. State and prove the analogue of Theorem 5.1.4 (iii) in this setting.

5.1.5.   A bilinear form $(\,,\,) : V \times V \to \mathbf{F}$ is called *nondegenerate* (or *nonsingular*) if it satisfies one of the following three equivalent conditions:
(a) Whenever $(v,v') = 0$ for all $v' \in V$, then necessarily $v = \vec{0}$.
(b) Whenever $(v,v') = 0$ for all $v \in V$, then necessarily $v' = \vec{0}$.
(c) $(\,,\,)$ has rank $n = dim(V)$.
Prove that these three conditions are equivalent.

5.1.6.   If $(\,,\,) : V \times V \to \mathbf{F}$ is a bilinear form on a finite-dimensional vector space V and W is a subspace of V, set

$$W^{\perp} = \{v \in V \mid (v,w) = 0 \text{ for all } w \in W\}$$

Show that $W^{\perp}$ is a subspace of V and that $dim(W) + dim(W^{\perp})$ is at least the rank of $(\,,\,)$. Give an example where $dim(W) + dim(W^{\perp})$ exceeds the rank of V.

5.1.7.   Describe (in terms of matrices) all bilinear forms on $\mathbf{F}^3$ which satisfy $(u,v) = -(v,u)$ for all $u,v \in \mathbf{F}^3$.

5.1.8.   What is the rank of the following bilinear forms:
(a) $(\,,\,) : \mathbf{F}^3 \times \mathbf{F}^3 \to \mathbf{F}$ defined by $((a,b,c),(d,e,f)) = ad - 2be + cd - 3ce$
(b) $(\,,\,) : \mathbf{F}^4 \times \mathbf{F}^4 \to \mathbf{F}$ defined by $((a,b,c,d),(e,f,g,h)) = af + bg + ch + de$
(c) An inner product on $\mathbf{R}^n$

5.1.9.   Describe all rank 1 bilinear forms on a finite-dimensional vector space V .

5.1.10.   If V is a finite-dimensional $\mathbf{F}$ vector space, denote by $Bil(V)$ the vector space of all ($\mathbf{F}$-valued) bilinear forms on V.
(a) If $dim(V) = n$, what is $dim(Bil(V))$?
(b) Let $T : V \to W$ be a linear transformation between finite-dimensional vector spaces. For $(\,,\,) \in Bil(W)$, define $T^*(\,,\,) : V \times V \to \mathbf{F}$ by $T^*(u,v) = (T(u),T(v))$ for all $u,v \in V$. Show that $T^*(\,,\,) \in Bil(V)$ and $T^* : Bil(W) \to Bil(V)$ is a homomorphism.

5.2   Congruence and the Diagonalizability of Symmetric Bilinear Forms

Theorem 5.1.5 shows how the matrices which represent a bilinear form are related when there is a change of basis. The proof of Theorem 5.1.5 was analogous to the proof of Theorem 3.3.7, describing the linear operator case. In that case, two matrices A and C represent the same linear operator, except in different bases, if there exists an invertible matrix P so that $C = P^{-1}AP$. Such A and C are called *similar* (or sometimes *conjugate*). When studying bilinear forms, one has the following definitions.

Definition 5.2.1 (i) Two $n \times n$ matrices A and A' are called *congruent* if there exists an invertible matrix P for which $A' = P^t AP$.
(ii) Two bilinear forms $(,) : V \times V \to \mathbf{F}$ and $(,)' : V' \times V' \to \mathbf{F}$ are called *isometric* if there is an isomorphism $T : V \to V'$ such that $(v, w) = (T(v), T(w))'$ for all $v, w \in V$.

The definition of isometry presented here agrees with Definition 4.5.3 of a real isometry given (which was given for arbitrary inner products). The relationship between the congruence of matrices and isometry of bilinear forms is the subject of the next lemma.

Lemma 5.2.2 *Suppose that A and A' are matrices with associated bilinear forms $(,)_A$ and $(,)_{A'}$. Then A and A' are congruent if and only if $(,)_A$ and $(,)_{A'}$ are isometric.*

Proof: If A and A' are congruent, then there exists an invertible matrix P for which $A' = P^t AP$. Let $v, w \in \mathbf{F}^n$. Then $(v, w)_{A'} = v^t A'w = v^t(P^t AP)w = (Pv)^t A(Pw) = (Pv, Pw)_A$. This shows that $(,)_{A'}$ and $(,)_A$ are isometric via the isomorphism $T_P : \mathbf{F}^n \to \mathbf{F}^n$. Conversely, assume that $(,)_{A'}$ and $(,)_A$ are isometric. Then there exists an isomorphism $T : \mathbf{F}^n \to \mathbf{F}^n$ for which $(T(v), T(w))_A = (v, w)_{A'}$ for all $v, w \in \mathbf{F}^n$. Set P to be the invertible matrix $[T]$ in the standard basis. Then $v^t A'w = (v, w)_{A'} = (Pv, Pw)_A = (Pv)^t A(Pw) = v^t(P^t AP)w$ for all $v, w \in \mathbf{F}^n$. By the uniqueness assertion of Theorem 5.1.4 (i), $A' = P^t AP$ follows. This proves the lemma. //

Examples 5.2.3 (i) Define $(,) : \mathbf{F}^3 \times \mathbf{F}^3 \to \mathbf{F}$ by $((a, b, c), (d, e, f)) = ae + bf - cd$. It is readily checked that $(,)$ is a bilinear form. Computing in the standard basis gives $((1,0,0),(1,0,0)) = 0$, $((1,0,0),(0,1,0)) = 1,\ldots$, which shows that the matrix corresponding to $(\ ,\ )$ is

$$A = \begin{pmatrix} 0 & 1 & 0 \\ 0 & 0 & 1 \\ -1 & 0 & 0 \end{pmatrix}$$

Set

$$B = \left\{ \begin{pmatrix} 1 \\ 1 \\ 1 \end{pmatrix}, \begin{pmatrix} 1 \\ 0 \\ 1 \end{pmatrix}, \begin{pmatrix} 0 \\ 1 \\ 1 \end{pmatrix} \right\}$$

The transition matrix

$$P = \begin{pmatrix} 1 & 1 & 0 \\ 1 & 0 & 1 \\ 1 & 1 & 1 \end{pmatrix}$$

changes B coordinates to standard coordinates. It follows that $C = P^t AP$ represents the bilinear form in the basis B. To illustrate this, one easily computes that

$$P^t AP = \begin{pmatrix} 1 & 0 & 2 \\ 0 & -1 & 1 \\ 0 & 0 & 1 \end{pmatrix}$$

According to the definition $((1,1,1),(0,1,1)) = 1+1-0 = 2$. Evidently we have

$$\begin{pmatrix} 1 \\ 1 \\ 1 \end{pmatrix}_B = \begin{pmatrix} 1 \\ 0 \\ 0 \end{pmatrix} , \begin{pmatrix} 0 \\ 1 \\ 1 \end{pmatrix}_B = \begin{pmatrix} 0 \\ 0 \\ 1 \end{pmatrix}$$

and

$$\begin{pmatrix} 1 & 0 & 0 \end{pmatrix} \begin{pmatrix} 1 & 0 & 2 \\ 0 & -1 & 1 \\ 0 & 0 & 1 \end{pmatrix} \begin{pmatrix} 0 \\ 0 \\ 1 \end{pmatrix} = (2)$$

as well; this shows how $P^t AP$ represents $(\ ,\ )$ in the basis B.

(ii)   In (i) above, we viewed the congruent matrices A and $P^t AP$ as representing the same bilinear form on $\mathbf{F}^3$ with respect to different bases. However, the matrices A and $P^t AP$ represent different bilinear forms with resect to the standard basis. What Lemma 5.2.2 shows is that the bilinear forms they represent are isometric. Spelled out explicitly, the bilinear form given by $P^t A P$ on $\mathbf{F}^3$ is described by $((a, b, c), (d, e, f))_{P^t A P} = a(d + 2f) + b(-e + f) + cf$. The linear isomorphism $T_P : \mathbf{F}^3 \to \mathbf{F}^3$ gives the isometry between $(\ ,\ )_A$ and $(\ ,\ )_{P^t A P}$. This is because $T_P(a, b, c) = (a + b, a + c, a + b + c)$ and $T_P(d, e, f) = (d + e, d + f, d + e + f)$, so that by substituting, $(T_P(a, b, c), T_P(d, e, f))_A = (a + b)(d + f) + (a + c)(d + e + f) - (a + b + c)(d + e) = ad + 2af - be + bf + cf = ((a, b, c), (d, e, f))_A$.

(iii) On $\mathbf{R}^2$ consider the symmetric bilinear form $(\ ,\ )_H$ given by the symmetric matrix

$$H = \begin{pmatrix} 0 & 1 \\ 1 & 0 \end{pmatrix}$$

Then, this bilinear form is explicitly given by $((a, b), (c, d))_H = ad + bc$. Observe that $((1, -1), (1, -1))_H = -2$, so the positive-definite property (required for inner products) fails for this symmetric bilinear form. Hence, $(\ ,\ )_H$ *cannot* be isometric to the usual inner product on $\mathbf{R}^2$. This shows that two symmetric bilinear forms on a vector space *need not* be isometric [in contrast to the situation for inner products described in Lemma 4.5.5 (ii)]. The bilinear form on $\mathbf{R}^2$ given by H gives what is called *hyperbolic space*. Hyperbolic space is extremely important in several branches of mathematics and physics (especially in relativity theory).

With the correspondence between bilinear forms and $n \times n$ matrices under control, we can proceed to the structure theory in the symmetric case. Our first result shows that an arbitrary symmetric bilinear form can be diagonalized. We emphasize that this diagonalization is that of a symmetric bilinear form and *not* that of a linear operator. Consequently (see Corollary 5.2.6), we will obtain that symmetric matrices are congruent (not necessarily similar) to diagonal matrices over any field $\mathbf{F}$ of characteristic different from 2.

Theorem 5.2.4 *Suppose that V is a finite-dimensional $\mathbf{F}$ vector space where $\mathbf{F}$ is a field of characteristic different from 2. Let $(\ ,\ ) : V \times V \to \mathbf{F}$ be a symmetric bilinear form on V. Then there is an ordered basis $\mathcal{B}$ for V for which the symmetric matrix $S_{\mathcal{B}}$ corresponding to $(\ ,\ )$ is diagonal.*

Proof: We proceed by induction on the dimension of V to show that there is an ordered basis $\mathcal{B} = \{u_1, u_2, \ldots, u_n\}$ for V with the property that $(u_i, u_j) = 0$ whenever $i \neq j$. In view of the definition of $S_{\mathcal{B}}$, the result will follow. The case of $n = 1$ is clear since 1×1 matrices are always diagonal.

We may now assume that the theorem is true for bilinear forms on $(n-1)$-dimensional vector spaces. First, in case $(\ ,\ )$ takes the value zero everywhere, the result is trivial, because then the matrix $S_{\mathcal{B}}$ is always zero. Thus we can assume that $(\ ,\ )$ is nonzero. By a direct calculation we see that for all $u, w \in V$, $(u, w) = (1/2)[(u + w, u+w) - (u,u) - (w,w)]$. Since the bilinear form $(\ ,\ )$ is nonzero, it follows that there is some $v \in V$ with $(v, v) \neq 0$. For such v we define $v^{\perp} = \ker(R_v : V \to F) \ (= \{w \in V \mid (v, w) = 0\})$. Since $R_v(v) \neq 0$, we see that R_v is onto $\mathbf{F}$ (which is one-dimensional), and hence $W = v^{\perp}$ is a $(n-1)$-dimensional subspace of V. The symmetric bilinear form $(\ ,\ )$ gives (via restriction) a symmetric bilinear form $(\ ,\ ) : W \times W \to \mathbf{F}$ (which we denote by the same parentheses). Applying our induction hypothesis, there is an ordered basis $\mathcal{B}' = \{u_2, u_3, \ldots, u_n\}$ for W for which $(u_i, u_j) = 0$ whenever $i \neq j$. As each such $u_i \in W$, we have $(v, u_i) = 0$ for $i = 2, 3, \ldots, n$. Setting $v = u_1$ gives the desired basis $\mathcal{B} = \{u_1, u_2, \ldots, u_n\}$ for V. The theorem is proved. //

Example 5.2.5 We consider the symmetric matrix

$$S = \begin{pmatrix} 2 & 1 & 3 \\ 1 & 0 & 1 \\ 3 & 1 & 3 \end{pmatrix}$$

We find a diagonal basis for $(\ ,\ )_S : \mathbf{F}^3 \times \mathbf{F}^3 \to \mathbf{F}$ by following the proceedure given in the proof of Theorem 5.2.4. First we observe that $((1,0,0), (1,0,0))_S = 2 \neq 0$. Thus we can take $u_1 = (1,0,0)$ as our first basis element. (There are many different choices for this first vector u_1.) Our choices of u_2 and u_3 must satisfy $(u_1, u_2)_S = 0 = (u_1, u_3)_S$. Since $((1,0,0), (a,b,c))_S = 2a + b + 3c$, this shows that the subspace

$$(1,0,0)^{\perp} = \{(a,b,c) \mid 2a + b + 3c = 0\}$$
$$= \operatorname{span} \left\{ \begin{pmatrix} 1 \\ -2 \\ 0 \end{pmatrix}, \begin{pmatrix} 3 \\ 0 \\ -2 \end{pmatrix} \right\}$$

Since $((1, -2, 0), (1, -2, 0))_S = -2 \neq 0$, we can choose our next basis vector to be $u_2 = (1, -2, 0)$. Finally, to find u_3, the condition that

$(u_2, u_3)_S = 0$ gives an equation $0 = ((1, -2, 0), (a, b, c))_S = b + c$. The vector u_3 can be chosen to be any nonzero solution to the system $2a + b + 3c = 0$, $b + c = 0$. One such solution is $u_3 = (-1, -1, 1)$. Our desired basis $\mathcal{B} = \{u_1, u_2, u_3\}$ has now been found. The transition matrix from $\mathcal{B}$ coordinates to standard coordinates is

$$P = \begin{pmatrix} 1 & 1 & -1 \\ 0 & -2 & -1 \\ 0 & 0 & 1 \end{pmatrix}$$

One readily computes that

$$P^t S P = \begin{pmatrix} 2 & 0 & 0 \\ 0 & -2 & 0 \\ 0 & 0 & 0 \end{pmatrix}$$

As observed in the preceding example, Theorem 5.2.4 gives the diagonalization of symmetric matrices (*up to congruence*). This is formulated in the following corollary.

Corollary 5.2.6 *Suppose that S is an $n \times n$ symmetric $\mathbf{F}$ matrix where $\mathbf{F}$ is a field of characteristic different from 2. Then there is an invertible $n \times n$ matrix P such that $P^t S P$ is diagonal (that is, S is congruent to a diagonal matrix).*

Proof: Applying Theorem 5.2.4, let $\mathcal{B} = \{u_1, u_2, \ldots, u_n\}$ be an ordered basis in which the matrix $S_{\mathcal{B}}$ corresponding to the symmetric bilinear form $(\ ,\ )_S$ is diagonal. Let P be the coordinate-change matrix from the basis $\mathcal{B}$ to the standard basis. According to Theorem 5.1.6 (iii), $S_{\mathcal{B}} = P^t S P$; this is the desired result. //

PROBLEMS

5.2.1. For each of the following symmetric matrices, find a congruent diagonal matrix and express this matrix in the form $P^t S P$.

(a) $\begin{pmatrix} 2 & 3 \\ 3 & 1 \end{pmatrix}$ (b) $\begin{pmatrix} 0 & 2 & 0 \\ 2 & 0 & 2 \\ 0 & 2 & 0 \end{pmatrix}$ (c) $\begin{pmatrix} 0 & 1 & 2 \\ 1 & 2 & 3 \\ 2 & 3 & 0 \end{pmatrix}$ (d) $\begin{pmatrix} 0 & 1 & 0 & 0 \\ 1 & 0 & 2 & 0 \\ 0 & 2 & 1 & 1 \\ 0 & 0 & 1 & 1 \end{pmatrix}$

5.2.2. For each of the following symmetric matrices, find a congruent diagonal matrix and express this matrix in the form $P^t S P$.

(a) $\begin{pmatrix} 1 & 2 \\ 2 & 1 \end{pmatrix}$ (b) $\begin{pmatrix} 1 & 1 & 1 \\ 1 & 1 & 1 \\ 1 & 1 & 1 \end{pmatrix}$ (c) $\begin{pmatrix} 0 & 1 & 0 \\ 1 & 0 & 1 \\ 0 & 1 & 0 \end{pmatrix}$ (d) $\begin{pmatrix} 0 & 1 & 0 & 0 \\ 1 & 0 & 0 & 0 \\ 0 & 0 & 0 & 1 \\ 0 & 0 & 1 & 0 \end{pmatrix}$

5.2.3. Prove that if a matrix is congruent to a diagonal matrix, then it is symmetric.

5.2.4. If $(\, , \,) : V \times V \to \mathbf{F}$ is a symmetric bilinear form on a finite-dimensional vector space V and W is a subspace of V, we define $W^{\perp} = \{u \in V \mid (w, u) = 0 \text{ for all } w \in W\}$.
(a) Show that $W^{\perp}$ is a subspace of V.
(b) Show that $V^{\perp} = \{\vec{0}\}$ if and only if the rank of $(\, , \,) = n$.
(c) Show that the rank of $(\, , \,)$ is $dim(V) - dim(V^{\perp})$.

5.2.5. Suppose a 2×2 real matrix A has two orthogonal (with respect to the dot product) eigenvectors. Show that A is symmetric.

5.3 Quadratic Forms

We now turn to the subject of quadratic forms. One reason for the study of quadratic forms is their geometric significance. The reader may have seen their use, for example, in multivariable calculus. Recall that it is the second-degree terms of the Taylor expansion of a function that determine such local behavior as maxima, minima, and curvature. For a real-valued function on $\mathbf{R}^n$, the sum of the second-degree terms in a Taylor expansion is a quadratic polynomial, which gives a quadratic form.

This section begins with the definition of quadratic forms, which are defined to be functions (instead of quadratic polynomials). The advantage of this approach is that it enables one to relate quadratic forms to symmetric bilinear forms immediately and it applies to the infinite-dimensional case as well. We continue to assume that the characteristic of $\mathbf{F}$ is different from 2 (that is, $1 + 1 \neq 0$ in $\mathbf{F}$).

Definition 5.3.1 Let V be a $\mathbf{F}$ vector space. A *quadratic form* is a function $Q : V \to \mathbf{F}$ with the properties:
(i) $Q(kv) = k^2 Q(v)$ for all $v \in V$ and all $k \in \mathbf{F}$.
(ii) $B_Q(v, w) = Q(v + w) - Q(v) - Q(w)$ is a symmetric bilinear form.

For any quadratic form, the definition includes a built-in symmetric bilinear form $B_Q(\ ,\ )$. We call $(1/2)B_Q(\ ,\ )$ the *bilinear form associated with Q*. The reason for the $1/2$ comes from the following theorem, which also provides the best source of examples of quadratic forms (in the sense of the above definition).

Theorem 5.3.2 *If* $(\ ,\ )$ *is a symmetric bilinear form on* V, *then* $Q(v) = (v, v)$ *is a quadratic form on* V. *Moreover, for this* Q *we have* $B_Q(v, w) = 2(v, w)$ *for all* $v, w \in V$.

Proof: To see $Q(v)$ is a quadratic form, note first that $Q(kv) = (kv, kv) = k(v, kv) = k^2(v, v) = k^2 Q(v)$. For $v, w \in V$ we also have $B_Q(v, w) = Q(v+w) - Q(v) - Q(w) = (v+w, v+w) - (v, v) - (w, w) = (v, v+w) + (w, v+w) - (v, v) - (w, w) = (v, v) + (v, w) + (w, v) + (w, w) - (v, v) - (w, w) = 2(v, w)$. Since $(\ ,\ )$ is a symmetric bilinear form, it follows that $B_Q(\ ,\ )$ is a symmetric bilinear form. This proves the theorem. $/\!/$

Whenever $(\ ,\ )$ is a symmetric bilinear form on V, we call the quadratic form Q defined by $Q(v) = (v, v)$ the *quadratic form associated with* $(\ ,\ )$. In case $(\ ,\ )$ is an inner product on a real vector space V, observe that $Q(v) = \|\, v \,\|^2$ is nothing other than the square of the associated norm.

Remark 5.3.3 Definition 5.3.1 makes sense even if the characteristic of $\mathbf{F}$ is 2. (In fact the theory of quadratic forms and symmetric bilinear forms in characteristic 2 is quite interesting.) However, for the correspondence between quadratic and symmetric bilinear forms, our assumption that the characteristic of $\mathbf{F}$ is not 2 is critical. This is because if $Q : V \to \mathbf{F}$ is a quadratic form, the symmetric bilinear form which gives Q is the form $(1/2)B_Q(\ ,\ )$. Hence, we must divide by the (nonzero) scalar $1 + 1 = 2$.

The next example and the lemma which follows show how quadratic forms (in the sense of Definition 5.3.1) are related to homogeneous quadratic polynomials.

Example 5.3.4 Suppose that $V = \mathbf{F}^n$ and let $q(X_1, X_2, \ldots, X_n)$ be a homogeneous polynomial of degree 2 over $\mathbf{F}$; that is, $q(X_1, \ldots, X_n) = \sum_{i,j} a_{ij} X_i X_j$ where $a_{ij} \in \mathbf{F}$. For $v = (b_1, b_2, \ldots, b_n) \in \mathbf{F}^n$, define

$Q(v) = q(b_1, b_2, \ldots, b_n)$. In other words, $Q(v)$ is the function defined by the homogeneous quadratic polynomial $q(X_1, \ldots, X_n)$. The next lemma shows that $Q(v)$ is a quadratic form in the sense of Definition 5.3.1. As a consequence of this calculation, we shall see that every quadratic form arises from a quadratic polynomial in this manner.

Lemma 5.3.5 (i) *For any homogeneous quadratic polynomial $q(X_1, \ldots, X_n) = \sum_{i,j} a_{ij} X_i X_j$ where $a_{ij} \in \mathbf{F}$, let S_q be the symmetric matrix $S_q = (s_{ij})$ with entries $s_{ij} = (1/2)(a_{ij} + a_{ji})$. Then for all $v = (b_1, b_2, \ldots, b_n) \in \mathbf{F}^n$, $q(b_1, \ldots, b_n) = (v, v)_{S_q}$. Consequently, the function $Q(v) = q(b_1, \ldots, b_n)$ is a quadratic form in the sense of Definition 5.3.1.*

(ii) *Let $Q : \mathbf{F}^n \to \mathbf{F}$ be a quadratic form, and assume $Q(v) = (v, v)_S$ for some symmetric matrix $S = (s_{ij})$. Let $q(X_1, \ldots X_n) = \sum s_{ij} X_i X_j$. Then for all $v = (b_1, \ldots, b_n) \in \mathbf{F}^n$, $Q(v) = q(b_1, \ldots, b_n)$; that is Q is the quadratic form associated with the quadratic polynomial $q(X_1, \ldots, X_n)$.*

Proof: (i) Calculating directly gives

$$
\begin{aligned}
q(b_1, b_2, \ldots, b_n) &= \sum_{i,j} a_{ij} b_i b_j \\
&= \sum_i a_{ii} b_i^2 + \sum_{i<j} (a_{ij} + a_{ji}) b_i b_j \\
&= \sum_i s_{ii} b_i^2 + \sum_{i<j} 2 s_{ij} b_i b_j \\
&= \sum_{i,j} s_{ij} b_i b_j \\
&= \sum_i b_i \left(\sum_j s_{ij} b_j \right) \\
&= \sum_i b_i S_q \begin{pmatrix} b_1 \\ \vdots \\ b_n \end{pmatrix} \\
&= \begin{pmatrix} b_1 & \cdots & b_n \end{pmatrix} S_q \begin{pmatrix} b_1 \\ \vdots \\ b_n \end{pmatrix} \\
&= (v, v)_{S_q}
\end{aligned}
$$

This establishes the first claim. To see that $Q(v)$ is a quadratic form in the sense of Definition 5.3.1, we first note that $Q(kv) = k^2 Q(v)$ by the homogeniety of $q(X_1, X_2, \ldots, X_n)$. That $B_Q(\ ,\ )$ associated with Q is a symmetric bilinear form follows from Lemma 5.3.2: as Q is the quadratic form associated to $(\ ,\ )_{S_q}$, it follows that $B_Q(\ ,\ ) = 2(\ ,\ )_{S_q}$.

(ii) This is a direct consequence of the calculation in the proof of (i). The matrix S_q in that calculation is precisely the matrix S, since the matrix $S = (s_{ij})$ is symmetric. (In other words, $s_{ij} = (1/2)(s_{ij} + s_{ji})$ always, since $s_{ij} = s_{ji}$.) This proves the lemma. //

Example 5.3.6: Consider the symmetric matrix from Example 5.2.5

$$S = \begin{pmatrix} 2 & 1 & 3 \\ 1 & 0 & 1 \\ 3 & 1 & 4 \end{pmatrix}$$

The symmetric bilinear form $(\ ,\ )_S : \mathbf{F}^3 \times \mathbf{F}^3 \to \mathbf{F}$ associated with S is computed as

$$
\begin{aligned}
((a, b, c), (d, e, f))_S &= (a\ \ b\ \ c) \begin{pmatrix} 2 & 1 & 3 \\ 1 & 0 & 1 \\ 3 & 1 & 4 \end{pmatrix} \begin{pmatrix} d \\ e \\ f \end{pmatrix} \\
&= (2a + b + 3c\ \ \ a + c\ \ \ 3a + b + 4c) \begin{pmatrix} d \\ e \\ f \end{pmatrix} \\
&= 2ad + bd + 3cd + ae + ce + 3af + bf + 4cf
\end{aligned}
$$

The quadratic form associated with S is then computed to be

$$
\begin{aligned}
Q(r, s, t) &= ((r, s, t), (r, s, t))_S \\
&= 2r^2 + 2rs + 6rt + 2st + 4t^2
\end{aligned}
$$

which, as a function, is specified by a homogeneous quadratic polynomial in r, s, t.

We have now completed our first task, which was to establish the equivalence of the concepts of symmetric bilinear forms, symmetric matrices, and quadratic forms (when the characteristic of $\mathbf{F}$ is not 2). In the remainder of this chapter we exploit this equivalence, using symmetric matrices to study quadratic forms and vice versa. The reader needs to become comfortable with the transitions given by these equivalences. In advanced treatments and applications of this material, it is customary to identify these equivalent concepts (and thereby blur the distinctions between them). The following table summarizes the relationships between symmetric bilinear forms, symmetric matrices, and quadratic forms. Also included are formulas for determining the (homogeneous) quadratic polynomial associated with a quadratic form, since it is through such polynomials that quadratic forms often arise.

TABLE 5.3.7

EQUIVALENCE	COMPUTATION
1. $(\, , \,) \mapsto S$	$S = (s_{ij})$ where $s_{ij} = (e_i, e_j)$
2. $S \mapsto (\, , \,)_S$	$(v, w)_S = v^t S w$
3. $(\, , \,) \mapsto Q$	$Q(v) = (v, v)$
4. $Q \mapsto (\, , \,)$	$(v, w) = (1/2) B_Q(v, w)$ $= (1/2)[Q(v + w) - Q(v) - Q(w)]$
5. $S \mapsto q(\vec{X})$	$q(X_1, \ldots, X_n) = \sum s_{ij} X_i X_j$ where $S = (s_{ij})$
6. $q(\vec{X}) \mapsto Q$	$Q(b_1 e_1 + \cdots + b_n e_n) = q(b_1, \ldots, b_n)$
7. $S \mapsto Q$	$Q(v) = v^t S v$
8. $Q \mapsto S$	4. followed by 1.

We now translate the notions of congruence for matrices and isometry of bilinear forms to quadratic forms. It will then be possible to apply Theorem 5.2.4 to quadratic forms as well.

Definition 5.3.8 Suppose that $Q : V \to \mathbf{F}$ and $Q' : V' \to \mathbf{F}$ are quadratic forms. Then Q and Q' are called *isometric* if there is an isomorphism $T : V \to V'$ such that $Q'(v) = Q(T(v))$ for all $v \in V$.

The ideas behind this next lemma are the same as those giving Lemma 5.2.2.

Lemma 5.3.9 *Suppose that S and S' are symmetric matrices with associated quadratic forms Q_S and $Q_{S'}$. Then S and S' are congruent if and only if Q_S and $Q_{S'}$ are isometric.*

Proof: By Lemma 5.2.2 we know that S and S' are congruent if and only if the associated symmetric bilinear forms $(\ ,\ )_S$ and $(\ ,\ )_{S'}$ are isometric. Suppose that S and S' are congruent, and let $T : \mathbf{F}^n \to \mathbf{F}^n$ be an isomorphism so that $(v, w)_{S'} = (T(v), T(w))_S$. Then $Q_{S'}(v) = (v, v)_{S'} = (T(v), T(v))_S = Q_S(T(v))$; this shows that $Q_{S'}$ and Q_S are isometric (via the same isomorphism T). Conversely, assume that $Q_{S'}$ and Q_S are isometric via an isomorphism $T : \mathbf{F}^n \to \mathbf{F}^n$. Since $Q(T(v)) = Q'(v)$, the definitions of B_{Q_S} and $B_{Q_{S'}}$ show that for all $v \in \mathbf{F}^n$, $B_{Q_{S'}}(v, w) = B_{Q_S}(T(v), T(w))$. According to Theorem 5.3.2 it follows that $(v, w)_{S'} = (T(v), T(w))_S$ for all $v, w \in \mathbf{F}^n$. Hence $(\ ,\ )_{S'}$ and $(\ ,\ )_S$ are isometric; this proves the lemma. //

The diagonalizability results from the preceding section can be interpreted in terms of quadratic polynomials. In order to spell this out, we give the following definitions.

Definition 5.3.10 (i) Two quadratic polynomials $q(X_1, \ldots, X_n)$ and $q'(X_1', \ldots, X_n')$ are called *isometric* if there is an invertible $n \times n$ matrix $P = (p_{ij})$ such that $q(X_1, \ldots, X_n) = q'(p_{11}X_1 + p_{12}X_2 + \cdots + p_{1n}X_n, \ldots, p_{n1}X_1 + p_{n2}X_2 + \cdots + p_{nn}X_n)$
(ii) We say that a homogeneous quadratic polynomial $q(X_1, \ldots, X_n)$ is *diagonal* if it has the form

$$q(X_1, \ldots, X_n) = a_1 X_1^2 + a_2 X_2^2 + \cdots + a_n X_n^2 \quad \text{where} \quad a_i \in \mathbf{F}$$

It is readily checked that two quadratic polynomials in n variables are isometric if and only if the quadratic forms they define on $\mathbf{F}^n$ are isometric. In turn, this is equivalent to their associated symmetric matrices being congruent. Furthermore, one notes that $q(X_1, X_2, \ldots, X_n)$ is diagonal if and only if the associated symmetric matrix is diagonal. Using this, one obtains the following consequence of Theorem 5.2.4; the proof is straightforward and is omitted.

Corollary 5.3.11 *Suppose* $\mathbf{F}$ *is a field and the characteristic of* $\mathbf{F}$ *is different from* 2. *Then every quadratic polynomial* $q(X_1, \ldots, X_n) = \sum a_{ij} X_i X_j$ *where* $a_{ij} \in \mathbf{F}$ *is isometric to a diagonal quadratic polynomial.*

Remark 5.3.12 Observe that the diagonal quadratic polynomial $r_1 X_1^2 + r_2 X_2^2 + \cdots + r_n X_n^2$ is isometric to the quadratic polynomial $r_1 s_1^2 Y_1^2 + r_2 s_2^2 Y_2^2 + \cdots + r_n s_n^2 Y_n^2$ for any $s_1, s_2, \ldots, s_r \in \mathbf{F}$ (set $Y_i = (1/s_i) X_i$). This shows that any coefficient of a diagonal quadratic polynomial can be multiplied by a square and the resulting polynomial will be isometric to the original.

Over the real numbers $\mathbf{R}$, every nonzero real number can be multiplied by a square to produce either 1 or -1. Thus, if the real diagonal quadratic polynomial $q(X_1, X_2, \ldots, X_n) = r_1 X_1^2 + r_2 X_2^2 + \cdots + r_n X_n^2$ has nonzero coefficients, there is some natural number s such that $q(X_1, X_2, \ldots, X_n)$ is isometric over $\mathbf{R}$ to $Y_1^2 + Y_2^2 + \cdots + Y_s^2 - Y_{s+1}^2 - Y_{s+2}^2 - \cdots - Y_n^2$. For the same reason, as every element of $\mathbf{C}$ is a square, any diagonal complex quadratic form with nonzero coefficients is isometric over $\mathbf{C}$ to $Y_1^2 + Y_2^2 + \cdots + Y_n^2$.

Remark 5.3.13 The two-variable case of the preceding corollary is a familiar result from high school algebra; it is nothing other than completing the square to eliminate mixed terms of a homogeneous quadratic. For example consider $q_1 = aX^2 + bXY + cY^2$. Then $q_1 = a[X + (b/2a)Y]^2 + [c - (b^2/4a^2)]Y^2$, which shows that the original polynomial is isometric to the polynomial $q_2 = aR^2 + (c - b^2/4a^2)S^2$. The matrix giving this isometry is

$$P = \begin{pmatrix} 1 & b/2a \\ 0 & 1 \end{pmatrix}$$

since

$$q_2 \left(P \begin{pmatrix} X \\ Y \end{pmatrix} \right) = q_1(X, Y)$$

Example 5.3.14 The homogeneous quadratic polynomial $X^2 + 2XY + 2XZ - 2YZ$ can be diagonalized by completing the square:

$$\begin{aligned} X^2 + 2XY + 2XZ - 2YZ &= (X + Y + Z)^2 - Y^2 - Z^2 - 4YZ \\ &= (X + Y + Z)^2 - (Y + 2Z)^2 + 3Z^2 \end{aligned}$$

This shows that the original polynomial in X, Y, Z is isometric to the quadratic polynomial $R^2 - S^2 + 3T^2$. In particular, this shows that the quadratic form $Q' : \mathbf{F}^3 \to \mathbf{F}$ defined by $Q'(a, b, c) = a^2 + 2ab + 2ac - 2bc$ is isometric to the quadratic form $Q : \mathbf{F}^3 \to \mathbf{F}$ defined by $Q(d, e, f) = d^2 - e^2 + 3f^2$. If we define the isomorphism $T : \mathbf{F}^3 \to \mathbf{F}^3$ by $T(u, v, w) = (u + v + w, v + 2w, w)$, then the previous calculation shows that $Q(T(u, v, w)) = Q'(u, v, w)$.

The symmetric matrices associated with Q and Q' are

$$S_{Q'} = \begin{pmatrix} 1 & 1 & 1 \\ 1 & 0 & -1 \\ 1 & -1 & 0 \end{pmatrix} \quad \text{and} \quad S_Q = \begin{pmatrix} 1 & 0 & 0 \\ 0 & -1 & 0 \\ 0 & 0 & 3 \end{pmatrix}$$

Set

$$P = \begin{pmatrix} 1 & 1 & 1 \\ 0 & 1 & 2 \\ 0 & 0 & 1 \end{pmatrix}$$

Since

$$Q(d, e, f) = (d \quad e \quad f) \, S_Q \begin{pmatrix} d \\ e \\ f \end{pmatrix}$$

we have

$$T(a, b, c) = P \begin{pmatrix} a \\ b \\ c \end{pmatrix}$$

and as $Q(T(u, v, w)) = Q'(u, v, w)$, we find that

$$Q'(a, b, c) = \left(P \begin{pmatrix} a \\ b \\ c \end{pmatrix} \right)^t S_Q \left(P \begin{pmatrix} a \\ b \\ c \end{pmatrix} \right) = (a \quad b \quad c) \, P^t S_Q P \begin{pmatrix} a \\ b \\ c \end{pmatrix}$$

As this equation is true for arbitrary $a, b, c \in \mathbf{F}$, we find by the uniqueness assertion of Theorem 5.1.6 (i) that $S_{Q'} = P^t S_Q P$ [or alternatively, $S_Q = (P^{-1})^t S_{Q'} (P^{-1})$].

The final example of this section ties together results described in the last three sections and reminds us that over the reals, symmetric matrices can be orthogonally diagonalized. Note that since an orthogonal matrix U satisfies $U^t = U^{-1}$, $U^t S U = U^{-1} S U$, so congruence by an orthogonal matrix is also a similarity.

Example 5.3.15 Consider the symmetric matrix

$$S = \begin{pmatrix} -1 & -2 & 4 \\ -2 & 2 & 2 \\ 4 & 2 & -1 \end{pmatrix}$$

Then S determines the bilinear form $(\, , \,)_S : \mathbf{R}^3 \to \mathbf{R}^3$

$$((a, b, c), (d, e, f))_S = (a \quad b \quad c) \begin{pmatrix} -1 & -2 & 4 \\ -2 & 2 & 2 \\ 4 & 2 & -1 \end{pmatrix} \begin{pmatrix} d \\ e \\ f \end{pmatrix}$$

$$= (-a - 2b + 4c \quad -2a + 2b + 2c \quad 4a + 2b - c) \begin{pmatrix} d \\ e \\ f \end{pmatrix}$$

$$= -ad - 2bd + 4cd - 2ae + 2be + 2ce + 4af + 2bf - cf$$

The quadratic form associated with this bilinear form is given by

$$\begin{aligned} Q_S((a, b, c)) &= -a^2 - 2ba + 4ca - 2ab + 2b^2 + 2cb + 4ac + 2bc - c^2 \\ &= -a^2 - 4ab + 2b^2 + 8ac + 4bc - c^2 \end{aligned}$$

The squares of the quadratic polynomial $q(X, Y, Z) = -X^2 - 4XY + 2Y^2 + 8XZ + 4YZ - Z^2$ can be completed by writing

$$\begin{aligned} q(X, Y, Z) &= -(X + 2Y - 4Z)^2 + 6Y^2 - 12YZ + 15Z^2 \\ &= -(X + 2Y - 4Z)^2 + 6(Y - Z)^2 + 9Z^2 \\ &= -U^2 + 6V^2 + 9W^2 \\ &= q'(U, V, W) \end{aligned}$$

where $U = X + 2Y - 4Z$, $V = Y - Z$, and $W = Z$. This diagonalization of the quadratic polynomial $q(X, Y, Z)$ also shows how to diagonalize S up to congruence. Suppose that

$$S' = \begin{pmatrix} -1 & 0 & 0 \\ 0 & 6 & 0 \\ 0 & 0 & 9 \end{pmatrix}$$

is the matrix associated with the diagonal polynomial $q'(U, V, W)$. Then,

$$(X \quad Y \quad Z) S \begin{pmatrix} X \\ Y \\ Z \end{pmatrix} = Q_S \left(\begin{pmatrix} X \\ Y \\ Z \end{pmatrix} \right)$$

$$= q(X, Y, Z)$$
$$= q'(X + 2Y - 4Z, Y - Z, Z)$$
$$= Q_{S'}\left(\begin{pmatrix} 1 & 2 & -4 \\ 0 & 1 & -1 \\ 0 & 0 & 1 \end{pmatrix}\begin{pmatrix} X \\ Y \\ Z \end{pmatrix}\right)$$

$$= (X \ \ Y \ \ Z)\begin{pmatrix} 1 & 0 & 0 \\ 2 & 1 & 0 \\ -4 & -1 & 1 \end{pmatrix} S'\begin{pmatrix} 1 & 2 & -4 \\ 0 & 1 & -1 \\ 0 & 0 & 1 \end{pmatrix}\begin{pmatrix} X \\ Y \\ Z \end{pmatrix}$$

Consequently,

$$S = \begin{pmatrix} 1 & 0 & 0 \\ 2 & 1 & 0 \\ -4 & -1 & 1 \end{pmatrix} S'\begin{pmatrix} 1 & 2 & -4 \\ 0 & 1 & -1 \\ 0 & 0 & 1 \end{pmatrix}$$

so, inverting, we obtain the congruence

$$\begin{pmatrix} -1 & 0 & 0 \\ 0 & 6 & 0 \\ 0 & 0 & 9 \end{pmatrix} = \begin{pmatrix} 1 & 0 & 0 \\ -2 & 1 & 0 \\ 2 & 1 & 1 \end{pmatrix}\begin{pmatrix} -1 & -2 & 4 \\ -2 & 2 & 2 \\ 4 & 2 & -1 \end{pmatrix}\begin{pmatrix} 1 & -2 & 2 \\ 0 & 1 & 1 \\ 0 & 0 & 1 \end{pmatrix}$$

which shows how to diagonalize our original matrix S (up to congruence).

The previous diagonalization (up to *congruence*) works over any field **F**. Over the real numbers **R**, our theorems assert that S is actually *orthogonally similar* to a diagonal matrix. To see this, we compute the eigenvalues and eigenspaces of S. By direct calculation we find $C_S(X) = (X - 3)^2(X + 6)$. The eigenspace E_3 associated with 3 is

$$ker\begin{pmatrix} -4 & -2 & 4 \\ -2 & -1 & 2 \\ 4 & 2 & -4 \end{pmatrix} = ker\begin{pmatrix} 2 & 1 & -2 \\ 0 & 0 & 0 \\ 0 & 0 & 0 \end{pmatrix}$$
$$= span\left\{\begin{pmatrix} 1 \\ 0 \\ 1 \end{pmatrix}, \begin{pmatrix} 1/2 \\ -1 \\ 0 \end{pmatrix}\right\}$$

and the eigenspace E_{-6} associated with -6 is

$$E_{-6} = ker\begin{pmatrix} 5 & -2 & 4 \\ -2 & 8 & 2 \\ 4 & 2 & 5 \end{pmatrix} = span\left\{\begin{pmatrix} 2 \\ 1 \\ -2 \end{pmatrix}\right\}$$

These bases for E_3 and E_{-6} can be orthonormalized by the Gram-Schmidt process to give an orthonormal basis for $\mathbf{R}^3$

$$\left\{ \begin{pmatrix} 1/\sqrt{2} \\ 0 \\ 1/\sqrt{2} \end{pmatrix}, \begin{pmatrix} -1/\sqrt{10} \\ 4/\sqrt{10} \\ 1/\sqrt{10} \end{pmatrix}, \begin{pmatrix} 2/3 \\ 1/3 \\ -2/3 \end{pmatrix} \right\}$$

Setting

$$U = \begin{pmatrix} 1/\sqrt{2} & -1/\sqrt{10} & 2/3 \\ 0 & 4/\sqrt{10} & 1/2 \\ 1/\sqrt{2} & 1/\sqrt{10} & -2/3 \end{pmatrix}$$

we have that

$$U^{-1}SU = \begin{pmatrix} 3 & 0 & 0 \\ 0 & 3 & 0 \\ 0 & 0 & -6 \end{pmatrix}$$

where $U^{-1} = U^t$ is orthogonal.

PROBLEMS

5.3.1. Find a diagonal quadratic polynomial isometric to
(a) $X^2 - XY + Y^2 + XZ - Z^2$
(b) $XY - XZ - YZ$
(c) $XZ + YW$

5.3.2. (Polarization Identity) Let $(\ ,\ )$ be a symmetric bilinear form on V with associated quadratic form Q. Prove for all $v, w \in V$,
$(v, w) = (1/4)(Q(v+w) + Q(v-w))$.

5.3.3. Find the symmetric matrix associated to each quadratic form in Prob. 5.3.1.

5.3.4. Find the symmetric bilinear form associated to each quadratic form in Prob 5.3.1.

5.3.5. Find the quadratic form (express as a quadratic polynomial) associated with each of the following symmetric matrices.

(a) $\begin{pmatrix} 2 & 4 \\ 4 & 1 \end{pmatrix}$ (b) $\begin{pmatrix} 1 & 2 & 1 \\ 2 & 0 & 3 \\ 1 & 3 & 2 \end{pmatrix}$ (c) $\begin{pmatrix} 1 & 3 & 4 \\ 3 & 0 & 1 \\ 4 & 1 & 2 \end{pmatrix}$

5.3.6. Let q be the quadratic form given by XY on $\mathbf{F}^2$. Find an invertible linear operator $T : \mathbf{F}^2 \to \mathbf{F}^2$ for which $q(T(a, b)) = a^2 - b^2$.

5.3.7. Find the symmetric bilinear forms on $\mathbf{F}^2$ associated with the following quadratic polynomials:

(a) $q(X, Y) = X^2$.
(b) $q(X, Y) = XY$.
(c) $q(X, Y) = X^2 - XY$.
(d) $q(X, Y) = X^2 + 3Y^2$.

5.3.8. Suppose that $Q : V \to F$ is a quadratic form, and let W be a subspace of V.
(a) Show that the restriction $Q|_W : W \to \mathbf{F}$ is a quadratic form.
(b) If Q is nondegenerate, is $Q|_W$ nondegenerate? Why or why not?

5.4   Real Quadratic Forms

When studying quadratic forms over the real field $\mathbf{R}$, it turns out that the two numerical invariants, the dimension (that is, the number of variables) and the signature (discussed next), suffice to determine if two real quadratic forms are isometric. Thus, we obtain a complete classification of real quadratic forms. As an application of this classification, we describe the relationship between quadratic forms and the local maxima/minima of real-valued functions alluded to earlier.

Definition 5.4.1 A quadratic form $Q : \mathbf{F}^n \to \mathbf{F}$ (respectively symmetric matrix, symmetric bilinear form) is called n-*dimensional*. An n-dimensional quadratic form is called *nonsingular* (or sometimes *nondegenerate*) if the associated symmetric matrix is nonsingular. This is the same as the associated quadratic polynomial being isometric to a diagonal quadratic polynomial with nonzero diagonal coefficients. If $Q : \mathbf{R}^n \to \mathbf{R}$ is a nondegenerate quadratic form, Q is called *positive definite* if $Q(v) > 0$ for all nonzero $v \in \mathbf{R}^n$. Analogously, Q is called *negative definite* if $Q(v) < 0$ for all nonzero $v \in \mathbf{R}^n$. If there exist $v, w \in \mathbf{R}^n$ with $Q(v) > 0 > Q(w)$, Q is called *indefinite*.

The nonsingular positive-definite real quadratic forms are precisely those quadratic forms whose associated symmetric bilinear forms are inner products. But the other nonsingular real quadratic forms are also quite important. According to Corollary 5.3.11, every nonsingular quadratic polynomial is isometric to a diagonal quadratic polynomial. Moreover, Remark 5.3.12 shows that every n-dimensional nonsingular real quadratic polynomial is isometric to a polynomial of the form $X_1^2 + X_2^2 + \cdots + X_r^2 - X_{r+1}^2 - X_{r+2}^2 - \cdots - X_n^2$ for some natural number r. This is needed to define the signature, which is given next. Note that the lemma which follows the definition assures that the concept of signature is well-defined.

Definition 5.4.2 Suppose $q(X_1, X_2, \ldots, X_n)$ is a nonsingular real quadratic homogeneous polynomial. If $q(X_1, X_2, \ldots, X_n)$ is isometric to the diagonal polynomial $X_1^2 + X_2^2 + \cdots + X_r^2 - X_{r+1}^2 - X_{r+2}^2 - \cdots - X_n^2$, then we say that $q(X_1, \ldots, X_n)$ has *signature* $r - (n - r)$. If $q(X_1, \ldots, X_n)$ has signature 0, then $q(X_1, \ldots, X_n)$ is called *hyperbolic*. The same termonology applies to real symmetric bilinear forms and real symmetric matrices.

The quadratic form q in the above definition can be diagonalized in many different ways. In order for the signature to be well-defined, we need to know that the number of positive coefficients in any two diagonalizations are the same. This is our next result.

Lemma 5.4.3 *The signature of a nonsingular real quadratic homogeneous polynomial is uniquely determined.*

Proof: We have to show that the signature of a real homogeneous quadratic polynomial (form) does not depend upon the choice of diagonalization. We study the associated quadratic form $Q : \mathbf{R}^n \to \mathbf{R}$. Let $v_1, \ldots, v_s, v_{s+1}, \ldots, v_n$ be a basis for $\mathbf{R}^n$ which diagonalizes Q and for which $Q(v_1), \ldots, Q(v_s) > 0$ and $Q(v_{s+1}), \ldots, Q(v_n) < 0$. Let $V^+ = span\{v_1, \ldots, v_s\}$ and $V^- = span\{v_{s+1}, \ldots, v_n\}$. Since this basis diagonalizes Q, we see that whenever $\vec{0} \neq w \in V^+$, $Q(w) > 0$ and whenever $\vec{0} \neq w' \in V^-$, $Q(w') < 0$.

Let $u_1, \ldots, u_t, u_{t+1}, \ldots, u_n$ be another basis for $\mathbf{R}^n$ which diagonalizes Q and for which $Q(u_1), \ldots, Q(u_t) > 0$ and $Q(u_{t+1}), \ldots, Q(u_n) < 0$. Let $U^+ = span\{u_1, \ldots, u_t\}$ and $U^- = span\{u_{t+1}, \ldots, u_n\}$. By the positivity conditions just observed, we must have that $V^+ \cap U^- = \{\vec{0}\}$ and $U^+ \cap V^- = \{\vec{0}\}$. Now $dim(V^+) = s$, $dim(V^-) = n - s$, $dim(U^+) = t$, and $dim(U^-) = n - t$. As $s + (n - t) = dim(V^+ \oplus U^-) \leq n$, and $t + (n - s) = dim(U^+ \oplus V^-) \leq n$, we conclude that $s = t$. From this the well-definition of the signature follows. //

Remark 5.4.4 In view of the well-definition of the signature of a real quadratic polynomial, we see that if $q(X_1, \ldots, X_n)$ has signature n, then $q(X_1, \ldots, X_n)$ is positive definite and if $q(X_1, \ldots, X_n)$ has signature $-n$, then $q(X_1, \ldots, X_n)$ is negative definite.

Examples 5.4.5 The signature of the positive-definite quadratic polynomial $X^2 + 2Y^2 + Z^2$ is 3, while the signature of the polynomial $X^2 - 2Y^2 + Z^2$ is 1. The matrix

$$S = \begin{pmatrix} -1 & -2 & 4 \\ -2 & 2 & 2 \\ 4 & 2 & -1 \end{pmatrix}$$

of Example 5.3.15 is congruent to

$$\begin{pmatrix} 3 & 0 & 0 \\ 0 & 3 & 0 \\ 0 & 0 & -6 \end{pmatrix}$$

This shows that the quadratic polynomial associated with S,

$$q(X, Y, Z) = -X^2 - 4XY + 2Y^2 + 8XZ + 4YZ - Z^2$$

has signature 1.

The signature of a real quadratic form is an invariant with many important geometric applications. Together with the dimension, the signature classifies all nondegenerate real quadratic forms up to isometry. This is proved next.

Theorem 5.4.6 (Sylvester) *Two nondegenerate real quadratic forms are isometric if and only if they have the same dimension and the same signature.*

Proof: Consider two real quadratic forms Q_1 and Q_2. Clearly, if they are isometric, they must have the same dimension. By the well-definition of the signature, they must have the same signature. Conversely, suppose that Q_1 and Q_2 have the same dimension and signature. Since Q_1 and Q_2 have the same dimension, they both are quadratic forms on $\mathbf{R}^n$ for some n. If the quadratic form $X_1^2 + \cdots + X_r^2 - X_{r+1}^2 - \cdots - X_n^2$ has signature s, then $s = r - (n - r) = 2r - n$. Thus the number r is completely determined by s and n. Since every n-dimensional real quadratic form is isometric to a unique form of this type, and Q_1 and Q_2 have the same signature, both Q_1 and Q_2 are isometric to the same form. Consequently, Q_1 and Q_2 are isometric. This proves the theorem. //

Discussion 5.4.7 Maxima and Minima of Real-Valued Functions

Recall the second-derivative test from elementary differential calculus: Suppose $\mathcal{U} \subseteq \mathbf{R}$ is some open interval and $f : \mathcal{U} \to \mathbf{R}$ has a continuous second derivative on $\mathcal{U}$. If $a \in \mathcal{U}$ is a critical point of f (that is, $f'(a) = 0$) and $f''(a) < 0$, then a is a local maximum of f. If $f''(a) > 0$, then a is a local minimum of f. [If $f''(a) = 0$, then the test does not apply.] In order to generalize the second-derivative test to functions of several variables, we need to replace the positivity (negativity) of the second derivative with the condition of positive (negative) definitness of an associated quadratic form.

Let $\mathcal{U} \subseteq \mathbf{R}^n$ be an open subset. Suppose that $f : \mathcal{U} \to \mathbf{R}$ is a function with continuous first and second partial derivatives. A critical point $(a_1, a_2, \ldots, a_n) \in \mathcal{U}$ is a point which satisfies the conditions

$$\frac{\partial f}{\partial x_1}(a_1, a_2, \ldots, a_n) = 0$$

$$\frac{\partial f}{\partial x_2}(a_1, a_2, \ldots, a_n) = 0$$

$$\vdots$$

$$\frac{\partial f}{\partial x_n}(a_1, a_2, \ldots, a_n) = 0$$

By advanced calculus, any local maximum or minimum for f must be a critical point. The second derivative test for critical points is the following:

Suppose that $(a_1, a_2, \ldots, a_n)$ is a critical point of f. Consider the symmetric matrix

$$S = \left(\frac{\partial^2 f}{\partial x_i \partial x_j}(a_1, a_2, \ldots, a_n) \right)$$

(that is, the matrix whose ijth entry is the ijth second-order partial derivative). That S is symmetric is a consequence of the equality of mixed second partial derivitives. The second-derivative test applies if the quadratic form Q associated with S is nonsingular. If Q is negative definite, then $(a_1, a_2, \ldots, a_n)$ is a local maximum, and if Q is positive definite, then $(a_1, a_2, \ldots, a_n)$ is a local minimum. In case the quadratic form Q is indefinite and nonsingular, then $(a_1, a_2, \ldots, a_n)$

is called a *saddle* point. [If $\mathcal{U} \subseteq \mathbf{R}^2$, then the graph of f is represented by a surface in $\mathbf{R}^3$. If (a_1, a_2) is a saddle point, then locally the point $(a_1, a_2, f(a_1, a_2))$ looks like the center of a saddle, hence the terminology.]

The reason this second derivative test works can be understood if one thinks in terms of the second-degree Taylor expansion of the function f around a critical point $(a_1, \ldots, a_n)$. Since the partials

$$\frac{\partial f}{\partial x_i}(a_1, a_2, \ldots, a_n) = 0$$

there are no linear terms, and this second-order Taylor polynomial is

$$f(a_1, a_2, \ldots, a_n) + \frac{1}{2}\sum_{i,j} \frac{\partial^2 f}{\partial x_i \partial x_j}(a_1, a_2, \ldots, a_n)(X_i - a_i)(X_j - a_j)$$

The sum of the second degree terms, is $1/2$ times the quadratic polynomial in the variables $(X_1 - a_1), \ldots, (X_n - a_n)$ associated with the symmetric matrix S described above. If this quadratic polynomial is positive definite, then it always takes positive values at points other than $(a_1, \ldots, a_n)$. Thus this second-degree approximation to f has a minimum at $(a_1, \ldots, a_n)$. Close to $(a_1, a_2, \ldots, a_n)$ the second-degree approximation to f is quite good; this means that f has a local minimum at the point $(a_1, a_2, \ldots, a_n)$. Similar reasoning illustrates why negative definiteness gives local maximums and why saddle points occur in the nonsingular indefinite case.

PROBLEMS

5.4.1. Find the signature of the quadratic form associated with the real symmetric matrices:

(a) $\begin{pmatrix} 1 & 2 \\ 2 & 3 \end{pmatrix}$

(b) $\begin{pmatrix} 2 & 0 & 1 \\ 0 & 2 & 0 \\ 1 & 0 & 1 \end{pmatrix}$

(c) $\begin{pmatrix} 1 & 1 & 0 \\ 1 & 0 & 1 \\ 0 & 1 & 1 \end{pmatrix}$

(d) $\begin{pmatrix} 0 & 1 & 0 & 0 \\ 1 & 0 & 0 & 0 \\ 0 & 0 & 0 & 1 \\ 0 & 0 & 1 & 0 \end{pmatrix}$

5.4.2. For each matrix S in Prob. 5.3.2, find an orthogonal matrix U for which $U^t S U$ is diagonal.

5.4.3. Show that the quadratic polynomial $q(X, Y) = aX^2 + 2bXY + CY^2$ gives a positive-definite quadratic form on $\mathbf{R}^2$ if and only if $ac - b^2 > 0$ and $a > 0$.

5.4.4. Suppose that P is a nonsingular, positive-definite real symmetric $n \times n$ matrix. If Q is a nonsingular real $n \times n$ matrix, prove that $Q^t P Q$ is nonsingular and posititve definite.

5.4.5. Suppose that S is a real symmetric $n \times n$ matrix with all eigenvalues positive. Define $<v, w> = w^t S v$ for all $v, w \in \mathbf{R}^n$. Prove that $< , >: \mathbf{R}^n \times \mathbf{R}^n \to \mathbf{R}$ is an inner product.

5.4.6. If P is a nonsingular, positive-definite real symmetric matrix, prove that $det(P) > 0$ and $tr(P) > 0$.

5.4.7 Consider the real matrix

$$A = \begin{pmatrix} 1 & 2 & 3 & 4 \\ 2 & 3 & 4 & 5 \\ 3 & 4 & 5 & 6 \\ 4 & 5 & 6 & 7 \end{pmatrix}$$

Find an invertible matrix C so that $C^t A C$ is diagonal, and find the rank and signature of A.

Chapter 6

The Structure of Linear Operators

6.1 Polynomial Actions on Matrices and Linear Operators; the Cayley-Hamilton Theorem

This chapter resumes the study of linear operators, picking up where Chap. 3 left off. In Sec. 3.5 we determined criteria for the diagonalizability of a linear operator. We saw that a linear operator $T : V \rightarrow V$ on a finite-dimensional vector space V was diagonalizable if and only if the vector space V had a basis of eigenvectors for T. Diagonalizable operators are easy to visualize geometrically. Such operators stretch or shrink the vector space along appropriately chosen axes (those coming from a basis of eigenvectors).

Unfortunately, not every linear operator can be diagonalized. In fact we have seen that there are operators on real vector spaces which do not have any real eigenvalues. In this chapter we study the nondiagonalizable case. We will show that the structure of a linear operator is largely determined by what happens when a polynomial (in one variable) is applied to the operator. This first section studies what it means to apply a polynomial to an operator.

In Sec. 6.2, we exhibit some special operators (that is, we specify some special matrices) which are the prototypes by which we understand general operators. We will identify specific properties of these operators. Afterwards, we will show that to a large extent, the behavior of a general operator can be understood by seeing how these same properties exhibit themselvēs. This task occupies the remainder of the chapter.

Suppose that A is an $n \times n$ matrix with coefficients in a field $\mathbf{F}$. Then, of course, the product of A with any other $n \times n$ matrix is also an $n \times n$ matrix. In view of this, it makes sense to substitute A into polynomials with coefficients from $\mathbf{F}$. To be precise, whenever $f(X) = k_n X^n + \cdots + k_1 X + k_0$ is a polynomial in the variable X with coefficients $k_i \in \mathbf{F}$, we define

$$f(A) = k_n A^n + \cdots + k_1 A + k_0 I_n$$

where multiplication is scalar multiplication of matrices by scalars from $\mathbf{F}$ and addition is addition of $n \times n$ matrices. Do not forget to multiply k_0 by I_n so that you obtain an $n \times n$ matrix!

For example, consider the matrix

$$A = \begin{pmatrix} 1 & 1 \\ 3 & -1 \end{pmatrix}$$

Let $f(X) = X^2 - X + 3$ and $g(X) = X^2 - 4$. Then we compute that

$$f(A) = \begin{pmatrix} 4 & 0 \\ 0 & 4 \end{pmatrix} - \begin{pmatrix} 1 & 1 \\ 3 & -1 \end{pmatrix} + \begin{pmatrix} 3 & 0 \\ 0 & 3 \end{pmatrix} = \begin{pmatrix} 6 & -1 \\ -3 & 8 \end{pmatrix}$$

and

$$g(A) = \begin{pmatrix} 4 & 0 \\ 0 & 4 \end{pmatrix} - \begin{pmatrix} 4 & 0 \\ 0 & 4 \end{pmatrix} = \begin{pmatrix} 0 & 0 \\ 0 & 0 \end{pmatrix}$$

[The fact that $g(A)$ is the zero matrix is not merely a coincidence. $g(X)$ is the characteristic polynomial of A, and as we shall see at the end of this section, the characteristic polynomial of a matrix always "annihilates" the matrix in this fashion.]

We next define how to substitute an operator into a polynomial. Addition and scalar multiplication for operators is the same as described in Sec. 3.4. But care is required when two operators are multiplied. Multiplication of operators is functional composition. The reason for this is that matrix multiplication corresponds to the composition of linear transformations (as pointed out in Theorems 3.1.7 and 3.3.6). What you *do not* want to think is that operator multiplication has something to do with products of images of elements. While something like this is true for addition, if you get confused and try this for multiplication, you will get totally fouled up. So be careful!

For emphasis, we record these concepts as a formal definition.

Definition 6.1.1 Whenever $T_1, T_2 : V \rightarrow V$ are linear operators and $a \in \mathbf{F}$, we can define new linear operators $T_1 + T_2$, aT_1 and $T_1 T_2$ by setting

$$
\begin{aligned}
(T_1 + T_2)(v) &= T_1(v) + T_2(v) \quad \text{for all } v \in V \\
(aT_1)(v) &= a[T_1(v)] \quad \text{for all } v \in V \\
(T_1 T_2)(v) &= T_1(T_2(v)) \quad \text{for all } v \in V \quad (\text{that is, } T_1 T_2 = T_1 \circ T_2)
\end{aligned}
$$

We omit the (extremely) easy verification that $T_1 + T_2$, aT_1, and $T_1 T_2$ are linear operators. We abbreviate TT by T^2, TTT by T^3, and so on. If $f(X) = a_n X^n + a_{n-1} X^{n-1} + \cdots + a_1 X + a_0$ is a polynomial, we define $f(T)$ to be the linear operator

$$
f(T) = a_n T^n + a_{n-1} T^{n-1} + \cdots + a_1 T + a_0 I
$$

Here, $I : V \rightarrow V$ is the identity operator [so, $I(v) = v$ for all $v \in V$].

The substitution of linear operators and matrices into polynomials is a key tool for developing the theory of linear operators. As we shall see, this tool is more than a computational device; an entirely new point of view is involved. We shall discuss this in greater detail as this chapter unfolds, so for now we make only a few comments. It is useful to think of the process of substituting an operator (or a matrix) into a polynomial as a kind of multiplication by the polynomial. Thus in the examples described above, we might say

$$
f(X)A = \begin{pmatrix} 6 & -1 \\ -3 & 8 \end{pmatrix} \quad \text{and} \quad g(X)A = \begin{pmatrix} 0 & 0 \\ 0 & 0 \end{pmatrix}
$$

We will *not* use this notation, but there is a setting (the development of the theory of modules) where this would be the preferred notation. Another way of saying this is to say that $g(X)$ acts on A and gives the zero matrix. We shall use this terminology of action throughout this chapter.

Corresponding to this new viewpoint is a change in one's approach to linear operators. In addition to studying how the operator acts on special vectors (eigenvectors for example) in the vector space, one is interested in which polynomials act in a special way upon the operator. Those that annihilate (give 0) are of paramount importance. One attempts to classify operators according to the behavior of certain sets of polynomials associated with them.

The two types of substitution defined above, one for matrices and the other for operators, are compatible. This means that when one computes the matrix for a polynomial action on an operator, the result is the same polynomial acting on the matrix that represents the original operator. This is an elementary but crucial observation. Another crucial fact is that when a polynomial factors as the product of two other polynomials, then the polynomial acts as the composition of the actions of its factors. Since these results are so often used, we highlight them and state them precisely in the next lemma.

Lemma 6.1.2 *Assume that $T : V \rightarrow V$ is a linear operator and $A = [T]_B$ where B is an ordered basis of V.*
(i) *For any polynomial $f(X)$, $[f(T)]_B = f(A)$.*
(ii) *If $f(X) = g(X)h(X)$, then $f(T) = g(T) \circ h(T)$.*

Proof: (i) Suppose that $B = \{v_1, v_2, \ldots, v_n\}$ and $f(X) = a_n X^n + \cdots + a_1 X + a_0$. For $v \in V$, $(f(T)(v))_B = (a_n T^n(v) + \cdots + a_1 T(v) + a_0(v))_B = (a_n T^n(v))_B + \cdots + (a_1 T(v))_B + a_0(v)_B = a_n A^n(v)_B + \cdots + a_1 A(v)_B + a_0(v)_B = f(A)(v)_B$. Thus, by the uniqueness of the matrix representation, $[f(T)]_B = f(A)$.

(ii) We know from Theorem 3.3.6 that matrix multiplication corresponds to functional composition. This shows that $[g(T) \circ h(T)]_B = [g(T)]_B [h(T)]_B$. By (i), $[g(T)]_B [h(T)]_B = g(A)h(A) = f(A) = [f(T)]_B$. Thus, $f(T) = g(T) \circ h(T)$ follows. //

Recall that if A is an $n \times n$ matrix, then $C_A(X) = det(XI_n - A)$ is the characteristic polynomial of A. Recall also that (using Lemma 3.4.8) the characteristic polynomial makes sense for operators. (The characteristic polynomial of an operator $T : V \rightarrow V$ is the characteristic polynomial of the matrix $[T]_B$ for any ordered basis B of V.)

Example 6.1.3: Let S be the standard basis of P^2, the three dimensional vector space of all polynomials of degree at most 2. We shall assume that $\mathbf{F} = \mathbf{R}$. Let $T : P^2 \rightarrow P^2$ be the operator defined by

$T(f(X)) = f''(X) - f(X)$. For such T we know that $ker(T) = \vec{0}$. [This is easily checked: Note that the differential equation $f'' - f = 0$ has as all its solutions the functions $A\sin x + B\cos x$, A, $B \in \mathbf{R}$, none of which lie in P^2.] It follows that T is an isomorphism. In the standard basis S we have

$$[T]_S = \begin{pmatrix} -1 & 0 & 2 \\ 0 & -1 & 0 \\ 0 & 0 & -1 \end{pmatrix}$$

(from which it is also evident that T is an isomorphism). The characteristic polynomial of T is the characteristic polynomial of $[T]_S$, which is $C_T(X) = (X + 1)^3$.

We compute $C_T(T)$, using Lemma 6.1.2. Observe that if we set $p(X) = (X + 1)$, then $p(T) = T + I$. Since $T(f(X)) = f''(X) - f(X)$, we find that $(T + I)(f(X)) = (f''(X) - f(X)) + f(X) = f''(X)$. In other words, $T + I$ is the operator which gives the second derivative of a polynomial in P^2. Lemma 6.1.2 (ii) now shows that $p(T)^2$ is the composite operator $(T+I) \circ (T+I)$. Thus, $p(T)^2(f(X)) = (f''(X))'' = f^{(4)}(X)$. Similarly, $C_T(T) = p(T)^3$ is the operator which gives the sixth derivative of any polynomial inside P^2. Since the third derivative of any polynomial inside P^2 is already 0, we conclude that both operators $p(T)^2$ and $C_T(T) = p(T)^3$ are 0.

Alternatively, one can use Lemma 6.1.2 (i) and compute directly with the matrix $[T]_S$ to see that

$$[p(T)]_S = [T]_S + I = \begin{pmatrix} -1 & 0 & 2 \\ 0 & -1 & 0 \\ 0 & 0 & -1 \end{pmatrix} + \begin{pmatrix} 1 & 0 & 0 \\ 0 & 1 & 0 \\ 0 & 0 & 1 \end{pmatrix} = \begin{pmatrix} 0 & 0 & 2 \\ 0 & 0 & 0 \\ 0 & 0 & 0 \end{pmatrix}$$

Consequently, both $[p(T)^2]_S = [p(T)]_S^2$ and $[C_T(T)]_S = [p(T)]_S^3$ are zero.

Example 6.1.4 In this example we continue our analysis of Example 6.1.3. Let $q(X) = a_n X^n + a_{n-1} X^{n-1} + \cdots + a_0$ be any polynomial. We are interested in determining when the operator $q(T) = 0$. Of course, we could substitute the matrix $[T]_S$ into $q(X)$ and try to unravel the computational mess that arises. Instead however, we observe that it is always possible to rewrite the polynomial $q(X)$ as a polynomial in $X + 1$; that is, there always exists an expression

$$q(X) = b_n(X + 1)^n + b_{n-1}(X + 1)^{n-1} + \cdots + b_0$$

(To see this, set $b_n = a_n$, then solve for b_{n-1} as $b_{n-1} = a_{n-1} - nb_n$, and so forth.) As $(T + I)^2 = 0$, so do all the other higher powers $(T + I)^n$ for $n \geq 2$. Consequently, $q(T) = b_1(T + I) + b_0 I = b_1 T + (b_1 + b_0)I$.

Evaluating this expression on a polynomial in P^2, we find that $q(T)(f(X)) = b_1 T(f(X)) + (b_1 + b_0)f(X) = b_1 f''(X) + (b_1 + b_0 - 1)f(X)$. Thus, $q(T)(X^2) = 2b_1 + (b_1 + b_0 - 1)X^2$ and $q(T)(1) = (b_1 + b_0 - 1)1$. It follows that in order for $q(T)$ to be the zero operator, we must have that $b_1 = b_0 = 0$. This shows that for such $q(X)$, we can write $q(X) = (X + 1)^2(b_n(X + 1)^{n-2} + b_{n-1}(X + 1)^{n-3} + \cdots + b_2)$. In other words, $(X + 1)^2$ is a factor of any polynomial $q(X)$ for which $q(T) = 0$.

We shall see at the end of this section that the phenomenon observed in Example 6.1.4 always occurs for any linear operator on a finite-dimensional vector space. It will turn out that any polynomial which annihilates an operator is a multiple of a fixed annihilating polynomial, called the *minimal polynomial*. The characteristic polynomial will always annihilate the operator, and thus the minimal poynomial will be a factor of the characteristic polynomial. This result is given in the next theorem, which proves to be a turning point in the study of the structure of linear operators. The first time is is read, it appears to be an interesting coincidence. The true depth of this result will become apparent as this chapter unfolds.

Theorem 6.1.5 (Cayley-Hamilton Theorem) *For any $n \times n$ matrix A, we have $C_A(A) = 0$. If V is finite-dimensional, then for any linear operator $T : V \to V$, $C_T(T) = 0$.*

Remark What follows is a computational and matrix-oriented proof of the Cayley-Hamilton theorem. It is a nice application of the formulas developed for the determinant in Chap. 2, but it is not well-motivated. An alternative proof is given in 6.3.11. The second proof uses a bit more abstract "machinery," but illustrates better why this result is true.

Proof: Fix an ordered basis $\mathcal{B}$ of V and set $A = [T]_{\mathcal{B}}$. Then we have that $C_T(X) = det(XI_n - A)$. By Lemma 6.1.2 (i), $[C_T(T)]_{\mathcal{B}} = C_A(A)$, so the second statement of the theorem follows from the first. We express the characteristic polynomial $C_T(X) = C_A(X) = X^n + b_{n-1}X^{n-1} + \cdots + b_1 X + b_0$ with $b_i \in \mathbf{F}$. We denote by $D(X)$ the matrix with polynomial entries $D(X) = adj(XI_n - A)$. Since each entry in $D(X)$ is the determinant of an $(n-1) \times (n-1)$ submatrix of $(XI_n - A)$, each entry of $D(X)$ is a polynomial of degree less than or equal to $n - 1$. It follows that there exist matrices $D_0, D_1, \ldots, D_{n-1}$ with entries from $\mathbf{F}$ such that $D(X) = D_{n-1}X^{n-1} + \cdots + D_1 X + D_0$. ($D_i X^i$ is an $n \times n$ matrix all of whose entries have the form dX^i for some $d \in \mathbf{F}$.) According to Theorem 2.3.7 we have the matrix equation

$$det(XI_n - A)I_n = (XI_n - A)adj(XI_n - A) = (XI_n - A)D(X)$$

Substituting $C_A(X) = det(XI_n - A)$, (and using the fact that scalars commute with matrices) gives

$$
\begin{aligned}
X^n I_n &+ b_{n-1}X^{n-1}I_n + \cdots + b_1 XI_n + b_0 I_n \\
&= C_A(X)I_n = det(XI_n - A)I_n \\
&= (XI_n - A)adj(XI_n - A) \\
&= (XI_n - A)(X^{n-1}D_{n-1} + \cdots + XD_1 + D_0) \\
&= X^n D_{n-1} - X^{n-1}AD_{n-1} + X^{n-1}D_{n-2} \qquad (*) \\
&\quad - X^{n-2}AD_{n-2} + \cdots + XD_0 - AD_0 \\
&= X^n D_{n-1} + X^{n-1}(-AD_{n-1} + D_{n-2}) \\
&\quad + \cdots + X(-AD_1 + D_0) - AD_0
\end{aligned}
$$

Since two polynomials are equal if and only if their coefficients are equal, we find that the coefficient matrices are equal; that is, $I_n = D_{n-1}$, $b_{n-1}I_n = (-AD_{n-1} + D_{n-2}), \ldots$, $b_1 I_n = -(AD_1 + D_0)$, and $b_0 I_n = -AD_0$. This means that we can substitute the matrix A for the variable X in the equations $(*)$ to conclude:

$$
\begin{aligned}
C_A(A) &= A^n + b_{n-1}A^{n-1} + \cdots + b_1 A + b_0 I_n \\
&= A^n D_{n-1} + A^{n-1}(-AD_{n-1} + D_{n-2}) + \cdots \\
&\quad + A(-AD_1 + D_0) - AD_0 \\
&= A^n D_{n-1} - A^n D_{n-1} + A^{n-1}D_{n-2} - A^{n-1}D_{n-2} + \cdots \\
&\quad + AD_0 - AD_0 = 0
\end{aligned}
$$

This proves the theorem. //

The next example shows how to use the Cayley-Hamilton theorem to invert matrices. The result is nice, but is not a serious computational tool. Our first important use of the Cayley-Hamilton theorem will be in the construction of the minimal polynomial.

Example 6.1.6 Let A be an $n \times n$ matrix, with characteristic polynomial $C_A(X) = X^n + b_{n-1}X^{n-1} + \cdots + b_1 X + b_0$. Then, by the Cayley-Hamilton theorem, $0 = C_A(A) = A^n + b_{n-1}A^{n-i} + \cdots + b_1 A + b_0 I_n$. Rearranging gives $-A(A^{n-1} + b_{n-1}A^{n-2} + \cdots + b_1 I_n) = b_0 I_n$. In particular, if $b_0 \neq 0$, $A^{-1} = -b_0^{-1}(A^{n-1} + b_{n-1}A^{n-2} + \cdots + b_1 I_n)$. [Note that if $0 = b_0 = det(A)$, then the matrix A is not invertible.] For example, if

$$A = \begin{pmatrix} 3 & 2 & 2 \\ 1 & 2 & 2 \\ -1 & -1 & 0 \end{pmatrix}$$

then we can compute $C_A(X) = X^3 - 5X^2 + 8X - 4$. We find

$$A^{-1} = (1/4)(A^2 - 5A + 8I)$$

$$= \frac{1}{4}\begin{pmatrix} 9 & 8 & 10 \\ 3 & 4 & 6 \\ -4 & -4 & -4 \end{pmatrix} - 5\begin{pmatrix} 3 & 2 & 2 \\ 1 & 2 & 2 \\ -1 & -1 & 0 \end{pmatrix} + 8\begin{pmatrix} 1 & 0 & 0 \\ 0 & 1 & 0 \\ 0 & 0 & 1 \end{pmatrix}$$

$$= \frac{1}{4}\begin{pmatrix} 2 & -2 & 0 \\ -2 & 2 & -4 \\ 1 & 1 & 4 \end{pmatrix} = \begin{pmatrix} 1/2 & -1/2 & 0 \\ -1/2 & 1/2 & -1 \\ 1/4 & 1/4 & 1 \end{pmatrix}$$

In order to obtain the important consequences of the Cayley-Hamilton theorem, we must digress slightly and discuss some results about polynomials. First, we give some definitions and notation for polynomials.

Definition 6.1.7 We denote by $\mathbf{F}[X]$ the set of all polynomials in the variable X with coefficients in $\mathbf{F}$. The degree of a polynomial $f(X) \in \mathbf{F}[X]$ is denoted by $deg(f(X))$. If $f(X) = h(X)g(X)$, we say that $g(X)$ *divides* $f(X)$ and $g(X)$ is a *factor* (or *divisor*) of $f(X)$. The relation "divides" is usually denoted by the bar: $g(X) \mid f(X)$. If the only factors $g(X)$ of the nonconstant polynomial $f(X)$ with $deg(g(X)) < deg(f(X))$ are constants (that is, such $g(X) = b \in \mathbf{F}$), then we say that $f(X)$ is *prime* (or *irreducible*). A polynomial $f(X) = X^n + b_{n-1}X^{n-1} + \cdots + b_1 X + b_0$ (whose leading coefficient is 1) is called a *monic* polynomial.

We now state some basic results from the theory of polynomials of one variable. These results should be familiar (at least for $\mathbf{F} = \mathbf{R}$) from some previous algebra course. They are the natural generalizations to polynomials of basic results about the division of natural numbers (found in grade school arithmetic). Their proofs, which are essentially long division for polynomials in one variable, are omitted.

Theorem 6.1.8 (i) (Division Algorithm) *Suppose that* $f(X)$, $g(X) \neq 0 \in \mathbf{F}[X]$ *with* $deg(g(X)) \leq deg(f(X))$. *Then there exist unique polynomials* $h(X), r(X) \in \mathbf{F}[X]$ *with* $deg(r(X)) < deg(g(X))$ *such that*

$$f(X) = h(X)g(X) + r(X)$$

(ii) (Unique Factorization or Fundamental Theorem of Arithmetic for Polynomials) *Every monic polynomial* $f(X) \in \mathbf{F}[X]$ *can be written*

$$f(X) = p_1(X)^{n_1} p_2(X)^{n_2} \ldots p_s(X)^{n_s}$$

where the $p_i(X)$ *are distinct monic prime polynomials. Moreover, the primes* $p_i(X)$ *with their exponents* n_i *are uniquely determined up to the order in which they are listed.*

As mentioned earlier in this section, is is important to view polynomials as acting upon linear operators. The Cayley-Hamilton theorem shows that the characteristic polynomial acts by taking an operator to zero. As exhibited in Example 6.1.3, the characteristic polynomial is not the only polynomial with this property. Further, whenever any polynomial acts by taking an operator to zero, so does any multiple of this polynomial [by Lemma 6.1.2 (ii)]. So in fact there are always infinitely many annihilating polynomials.

We next define the minimal polynomial. It will be a consequence of the division algorithm that every annihilating polynomial is a multiple of the minimal polynomial (see Theorem 6.1.10). In general, the minimal polynomial may or may not be the same as the characteristic polynomial.

Definition 6.1.9 Suppose that A is an $n \times n$ matrix. We define $M_A(X)$, called the *minimal polynomial of A*, to be the (nonzero) monic polynomial of least degree for which $M_A(A) = 0$. If $T : V \to V$ is a linear operator, then we define $M_T(X)$, called the *minimal polynomial of T*, to be the monic polynomial of least degree for which $M_T(T) = 0$.

The Cayley-Hamilton theorem ensures that the minimal polynomial exists. We know that $C_A(A) = 0$, so there is some monic polynomial which annihilates A. Moreover, the monic polynomial of least degree must be unique. (If there were two distinct monic polynomials of least degree annihilating A, then their difference would be a nonzero polynomial of lower degree annihilating A, a scalar multiple of which would be monic.)

As an example, consider the 3×3 matrix

$$M = \begin{pmatrix} 2 & 3 & 1 \\ 0 & 2 & 0 \\ 0 & 0 & 2 \end{pmatrix}$$

In view of the Cayley-Hamilton theorem, the minimal polynomial of M has degree at most three. One way to find the minimal polynomial is to compute matrix powers of M: M^0, M^1, M^2, and M^3 and look for a linear dependence relation. These powers are

$$\begin{pmatrix} 1 & 0 & 0 \\ 0 & 1 & 0 \\ 0 & 0 & 1 \end{pmatrix}, \begin{pmatrix} 2 & 3 & 1 \\ 0 & 2 & 0 \\ 0 & 0 & 2 \end{pmatrix}, \begin{pmatrix} 4 & 12 & 4 \\ 0 & 4 & 0 \\ 0 & 0 & 4 \end{pmatrix}, \begin{pmatrix} 8 & 96 & 32 \\ 0 & 8 & 0 \\ 0 & 0 & 8 \end{pmatrix}$$

Inspecting these powers of M, one can see that I and M are linearly independent (so the degree of the minimal polynomial is at least two) and I, M, and M^2 are linearly dependent. In particular $M^2 - 4M = 4I_3$. This shows that the minimal polynomial of M is $X^2 - 4X + 4 = (X - 2)^2$.

Note that whenever $T : V \to V$ is a linear operator and B is some ordered basis for V, Lemma 6.1.2 (i) guarantees $M_T(X) = M_A(X)$ where $A = [T]_B$. Suppose that P is an invertible $n \times n$ matrix. Then for any polynomial $f(X)$, $f(P^{-1}AP) = P^{-1}f(A)P$ (an easy calculation). From this we see that $f(A) = 0$ if and only if $f(P^{-1}AP) = 0$. This shows directly that like the characteristic polynomials, the minimal polynomials for similar matrices are the same. Of course this better be true, for the minimal polynomial is defined for an operator, and we noted that the minimal polynomial of an operator agrees with the minimal polynomial of the matrix representing the operator (in any basis). We remark that whenever $f(X)$ is a nonzero polynomial and $deg(f(X)) < deg(M_A(X))$, then the definition of the minimal polynomial guarantees that $f(A) \neq 0$.

As we shall see throughout this chapter, the interplay between the minimal polynomial and the characteristic polynomial largely determines the structure of a linear operator. The final result of this section is an application of the division algorithm. Because $M_T(X) \mid C_T(X)$ is so closely tied to the Cayley-Hamilton theorem, it is often referred to as the Cayley-Hamilton theorem.

Theorem 6.1.10 *Suppose that $T : V \to V$ is a linear operator and $f(X)$ is a nonzero polynomial for which $f(T) = 0$. Then $M_T(X) \mid f(X)$. In particular, $M_T(X) \mid C_T(X)$.*

Proof: We know by definition that $deg(M_T(X)) \leq deg(f(X))$. We apply the division algorithm to write

$$f(X) = h(X)M_T(X) + r(X)$$

where $h(X)$ and $r(X)$ are polynomials and $deg(r(X)) < deg(M_T(X))$ Observe that $r(T) = f(T) - h(T)M_T(T) = 0 - 0 = 0$, by hypothesis and by the definition of $M_T(X)$. But since $deg(r(X)) < deg(M_T(X))$, we see that necessarily $r(X) = 0$. This shows that $M_T(X) \mid f(X)$, and this proves the theorem. //

Theorem 6.1.10 is important for the computation of the minimal polynomial. It shows that once the characteristic polynomial is computed, to find the minimal polynomial, one must find the factor of least degree which annihilates the operator. For example, if the characteristic polynomial is irreducible, then the minimal polynomial is necessarily the same as the characteristic polynomial. We shall see somewhat later, in Corollary 6.6.8, that the minimal polynomial and the characteristic polynomial always have the same prime factors. Consequently, if the characteristic polynomial factors into distinct primes, then again the characteristic polynomial and the minimal polynomial are the same.

We conclude this section with an additional example. This example shows that sometimes a small change in a matrix can greatly effect the relationship between the minimal and characteristic polynomials.

Example 6.1.11 Consider

$$R = \begin{pmatrix} 2 & 0 & 0 \\ 0 & 2 & 0 \\ 0 & 0 & 3 \end{pmatrix} \quad \text{and} \quad S = \begin{pmatrix} 2 & 1 & 0 \\ 0 & 2 & 0 \\ 0 & 0 & 3 \end{pmatrix}$$

We have $C_S(X) = C_R(X) = (X-2)^2(X-3)$. By the preceding remarks, the possible minimal polynomials for R and S are $g(X) = (X-2)(X-3)$ and $(X-2)^2(X-3)$. A direct calculation reveals that $g(R)$ is zero, while $g(S)$ is not. Hence, $M_R(X) = g(X) \neq C_R(X)$, while $M_S(X) = C_S(X)$. The key reason these two matrices have different minimal polynomials is that T_R has a two-dimensional eigenspace associated with the eigenvalue 2, while T_S has only a one-dimensional eigenspace associated with the eigenvalue 2. This will be further illuminated in the sections that follow.

PROBLEMS

6.1.1. Find 3×3 real matrices with minimal polynomials X, X^2, and X^3.

6.1.2. Use the Cayley-Hamiton theorem to show that if $T : V \to V$ is a linear operator on an n-dimensional vector space V and $T^k = 0$ for some k, then $T^n = 0$.

6.1.3. Find the characteristic and minimal polynomials of the following matrices:

(a) $\begin{pmatrix} 0 & 0 & a \\ 1 & 0 & b \\ 0 & 1 & c \end{pmatrix}$ (b) $\begin{pmatrix} 0 & 1 & 0 \\ 1 & 0 & 0 \\ 0 & 0 & 1 \end{pmatrix}$ (c) $\begin{pmatrix} 0 & 1 & 0 & 1 \\ 1 & 0 & 1 & 0 \\ 0 & 1 & 0 & 1 \\ 1 & 0 & 1 & 0 \end{pmatrix}$

6.1.4. Find the characteristic and minimal polynomials of the following matrices. Use the Cayley-Hamilton theorem to invert each.

(a) $\begin{pmatrix} 4 & -5 & 3 \\ 2 & -3 & 2 \\ -1 & 1 & 0 \end{pmatrix}$ (b) $\begin{pmatrix} -2 & -6 & -9 \\ 3 & 7 & 9 \\ -1 & -2 & -2 \end{pmatrix}$ (c) $\begin{pmatrix} 0 & 0 & 0 & 1 \\ 0 & 0 & 1 & 0 \\ 0 & 1 & 0 & 0 \\ 1 & 0 & 0 & 0 \end{pmatrix}$

6.1.5. Prove that for any square matrix A, A and A^t have the same minimal polynomial.

6.1.6. Suppose that $T : M_{n \times n}(\mathbf{F}) \to M_{n \times n}(\mathbf{F})$ is the linear transformation defined by $T(A) = AB$ for some fixed $n \times n$ matrix B. Show that the minimal polynomial of T is the same as the minimal polynomial of the matrix B.

6.1.7. What is the minimal polynomial of the differentiation operator $D : P^n \to P^n$?

6.1.8. What is the minimal polynomial of the operator $S : P^n \to P^n$ defined by $S(f(X)) = f(X) - Xf'(X)$?

6.1.9.   Suppose that $Proj_W : \mathbf{R}^n \to W$ is the orthogonal projection, where W is an m-dimensional subspace of $\mathbf{R}^n$. Find the minimal and characteristic polynomials of $Proj_W$.

6.1.10.   (a) Suppose that a 5×5 matrix A is diagonalizable and has characteristic polynomial $(X - 2)^3(X + 1)^2$. What is $M_A(X)$?
(b) If $C_A(X) = (X - 2)^3(X + 1)^2$, give examples of matrices with all possible minimal polynomials.

6.1.11.   Let $T : \mathbf{F}^n \to \mathbf{F}^n$ be a linear operator. Let $f(X) \in \mathbf{F}[X]$ be an arbitrary polynomial. If c is an eigenvalue of T, prove that $f(c)$ is an eigenvalue of $f(T)$.

6.2   Two Important Examples

This section is devoted to the close examination of two examples. During the remainder of the chapter we will study to what extent any linear operator can be interpreted as looking like these examples. Both the minimal and characteristic polynomials will be key tools in this analysis. In order to use these tools effectively, we need to understand how both polynomials are determined when the operator acts in a special way on some ordered basis. Our first task is to explain this relationship for these examples.

The first example leads to the so-called Jordan form. In the following definition, we give the building block from which the Jordan form is constructed.

Definition 6.2.1 A square matrix J is called an *Jordan block matrix of eigenvalue c* if it is of the following form:

$$
J = \begin{pmatrix}
c & 0 & 0 & \cdots & 0 & 0 \\
1 & c & 0 & \cdots & 0 & 0 \\
0 & 1 & c & \cdots & 0 & 0 \\
\vdots & \vdots & \vdots & \ddots & \vdots & \vdots \\
0 & 0 & 0 & \cdots & c & 0 \\
0 & 0 & 0 & \cdots & 1 & c
\end{pmatrix}
$$

Evidently, Jordan block matrices are lower triangular. This makes the computation of the characteristic polynomial of a Jordan block matrix trivial; in the above we have $C_J(X) = (X - c)^n$ (where J is $n \times n$). Moreover, it turns out that for this matrix, the minimal polynomial must be the same as the characteristic polynomial. This is given next.

Lemma 6.2.2 *Suppose that J is an $n \times n$ Jordan block matrix of eigenvalue c. Then $C_J(X) = M_J(X) = (X - c)^n$.*

Proof: That $C_J(X) = (X - c)^n$ is clear. To see that $M_J(X) = (X - c)^n$ as well follows from the calculation

$$(J - cI)^{n-1} = \begin{pmatrix} 0 & 0 & 0 & \cdots & 0 & 0 \\ 1 & 0 & 0 & \cdots & 0 & 0 \\ 0 & 1 & 0 & \cdots & 0 & 0 \\ \vdots & \vdots & \vdots & \ddots & \vdots & \vdots \\ 0 & 0 & 0 & \cdots & 0 & 0 \\ 0 & 0 & 0 & \cdots & 1 & 0 \end{pmatrix}^{n-1}$$

$$= \begin{pmatrix} 0 & 0 & 0 & \cdots & 0 & 0 \\ 0 & 0 & 0 & \cdots & 0 & 0 \\ 0 & 0 & 0 & \cdots & 0 & 0 \\ \vdots & \vdots & \vdots & \ddots & \vdots & \vdots \\ 0 & 0 & 0 & \cdots & 0 & 0 \\ 1 & 0 & 0 & \cdots & 0 & 0 \end{pmatrix}$$

This shows that $(X - c)^{n-1}$ cannot annihilate the matrix J. By Theorem 6.1.10 we know that $M_T(X)$ must be a factor of $(X - c)^n$, and we have shown that no factor of degree less than n can annihilate J. Hence $M_J(X) = (X - c)^n$. //

Jordan block matrices have some special properties. First, if J is an $n \times n$ Jordan block matrix of eigenvalue c, then J has only the single eigenvalue c. In addition, the eigenspace E_c of J is one-dimensional, because by inspection $rk(J - cI) = n - 1$. Such a matrix J is far from being diagonalizable, in fact you might say "farthest possible," among matrices that are triangular.

In contrast, consider the matrix

$$\begin{pmatrix} 3 & 0 & 0 \\ 2 & 2 & 0 \\ 5 & 2 & 9 \end{pmatrix}$$

Easily, one sees $C_A(X) = (X - 3)(X - 2)(X - 9)$. Moreover, by substituting A into the possible factors of $C_A(X)$, one finds that $M_A(X) =$

$C_A(X)$. [By the forthcoming Corollary 6.6.8, the minimal polynomial and the charactristic polynomial have the same prime factors, and this shows immediately that $M_A(X) = C_A(X)$ in this case.] Since A has three distinct eigenvalues, A is diagonalizable.

Why are some matrices diagonalizable, and why does a Jordan block matrix fail to be diagonalizable? The two examples above illustrate that the number of zeros in the original matrix is irrelevant. Jordan block matrices have so many zeros that at first glance they may appear to be diagonal. On the other hand the triangular matrix A has no nonzero entries below the diagonal. It turns out that the key is the minimal polynomial. The next result deals with the special case of a single eigenvalue.

Lemma 6.2.3   *Suppose that A is an $n \times n$ matrix and $C_A(X) = (X - c)^n$ for some $c \in \mathbf{F}$. Then A is diagonalizable if and only if $M_A(X) = (X - c)$. In this case, $A = cI_n$.*

Proof: Suppose that $M_A(X) = (X - c)$. Then since $M_A(A) = 0$, we find that $A - cI_n = 0$; that is, $A = cI_n$ is diagonal. Conversely, if A is diagonalizable, since $C_A(X) = (X - c)^n$, A has only the single eigenvalue c. It follows that A is similar to the matrix cI_n, whose minimal polynomial is $(X - c)$. However, if $A = PcI_nP^{-1}$, then as cI_n commutes with all matrices, we find that $A = cI_n$ as required. //

The lemma shows that diagonalizability corresponds (in the above case) to the minimal polynomial being as "small" as possible. In fact this generalizes: whenever the minimal polynomial is a product of distinct linear factors, the operator will be diagonalizable. This is proved later (see Corollary 6.5.8).

We next illustrate how the linear operator associated with a Jordan block matrix can be visualized. This will provide the motivation for much of the work that follows. The situation is the following: Suppose that J is the $n \times n$ Jordan block matrix of eigenvalue c. We shall investigate how T_J acts on the standard basis for $\mathbf{F}^n$. Remember, the columns of A are precisely the images of the standard basis $e_1, e_2, \ldots, e_n$ of $\mathbf{F}^n$. From this we see that $T_J(e_1) = ce_1 + e_2$, $T_J(e_2) = ce_2 + e_3, \ldots, T_J(e_{n-1}) = ce_{n-1} + e_n$, and $T_J(e_n) = ce_n$. Immediately, we find that $E_c = span\{e_n\}$ is the eigenspace of J (which previously was observed to be one-dimensional). The remaining basis elements

$e_1, e_2, \ldots, e_{n-1}$ try hard to be eigenvectors, but do not quite make it. Observe however, that the "amount" by which each one fails is precisely the next basis element. This turns out to be a very useful feature of the chosen basis.

The preceding description motivates the following definition.

Definition 6.2.4 Suppose that $T : V \to V$ is an operator, and let $v_n \in V$ be an eigenvector for T of eigenvalue c. We say that $v_1, v_2, \ldots, v_{n-1}, v_n$ is a sequence of *generalized eigenvectors* if $T(v_1) = cv_1 + v_2$, $T(v_2) = cv_2 + v_3$, $\ldots$, $T(v_{n-1}) = cv_{n-1} + v_n$, and $T(v_n) = cv_n$.

We show next that if an n-dimensional vector space has a sequence $v_1, \ldots, v_{n-1}, v_n$ of generalized eigenvectors for a linear operator T, then T is represented by a Jordan block matrix in the ordered basis given by the sequence.

Lemma 6.2.5 *Suppose that $T : V \to V$ is a linear operator on a vector space V. Suppose that $v_n \neq \vec{0}$ is an eigenvector of T of eigenvalue c and $v_1, v_2, \ldots, v_{n-1}, v_n$ is a sequence of generalized eigenvectors. Then $v_1, v_2, \ldots, v_{n-1}, v_n$ are linearly independent in V. If $dim(V) = n$ and we set $\mathcal{B}$ to be the ordered basis $\{v_1, v_2, \ldots, v_n\}$ of V, then $[T]_\mathcal{B}$ is the $n \times n$ Jordan block matrix of eigenvalue c.*

Proof: That $[T]_\mathcal{B}$ is as described follows immediately from the definition of the representing matrix of a linear transformation. Thus, we need only establish the linear independence of the sequence of generalized eigenvectors $v_1, v_2, \ldots, v_{n-1}, v_n$. To do this, we study the operators $(T - cI), (T - cI)^2, \ldots, (T - cI)^{n-1}$. First, as v_n is an eigenvector for T of eigenvalue c, we see that $(T - cI)(v_n) = \vec{0}$. Moreover, from this it follows that $(T - cI)^i(v_n) = \vec{0}$ for all $i \geq 1$. Next we compute that $(T - cI)(v_{n-1}) = v_n \neq \vec{0}$. From this $(T - cI)^2(v_{n-1}) = (T - cI)(v_n) = \vec{0}$, and consequently, $(T - cI)^i(v_{n-1}) = \vec{0}$ for all $i \geq 2$. Continuing in this fashion, we obtain that $(T - cI)^{n-k}(v_k) = v_n \neq \vec{0}$ and that $(T - cI)^j(v_k) = \vec{0}$ whenever $j > n - k$.

We now can establish the linear independence of $v_1, v_2, \ldots, v_{n-1}, v_n$. Suppose that $a_1 v_1 + a_2 v_2 + \cdots + a_n v_n = \vec{0}$. Choose k to be the least subscript for which $a_k \neq 0$. Then we compute that

$$
\begin{aligned}
\vec{0} &= (T - cI)^{n-k}(\vec{0}) \\
&= (T - cI)^{n-k}(a_k v_k + a_{k+1} v_{k+1} + \cdots + a_n v_n)
\end{aligned}
$$

$$\begin{aligned} &= a_k(T - cI)^{n-k}(v_k) + \cdots + a_n(T - cI)^{n-k}(v_n) \\ &= a_k v_n + \vec{0} + \cdots + \vec{0} = a_k v_n \end{aligned}$$

Since $v_k \neq \vec{0}$, this shows that $a_k = 0$, and the linear independence of $v_1, v_2, \ldots, v_{n-1}, v_n$ follows. This proves the lemma. //

The general question addressed in Sec. 6.4 is that of determining when an operator can be represented by a Jordan block matrix or, more generally, a matrix built from Jordan block matrices (a Jordan matrix). Such a representation will be called the *Jordan form* of the operator. It will turn out that Jordan forms will exist whenever the characteristic polynomial factors completely into linear (degree one) terms. Consequently, over any algebraically closed field, for example the complex numbers **C**, linear operators always have a Jordan form. The powerful tool of the Jordan form will be exploited in Sec. 7.2 in the study of systems of linear differential equations.

Unfortunately, not every matrix has a Jordan form. For example, the real matrix

$$\begin{pmatrix} 0 & -1 \\ 1 & 0 \end{pmatrix}$$

has characteristic polynomial $X^2 + 1$, which does not factor into linear terms (with real coefficients). This brings us to the second example of this section. While reading the next definition, note the similarity between companion matrices and Jordan block matrices. But also note the difference. Structurally they are *quite* different, as we shall see.

Definition 6.2.6 Let $f(X) = X^n + a_{n-1}X^{n-1} + \cdots + a_1 X + a_0$. The $n \times n$ matrix Q (below) is called the *companion matrix for* $f(X)$. A matrix Q is called a companion matrix if it is the companion matrix for some monic polynomial $f(X)$.

$$Q = \begin{pmatrix} 0 & 0 & \cdots & 0 & 0 & -a_0 \\ 1 & 0 & \cdots & 0 & 0 & -a_1 \\ 0 & 1 & \cdots & 0 & 0 & -a_2 \\ \vdots & \vdots & \ddots & \vdots & \vdots & \vdots \\ 0 & 0 & \cdots & 1 & 0 & -a_{n-2} \\ 0 & 0 & \cdots & 0 & 1 & -a_{n-1} \end{pmatrix}$$

Companion matrices have a simple appearance, except for the last column. The diagonal of ones just below the main diagonal has a nice interpretation. Suppose $[T]_\mathcal{B} = Q$ where Q is the above companion matrix and $\mathcal{B} = \{v_1, v_2, \ldots, v_n\}$ is some ordered basis. The diagonal of ones shows that $T(v_1) = v_2$, $T(v_2) = v_3, \ldots$, and $T(v_{n-1}) = v_n$. Also, $T(v_n) = -a_{n-1}v_n - a_{n-2}v_{n-1} - \cdots - a_0 v_1$, so this last basis element is "spread around" by T according to the various a_i which comprise the last column. We begin by putting this observation to use explicitly and compute the characteristic and minimal polynomials of a companion matrix.

Lemma 6.2.7 *If Q is the $n \times n$ companion matrix for the monic polynomial $f(X)$, then $C_Q(X) = f(X) = M_Q(X)$.*

Proof: This proof can be simplified if one uses the Cayley-Hamilton theorem (see the Remark below). However, we want to use this result in an alternative proof of the Cayley-Hamilton theorem (see 6.3.11), so we give a proof that uses only the definitions. We proceed by induction on n. If $n = 1$, the result is trivial. To evaluate $C_Q(X)$, we compute $det(XI_n - I)$ by a cofactor expansion along the first row, applying the $(n-1) \times (n-1)$ case to the first summand:

$$C_Q(X) = det \begin{pmatrix} X & 0 & \cdots & 0 & 0 & a_0 \\ -1 & X & \cdots & 0 & 0 & a_1 \\ 0 & -1 & \cdots & 0 & 0 & a_2 \\ \vdots & \vdots & \ddots & \vdots & \vdots & \vdots \\ 0 & 0 & \cdots & -1 & X & a_{n-2} \\ 0 & 0 & \cdots & 0 & -1 & X + a_{n-1} \end{pmatrix}$$

$$= X \cdot det \begin{pmatrix} X & \cdots & 0 & 0 & a_1 \\ -1 & \cdots & 0 & 0 & a_2 \\ \vdots & \ddots & \vdots & \vdots & \vdots \\ 0 & \cdots & -1 & X & a_{n-2} \\ 0 & \cdots & 0 & -1 & X + a_{n-1} \end{pmatrix}$$

$$+ a_0(-1)^{n+1} \cdot det \begin{pmatrix} -1 & X & \cdots & 0 & 0 \\ 0 & -1 & \cdots & 0 & 0 \\ \vdots & \vdots & \ddots & \vdots & \vdots \\ 0 & 0 & \cdots & -1 & X \\ 0 & 0 & \cdots & 0 & -1 \end{pmatrix}$$

$$= (X)(X^{n-1} + a_{n-1}X^{n-2} + \cdots + a_1) + a_0(-1)^{n+1}(-1)^{n-1}$$
$$= (X^n + a_{n-1}X^{n-1} + \cdots + a_1 X + a_0) = f(X)$$

as desired.

To see that $M_Q(X) = f(X)$, we first note that as Q is a companion matrix, $e_2 = Qe_1$, $e_3 = Qe_2 = Q^2e_1$, $\ldots$, $e_n = Qe_{n-1} = Q^{n-1}e_1$. In particular, $e_1, Qe_1, Q^2e_1, \ldots, Q^{n-1}e_1$ are all linearly independent. Thus, for any nonzero polynomial $g(X) = b_0 + b_1X + \cdots + b_{n-1}X^{n-1}$, we find that $g(Q)e_1 = b_0e_1 + b_1T_Qe_1 + \cdots + b_{n-1}T_Q^{n-1}e_1 \neq \vec{0}$. This shows that $g(Q)$ cannot be the zero operator; that is, no polynomial of degree less than n can annihilate Q.

Next we note that

$$
\begin{aligned}
Q^n e_1 &= Qe_{n-1} \\
&= -a_{n-1}e_{n-1} - a_{n-2}e_{n-2} - \cdots - a_0e_0 \\
&= -a_{n-1}Q^{n-1}e_1 - a_{n-2}Q^{n-2}e_1 - \cdots - a_0e_1
\end{aligned}
$$

This shows that $f(Q)e_1 = \vec{0}$. From this, $\vec{0} = Q\vec{0} = Qf(Q)e_1 = f(Q)Qe_1 = f(Q)e_2$, and likewise $\vec{0} = f(Q)e_3, \ldots, \vec{0} = f(Q)e_{n-1}$. Since $f(Q)$ is zero on a basis for $\mathbf{F}^n$, we find that $f(Q) = 0$. It follows that $f(X)$ is a monic polynomial of least degree annihilating Q. This shows that $f(X)$ is the minimal polynomial $M_Q(X)$, and this proves the lemma. //

Remark If one assumes the Cayley-Hamilton theorem, the computation in the first paragraph of the above proof can be eliminated. The second and third paragraphs show that the monic polynomial $f(X)$ is the minimal polynomial of Q. Thus, as the minimal polynomial $M_Q(X)$ must be a factor of the monic polynomial $C_Q(X)$ and $n = deg(C_Q(X)) = deg(f(X))$, we conclude that necessarily $C_Q(X) = f(X)$.

The preceding proof gives one of the critical properties of the linear operator associated with a companion matrix. If Q is a companion matrix, then e_1, $Qe_1 = e_2, \ldots$, $Q^{n-1}e_1 = e_n$ are linearly independent and hence form a basis for $\mathbf{F}^n$. We look for such vectors when we turn the problem around and try to find a basis in which a given linear operator would be represented by a companion matrix. This motivates the following definition.

Definition 6.2.8 Suppose that $T : V \to V$ is a linear operator on a finite-dimensional vector space V. We say that $v \in V$ is a *cyclic vector for T* if V is spanned by the set of vectors $\{v, T(v), T^2(v), T^3(v), \ldots\}$.

We now show that cyclic vectors give rise to companion matrix representations of linear operators. The idea behind the proof is much the same as Lemma 6.2.5. The key point is verifying that the appropriate vectors do indeed form a basis for V. The rest then follows from the definition of the matrix of an operator with respect to a basis. We point out that like the proof of Lemma 6.2.7, the proof of this lemma does not use the Cayley-Hamilton theorem.

Lemma 6.2.9 *Suppose that $T : V \to V$ has a cyclic vector v, where V is n-dimensional. Then:*
(i) $\mathcal{B} = \{v, T(v), T^2(v), \ldots, T^{n-1}(v)\}$ *is an ordered basis for V.*
(ii) *The matrix $[T]_\mathcal{B}$ is the companion matrix for $C_T(X)$, and consequently $M_T(X) = C_T(X)$.*

The basis $\mathcal{B}$ given in part (i) of the lemma is called a *cyclic* basis for V.

Proof: (i) By hypothesis, the (possibly infinite) collection of vectors $v, T(v)$, $T^2(v)$, $T^3(v)$, $\ldots$ span the vector space V. Since V is n-dimensional, the $n + 1$ vectors $v, T(v), \ldots, T^n(v)$ are linearly dependent. Thus for some $j \leq n$, $T^j(v)$ is a linear combination of $v, T(v), \ldots$, $T^{j-1}(v)$. If $T^j(v) = a_{j-1}T^{j-1}(v) + \cdots + a_1 T(v) + a_0 v$, we find $T^{j+1}(v) = T(T^j(v)) = a_{j-1}T_j(v) + \cdots + a_1 T^2(v) + a_0 T(v)$. Consequently, as $T^j(v)$ is a linear combination of $T^{j-1}(v), \ldots, T(v), v$, we find that $T^{j+1}(v)$ is also a linear combination of $T^{j-1}(v), \ldots, T(v), v$. Iterating this argument, we find that $T^k(v)$ is a linear combination of $T^{j-1}(v), \ldots$, $T(v), v$ for all $k \geq j$. This shows that $V = span\{v, T(v), T^2(v), \ldots\} = span\{v, T(v), \ldots, T^{j-1}(v)\}$. Since V is n-dimensional, we conclude that $j \geq n$. As $j \leq n$, we obtain that $j = n$, and consequently the n vectors $\{v, T(v), \ldots, T^{n-1}(v)\}$ must be a basis for V. This proves (i).

(ii) We determine $[T]_\mathcal{B}$ by directly computing the action of T on the basis elements in $\mathcal{B}$: $T(v) = T(v) \in \mathcal{B}$, $T(T(v)) = T^2(v) \in \mathcal{B}$, $T(T^2(v)) = T^3(v) \in \mathcal{B}, \ldots, T(T^{n-2}(v)) = T^{n-1}(v) \in \mathcal{B}$, and finally $T(T^{n-1}(v)) = T^n(v) = -a_{n-1}T^{n-1}(v) - \cdots - a_1 T(v) - a_0 v$, expressed as a linear combination of elements of $\mathcal{B}$. Since the columns of $[T]_\mathcal{B}$ are given by the $\mathcal{B}$ coordinates of these images, we see that $[T]_\mathcal{B}$ is the companion matrix for $C_T(X)$. We conclude that $M_T(X) = C_T(X)$, since this is true for companion matrices by Lemma 6.2.7. This proves the result. //

We conclude this section by noting that there is one special kind of $n \times n$ matrix that is both a Jordan block matrix and a companion matrix. This matrix looks like

$$N = \begin{pmatrix} 0 & 0 & \cdots & 0 & 0 \\ 1 & 0 & \cdots & 0 & 0 \\ 0 & 1 & \cdots & 0 & 0 \\ \vdots & \vdots & \ddots & \vdots & \vdots \\ 0 & 0 & \cdots & 1 & 0 \end{pmatrix}$$

This $n \times n$ matrix is a Jordan block matrix of eigenvalue 0 and is also the companion matrix for the polynomial X^n. It follows from both the results of this section that this matrix has X^n as its characteristic and minimal polynomial. This particular kind of matrix will be useful in establishing the existence of the Jordan form (see Sec. 6.4). A matrix N with the property that $N^k = 0$ for some power k is called *nilpotent*.

PROBLEMS

6.2.1. Find a 5×5 rank 3 matrix N for which $N^3 = 0$ but $N^2 \neq 0$. Can such a matrix have rank 4?

6.2.2. Suppose that $T_A : \mathbf{F}^n \to \mathbf{F}^n$ and $T_B : \mathbf{F}^n \to \mathbf{F}^n$ both have a cyclic vector. Show that the $n \times n$ matrices A and B are similar if and only if $C_A(X) = C_B(X)$.

6.2.3. Show that the real matrices

$$A = \begin{pmatrix} 0 & 0 & 0 \\ 3 & 0 & 0 \\ 1 & 4 & 0 \end{pmatrix} \quad \text{and} \quad B = \begin{pmatrix} 0 & 1 & 0 \\ 0 & 0 & 1 \\ 0 & 0 & 0 \end{pmatrix}$$

are similar.

6.2.4. Which of the following real matrices are similar to another?

$$\begin{pmatrix} 2 & 0 & 0 \\ 1 & 2 & 0 \\ 0 & 1 & 2 \end{pmatrix}, \quad \begin{pmatrix} 2 & 1 & 0 \\ 0 & 2 & 0 \\ 0 & 0 & 2 \end{pmatrix}, \quad \begin{pmatrix} 2 & 0 & 0 \\ 1 & 2 & 0 \\ 1 & 0 & 2 \end{pmatrix}$$

$$\begin{pmatrix} 2 & 0 & 0 \\ 1 & 2 & 0 \\ 1 & 1 & 2 \end{pmatrix}, \quad \begin{pmatrix} 2 & 0 & 0 \\ 3 & 2 & 0 \\ 5 & 7 & 2 \end{pmatrix}$$

6.2.5. Find a sequence of generalized eigenvectors of maximal length for

$$M = \begin{pmatrix} 2 & 2 & 3 \\ 0 & 2 & 1 \\ 0 & 0 & 2 \end{pmatrix}$$

6.2.6. Find two sequences of generalized eigenvectors of maximal length with distinct eigenvalues for

$$Q = \begin{pmatrix} 3 & 1 & 0 & 2 \\ 0 & 3 & 1 & 6 \\ 0 & 0 & 2 & 1 \\ 0 & 0 & 0 & 2 \end{pmatrix}$$

6.2.7. If $T_A : \mathbf{F}^3 \to \mathbf{F}^3$ is given by

$$A = \begin{pmatrix} 0 & 1 & 1 \\ 2 & 3 & 0 \\ 2 & 0 & 1 \end{pmatrix}$$

find a cyclic basis for T_A.

6.2.8. Determine if the following matrices have cyclic vectors.

(a) $\begin{pmatrix} 0 & 1 & 0 & 1 \\ 1 & 0 & 1 & 0 \\ 0 & 1 & 0 & 1 \\ 1 & 0 & 1 & 0 \end{pmatrix}$ (b) $\begin{pmatrix} 1 & 0 & 1 \\ 2 & 3 & 1 \\ 1 & 0 & 1 \end{pmatrix}$ (c) $\begin{pmatrix} 1 & 0 & 0 \\ 0 & 1 & 0 \\ 0 & 0 & 2 \end{pmatrix}$

6.2.9. When does a diagonalizable operator $T : V \to V$ have a cyclic basis?

6.2.10. Which companion matrix is similar to the Jordan block matrix

$$J = \begin{pmatrix} 2 & 0 & 0 \\ 1 & 2 & 0 \\ 0 & 1 & 2 \end{pmatrix}$$

6.2.11. Let

$$A = \begin{pmatrix} 1 & 0 \\ 1 & 1 \end{pmatrix} \quad \text{and} \quad B = \begin{pmatrix} 0 & -1 \\ 1 & 2 \end{pmatrix}$$

Find an invertible matrix C such that $A = C^{-1}BC$.

6.3 Direct-Sum Decompositions, Invariant Subspaces, and Quotient Spaces

In this section we study direct-sum decompositions of a vector space and use them to study linear operators. We have two tasks. First, we need to formulate the definition of multiple direct sums. This definition generalizes the definition of the direct sum of two vector spaces given in Definition 4.3.4; but some caution is necessary. For emphasis, we will recall the earlier results on direct sums of two vector spaces as we proceed.

Our second task is, given a linear operator, to define and study invariant direct sums. The point is that for an invariant direct sum, the given operator induces (via restriction) operators on the summands. Thus one reduces the problem of understanding the original operator to the problem of understanding restriction operators on smaller vector spaces. In terms of matrices, invariant direct-sum decompositions correspond to block-diagonal representing matrices for the operator.

The definition of direct sums, like the definition of a basis, has two parts. First there is a linear independence criterion, and second there is a spanning criterion. The definitions, in fact, are closely related. In a basis one has a collection of spanning vectors that point in independent directions. In a direct-sum decomposition, one has a collection of spanning subspaces with independent directions. For example, if one replaces each element in a basis by the the one-dimensional space it spans, a direct-sum decomposition is obtained (see Example 6.3.2). In this example one is visualizing the vector space as a direct sum of coordinate axes. Of course in general, summands in a direct sum have arbitrary dimension. We will establish in Corollary 6.3.5 that their dimensions sum to the dimension of the vector space.

Definition 6.3.1 Suppose that $W_1, W_2, \ldots, W_n$ are subspaces of a vector space V.

(i) $W_1, W_2, \ldots, W_n$ are called *independent* if whenever $w_1 \in W_1$, $w_2 \in W_2, \ldots, w_n \in W_n$, and $w_1 + w_2 + \cdots + w_n = \vec{0}$, then $w_1 = \vec{0}$, $w_2 = \vec{0}, \ldots, w_n = \vec{0}$.

(ii) $V = W_1 + W_2 + \cdots + W_n$ if every $v \in V$ can be expressed as $v = w_1 + w_2 + \cdots + w_n$ where $w_1 \in W_1, w_2 \in W_2, \ldots, w_n \in W_n$.

(iii) We say that V is the *direct sum* of $W_1, W_2, \ldots, W_n$, denoted $V = W_1 \oplus W_2 \oplus \cdots \oplus W_n$ if $W_1, W_2, \ldots, W_n$ are independent and

$$V = W_1 + W_2 + \cdots + W_n.$$

Whenever $W_1, W_2, \ldots, W_n$ are nonzero subspaces of V, the independence of $W_1, W_2, \ldots, W_n$ is equivalent to the following:

(i') If $w_1 \in W_1$, $w_2 \in W_2, \ldots$, $w_n \in W_n$ are each nonzero, then w_1, $w_2, \ldots, w_n$ are linearly independent.

To see this equivalence, we see easily (i) is a consequence of (i'). Conversely, assuming (i), suppose that $k_1 w_1 + k_2 w_2 + \cdots + k_n w_n = \vec{0}$ for $w_i \neq \vec{0} \in W_i$. Then $k_i w_i \in W_i$, so by (i) we have $k_i w_i = \vec{0}$ for all i. As each $w_i \neq \vec{0}$, we conclude that $k_i = 0$; this gives (i').

It is worthwhile to isolate the meaning of the direct sum in the case where n = 2. In this case we obtain Definition 4.3.4.

Definition 6.3.1' Suppose that W_1, W_2 are subspaces of V. Then $V = W_1 \oplus W_2$ if
(i) $W_1 \cap W_2 = \{\vec{0}\}$.
(ii) $V = W_1 + W_2$; that is, every $v \in V$ can be expressed as $v = w_1 + w_2$ where $w_1 \in W_1$ and $w_2 \in W_2$.

The first condition of 6.3.1' just given follows easily from the first condition of the original Definition 6.3.1. If $w \in W_1 \cap W_2$, then $w - w = \vec{0}$ gives $w = \vec{0}$. Conversely, if $W_1 \cap W_2 = \{\vec{0}\}$, $w_1 \in W_1$, and $w_2 \in W_2$, then $w_1 + w_2 = \vec{0}$ gives that $w_1 = -w_2 \in W_1 \cap W_2$, so $w_1 = -w_2 = \vec{0}$ follows. Note however, that the first condition given in 6.3.1' cannot be directly generalized to describe the general case of independent subspaces. For example, if one sets

$$W_1 = span\left\{\begin{pmatrix} 1 \\ 0 \end{pmatrix}\right\}, \ W_2 = span\left\{\begin{pmatrix} 0 \\ 1 \end{pmatrix}\right\}, \ W_3 = span\left\{\begin{pmatrix} 1 \\ 1 \end{pmatrix}\right\} \subseteq \mathbf{F}^2$$

then easily $W_1 \cap W_2 = W_1 \cap W_3 = W_2 \cap W_3 = \{\vec{0}\}$. However, clearly W_1, W_2, W_3 are not independent subspaces of $\mathbf{F}^2$.

Examples 6.3.2 Suppose that $v_1, v_2, \ldots, v_{2n}$ form a basis for a vector space V.

(i) Set $W_i = span\{v_i\}$ for $i = 1, 2, \ldots, 2n$. Then one has that $V = W_1 \oplus W_2 \oplus \cdots \oplus W_{2n}$.

(ii) Set $U_j = span\{v_{2j-1}, v_{2j}\}$ for $j = 1, 2, \ldots, n$. Then $V = U_1 \oplus U_2 \oplus \cdots \oplus U_n$ is easily checked.

(iii)   Let $V_1, V_2, \ldots, V_s$ be arbitrary $\mathbf{F}$ vector spaces. Form a new vector space V from these vector spaces as follows: The elements of V are n-tuples $(v_1, v_2, \ldots, v_s)$ where each $v_i \in V_i$. (In other words V is the cartesian product $V = V_1 \times V_2 \times \cdots \times V_s$.) Addition and scalar multiplication in V is defined entrywise (exactly as in $\mathbf{F}^n$). We identify each vector space V_i with the subspace $\{(0, \ldots, 0, v_i, 0, \ldots, 0) \in V \mid v_i \in V_i\}$ consisting of all n-tuples which are zero in all except for the ith entry. It is easily checked that with this identification $V = V_1 \oplus V_2 \oplus \cdots \oplus V_s$. Often the compact notation $V = \bigoplus_{i=1}^{s} V_i$ is used to denote this construction of V.

The $n = 2$ case of Definition 6.3.1, reformulated as in Definition 6.3.1', provides an inductive base for the study of general direct-sum decompositions. This is the content of the next result, which is usually referred to as the *associativity* of the direct sum.

Lemma 6.3.3   (i) *Suppose that $W_1, W_2, \ldots, W_n, W$ are subspaces of the vector space V with $W_1, W_2, \ldots, W_{n-1} \subseteq W$. If*

$$W = W_1 \oplus W_2 \oplus \cdots \oplus W_{n-1}$$

and $V = W \oplus W_n$, then

$$V = W_1 \oplus W_2 \oplus \cdots \oplus W_n$$

In particular, the assertion $V = W_1 \oplus W_2 \oplus \cdots \oplus W_n$ is the same as saying that $V = ((\cdots (W_1 \oplus W_2) \oplus \cdots \oplus W_{n-1}) \oplus W_n)$.
(ii) *Suppose that $V = W_1 + W_2 + \cdots + W_n$. If for each $i = 2, 3, \ldots, n$, $W_i \cap (W_1 + W_2 + \cdots + W_{i-1}) = \{\vec{0}\}$, then $V = W_1 \oplus W_2 \oplus \cdots \oplus W_n$.*

Proof: (i)   That $V = W_1 + W_2 + \cdots + W_n$ is clear. Now suppose that $w_1 \in W_1, \ldots, w_n \in W_n$, and $w_1 + w_2 + \cdots + w_{n-1} + w_n = \vec{0}$. Since $V = W \oplus W_n$, we conclude that $w_1 + w_2 + \cdots + w_{n-1} = \vec{0}$ and $w_n = \vec{0}$. As $W_1, W_2, \ldots, W_{n-1}$ are independent, we see that $w_1 = \vec{0}, \ldots, w_{n-1} = \vec{0}$ as well; this gives the independence of $W_1, W_2, \ldots, W_n$. This shows that $V = W_1 \oplus W_2 \oplus \ldots \oplus W_n$. Part (ii) now follows from (i) by induction, using the description of the $n = 2$ case of the direct sum as given in Definition 6.3.1'. //

The next two results show that direct-sum decompositions always arise in (essentially) the same manner as did the examples of 6.3.2.

Lemma 6.3.4 is precisely Lemma 4.3.6, which we restate without proof. The corollary which follows is the generalization of this result.

Lemma 6.3.4 *Suppose that $V = W_1 \oplus W_2$. Then*
(i)   *Every $v \in V$ can be expressed uniquely as $v = w_1 + w_2$ where $w_1 \in W_1$ and $w_2 \in W_2$.*
(ii) *If $\{v_1, \ldots, v_r\}$ is a basis for W_1 and $\{u_1, \ldots, u_s\}$ is a basis for W_2, then $\{w_1, \ldots, w_r, u_1, \ldots, u_s\}$ is a basis for V. In particular, $dim(V) = dim(W_1) + dim(W_2)$.*

Applying Lemma 6.3.3 inductively we obtain the following.

Corollary 6.3.5 *Suppose that $V = W_1 \oplus W_2 \oplus \cdots \oplus W_n$. Then*
(i)   *Every $v \in V$ can be expressed uniquely as $v = w_1 + w_2 + \cdots + w_n$ where $w_1 \in W_1$, $w_2 \in W_2, \ldots, w_n \in W_n$.*
(ii) *$dim(V) = dim(W_1) + dim(W_2) + \cdots + dim(W_n)$; in particular the union of a collection of bases, one for each W_i, is a basis of V.*
(iii) *Suppose that $W_1, W_2, \ldots, W_n$ are independent subspaces of V. If $dim(V) = dim(W_1) + dim(W_2) + \cdots + dim(W_n)$, then $V = W_1 \oplus W_2 \oplus \cdots \oplus W_n$.*

Proof: (i) We use Lemma 6.3.3 and proceed by induction on n. When $n = 2$, the result is Lemma 6.3.4. By Lemma 6.3.3 we write $V = (W_1 \oplus \cdots \oplus W_{n-1}) \oplus W_n$, and we abbreviate $W = W_1 \oplus \cdots \oplus W_{n-1}$. For $v \in V$ we express uniquely $v = w + w_n$ where $w \in W$ and $w_n \in W_n$. By induction, we express uniquely $w = w_1 + \cdots + w_{n-1}$ where $w_i \in W_i$. This gives $v = w_1 + \cdots + w_{n-1} + w_n$, where the w_i are each uniquely determined. This proves (i).

(ii) Consider a collection of bases, one for each subspace W_i. By induction, the union of this collection for the $n - 1$ subspaces $W_1, \ldots, W_{n-1}$ is a basis for W. By Lemma 6.3.4 (ii), the union of the total collection must be a basis of V. This establishes (ii).

(iii) Let U be the subspace of V spanned by the W_i; that is, $U = W_1 + W_2 + \cdots + W_n \subseteq V$. By the independence assumption we conclude that $U = W_1 \oplus W_2 \oplus \cdots \oplus W_n$. Consequently, part (ii) of this corollary applies to give that $dim(U) = dim(W_1) + dim(W_2) + \cdots + dim(W_n) = dim(V)$. As $U \subseteq V$, it follows that $V = W$; this proves what was required. //

We now turn to the subject of invariant direct sums, which is crucial to the study of the structure of linear operators. As remarked earlier, invariant direct sums correspond to block-diagonal decompositions of a representing matrix for the operator. The remainder of this section will be devoted to developing these points in detail. First, we give the definition of invariant subspaces.

Definition 6.3.6 Suppose that $T : V \to V$ is a linear operator and W is a subspace of V. W is called a *T-invariant subspace of V* (or simply *T-invariant*) if for all $w \in W$, $T(w) \in W$.

In order to tell if a given subspace W of V is T-invariant, it suffices to show for some basis B for W (or spanning set of W) that whenever $v \in B$, then $T(v) \in W$. This is because any element $w \in W$ can be expressed as $w = a_1 v_1 + a_2 v_2 + \cdots + a_n v_n$ with $a_i \in F$ and $v_i \in B$, and then $T(v) = a_1 T(v_1) + a_2 T(v_2) + \cdots + a_n T(v_n) \in W$ since each $T(v_i) \in W$. This easy observation simplifies testing for the T-invariance of a subspace.

Observe that if $T : V \to V$ is a linear operator, then the subspaces $\{\vec{0}\}$, $V \subseteq V$ are always T-invariant (for trivial reasons). A main task in the study of an operator T is to determine nontrivial T-invariant subspaces. One process for obtaining T-invariant subspaces, is to expand upon the idea of a cyclic vector given in Definition 6.2.8. For any vector $v \in V$, consider the subspace of V spanned by $\{v, T(v), T^2(v), T^3(v), \ldots\}$. [Subsequently this subspace will be denoted by $Z(T, v)$ and called the T-cyclic subspace of V generated by v.] Since $T(T^i(v)) = T^{i+1}(v)$ again lies in the spanning set, we see that this subspace is T-invariant. The only problem with this process is that given an arbitrary vector v, it is nearly impossible to predict how large a subspace it will generate. Often it will be the entire vector space V. We shall study this in much greater detail in Sec. 6.6.

As another example, we consider the vector space $V = C^\infty(\mathbf{R}, \mathbf{R})$ of infinitely differentiable functions on $\mathbf{R}$. Let $D : V \to V$ be the differentiation linear operator. Then, since the derivative of a polynomial function is also a polynomial function, the subspace $P \subseteq V$ of all polynomial functions is D-invariant. More generally, consider a polynomial applied to D, say $T = a_n D^n + a_{n-1} D^{n-1} + \cdots + a_0 I$. (Such T is the linear operator associated with a linear differential equation.) P is also a T-invariant subspace of V. Although not usually described in this language, this observation is useful in the study of such differential equations. One often approximates functions by polynomials and then studies the behavior of the differential operator on these functions. In effect, one is restricting attention to an invariant subspace, in order later to pull the information obtained back to V.

For a matrix example, consider

$$A = \begin{pmatrix} 2 & 0 & 1 & 0 & 1 \\ 0 & 5 & 0 & 7 & 0 \\ 1 & 0 & 3 & 0 & 1 \\ 0 & 8 & 0 & 5 & 0 \\ 2 & 0 & 2 & 0 & 6 \end{pmatrix}$$

Then, it is easy to see that $\mathbf{F}^5$ has as T_A-invariant subspaces $W_1 = span\{e_1, e_3, e_5\}$ and $W_2 = span\{e_2, e_4\}$. By inspecting A one can see that $T(e_1)$, $T(e_3)$, and $T(e_5)$ are linear combinations of e_1, e_3, and e_5 and that $T(e_2)$ and $T(e_4)$ are linear combinations of e_2 and e_4. Further, one can compute that $T_A(e_1) = (2, 0, 1, 0, 2)$ and $T_A(2, 0, 1, 0, 2) = (6, 0, 6, 0, 12) = -6e_1 + 6(2, 0, 1, 0, 2)$. It follows that

$$W_3 = span\{(1, 0, 0, 0, 0), (2, 0, 1, 0, 2)\}$$

is also a T-invariant (and T-cyclic) subspace of $\mathbf{F}^5$. (In fact, $W_3 \subseteq W_1$.)

We consider the ordered basis $\mathcal{B} = \{e_1, e_3, e_5, e_2, e_4\}$ obtained from the standard basis by reordering. Then

$$[T_A]_{\mathcal{B}} = \begin{pmatrix} 2 & 1 & 1 & 0 & 0 \\ 1 & 3 & 1 & 0 & 0 \\ 2 & 2 & 2 & 0 & 0 \\ 0 & 0 & 0 & 5 & 7 \\ 0 & 0 & 0 & 8 & 5 \end{pmatrix}$$

which is a *block-diagonal matrix*. The operator T_A (and hence the matrix A) can now be studied by looking at operators on the smaller vector spaces W_1 and W_2. Note that $V = W_1 \oplus W_2$ and that each of W_1, W_2 are T-invariant, giving a so-called T-invariant direct-sum decomposition of $\mathbf{F}^5$. The next definition explains which operators on W_1 and W_2 we are referring to.

Definition 6.3.7 Suppose that $T : V \to V$ is a linear operator and W is a T-invariant subspace of V. The *restriction of T to W* is the linear operator $T|_W : W \to W$ defined by $T|_W(w) = T(w)$ for all $w \in W$.

In other words, the restriction $T|_W$ is exactly the same as T, except we now view the domain and range to be the subspace W (instead of V). It is important to note that if W was not T-invariant, then the restriction of T to W would not make sense as an operator. We need to be sure that whenever $v \in W$, the image $T|_W(v) = T(v)$ is also an element of W. (For an arbitrary subspace $W \subseteq V$ we can consider a restriction linear transformation $T|_W \colon W \to V$, but this has little use for us in this chapter where we need to work with operators.)

In the above example, one can easily understand the the restrictions of T_A to W_1 and W_2 by the matrix representation $[T_A]_B$. The upper left block describes how T_A acts on W_1 in terms of the ordered basis $\mathcal{B}_1 = \{e_1, e_3, e_5\}$ of W_1. The columns in this block give the $\mathcal{B}_1$ coordinates of $T_A(e_1)$, $T_A(e_3)$, and $T_A(e_5)$ respectively. In other words

$$[T_A|_{W_1}]_{\mathcal{B}_1} = \begin{pmatrix} 2 & 1 & 1 \\ 1 & 3 & 1 \\ 2 & 2 & 2 \end{pmatrix}$$

Similarly,

$$[T_A|_{W_2}]_{\mathcal{B}_2} = \begin{pmatrix} 5 & 7 \\ 8 & 5 \end{pmatrix}$$

where $\mathcal{B}_2$ is the ordered basis $\{e_2, e_4\}$ of W_2.

Although the idea should be clear from the previous discussions, we formally give the definition of a block-diagonal matrix. Recall that the definition of a partitioned matrix was given in Definition 0.3.9.

Definition 6.3.8 An $n \times n$ *block-diagonal matrix* A is a partitioned matrix $A = (A_{ij})$ for which

(i)   Each A_{ii} is square.

(ii) If $i \neq j$, A_{ij} is a matrix with all entries 0.

When $A = (A_{ij})$ is block-diagonal, we use the notation $A = A_{11} \oplus A_{22} \oplus \cdots \oplus A_{tt}$ (also called the *direct sum* of A_{11}, A_{22}, ..., A_{tt}).

The next result explains why invariant direct sum decompositons are important. It generalizes what we saw in the above example and shows that when there is an invariant direct sum decomposition, the operator can be represented by a block-diagonal matrix. In other words, the operator can be viewed as the (direct) sum of operators, one on each subspace of the decomposition. For simplicity we give the result only for direct sums of two vector spaces. The result general-izes (by an easy induction) in the obvious manner to invariant direct sums of more than two vector spaces; the precise statement of this generalization is omitted.

Lemma 6.3.9 *Suppose that $T : V \to V$ is a linear operator and $V = W_1 \oplus W_2$ where W_1 and W_2 are T-invariant subspaces of V. Let $B_1 = \{v_1, \ldots, v_r\}$ be an ordered basis for W_1 and $B_2 = \{u_1, \ldots, u_s\}$ be an ordered basis for W_2. Set $B = \{v_1, \ldots, v_r, u_1, \ldots, u_s\}$, an ordered basis for V. If $[T|_{W_1}]_{B_1} = A_1$ and $[T|_{W_2}]_{B_2} = A_2$ then*

$$[T]_B = \begin{pmatrix} A_1 & 0 \\ 0 & A_2 \end{pmatrix} = A_1 \oplus A_2$$

Proof: Suppose that $A_1 = (a_{ij})$. This means that $T|_W(v_j) = a_{1j}v_1 + \cdots + a_{rj}v_r$. Hence $T(v_j) = a_{1j}v_1 + \cdots + a_{rj}v_r + 0u_1 + \cdots + 0u_s$. This shows that the jth column of $[T]_B$ is precisely the jth column of $A_1 \oplus A_2$ where $j \leq r$. When $j > r$, the result is analogous. This gives the lemma. //

Examples 6.3.10 (i) Suppose that $T : V \to V$ is a linear operator, and assume that $B = \{v_1, \ldots, v_n\}$ is a basis of eigenvectors for T. Such T is diagonalizable, and if the eigenvalue of v_i is c_i, we know that

$$[T]_B = \begin{pmatrix} c_1 & 0 & \cdots & 0 \\ 0 & c_2 & \cdots & 0 \\ \vdots & \vdots & \ddots & \vdots \\ 0 & 0 & \cdots & c_n \end{pmatrix}$$

In this case we have decomposed $V = W_1 \oplus \cdots \oplus W_n$ where each $W_i = span\{v_i\}$ is a one-dimensional T-invariant subspace of V. The matrix $[T]_B$ can be viewed as a block-diagonal matrix with n 1×1 blocks along the diagonal.

(ii) Consider the operator $T_D : \mathbf{R}^3 \to \mathbf{R}^3$ where

$$D = \begin{pmatrix} 1 & 2 & 3 \\ 2 & 1 & 3 \\ 0 & 0 & 6 \end{pmatrix}$$

Then $T(e_1) = e_1 + 2e_2$, $T(e_2) = 2e_1 + e_2$, and $T(e_3) = 3e_1 + 3e_2 + 6e_3$. Setting $W_1 = span\{e_1, e_2\}$, we see that as $T(e_1), T(e_2) \in W_1$, W_1 is a

T-invariant subspace of $\mathbf{F}^3$. Note that $T(e_3) \notin span\{e_3\}$, so $span\{e_3\}$ is *not* a T-invariant subspace of $\mathbf{R}^3$. However $T(e_1 + e_2 + e_3) = 6e_1 + 6e_2 + 6e_3 = 6(e_1+e_2+e_3)$. Thus $W_2 = span\{e_1+e_2+e_3\}$ is a T-invariant subspace of $\mathbf{R}^3$. Set $\mathcal{B} = \{e_1, e_2, e_1 + e_2 + e_3\}$. Evidently, $\mathcal{B}$ is a basis of $\mathbf{R}^3$, and $\mathbf{R}^3 = W_1 \oplus W_2$ is a T-invariant direct sum decomposition of $\mathbf{R}^3$. The matrix for T_D in the basis $\mathcal{B}$ is the block-diagonal matrix

$$[T_D]_\mathcal{B} = \begin{pmatrix} 1 & 2 & 0 \\ 2 & 1 & 0 \\ 0 & 0 & 6 \end{pmatrix}$$

In the latter example we came up with an invariant direct-sum decomposition for an operator, but we gave no indication of how to find this decomposition. In general this is a difficult problem. A key is the relationship between the minimal polynomial and the characteristic polynomial. This is the subject of the remainder of this chapter.

We are now ready to give the alternative proof of the Cayley-Hamilton theorem promised earlier. This proof is a nice application of the concept of an invariant subspace and relies upon the relationship between invariant subspaces and the associated matrix representations of an operator. In contrast to the earlier proof, which was a mysterious "brute-force" matrix computation, this proof utilizes the function theoretic viewpoint of an operator. For this reason it gives a better illustration of why the Cayley-Hamilton theorem is true.

Theorem 6.3.11 (Cayley-Hamilton Theorem) *For any linear operator $T : V \to V$ on a finite-dimensional vector space V, $C_T(T) = 0$.*

Proof: We use Lemmas 6.2.7 and 6.2.9, whose proofs did not depend upon the Cayley-Hamilton theorem! Consider any $v \in V$. We let W be the subspace of V spanned by v, $T(v)$, $T^2(v), \ldots$. Then, as observed earlier, W is a T-invariant subspace of V with cyclic vector v. If $dim(W) = r$, then by Lemma 6.2.9 (i), W has the basis $\mathcal{B}_1 = \{v, T(v), \ldots, T^{r-1}(v)\}$. By Lemma 6.2.9 (ii), $[T|_W]_{\mathcal{B}_1}$ is the companion matrix for some monic polynomial $f(X)$. By Lemma 6.2.7 we find that $C_{T|_W}(X) = f(X)$ and $M_{T|_W}(X) = f(X)$. In particular, $C_{T|_W}(T)(v) = \vec{0}$.

To conclude the proof, we apply the basis-extension theorem to extend $\mathcal{B}_1$ to an ordered basis $\mathcal{B}$ for all of V. When this is done, it follows that the matrix $[T]_{\mathcal{B}}$ has the block-triangular form

$$\begin{pmatrix} K & R \\ 0 & S \end{pmatrix}$$

where K is the companion matrix $[T|_W]_{\mathcal{B}_1}$. Since this matrix is block triangular, by applying Theorem 2.2.10 it follows that its characteristic polynomial is $C_K(X)C_S(X) = C_T(X)$. We have already shown that $C_K(T)(v) = \vec{0}$, so it follows that $C_T(T)(v) = \vec{0}$. Since $v \in V$ was an arbitrary vector, we have $C_T(T)(v) = v$ for all $v \in V$; this proves the Cayley-Hamilton theorem. //

We conclude this section with another use of T-invariant subspaces, the operator induced on quotient spaces. This construction is extremely important for the proofs in Sec. 6.9, and the reading of this section could be postponed until that time. However, these results also serve to unify a number of the ideas developed earlier in this section. The definition of the quotient space is given next.

Definition 6.3.12 Let V be a vector space and W a subspace of V. The quotient space V/W is defined to be the following vector space: The elements of V/W are subsets of the form $(v + W) \subseteq V$, where $v \in V$. Such subsets are called *cosets* of W in V. Addition of cosets is given by the formula $(v + W) + (v' + W) = (v + v' + W)$, and scalar multiplication is given by $k(v + W) = (kv + W)$ for all $k \in \mathbf{F}$.

Note that $(v + W) = (v' + W)$ if and only if $v - v' \in W$. This is true because whenever $v - v' \in W$, any $u = v + w \in (v + W)$ can be expressed as $u = v' + (v - v') + w \in (v' + W)$, so $(v + W) = (v' + W)$ follows. Conversely, if $(v + W) = (v' + W)$, then $v \in (v' + W)$, so writing $v = v' + w$ with $w \in W$ shows that $v - v' \in W$, as required. With this observation, it is straightforward to verify all the axioms of a vector space for V/W. The details will be omitted. Observe that the zero vector in V/W is the coset $(\vec{0} + W) = W$.

Whenever studying quotient spaces, it is very convenient to consider a projection linear transformation. This is the function $\pi : V \to V/W$ defined by $\pi(v) = (v + W)$ for all $v \in W$. Evidently, π is a linear transformation, which is surjective, and $\pi(v) = \vec{0} = W \in V/W$ if and only if $v \in W$ (that is, $ker(\pi) = W$). A consequence of this observation is the following.

Corollary 6.3.13 (i) *If V is an n-dimensional vector space and W is a r-dimensional subspace of V, then V/W is an $(n-r)$-dimensional vector space.*

(ii) *Suppose that $L : V \to W$ is a surjective linear transformation. Then the function $\overline{L} : V/(ker(L)) \to W$ defined by $\overline{L}(v + ker(L)) = L(v)$ is a well-defined vector space isomorphism.*

Proof: (i) The linear transformation $\pi : V \to V/W$ is surjective with kernel W. The desired result now follows from the rank plus nullity theorem (Theorem 3.2.4).

(ii) We first show that $\overline{L}(v + ker(L))$ does not depend upon the choice of the element v which represents the coset $v + ker(L)$. In other words, we must see that in case $(v + ker(L)) = (v' + ker(L))$, then the images $\overline{L}(v + ker(L)) = L(v)$ and $\overline{L}(v' + ker(L)) = L(v')$ are the same. But, in this situation, $(v + ker(L)) = (v' + ker(L))$ gives that $v - v' \in ker(L)$ so that $L(v) = L(v')$. This shows that $\overline{L}$ is a well-defined function.

It is readily checked that $\overline{L}$ is a linear transformation. Necessarily $\overline{L}$ is surjective as L is. Finally, if $\overline{L}(v + ker(L)) = \vec{0}$, then $L(v) = \vec{0}$; that is, $v \in ker(L)$. This shows that $ker(\overline{L}) = \{(ker(L))\} = \{\vec{0}\} \subseteq V/ker(L)$. Hence, $\overline{L}$ is injective and the corollary is proved. //

The important connection between T-invariant subspaces and quotient spaces is given in the following lemma.

Lemma 6.3.14 *Suppose that W is a T-invariant subspace of V. Then $\overline{T} : V/W \to V/W$ defined by $\overline{T}(v + W) = (T(v) + W)$ is a well-defined linear operator on V/W.*

Proof: First we show that $\overline{T}$ is well-defined. We assume that $v + W = v' + W$. Then $v - v' \in W$. Since W is T-invariant, we know that $T(v - v') = T(v) - T(v') \in W$ as well. Hence, $(T(v) + W) = (T(v') + W)$; this gives the well-definition of the function $\overline{T}$.

To see that $\overline{T}$ is linear we note that $\overline{T}((v + W) + (v' + W)) = \overline{T}(v + v' + W) = (T(v + v') + W) = (T(v) + T(v') + W) = (T(v) + W) + (T(v') + W) = \overline{T}((v + W)) + \overline{T}((v' + W))$ and $k\overline{T}(v + W) = k(T(v) + W) = (kT(v) + kW) = (T(kv) + W) = \overline{T}((kv + W)) = \overline{T}(k(v + W))$, as required. This gives the linearity of $\overline{T}$ and proves the lemma. //

Example 6.3.15 Let $T_D : \mathbf{R}^3 \to \mathbf{R}^3$ be the linear transformation considered in Example 6.3.10 (ii) where

$$D = \begin{pmatrix} 1 & 2 & 3 \\ 2 & 1 & 3 \\ 0 & 0 & 6 \end{pmatrix}$$

Recall that $\mathbf{R}^3$ had a T_D-invariant decomposition $\mathbf{R}^3 = W_1 \oplus W_2$ where W_1 has as a basis $\mathcal{B}_1 = \{e_1, e_2\}$ and W_2 has as a basis $\mathcal{B}_2 = \{e_1 + e_2 + e_3\}$. Consider the quotient space V/W_2. By Corollary 6.3.13, we see that V/W_2 is a two-dimensional vector space.

Clearly $(e_1 + W_2)$ and $(e_2 + W_2)$ span V/W_2, so they form a basis which we label $\mathcal{B}_1' = \{(e_1 + W_2), (e_2 + W_2)\}$. Let $\pi : V \to V/W_2$ be the projection linear transformation. Since $\pi(e_1) = (e_1 + W_2)$ and $\pi(e_2) = (e_2 + W_2)$, it follows that the restriction of π to W_1, $\pi|_{W_1} : W_1 \to V/W_2$ is an isomorphism. Thus in a natural way, V/W_2 can be identified with W_1.

This identification can be carried one step farther. The two operators $T_D|_{W_1} : W_1 \to W_1$ and $\overline{T_D} : V/W_2 \to V/W_2$ actually coincide when W_1 and V/W_2 are identified. This follows from the fact that whenever $v, v' \in W_1$, $T_D(v) = v'$ if and only if $\overline{T_D}((v + W_2)) = (v' + W_2)$. Another way to look at this equivalence is to see that the representing matrix

$$[\overline{T_D}]_{\mathcal{B}'} = \begin{pmatrix} 2 & 1 \\ 1 & 2 \end{pmatrix}$$

which is precisely $[T_D]_{\mathcal{B}_1}$.

The type of identification exhibited in this example will be useful later in the study of decompositions of linear operators.

PROBLEMS

6.3.1. Suppose that V is a finite-dimensional vector space and $W_1 \subseteq V$ is a subspace. Prove that there exists some $W_2 \subseteq V$ such that $V = W_1 \oplus W_2$. Is W_2 unique?

6.3.2. Suppose that V is a finite-dimensional vector space and W_1, W_2 are subspaces of V with $V = W_1 + W_2$. Prove there is a subspace $W_2' \subseteq W_2$ such that $V = W_1 \oplus W_2'$.

6.3.3. Consider $T_A : \mathbf{F}^2 \to \mathbf{F}^2$ given by

$$A = \begin{pmatrix} 1 & 1 \\ 0 & 1 \end{pmatrix}$$

Show that

$$W = span\left\{\begin{pmatrix} 1 \\ 0 \end{pmatrix}\right\}$$

is T_A-invariant. Show that there are no other one-dimensional T_A-invariant subspaces of $\mathbf{F}^2$.

6.3.4. Let $V = \mathcal{C}([-1, 1], \mathbf{R})$ be the vector space of continuous real-valued functions defined on the interval $[-1, 1]$. Let $W_e \subseteq V$ be the subspace of even functions [that is, those which satisfy $f(-x) = f(x)$] and let $W_o \subseteq V$ be the subspace of odd functions [that is, those which satisfy $f(-x) = -f(x)$]. Prove that $V = W_e \oplus W_o$.

6.3.5. Prove that any eigenspace of a linear operator T is always T-invariant. If E is the eigenspace of some linear operator $T : V \to V$, what is the restriction of T to E?

6.3.6. Suppose that $T : V \to V$ is a linear operator and every subspace of V is T-invariant. Show that $T = kI$ for some scalar k.

6.3.7. Give an example of a linear operator $T : \mathbf{R}^2 \to \mathbf{R}^2$ whose only T-invariant subspaces are $\{\vec{0}\}$ and $\mathbf{R}^2$. Can you do this for $\mathbf{R}^3$ or $\mathbf{C}^2$?

6.3.8. Suppose that $T : V \to V$ satisfies $T^2 = T$ (that is, T is *idempotent*) where V is finite-dimensional.
(a) Show that $im(T^2) = im(T)$.
(b) Show that $V = ker(T) \oplus im(T)$.
(c) Show that both $im(T)$ and $ker(T)$ are T-invariant, and find the corresponding block-diagonal representation.

6.3.9. Suppose that $T : \mathbf{R}^n \to \mathbf{R}^n$ is a linear operator with the property that $T^2 = I$. Let E_1 be the eigenspace associated with the eigenvalue 1 and E_{-1} be the eigenspace associated with the eigenvalue -1. Prove that $\mathbf{R}^n = E_1 \oplus E_{-1}$.

6.3.10. If $T : V \to V$ is diagonalizable and $W \subseteq V$ is a T-invariant subspace of V, show that the induced operator $\overline{T} : (V/W) \to (V/W)$ is also diagonalizable.

6.3.11. If $T : V \to V$ and V has a T-invariant direct-sum decomposition $V = W_1 \oplus W_2 \oplus \cdots \oplus W_n$ where $dim(W_i) = 1$ for all i, show that T is diagonalizable.

6.3.12. Show that for any pair of polynomials $f(X)$ and $g(X)$ with $g(X)$ dividing $f(X)$ and having the same prime factors, there exists a matrix A with $C_A(X) = f(X)$ and $M_A(X) = g(X)$.

6.4   The Jordan Form

In this section we show that if the minimal (or characteristic) polynomial of a linear operator factors completely over **F** (that is, into linear terms), then the operator can be represented (over **F**) by a Jordan matrix. It follows, since Jordan matrices are triangular, that whenever the minimal polynomial factors completely, the operator is represented by a triangular matrix. Conversely, since the characteristic polynomial of a triangular matrix obviously factors into linear terms, we obtain a complete characterization of those operators which can be "triangulated."

The key to finding Jordan matrix representations is finding sequences of generalized eigenvectors which together form a basis of the vector space. To do this, we must start with appropriate bases of the eigenspaces of the operator and then locate the generalized eigenvectors. We must address two questions. The first question is how to find such sequences computationally. For this, the idea is to use what was observed in the proof of Lemma 6.2.5. The generalized eigenvectors lie inside the kernels of the operators $(T - cI)^k$ for $k = 1, 2, \ldots$. The second question is why these sequences of generalized eigenvectors exist. They do in fact always exist, but the proof does not resemble the computational techniques used to actually find them.

We begin with the (full) definition of Jordan matrices.

Definitions 6.4.1 A matrix Q is called a *Jordan matrix of eigenvalue* c if it is a block-diagonal matrix

$$Q = \begin{pmatrix} J_1 & 0 & \cdots & 0 \\ 0 & J_2 & \cdots & 0 \\ \vdots & \vdots & \ddots & \vdots \\ 0 & 0 & \cdots & J_s \end{pmatrix}$$

where each J_i is an $n_i \times n_i$ Jordan block matrix of eigenvalue c and $n_1 \geq n_2 \geq \ldots \geq n_s$. A matrix D is called a *Jordan matrix* if it is a block-diagonal matrix of the form

$$D = \begin{pmatrix} Q_1 & 0 & \cdots & 0 \\ 0 & Q_2 & \cdots & 0 \\ \vdots & \vdots & \ddots & \vdots \\ 0 & 0 & \cdots & Q_s \end{pmatrix}$$

where each Q_i is a Jordan matrix of eigenvalue c_i and the scalars c_1, $c_2, \ldots, c_s$ are distinct.

For example,

$$\begin{pmatrix} 2 & 0 & 0 & 0 & 0 \\ 1 & 2 & 0 & 0 & 0 \\ 0 & 1 & 2 & 0 & 0 \\ 0 & 0 & 0 & 3 & 0 \\ 0 & 0 & 0 & 0 & 3 \end{pmatrix} \quad \text{and} \quad \begin{pmatrix} 2 & 0 & 0 & 0 & 0 \\ 0 & 2 & 0 & 0 & 0 \\ 0 & 0 & 2 & 0 & 0 \\ 0 & 0 & 0 & 3 & 0 \\ 0 & 0 & 0 & 1 & 3 \end{pmatrix}$$

are Jordan matrices with characteristic polynomial $(X - 2)^3(X - 3)^2$. The first matrix has minimal polynomial $(X - 2)^3(X - 3)$, while the second has minimal polynomial $(X - 2)(X - 3)^2$. The minimal polynomials are easy to determine by inspection, using Lemma 6.2.2. This is because the minimal polynomial will be the least common multiple of all the minimal polynomials of the individual Jordan blocks which constitute the entire Jordan matrix. In the first matrix, the Jordan block of eigenvalue 2 is 3×3, so it has minimal poynomial $(X - 2)^3$, while both Jordan blocks of eigenvalue 3 are 1×1 and have minimal polynomial $(X - 3)$. Consequently the first matrix has as minimal polynomial the product $(X - 2)^3(X - 3)$.

If J is a Jordan matrix, then corresponding to each Jordan block of J, there is a sequence of generalized eigenvectors for T_A. For instance, in the second example above, the first three standard basis elements $\{e_1, e_2, e_3\}$ form a basis of the eigenspace E_2 of the associated linear operator. Individually, each of these vectors is a sequence of length 1. The fourth basis vector e_4 is not an eigenvector, but e_5 is, and the sequence e_4, e_5 is a sequence of generalized eigenvectors (with associated eigenvalue 3). One has a sequence of two vectors, since the single Jordan block of eigenvalue 3 is a 2×2 matrix. The next result generalizes Lemma 6.2.5 and characterizes the type of basis necessary to obtain Jordan-form representations of a linear operator. We give a definition first.

Definition 6.4.2 Let $T : V \to V$ be a linear operator. An ordered basis $\mathcal{B} = \{v_1, \ldots, v_n\}$ for V is called a *Jordan basis* for T if each v_i is part of a sequence of generalized eigenvectors, the entire sequence of which lies in $\mathcal{B}$.

Lemma 6.4.3 *Suppose that* $T : V \to V$ *is a linear operator on a finite-dimensional vector space V. Then T can be represented by a Jordan matrix if and only if V has a Jordan basis for T.*

Proof: If $[T]_\mathcal{B}$ is a Jordan matrix, then the basis elements which correspond to each Jordan block of $[T]_\mathcal{B}$ will give a sequence of generalized eigenvectors for T. Conversely, if we have a Jordan basis for T, we order the basis in such a way that the sequences of generalized eigenvectors are listed in order and the sequences with common eigenvalue are grouped together with the longer sequences preceding the shorter. Each sequence of generalized eigenvectors will span a T-invariant subspace of V, and the corresponding block-diagonal matrix $[T]_\mathcal{B}$ will consist of Jordan blocks by Lemma 6.2.5. The lemma follows.
//

We next give a few examples to illustrate how the Jordan form arises.

Examples 6.4.4 (i) Set

$$G = \begin{pmatrix} -16 & 0 & 30 \\ 3 & -1 & -5 \\ -9 & 0 & 17 \end{pmatrix}$$

Then one may compute that $C_G(X) = (X + 1)^2(X - 2)$. Since the $rk(G + I) = 2$, the eigenspace E_{-1} of T_G of eigenvalue -1 must be one-dimensional. It follows that T_G cannot be diagonalizable, since the algebraic multiplicity of the eigenvalue -1 is 2. Consequently, the Jordan form of G must be

$$\begin{pmatrix} -1 & 0 & 0 \\ 1 & -1 & 0 \\ 0 & 0 & 2 \end{pmatrix}$$

The vector

$$e_2 = \begin{pmatrix} 0 \\ 1 \\ 0 \end{pmatrix}$$

is a eigenvector with eigenvalue -1. Moreover,

$$G \begin{pmatrix} 2 \\ 0 \\ 1 \end{pmatrix} = (-1) \begin{pmatrix} 2 \\ 0 \\ 1 \end{pmatrix} + \begin{pmatrix} 0 \\ 1 \\ 0 \end{pmatrix} \quad \text{and} \quad \begin{pmatrix} 5 \\ 0 \\ 3 \end{pmatrix}$$

is an eigenvector with eigenvalue 2. Consequently, the basis

$$\mathcal{B} = \left\{ \begin{pmatrix} 2 \\ 0 \\ 1 \end{pmatrix}, \begin{pmatrix} 0 \\ 1 \\ 0 \end{pmatrix}, \begin{pmatrix} 5 \\ 0 \\ 3 \end{pmatrix} \right\}$$

for $\mathbf{F}^3$ is a basis for which $[T_G]_{\mathcal{B}}$ is the Jordan form of G.

(ii) We consider the following matrix:

$$A = \begin{pmatrix} 0 & -1 & 2 & 0 \\ -3 & 0 & 0 & 2 \\ -2 & 0 & 0 & 1 \\ 0 & -2 & 3 & 0 \end{pmatrix}$$

Then we can can easily compute that $C_A(X) = (X^2+1)^2$ and $M_A(X) = X^2 + 1$. Over $\mathbf{R}$, these polynomials do not factor into linear factors. Hence A does not have a real Jordan form. However, working over the complex numbers $\mathbf{C}$, $C_A(X) = (X + i)^2(X - i)^2$ and $M_A(X) = (X+i)(X-i)$. Hence A is diagonalizable over $\mathbf{C}$, and its diagonalization is the Jordan-form matrix

$$A = \begin{pmatrix} i & 0 & 0 & 0 \\ 0 & i & 0 & 0 \\ 0 & 0 & -i & 0 \\ 0 & 0 & 0 & -i \end{pmatrix}$$

We next describe how to compute sequences of generalized eigenvectors for a given $n \times n$ matrix A. The procedure is the following: Once the eigenvalues have been determined, each eigenspace is found as usual; that is, the eigenspace E_c is $ker(A - cI_n)$. If $dim(E_c)$ equals the algebraic multiplicity of c, then there is nothing left to do as each sequence will consist of a single eigenvector.

In case $dim(E_c)$ is less that the algebraic multiplicity of c, then we next must find the generalized eigenvectors v for which $(A - cI_n)v \in E_c$. To do this, let $w_1, w_2, \ldots, w_s \in E_c$ be a basis for E_c. Let R be the matrix whose columns are the w_i. We must solve the matrix equation $(A - cI_n)v = Ru$ for all $(n \times 1)$ vectors v and all possible $s \times 1$ matrices u. (Ru is a linear combination of $w_1, w_2, \ldots, w_s$ and hence is an element of E_c.) The reason for this is that there is no way to tell in advance which eigenvectors in E_c can be the final element in some chain of generalized eigenvectors. For every such solution v and u, the vectors v, $(A - cI_n)v = Ru$ are the last two elements in a sequence of generalized eigenvectors. Note that $D_c = \{Ru \mid \text{there exists } v \in V$ with $(A - cI_n)v = Ru\}$ is a subspace of E_c. Let $Ru_1, Ru_2, \ldots, Ru_t$ be a basis for D_c and choose $z_1, z_2, \ldots, z_t \in V$ such that $(A - cI_n)z_i = Ru_i$. We denote by $E_c^{(1)}$ the $span\{z_1, z_2, \ldots, z_t\}$. The collection of second to last elements in the desired sequence of generalized eigenvectors will all lie inside $E_c^{(1)}$.

If $dim(E_c^{(1)}) + dim(E_c)$ equals the algebraic multiplicity, then the process is complete. Then the basis $\{z_1, z_2, \ldots, z_t\}$ of $E_c^{(1)}$, together with the corresponding linearly independent elements

$$\{(A - cI_n)z_1, (A - cI_n)z_2, \ldots, (A - cI_n)z_t\} \subset E_c$$

will give the desired length 2 sequences of generalized eigenvectors. Lastly, to obtain all the vectors required for the Jordan form, one must extend the linearly independent vectors $(A - cI_n)z_1, (A - cI_n)z_2$, $\ldots, (A - cI_n)z_t \in E_c$ to a basis of E_c.

In case $dim(E_c^{(1)}) + dim(E_c)$ is less than the algebraic multiplicity of c, the process is not complete. This means there are some sequences of generalized eigenvectors of length at least 3. To continue, one must study the possible solutions of $(A - cI_n)v \in E_c^{(1)}$. The process continues in direct analogy to what was previously described, until sequences of generalized eigenvectors are produced spanning a subspace of dimension equal to the algebraic multiplicity of c.

In order to understand the (somewhat longwinded) process just described better, we first carry out this process for a matrix already in Jordan form. It should then be clear why it is necessary to proceed one layer at a time in producing the sequences. The second example also illustrates this computational proceedure.

Examples 6.4.5 (i) Consider the Jordan matrix

$$J = \begin{pmatrix} 4 & 0 & 0 & 0 & 0 & 0 & 0 \\ 1 & 4 & 0 & 0 & 0 & 0 & 0 \\ 0 & 1 & 4 & 0 & 0 & 0 & 0 \\ 0 & 0 & 0 & 4 & 0 & 0 & 0 \\ 0 & 0 & 0 & 1 & 4 & 0 & 0 \\ 0 & 0 & 0 & 0 & 0 & 4 & 0 \\ 0 & 0 & 0 & 0 & 0 & 0 & 5 \end{pmatrix}$$

The eigenvalue 4 is the interesting one. Since J is already in Jordan form, we can see that $E_4 = span\{e_3, e_5, e_6\}$. In the process above, we must solve the equation $(J - 4I_7)v = Ru$, where R is the 7×3 matrix with columns e_3, e_5, and e_6. Note that the the image of the operator $(J - 4I_7)$ is $span\{e_2, e_3, e_5, e_7\}$. Since the intersection $im(J - 4I_7) \cap E_4 = span\{e_3, e_5\}$ is two-dimensional, we conclude that there are two sequences of generalized eigenvectors of length greater than 1. As $(J - 4I_7)e_2 = e_3$ and $(J - 4I_7)e_4 = e_5 \in E_4$, we obtain $E_4^{(1)} = span\{e_2, e_4\}$. (Observe that any element of $E_4^{(1)}$ is the first element of a two-element sequence of generalized eigenvectors.)

The process concludes by observing that $im(I - 4I_7) \cap E_4^{(1)} = span\{e_2\}$ is one-dimensional. Since $(J - 4I_7)e_1 = e_2 \in E_4^{(1)}$, we set $E_4^{(2)} = span\{e_1\}$. The vector e_1 is the first vector in a sequence of three generalized eigenvectors. As $dim(E_4^{(2)}) + dim(E_4^{(1)}) + dim(E_4) = 1 + 2 + 3 = 6$, which is the algebraic multiplicity of 4, the process is complete. For a Jordan basis we start with e_1, $(J - 4I_7)e_1 = e_2$, $(J - 4I_7)e_2 = e_3 \in E_4$, then as $E_4^{(2)}$ is exhausted, we choose some element, say $e_4 \in E_4^{(1)}$, not yet spanned and form the sequence e_4, $(J - 4I_4)e_4 = e_5$. Finally, as $E_4^{(1)}$ has been spanned, we add one additional eigenvector, say e_6, to complete our collection of sequences of generalized eigenvectors associated with the eigenvalue 4.

Observe that in this process the choices of starting vectors for the sequences are not unique. For example, instead of e_4, any element of the form $re_2 + se_4 \in E_4^{(1)}$ with $s \neq 0$ could have been chosen to start the second sequence.

(ii) Set

$$K = \begin{pmatrix} 1 & 2 & 1 \\ 0 & 1 & 3 \\ 0 & 0 & 1 \end{pmatrix}$$

One easily checks that $C_K(X) = (X - 1)^3$ and $M_K(X) = (X - 1)^3$. Since $rk(K - I) = 2$, we see that the Jordan form of K must be

$$\begin{pmatrix} 1 & 0 & 0 \\ 1 & 1 & 0 \\ 0 & 1 & 1 \end{pmatrix}$$

To find the basis in which T_K is represented by this Jordan block matrix, we must find a sequence of generalized eigenvectors for T_K.

First the eigenspace E_1 is found to be

$$E_1 = ker \begin{pmatrix} 0 & 2 & 1 \\ 0 & 0 & 3 \\ 0 & 0 & 0 \end{pmatrix} = span \left\{ \begin{pmatrix} 1 \\ 0 \\ 0 \end{pmatrix} \right\}$$

We let

$$v_3 = \begin{pmatrix} 1 \\ 0 \\ 0 \end{pmatrix}$$

which is our third basis vector for the Jordan basis. The second basis vector v_2 must satisfy the condition that $Kv_2 = v_2 + v_3$. In other words, $(K - I_3)v_2 = v_3$. Hence we must solve

$$\begin{pmatrix} 0 & 2 & 1 \\ 0 & 0 & 3 \\ 0 & 0 & 0 \end{pmatrix} \begin{pmatrix} X \\ Y \\ Z \end{pmatrix} = \begin{pmatrix} 1 \\ 0 \\ 0 \end{pmatrix}$$

to find v_2. We see that

$$v_2 = \begin{pmatrix} 0 \\ 1/2 \\ 0 \end{pmatrix}$$

is such a vector. Finally, the first basis vector v_1 must satisfy the equation $(K - I_3)v_1 = v_2$. Solving

$$\begin{pmatrix} 0 & 2 & 1 \\ 0 & 0 & 3 \\ 0 & 0 & 0 \end{pmatrix} \begin{pmatrix} X \\ Y \\ Z \end{pmatrix} = \begin{pmatrix} 0 \\ 1/2 \\ 0 \end{pmatrix}$$

we find a solution

$$v_1 = \begin{pmatrix} 0 \\ -1/12 \\ 1/6 \end{pmatrix}$$

Setting $\mathcal{B} = \{v_1, v_2, v_3\}$, we have that $[T_K]_{\mathcal{B}}$ is the Jordan form of K.

The next result is the generalization of Lemma 6.2.5 to collections of sequences of generalized eigenvectors. It is crucial to the proof given in this section for the existence of the Jordan form. We remark that the numbering system for sequences of generalized eigenvectors in this lemma is reversed from that of the notation in Lemma 6.2.5. This is merely a technical convenience.

Lemma 6.4.6 *Suppose that $T : V \to V$ is a linear operator and $v_{1,1}, v_{2,1}, \ldots, v_{r,1}$ is a collection of r linearly independent eigenvectors for T. Suppose that for each $i = 1, 2, \ldots, r$, $v_{i,s_i}, v_{i,s_i-1}, \ldots, v_{i,1}$ is a sequence of generalized eigenvectors. Then the collection of all the generalized eigenvectors $\{v_{i,j}\}$, where $1 \le i \le r, 1 \le j \le s_i$, is linearly independent.*

Proof: If the collection $\{v_{i,j}\}$ is not linearly independent, there is a nontrivial relation $\sum a_{i,j} v_{i,j} = \vec{0}$. Denote by k the largest subscript j for which some $a_{i,j} \ne 0$. (So, for this k there is some i with $a_{ik} \ne 0$, but $a_{ij} = 0$ whenever $j > k$.) Among all possible nontrivial such relations, we choose one with this number k being minimal. Since $v_{1,1}, \ldots, v_{r,1}$ are linearly independent, we must have $k \ge 2$. Moreover, we may additionally assume that (among such relations) our relation is chosen to have a minimal number of nonzero $a_{i,k}$ $(i = 1, 2, \ldots)$. We denote by c_i the eigenvalue of the eigenvector $v_{i,1}$.

Choose some p with $a_{p,k} \ne 0$. We apply the operator $(T - c_p I)$ to our relation. Using the facts that $T(v_{i,j}) = c_i v_{i,j} + v_{i,j-1}$ for $1 < j < s_j$ and $T(v_{j,1}) = c_j v_{j,1}$, we obtain that

$$
\begin{aligned}
\vec{0} &= (T - c_p I)\left(\sum a_{i,j} v_{i,j}\right) \\
&= \sum_{i=1}^{n} [a_{i,k}(c_i - c_p)v_{i,k} + a_{i,k}v_{i,k-1} + a_{i,k-1}(c_i - c_p)v_{i,k-1}] + (\cdots)
\end{aligned}
$$

where $(\cdots)$ denotes other terms which involve the vectors $v_{i,j}$ with $j < k - 1$. In the case where there are at least two distinct c_i among values of i for $a_{i,k} \ne 0$, the new relation is a nontrivial relation among the $v_{i,j}$ which has fewer nonzero coefficients for the $v_{i,k}$. This contradicts our second minimality assumption. In the case where there is only one eigenvalue c_i among the values of i for which $a_{i,k} \ne 0$, then there are no $v_{i,k}$ terms in this new relation, and the coefficient of $v_{p,k-1}$ is $a_{p,k} \ne 0$. We have a nontrivial relation with smaller associated value k. This contradicts the minimality of k. From these two contradictions the lemma follows. //

We now give Fillipov's proof of the existence theorem for the Jordan form. The beauty of this proof is that it uses induction (on the dimension) in a clever way.

Theorem 6.4.7   (i)   *Suppose* $T : V \to V$ *is a linear operator on a finite-dimensional* **F** *vector space, whose minimal polynomial factors into a product of linear factors. Then there is an ordered basis* $\mathcal{B}$ *of* V *for which* $[T]_{\mathcal{B}}$ *is a Jordan matrix. Moreover, the Jordan matrix* $[T]_{\mathcal{B}}$ *is uniquely determined up to the order in which the eigenvalues are listed (although the basis* $\mathcal{B}$ *need not be).*
(ii)   *Every* $n \times n$ *matrix* A, *with coefficients in a field* **F**, *whose minimal polynomial factors into a product of linear factors, is similar to a unique Jordan matrix (up to the ordering of the eigenvalues).*
(iii)   *Every* $n \times n$ *matrix with coefficients in an algebraically closed field (for example,* **C**) *is similar to a unique Jordan matrix (up to the ordering of the eigenvalues).*

Proof: (i) This follows from part (ii) since one can replace the operator T by any matrix representation.

(iii) This is also an immediate consequence of part (ii), since by definition, in an algebraically closed field every polynomial factors into a product of linear factors.

(ii) We begin with the existence assertion and proceed by induction on n. If $n = 1$, the result is trivial. Thus we assume the result for all $s \times s$ matrices whenever $s < n$. We know that A has some eigenvalue c. Replacing the matrix A by $A - cI_n$, we have a singular matrix. If we can show that the matrix $A - cI_n$ has a Jordan form, then we will obtain a Jordan form for the original A by adding cI_n to the Jordan form of $A - cI_n$. With this reduction we can assume that A is singular.

Set $V = col(A)$, the column space of A. V is a T_A-invariant subspace of $\mathbf{F}^n$, and $rk(A) = dim(V) = s < n$ as A is singular. The restriction of T_A to V, $T_A|_V$, is an operator to which we may apply our induction hypothesis. [Since the minimal polynomial $M_A(X)$ annihilates T_A, that is, $M_A(T_A) = 0$, we conclude that $M_A(X)$ annihilates $T_A|_V$. From this we conclude that the minimal polynomial of $T_A|_V$, which is a factor of the annihilating polynomial $M_A(X)$, must factor into linear factors.]

Our induction hypothesis gives us a basis for V, $\{v_1, \ldots, v_s\}$ where each basis element is part of a sequence of generalized eigenvectors all of which occur in this basis. We next consider the nullspace $K = ker(A)$. We know that $dim(K) = n - s$. Set $W = V \cap K$ and suppose that $dim(W) = r \leq s$. This means that V contains an r-dimensional subspace W of the eigenspace E_0 $(= K)$ of T corresponding to the eigenvalue 0. W is the eigenspace corresponding to the eigenvalue 0 of the restricted operator $T_A|_V$. From this it follows that among our Jordan basis $\{v_1, \ldots, v_s\}$ of V, there is a subcollection which is a basis for W. For convenience we shall assume these are the vectors $v_1, \ldots, v_r$.

The vectors $v_1, \ldots, v_r$ are linearly independent in the $(n-s)$-dimensional vector space K. We apply the basis-extension theorem and obtain $n-s-r$ new vectors $u_1, \ldots, u_{n-s-r}$ so that $v_1, \ldots, v_r, u_1, \ldots, u_{n-s-r}$ form a basis for K. For each of $i = 1, 2, \ldots, r$ we denote by v_i' the first element in the sequence of generalized eigenvectors of $T_A|_V$ (among $\{v_1, \ldots, v_s\}$) that ends with v_i. As each of these v_i' lies inside $col(A) = im(T_A)$, it follows that there is some $w_i \in \mathbf{F}^n$ for which $Aw_i = v_i'$. Observe that as the eigenvalue of v_i $(1 \le i \le r)$ is 0, the equation $Aw_i = 0w_i + v_i'$ shows that each sequence of generalized eigenvectors starting with v_i has been extended by one new element w_i.

Altogether we have a set of n vectors $v_1, \ldots, v_s$, $u_1, \ldots, u_{n-s-r}$, $w_1, \ldots, w_r$, each of which is either an eigenvector for T_A or part of a sequence of generalized eigenvectors entirely contained in this set. Once we know that this set is linearly independent, we will have a Jordan basis of $\mathbf{F}^n$ for T_A, and the existence assertion will be proved. To see this, it suffices by Lemma 6.4.6 to show that the eigenvectors in this list are linearly independent.

The eigenvectors in our list are the eigenvectors $v_1, \ldots, v_s$, together with $u_1, \ldots, u_{n-s-r}$. To see that a collection of eigenvectors is linearly independent, it suffices to show, by Lemma 3.5.2, that each subcollection of eigenvectors of a fixed eigenvalue is linearly independent. For the nonzero eigenvalues, all the eigenvectors occur among the $v_1, \ldots, v_s$, which are linearly independent by the induction hypothesis. For the eigenvalue 0, the eigenvectors are the vectors $v_1, \ldots, v_r$ together with $u_1, \ldots, u_{n-s-r}$. By the construction of the u_i these are linearly independent. This gives the existence of the Jordan basis for T_A.

The uniqueness of the Jordan form of A remains. We must show that the number of $m \times m$ Jordan block matrices of a fixed eigenvalue in any two Jordan forms of A is the same. This is an immediate consequence of the following lemma.

Lemma 6.4.8 *Let A be an $n \times n$ matrix and suppose that $M_A(X)$ factors completely. Suppose that c is an eigenvalue of A. Set $r_j =$*

$rk((A - cI_n)^j)$ for $j \geq 0$, (so by convention, $r_0 = n$). Then there are precisely $r_{m+1} - 2r_m + r_{m-1}$ $m \times m$ Jordan block matrices of eigenvalue c in any Jordan form of A.

Proof: Suppose that A is similar to the Jordan matrix J. Then $J = PAP^{-1}$ for some invertible $n \times n$ matrix P. Using this we see that $P(A - cI_n)^j P^{-1} = (PAP^{-1} - PcI_nP^{-1})^j = (J - cI_n)^j$, so in particular $r_j = rk((A - cI_n)^j) = rk((J - cI_n)^j)$. Thus, in order to prove the lemma, it suffices to verify the result in the case of Jordan matrices. For this we first show that the number of $s \times s$ Jordan blocks of eigenvalue c with $s \geq j$ in an $n \times n$ Jordan matrix J is $r_{j-1} - r_j$.

If J_ℓ is an $s \times s$ Jordan block matrix of eigenvalue c, then it is readily checked (see the proof of Lemma 6.2.2) that the rank $rk((J_\ell - cI_s)^j) = s - j$ for all $j \leq s$ and $rk((J_\ell - cI_s)^j) = 0$ for all $j \geq s$. Observe also that if J_ℓ is an $s \times s$ Jordan block matrix with eigenvalue $c' \neq c$, then $rk((J_\ell - cI_s)^j) = s$ for all j. By definition, if J is an $n \times n$ Jordan matrix, then it is a block-diagonal matrix with diagonal blocks that are Jordan block matrices. Consequently the matrix $(J - cI_n)^j$ is a block-diagonal matrix with diagonal blocks having the form $(J_\ell - cI_s)^j$ for the various J_ℓ and s. Since the rank of the matrix $(J - cI_n)^j$ is the sum of the ranks of its diagonal blocks, we see that the difference in ranks $rk((J - cI_n)^{j-1}) - rk((J - cI_n)^j) = r_{j-1} - r_j$ is precisely the number of $s \times s$ Jordan blocks J_ℓ of eigenvalue c in J where $s \geq j$.

For any m there are $r_{m-1} - r_m$ $s \times s$ Jordan blocks of eigenvalue c with $s \geq m$ and $r_m - r_{m+1}$ $s \times s$ Jordan blocks of eigenvalue c with $s > m$. Consequently there are $(r_{m-1} - r_m) - (r_m - r_{m+1}) = r_{m+1} - 2r_m + r_{m-1}$ $m \times m$ such blocks. This proves the lemma. //

Often it is possible to determine what the Jordan form should be knowing only the characteristic and minimal polynomials. Other times this is not possible, but a complete list of all the possible Jordan forms can always be written down. We conclude with some examples which illustrate how the characteristic and minimal polynomials determine the possible Jordan forms that can occur for a given matrix.

Examples 6.4.9 (i) In case U has characteristic polynomial $C_U(X) = X^2(X - 3)^3$ and minimal polynomial $M_U = X(X - 3)^3$, then T_U must have a two-dimensional eigenspace E_0, while only a one-dimensional eigenspace E_3. Consequently, (listing the eigenvalue 0 first) the Jordan form of U must be the matrix

$$\begin{pmatrix} 0 & 0 & 0 & 0 & 0 \\ 0 & 0 & 0 & 0 & 0 \\ 0 & 0 & 3 & 0 & 0 \\ 0 & 0 & 1 & 3 & 0 \\ 0 & 0 & 0 & 1 & 3 \end{pmatrix}$$

(ii) Suppose that a matrix B satisfies $C_B(X) = (X - 2)^4(X - 3)^2$ and $M_B(X) = (X - 2)^2(X - 3)$. Then, assuming the eigenvalue 2 is listed first, there are two possible Jordan forms for B, depending upon the types of sequences of generalized eigenvectors of eigenvalue 2 that occur. This four-dimensional subspace can decompose into two subspaces, each with minimal polynomial $(X-2)^2$; or it can decompose into three subspaces, the first with minimal polynomial $(X-2)^2$ and the second two with minimal polynomial $(X - 2)$. Note that the product of these minimal polynomials must equal $(X-2)^4$ and their exponents must decrease in size.

The two possible Jordan forms are

$$\begin{pmatrix} 2 & 0 & 0 & 0 & 0 & 0 \\ 1 & 2 & 0 & 0 & 0 & 0 \\ 0 & 0 & 2 & 0 & 0 & 0 \\ 0 & 0 & 1 & 2 & 0 & 0 \\ 0 & 0 & 0 & 0 & 3 & 0 \\ 0 & 0 & 0 & 0 & 0 & 3 \end{pmatrix} \quad \text{and} \quad \begin{pmatrix} 2 & 0 & 0 & 0 & 0 & 0 \\ 1 & 2 & 0 & 0 & 0 & 0 \\ 0 & 0 & 2 & 0 & 0 & 0 \\ 0 & 0 & 0 & 2 & 0 & 0 \\ 0 & 0 & 0 & 0 & 3 & 0 \\ 0 & 0 & 0 & 0 & 0 & 3 \end{pmatrix}$$

PROBLEMS

6.4.1. For each matrix below, find the Jordan form over $\mathbf{R}$, if it exists.

(a) $\begin{pmatrix} 0 & 0 & 1 \\ 1 & 0 & -1 \\ 0 & 1 & 1 \end{pmatrix}$ (b) $\begin{pmatrix} 3 & 0 & 0 \\ 2 & 3 & -1 \\ 0 & 0 & 3 \end{pmatrix}$ (c) $\begin{pmatrix} 2 & -3 & 1 \\ 1 & -2 & 1 \\ 1 & -3 & 2 \end{pmatrix}$

6.4.2. For each matrix below, find the Jordan form over $\mathbf{R}$, if it exists.

(a) $\begin{pmatrix} 0 & 2 & -1 \\ 2 & 3 & -2 \\ -1 & -2 & 0 \end{pmatrix}$ (b) $\begin{pmatrix} 3 & 1 & 0 \\ 0 & 3 & 9 \\ 0 & 0 & 5 \end{pmatrix}$ (c) $\begin{pmatrix} 3 & 0 & 0 \\ 0 & 3 & 9 \\ 0 & 0 & 5 \end{pmatrix}$

6.4.3.   Prove that whenever the characteristic polynomial of a linear operator T factors completely, there is an ordered basis $\mathcal{B}$ for which $[T]_\mathcal{B} = N + D$ where D is diagonal and N is nilpotent.

6.4.4. Prove directly from the definitions that if a 3×3 matrix has a Jordan form, then such Jordan form is unique.

6.4.5. Are the following real matrices similar?

(a) $\begin{pmatrix} 1 & -1 \\ 4 & -3 \end{pmatrix}$ and $\begin{pmatrix} -1 & 0 \\ 1 & -1 \end{pmatrix}$

(b) $\begin{pmatrix} 3 & 0 & 0 \\ 1 & 3 & 0 \\ 1 & 0 & 3 \end{pmatrix}$ and $\begin{pmatrix} 3 & 0 & 0 \\ 1 & 3 & 0 \\ 0 & 1 & 3 \end{pmatrix}$

(c) $\begin{pmatrix} 1 & 3 & 5 \\ 3 & 9 & 15 \\ -2 & -6 & -10 \end{pmatrix}$ and $\begin{pmatrix} 0 & 0 & 0 \\ 1 & 0 & 0 \\ 0 & 0 & 0 \end{pmatrix}$

6.4.6. (a) Find a matrix A with characteristic polynomial $C_A(X) = (X-2)^3(X-3)^4$ and minimal polynomial $M_A(X) = (X-2)^2(X-3)^2$.
(b) Find all Jordan-form matrices J with $C_J(X) = (X-2)^5(X+5)^4$ and $M_J(X) = (X-2)^3(X+5)^2$.

6.4.7. For each of the following matrices determine a matrix P for which $P^{-1}MP$ is a Jordan-form matrix.

(a) $M = \begin{pmatrix} 2 & 1 \\ -1 & 4 \end{pmatrix}$

(b) $M = \begin{pmatrix} 2 & 1 & 0 & 0 \\ 1 & 2 & 0 & 0 \\ 0 & 0 & 2 & 5 \\ 0 & 0 & 0 & 2 \end{pmatrix}$

(c) $M = \begin{pmatrix} 0 & 0 & 8 \\ 1 & 0 & -12 \\ 0 & 1 & 6 \end{pmatrix}$

6.4.8. For each of the following matrices determine a matrix P for which $P^{-1}MP$ is a Jordan-form matrix.

(a) $M = \begin{pmatrix} 1 & 2 & 0 & 0 \\ 0 & 1 & 3 & 1 \\ 0 & 0 & 1 & 2 \\ 0 & 0 & 0 & 2 \end{pmatrix}$

(b) $M = \begin{pmatrix} 5 & -1 & 3 \\ -6 & 4 & -6 \\ -6 & 2 & -4 \end{pmatrix}$

(c) $M = \begin{pmatrix} 2 & 0 & 1 \\ 0 & 3 & 0 \\ 1 & 0 & 2 \end{pmatrix}$

6.4.9. Suppose that $\mathbf{F}$ is algebraically closed (for example $\mathbf{F} = \mathbf{C}$) and $f(X) \in \mathbf{F}[X]$. For any $n \times n$ $\mathbf{F}$ matrix A and any $c \in \mathbf{F}$, show that whenever $f(c)$ is an eigenvalue of $f(A)$, necessarily c is an eigenvalue of A. (This is the converse to Prob. 6.1.11.)

6.4.10. Let

$$A = \begin{pmatrix} -4 & -2 & -4 \\ 4 & 2 & 4 \\ 2 & 1 & 2 \end{pmatrix}$$

Show directly that A is nilpotent. From this determine the minimal polynomial of A, the characteristic polynomial of A, the Jordan form of A, and a Jordan basis for T_A.

6.5   The Primary-Decomposition Theorem

The Jordan form of an operator always exists when the characteristic (and hence minimal) polynomial factors completely. However, in many cases this does not occur. The remaining sections of this chapter are devoted to this situation. This section shows that there is always an invariant direct sum decompositon of an operator which corresponds to the factorization of the characteristic polynomial into irreducible factors. This decomposition is called the *primary decomposition.*

As exhibited in our study of the Jordan form, the relationship between the minimal polynomial and the characteristic polynomial determines many of the key properties of the operator. We shall see that the minimal polynomial, which is a factor of the characteristic polynomial, always has the same prime factors as the characteristic polynomial. Thus, the difference between these polynomials is the degree to which each prime factor divides the polynomial. We saw in the case of the Jordan form, that the nicest Jordan forms occurred where the minimal polynomial was a product of distinct linear factors. In this case the Jordan blocks were as small as possible, that is, 1×1. (In other words, the operator was diagonalizable.) This phenomenon is also true more generally; if the minimal polynomial has low degree in comparison with the characteristic polynomial, then the operator can be decomposed into a large number of smaller blocks.

A particularly easy case to understand is that of a diagonalizable operator. As previously mentioned, diagonalizability is equivalent to the minimal polynomial factoring into distinct linear terms. We will finally give the full proof of this result in this section. Half of this result is given in the next theorem.

Theorem 6.5.1 *If a linear operator $T : V \to V$ is diagonalizable, then $M_T(X) = (X - a_1)(X - a_2) \cdots (X - a_s)$ where $a_1, a_2, \ldots, a_s \in \mathbf{F}$ are the distinct eigenvalues of T.*

Proof: Let $\mathcal{B} = \{v_1, v_2, \ldots, v_n\}$ be an ordered basis of eigenvectors for T, where the associated eigenvalue of v_i is r_i (the r_i need not be distinct). Let $a_1, a_2, \ldots, a_s$ denote the distinct eigenvalues of T (so in fact, as sets $\{a_1, a_2, \ldots, a_s\} = \{r_1, r_2, \ldots, r_n\}$). Observe that $(T - a_i I)(v_j) = T(v_j) - a_i I(v_j) = r_j v_j - a_i v_j = (r_j - a_j)v_j$. From this it follows that the composition $(T - a_1 I) \circ (T - a_2 I) \circ \cdots \circ (T - a_s I)(v_j) = (r_j - a_1)(r_j - a_2) \cdots (r_j - a_s)v_j$. However, $r_j = a_i$ for some i, so

$$(T - a_1 I) \circ (T - a_2 I) \circ \cdots \circ (T - a_s I)(v_j) = \vec{0}$$

Since $\{v_1, v_2, \ldots, v_n\}$ is a basis of V, it follows that the composition $(T - a_1 I) \circ (T - a_2 I) \circ \cdots \circ (T - a_s I) = 0$. This shows that for $g(X) = (X - a_1)(X - a_2) \cdots (X - a_s)$, $g(T) = 0$.

It follows from Corollary 6.1.10 that $M_T(X) \mid g(X)$. If $M_T(X) \neq g(X)$, then some factor of $g(X)$, say $(X - a_1)$, does not divide $M_T(X)$. If this is the case, we find that $M_T(X) \mid (X - a_2)(X - a_3) \cdots (X - a_s)$. We can assume that $r_1 = a_1$. Then, for the same reasons as above, $(T - a_2 I)(T - a_3 I) \cdots (T - a_s I)(v_1) = (r_1 - a_2)(r_1 - a_2) \cdots (r_1 - a_s)v \neq \vec{0}$. This is a contradiction to Corollary 6.1.10, and the theorem is proved. //

To complete the characterization of diagonalizability in terms of the minimal polynomial, we must establish the converse of Theorem 6.5.1. We will set this aside for the moment, as it is a consequence of a more general result, namely the primary-decomposition theorem. To prove the primary-decomposition theorem, we need some more results about polynomials.

Definition 6.5.2 A subset $I \subseteq \mathbf{F}[X]$ is called an *ideal* in $\mathbf{F}[X]$ if
(i) Whenever $g(X), h(X) \in I$, then $g(X) + h(X) \in I$.
(ii) Whenever $s(X) \in \mathbf{F}[X]$ and $g(X) \in I$, then $s(X)g(X) \in I$.

The easiest way to construct examples of polynomial ideals is to take a single polynomial $f(X)$ and consider the collection of all multiples of $f(X)$. In other words, if we define $(f(X)) = \{g(X)f(X) \mid g(X) \in \mathbf{F}[X]\}$, then $(f(X))$ is an ideal. To check this, note that if $g(X)f(X), h(X)f(X) \in (f(X))$, then $g(X)f(X) + h(X)f(X) = (g(X) + h(X))f(X) \in (f(X))$ and if $s(X) \in \mathbf{F}[X]$ with $g(X)f(X) \in (f(X))$, then $s(X)(g(X)f(X)) = (s(X)g(X))f(X) \in (f(X))$. The ideal $(f(X))$ is called the *principal ideal* generated by $f(X)$.

One can generalize the concept of a principal ideal in an obvious way. Suppose $g_1(X), g_2(X), \ldots, g_s(X) \in \mathbf{F}(X)$ are polynomials. We denote by $(g_1(X), g_2(X), \ldots, g_s(X))$ the subset of all polynomials of the form

$$f_1(X)g_1(X) + f_2(X)g_2(X) + \cdots + f_s(X)g_s(X)$$

where $f_1(X), f_2(X), \ldots, f_s(X) \in \mathbf{F}[X]$ are arbitrary. As in the previous case, it is easily checked that $(g_1(X), g_2(X), \ldots, g_s(X))$ is an ideal, called the *ideal generated by* $g_1(X), g_2(X), \ldots, g_s(X)$.

It turns out that in the study of polynomials of *one* variable over a field, the only ideals are the principal ideals. This is given in the next theorem. This result should be thought of as the polynomial version of the famous result of grade school arithmetic that says "any finite collection of natural numbers has a greatest common divisor." The proof is similar to that of Theorem 6.1.10 and uses only the division algorithm. (Note that the proof applies verbatim to give the just stated g.c.d. result from elementary arithmetic.)

Theorem 6.5.3 (Principal-Ideal Theorem) *Suppose I is a nonzero ideal in $\mathbf{F}[X]$. Then there is a unique monic polynomial $g(X)$ such that $I = (g(X))$. The polynomial $g(X)$ is the nonzero monic polynomial of least degree in I.*

Proof: Choose $g(X) \in I$ to be a nonzero polynomial of smallest degree in I. Multiplying $g(X)$ by a scalar, we can assume that $g(X)$ is a monic polynomial. Let $0 \neq f(X) \in I$. Then $deg(f(X)) \geq deg(g(X))$. Applying the division algorithm, there are polynomials $h(X), r(X) \in \mathbf{F}[X]$ so that $f(X) = h(X)g(X) + r(X)$, where $deg(r(X)) < deg(g(X))$. We note that both $f(X), h(X)g(X) \in I$. It follows that $r(X) \in I$. But $g(X)$ was chosen to have the smallest degree among the nonzero polynomials in I, and $deg(r(X)) < deg(g(X))$. Thus $r(X) = 0$. From this we see that $f(X) = h(X)g(X)$, which shows that $I = (g(X))$.

To see that $g(X)$ is uniquely determined, we suppose that $I = (g_1(X)) = (g_2(X))$ where $g_1(X)$ and $g_2(X)$ are monic polynomials. Since $g_2(X) \in I = (g_1(X))$, we find that $g_1(X) \mid g_2(X)$. By the same reason, $g_2(X) \mid g_1(X)$. Hence, $deg(g_1(X)) = deg(g_2(X))$. Since both $g_1(X)$ and $g_2(X)$ are monic, we find that $deg(g_1(X) - g_2(X)) < deg(g_1(X))$. But, as $g_1(X) - g_2(X) \in I$, necessarily $g_1(X) - g_2(X) = 0$. This shows that $g_1(X) = g_2(X)$ and proves the theorem. //

Suppose that $g_1(X), g_2(X), \ldots, g_s(X) \in \mathbf{F}[X]$ are polynomials. Then, for some polynomial $g(X) \in \mathbf{F}[X]$, we have that

$$(g_1(X), g_2(X), \ldots, g_s(X)) = (g(X))$$

It follows, as each $g_i(X)$ is a multiple of $g(X)$, that $g(X)$ is a *common divisor* of $g_1(X), g_2(X), \ldots, g_s(X)$. Moreover, such $g(X)$ is a *greatest* common divisor, since any other common divisor of $g_1(X), g_2(X), \ldots, g_s(X)$ must be a divisor of $g(X)$. If the greatest common divisor of $g_1(X), g_2(X), \ldots, g_s(X)$ is 1, then we say that $g_1(X), g_2(X), \ldots, g_s(X)$ are *relatively prime*. The next corollary deals with this situation. This result is necessary for the proof of the Primary-Decomposition Theorem 6.5.7.

Corollary 6.5.4 *Suppose that $g_1(X), g_2(X), \ldots, g_s(X)$ are polynomials in $\mathbf{F}[X]$ with no (collective) common factors other than constant polynomials. Then $(g_1(X), g_2(X), \ldots, g_s(X)) = \mathbf{F}[X]$.*

Proof: By the principal-ideal theorem, we can write

$$I = (g_1(X), g_2(X), \ldots, g_s(X)) = (g(X))$$

for some monic $g(X) \in \mathbf{F}[X]$. Since $g_i(X) \in I$ for each i, we have that $g_i(X) = h_i(X)g(X)$ for appropriate polynomials $h_i(X) \in \mathbf{F}[X]$. But then $g(X)$ is a monic common factor of $g_1(X), g_2(X), \ldots, g_s(X)$; so by hypothesis such $g(X) = 1$. It follows that $I = (1) = \mathbf{F}[X]$. //

We are now ready to use the theory of polynomial ideals as a tool for studying linear operators. Given a linear operator, the collection of all polynomials which annihilate the operator forms an ideal (see Lemma 6.5.6). The minimal polynomial is precisely the monic generator of this (necessarily) principal ideal.

Definition 6.5.5 Suppose that $T : V \to V$ is a linear operator. We define $An(T) \subseteq \mathbf{F}[X]$, the *annihilator* of T, by

$$An(T) = \{f(X) \in \mathbf{F}[X] \mid f(T) = 0\}$$

Lemma 6.5.6 *Suppose that V is a finite-dimensional vector space over $\mathbf{F}$. For any $T : V \to V$, $An(T)$ is a nonzero ideal of $\mathbf{F}[X]$. In fact, $An(T)$ is the principal ideal $(M_T(X))$.*

Proof: Suppose that $g_1(X), g_2(X) \in An(T)$. Then for $g(X) = g_1(X) + g_2(X)$, we have $g(T) = g_1(T) + g_2(T) = 0 + 0 = 0$, so that $g(X) \in An(T)$. Also, for $f(X) \in \mathbf{F}[X]$, $g(X) \in An(T)$ and $h(X) = f(X)g(X)$, so by Lemma 6.1.2 (ii) we have that $h(T) = f(T) \circ g(T) = f(T) \circ 0 = 0$. This shows that $An(T)$ is an ideal. By the Cayley-Hamilton theorem, $C_T(X) \in An(T)$, so $An(T) \neq \{0\}$. Finally, by Theorem 6.1.10, every element of $An(T)$ is a multiple of $M_T(X)$; that is, $An(T) = (M_T(X))$. This proves the lemma. $//$

We now give the primary-decomposition theorem. The proof requires the machinery of polynomial ideals just introduced. The converse to Theorem 6.5.1 will follow as as a corollary.

Theorem 6.5.7 (Primary-Decomposition Theorem) *Suppose that $T : V \to V$ is a linear operator. We assume that*

$$M_T(X) = p_1(X)^{n_1} p_2(X)^{n_2} \cdots p_s(X)^{n_s}$$

where $p_1(X)$, $p_2(X), \cdots, p_s(X)$ are distinct monic prime polynomials. Then there is a direct-sum decomposition

$$V = V_1 \oplus V_2 \oplus \cdots \oplus V_s$$

such that
(i)   *Each V_i is a T-invariant subspace of V.*
(ii)   *The minimal polynomials $M_{T|_{V_i}} = p_i(X)^{n_i}$. In particular, the minimal polynomial of T is the product of the minimal polynomials of the $T|_{V_i}$.*
(iii)   *Such a direct-sum decomposition of V is uniquely determined by T, conditions (i) and (ii), and the ordering of $p_1(X)$, $p_2(X), \ldots, p_s(X)$. In this unique decomposition we have $V_i = ker(p_i(T)^{n_i})$.*

Proof: (i) and (ii) We define the polynomials $g_i(X) = M_T(X)/p_i(X)^{n_i}$ for each $i = 1, 2, \ldots, s$. The collection of polynomials $g_1(X)$, $g_2(X)$, $\ldots$, $g_s(X)$ have no common factors other than constants. It follows from Corollary 6.5.4 that the ideal $(g_1(X), g_2(X), \ldots, g_s(X)) = \mathbf{F}[X]$. Thus there exist polynomials $h_1(X), h_2(X), \ldots, h_s(X)$ such that

$$h_1(X)g_1(X) + h_2(X)g_2(X) + \cdots + h_s(X)g_s(X) = 1 \in \mathbf{F}[X]$$

Set $V_i = im(h_i(T) \circ g_i(T))$. We will show that $V = V_1 \oplus V_2 \oplus \cdots \oplus V_s$ and these V_i satisfy conditions (i) and (ii).

First, the polynomial equation

$$h_1(X)g_1(X) + h_2(X)g_2(X) + \cdots + h_s(X)g_s(X) = 1$$

gives that $h_1(T)g_1(T) + h_2(T)g_2(T) + \cdots + h_s(T)g_s(T) = I$. Let $v \in V$. It follows that $h_1(T)g_1(T)(v) + h_2(T)g_2(T)(v) + \cdots + h_s(T)g_s(T)(v) = I(v) = v$. As $h_i(T)g_i(T)(v) \in V_i$, we obtain that $V_1 + V_2 + \cdots + V_s = V$.

Now suppose that some $v \in V_1 \cap (V_2 + V_3 + \cdots + V_s)$. We write $v = h_1(T)g_1(T)(u_1)$ and $v = \sum_{i=2}^{s} h_i(T)g_i(T)(u_i)$. According to the definitions, the polynomial $M_T(X)$ divides both $p_1(X)^{n_1}h_1(X)g_1(X)$ and $p_2(X)^{n_2} \cdots p_s(X)^{n_s}h_i(X)g_i(X)$ for each $i > 1$. It follows that $\vec{0} = p_1(T)^{n_1}h_1(T)g_1(T)(u_1) = p_1(T)^{n_1}(v)$ and

$$
\begin{aligned}
\vec{0} &= \sum_{i=2}^{s} p_2(T)^{n_2} \cdots p_s(T)^{n_s} h_i(T)g_i(T)(u_i) \\
&= p_2(T)^{n_2} \cdots p_s(T)^{n_s} \left(\sum_{i=2}^{s} h_i(T)g_i(T)(u_i) \right) \\
&= p_2(T)^{n_2} \cdots p_s(T)^{n_s}(v)
\end{aligned}
$$

As $p_1(X)^{n_1}$ and $p_2(X)^{n_2} \cdots p_s(X)^{n_s}$ have no common factors, Corollary 6.5.4 gives $(p_1(X)^{n_1}, p_2(X)^{n_2} \cdots p_s(X)^{n_s}) = \mathbf{F}[X]$. Hence, there is an expression $1 = r_1(X)p_1(X)^{n_1} + r_2(X)p_2(X)^{n_2} \cdots p_s(X)^{n_s}$ where $r_1(X), r_2(X) \in \mathbf{F}[X]$. This gives that $v = I(v) = r_1(T)p_1(T)^{n_1}(v) + r_2(T)p_2(T)^{n_2} \cdots p_s(T)^{n_s}(v) = r_1(T)(\vec{0}) + r_2(T)(\vec{0}) = \vec{0}$. We have established $V_1 \cap (V_2 + \cdots + V_s) = \{\vec{0}\}$. By the same reasoning $V_i \cap (\sum_{j \neq i} V_j) = \{\vec{0}\}$ for $i \neq j$, and it follows from Lemma 6.3.3 (ii) that $V = V_1 \oplus V_2 \oplus \cdots \oplus V_s$.

We next show that each V_i is T-invariant. For any $w \in V_i$, we can express $w = h_i(T)g_i(T)(v) \in V_i$, where $v \in V$. We use the equality of polynomials $Xh_i(X)g_i(X) = h_i(X)g_i(X)X$ and apply Lemma 6.1.2 (ii). Thus,

$$
\begin{aligned}
T(w) &= T(h_i(T)g_i(T)(v)) = Th_i(T)g_i(T)(v) \\
&= h_i(T)g_i(T)T(v) = h_i(T)g_i(T)(T(v)) \in V_i
\end{aligned}
$$

This shows that V_i is T-invariant.

We saw above that for any $v \in V_i$, $p_i(T)^{n_i}(v) = \vec{0}$. This shows that $M_{T|_{V_i}}(X) \mid p_i(X)^{n_i}$. Consider $f(X) = p_i(X)^{n_i-1}h_i(X)g_i(X)$. Observe that $p_i(X)$ fails to divide $h_i(X)g_i(X)$, because otherwise $p_i(X)$ would divide

$$h_1(X)g_1(X) + h_2(X)g_2(X) + \cdots + h_s(X)g_s(X) = 1$$

From this we see that $p_i(X)^{n_i}$ does not divide $f(X)$. As $p_i(X)^{n_i} \mid M_T(X)$, we see that $M_T(X)$ does not divide $f(X)$; that is, $f(X) \notin An(T)$. We choose some $v \in V$ with $f(T)(v) \neq \vec{0}$. We have that $f(T)(v) = p_i(T)^{n_i-1}h_i(T)g_i(T)(v) = p_i(T)^{n_i-1}(h_i(T)g_i(T)(v))$, where $h_i(T)g_i(T)(v) \in V_i$. Thus, $p_i(X)^{n_i-1} \notin An(T \mid_{V_i})$. It follows that $M_{T|_{V_i}}(X) = p_i(X)^{n_i}$; this shows that parts (i) and (ii) of the theorem are satisfied.

(iii) Suppose that $V = V_1 \oplus V_2 \oplus \cdots \oplus V_s$ is a direct-sum decomposition satisfying (i) and (ii). If we show that $V_i = ker(p_i(T)^{n_i})$, then the uniqueness assertion will follow. Since $M_{T|_{V_i}} = p_i(X)^{n_i}$, we necessarily have that $V_i \subseteq ker(p_i(T)^{n_i})$. To obtain the reverse inclusion, we observe that the argument given in the third paragraph of this proof applies when we replace V_i with $ker(p_i(T)^{n_i})$. Thus V is the direct sum of these kernels. This proves the theorem. //

As promised, we give the the converse to Theorem 6.5.1 and thereby complete the characterization of diagonalizable operators in terms of the minimal polynomial.

Corollary 6.5.8 *A linear operator $T : V \to V$ is diagonalizable if and only if $M_T(X) = (X - a_1)(X - a_2) \cdots (X - a_s)$ where $a_1, a_2, \ldots, a_s \in \mathbf{F}$ are the distinct eigenvalues of T.*

Proof: The only-if part is Theorem 6.5.1. Conversely, assume that $M_T(X) = (X-a_1)(X-a_2) \cdots (X-a_s)$. By the primary-decomposition theorem, $V = V_1 \oplus V_2 \oplus \cdots \oplus V_s$ where $M_{T|_{V_i}}(X) = (X-a_i)$. This means that for each $v \in V_i$, $T(v) = a_i v$; that is, V_i consists of eigenvectors for T with associated eigenvalues a_i. If $\mathcal{B}_i$ is a basis for V_i, then by Corollary 6.3.4 (ii) we know that $\mathcal{B}_1 \cup \mathcal{B}_2 \cup \cdots \cup \mathcal{B}_s$ is a basis for V. Thus V has a basis of eigenvectors for T. //

The proof of the primary-decomposition theorem gives a procedure for finding the summands that occur in the primary decomposition of a linear operator. The process is remarkably like that of finding eigenspaces. (In fact it reduces to the eigenspace case whenever a linear factor divides the minimal polynomial with multiplicity one.) The theorem says that if $p(X)^n$ is the exact power of the irreducible polynomial $p(X)$ dividing $M_T(X)$, then the summand associated with $p(X)$ is $ker(p(T)^n)$. So, in the matrix case, to find the invariant subspace associated with some prime power $p(X)^r$, one simply computes the kernel of $p(A)^r$. The next examples illustrate this procedure.

Examples 6.5.9 (i) Consider the real matrix

$$A = \begin{pmatrix} 0 & -2 & 1 \\ 1 & -2 & 2 \\ 1 & -4 & 4 \end{pmatrix}$$

Then we may compute that $C_A(X) = (X^2 + 1)(X - 2)$. To apply the primary-decomposition theorem, we compute the two subspaces $ker(A^2 + I)$ and $ker(A - 2I)$ which correspond to our given factorization of $C_A(X) = M_A(X)$. As

$$A^2 = \begin{pmatrix} -1 & 0 & 0 \\ 0 & -6 & 5 \\ 0 & -10 & 9 \end{pmatrix}$$

we find that $W_1 = ker(A^2 + I)$ is

$$ker \begin{pmatrix} 0 & 0 & 0 \\ 0 & -5 & 5 \\ 0 & -10 & 10 \end{pmatrix} = span \left\{ \begin{pmatrix} 1 \\ 0 \\ 0 \end{pmatrix}, \begin{pmatrix} 0 \\ 1 \\ 1 \end{pmatrix} \right\}$$

Similarly, we find that $W_2 = ker(A - 2I)$ is

$$ker \begin{pmatrix} -2 & -2 & 1 \\ 1 & -4 & 2 \\ 1 & -4 & 2 \end{pmatrix} = ker \begin{pmatrix} 1 & 1 & -1/2 \\ 0 & -5 & -5/2 \\ 0 & 0 & 0 \end{pmatrix} = span \left\{ \begin{pmatrix} 0 \\ 1 \\ 2 \end{pmatrix} \right\}$$

(which is the eigenspace associated with the eigenvalue 2). As

$$T_A \begin{pmatrix} 1 \\ 0 \\ 0 \end{pmatrix} = \begin{pmatrix} 0 \\ 1 \\ 1 \end{pmatrix} \quad \text{and} \quad T_A \begin{pmatrix} 0 \\ 1 \\ 1 \end{pmatrix} = \begin{pmatrix} -1 \\ 0 \\ 0 \end{pmatrix}$$

we consider the ordered basis

$$B = \left\{ \begin{pmatrix} 1 \\ 0 \\ 0 \end{pmatrix}, \begin{pmatrix} 0 \\ 1 \\ 1 \end{pmatrix}, \begin{pmatrix} 0 \\ 1 \\ 2 \end{pmatrix} \right\}$$

(corresponding to the invariant direct-sum decomposition $W_1 \oplus W_2$).
We find that

$$[T_A]_B = \begin{pmatrix} 0 & -1 & 0 \\ 1 & 0 & 0 \\ 0 & 0 & 2 \end{pmatrix}$$

which is a block-diagonal matrix.

(ii) Set

$$H = \begin{pmatrix} 1 & 3 & 1 \\ 0 & 1 & 4 \\ 0 & 0 & 2 \end{pmatrix}$$

Then $C_H(X) = (X-1)^2(X-2)$. Set $f(X) = (X-1)(X-2)$. Then

$$f(H) = \begin{pmatrix} 0 & 3 & 1 \\ 0 & 0 & 4 \\ 0 & 0 & 1 \end{pmatrix} \begin{pmatrix} -1 & 3 & 1 \\ 0 & -1 & 4 \\ 0 & 0 & 0 \end{pmatrix} = \begin{pmatrix} 0 & -3 & 12 \\ 0 & 0 & 0 \\ 0 & 0 & 0 \end{pmatrix} \neq 0$$

This shows that $M_H(X) \neq f(X)$ and thus $M_H(X) = (X-1)^2(X-2)$.
In particular we see that H is not diagonalizable. To compute the
primary decomposition of H, we must compute the kernels of $(H-I_3)^2$
and $(H-2I_3)$. [Note: The second power of $(A-I_3)$ must be used in
the computation of the primary decomposition since the second power
of $(X-1)$ occured in $M_H(X)$.] We find:

$$V_1 = ker(H-I_3)^2 = ker \begin{pmatrix} 0 & 0 & 13 \\ 0 & 0 & 4 \\ 0 & 0 & 1 \end{pmatrix} = span \left\{ \begin{pmatrix} 1 \\ 0 \\ 0 \end{pmatrix}, \begin{pmatrix} 0 \\ 1 \\ 0 \end{pmatrix} \right\}$$

and

$$V_2 = ker(H-2I_3) = ker \begin{pmatrix} -1 & 3 & 1 \\ 0 & -1 & 4 \\ 0 & 0 & 0 \end{pmatrix}$$

$$= ker \begin{pmatrix} -1 & 0 & -13 \\ 0 & 1 & -4 \\ 0 & 0 & 0 \end{pmatrix} = span \left\{ \begin{pmatrix} 13 \\ 4 \\ 1 \end{pmatrix} \right\}$$

Setting

$$B = \left\{ \begin{pmatrix} 1 \\ 0 \\ 0 \end{pmatrix}, \begin{pmatrix} 0 \\ 1 \\ 0 \end{pmatrix}, \begin{pmatrix} 13 \\ 4 \\ 1 \end{pmatrix} \right\}$$

we find that

$$[T_H]_B = \begin{pmatrix} 1 & 3 & 0 \\ 0 & 1 & 0 \\ 0 & 0 & 2 \end{pmatrix}$$

which is the block-diagonal matrix corresponding to the T_H-invariant direct-sum decomposition $\mathbf{F}^3 = V_1 \oplus V_2$.

PROBLEMS

6.5.1. Find the primary decompositions of $\mathbf{R}^n$ for the operators associated with the following matrices.

(a) $\begin{pmatrix} 4 & 0 & 0 \\ 2 & 4 & 0 \\ 3 & 1 & 1 \end{pmatrix}$ (b) $\begin{pmatrix} 0 & -1 & 0 & 0 \\ 1 & 0 & 0 & 0 \\ -2 & -2 & 2 & 1 \\ 1 & 1 & -1 & 0 \end{pmatrix}$ (c) $\begin{pmatrix} 3 & 2 & -1 \\ -2 & 3 & 0 \\ -4 & 0 & 3 \end{pmatrix}$

6.5.2. Find the primary decompositions of $\mathbf{R}^n$ for the operators associated with the following matrices.

(a) $\begin{pmatrix} 0 & -1 & 1 & 1 \\ 1 & 0 & 2 & 1 \\ 0 & 0 & 1 & 1 \\ 0 & 0 & 0 & 1 \end{pmatrix}$ (b) $\begin{pmatrix} 0 & 1 & 1 \\ 1 & 0 & 1 \\ 1 & 1 & 0 \end{pmatrix}$ (c) $\begin{pmatrix} 3 & 0 & 2 \\ 1 & 2 & 3 \\ 2 & -2 & -1 \end{pmatrix}$

6.5.3. Let V be a finite-dimensional $\mathbf{F}$ vector space and $T : V \to V$ a linear operator. Suppose that V has no nontrivial T-invariant subspaces.
(a) Use the primary-decompositon theorem to show that $M_T(X)$ is a power of a prime polynomial.
(b) Now show that $M_T(X)$ must be a prime polynomial by considering some cyclic subspaces.

6.5.4. (a) Suppose that $V = V_1 \oplus V_2 \oplus \cdots \oplus V_n$ is a direct-sum decomposition of a finite-dimensional vector space V. Prove there exist linear operators $E_i : V \to V$ which satisfy $E_i(v_i) = v_i$ for all $v_i \in V_i$, $E_i(v_j) = \vec{0}$ whenever $i \neq j$ and $v_j \in V_j$, $E_i E_j = E_j E_i$, $E_i^2 = E_i$, and $I = E_1 + E_2 + \cdots + E_n$. The E_i are called *projection operators*.

(b) Establish a converse to (i); that is, show that whenever one has a sequence of operators $E_1, E_2, \ldots, E_n$ satisfying $E_i E_j = E_j E_i$, $E_i^2 = E_i$ and $I = E_1 + E_2 + \cdots + E_n$, then setting $V_i = im(E_i)$ gives a direct sum decompositon $V = V_1 \oplus V_2 \oplus \cdots \oplus V_n$.

6.5.5. Suppose that V is a finite-dimensional $\mathbf{F}$ vector space.
(a)   Assume that $T : V \to V$ has minimal polynomial $p(X)^r$ for some prime polynomial $p(X)$. Show there is some vector $v \in V$ with $p(T)^{r-1}(v) \neq \vec{0}$.
(b) If $T : V \to V$ has minimal polynomial $f(X)$, show using (i) and the primary-decomposition theorem, there exists some $v \in V$ with $g(T)(v) \neq \vec{0}$ for all $g(X)$ with $deg(g(X)) < deg(f(X))$.

6.5.6.   Show that whenever $T_1, T_2, \ldots, T_r : V \to V$ are all diagonalizable operators on a finite-dimensional vector space V for which $T_i T_j = T_j T_i$, then there is a basis $\mathcal{B}$ of V for which $[T_i]_\mathcal{B}$ is diagonal for all i (that is, the T_i are simultaneously diagonalizable). Hints: Induct on $n = dim(V)$. Apply the primary-decomposition theorem to T_1. Consider linear operators E_i as given in Prob. 6.5.4. Show that the hypotheses guarantee that $T_j E_k = E_k T_j$ for all j, k, and derive the desired conclusion.

6.5.7.   What is the characteristic polynomial of a nilpotent matrix?

6.5.8.   Let $V \subset C(\mathbf{R}, \mathbf{R})$ be the n-dimensional vector space of solutions to the linear differential equation $f^{(n)} + a_{n-1} f^{(n-1)} + \cdots + a_1 f' + a_0 f = 0$. Let $D : C(\mathbf{R}, \mathbf{R}) \to C(\mathbf{R}, \mathbf{R})$ be the differetiation operator.
(a) Show that V is D-invariant.
(b)   Apply the primary-decomposition theorem to $D : V \to V$ and show that V is spanned by solutions to differential equations of the form $(f' - af)^n = 0$ and $(f'' - af' - bf)^n = 0$.

6.5.9.   Suppose that T is a linear opeartor on a finite-dimensional real vector space for which $T^2 = 3T$. Prove that T is diagonalizable.

6.6   Cyclic Subspaces and the Cyclic-Decomposition Theorem

We continue to study the general situation where the minimal polynomial of an operator need not factor into linear terms. Given a linear operator $T : V \to V$, we seek T-invariant subspaces of V for which the restriction of T to this subspace is represented by a companion matrix. In Lemma 6.2.9 we saw that the existence of a cyclic vector for a vector space is sufficient to guarantee that the operator is represented by a companion matrix. The next definition gives the appropriate terminology in the context of (invariant) subspaces.

Definition 6.6.1 Suppose that $T : V \to V$ is a linear operator. Let $v \in V$. The span of the vectors v, $T(v)$, $T^2(v)$, $T^3(v), \ldots$ inside V is called the *T-cyclic subspace generated by* v and is denoted by $Z(T, v)$. A vector $v \in V$ is called a *cyclic vector* for T if $Z(T, v) = V$. If V has a cyclic vector for T, then V is called *T-cyclic*.

Evidently, $Z(T, v)$ is always a T-invariant subspace of V, for which v is a cyclic vector. This will be exploited in the sequel. One can see from the definition that $Z(T, v)$ is in fact the smallest T-invariant subspace of V which contains v.

As as an example consider the matrix

$$M = \begin{pmatrix} 1 & 2 & -2 \\ 3 & 0 & 3 \\ 1 & 1 & -2 \end{pmatrix}$$

Then, by direct calculation

$$T_M \begin{pmatrix} 1 \\ 0 \\ 0 \end{pmatrix} = \begin{pmatrix} 1 \\ 3 \\ 1 \end{pmatrix} \quad \text{and} \quad T_M \begin{pmatrix} 1 \\ 3 \\ 1 \end{pmatrix} = \begin{pmatrix} 5 \\ 6 \\ 2 \end{pmatrix}$$

However,

$$\begin{pmatrix} 5 \\ 6 \\ 2 \end{pmatrix} = 3 \begin{pmatrix} 1 \\ 0 \\ 0 \end{pmatrix} + 2 \begin{pmatrix} 1 \\ 3 \\ 1 \end{pmatrix} \in span \left\{ \begin{pmatrix} 1 \\ 0 \\ 0 \end{pmatrix}, \begin{pmatrix} 1 \\ 3 \\ 1 \end{pmatrix} \right\}$$

It follows that the T-cyclic subspace generated by

$$\begin{pmatrix} 1 \\ 0 \\ 0 \end{pmatrix}$$

is this two-dimensional subspace of $\mathbf{F}^3$. In contrast, we compute that

$$T_M \begin{pmatrix} 1 \\ 1 \\ 0 \end{pmatrix} = \begin{pmatrix} 3 \\ 3 \\ 2 \end{pmatrix} \quad \text{and} \quad T_M \begin{pmatrix} 3 \\ 3 \\ 2 \end{pmatrix} = \begin{pmatrix} 5 \\ 15 \\ 2 \end{pmatrix}$$

These three vectors are linearly independent, so it follows that

$$\begin{pmatrix} 1 \\ 1 \\ 0 \end{pmatrix}$$

generates a three-dimensional cyclic subspace, that is, all of $\mathbf{F}^3$.

T-cyclic subspaces also arise naturally in the context of function spaces. For example, if $V = C^\infty(\mathbf{R}, \mathbf{R})$ is the vector space of infinitely differentiable functions on $\mathbf{R}$, let $D : V \to V$ be the differentiation operator. Consider the vector $v = \sin x \in V$. Then the cyclic subspace generated by v is the two-dimensional subspace $Z(D, v) = span\{\sin x, \cos x\}$. Evidently, if we take $\mathcal{B} = \{\sin x, \cos x\}$ as an ordered basis of $Z(D, v)$, we have that the matrix

$$[T\,|\,Z(D,v)]_{\mathcal{B}} = \begin{pmatrix} 0 & -1 \\ 1 & 0 \end{pmatrix}$$

We next generalize the concept of the minimal polynomial. Instead of considering polynomials which annihilate the entire operator, we restrict our attention to cyclic subspaces. This gives rise to the concept of a minimal polynomial of a vector under an operator.

Definition 6.6.2 Suppose that $T : V \to V$ is a linear operator and $v \in V$. The *minimal polynomial of v under T*, denoted $M_{T,v}(X)$, is defined to be the minimal polynomial of $T\,|\,Z(T,v) : Z(T, v) \to Z(T, v)$ [the restriction of the operator T to the T-invariant subspace $Z(T, v)$].

Alternatively, the minimal polynomial of v under T is the monic polynomial $f(X)$ of smallest degree for which $f(T)(v) = \vec{0}$. The point is that if such $f(X)$ has degree d, then by the minimality of d, the vectors $v, T(v), T^2(v), \ldots, T^{d-1}(v)$ will be linearly independent. Moreover, as $f(T)(v) = \vec{0}$, $T^d(v)$ is a linear combination of these vectors, so they span $Z(T, v)$. It then follows that $f(X)$ is the minimal polynomial of $T\,|\,Z(T,v)$ by Lemma 6.2.9.

For example, if $T : V \to V$ is a linear operator and $v \in V$ is an eigenvector with eigenvalue c, then since $T(v) = cv \in span\{v\}$, $Z(T, v) = span\{v\}$ is a one-dimensional subspace of V. The restriction of T to this subspace has minimal (and also characteristic) polynomial $X - c$. This is because for such an eigenvector, $(T - cI)(v) = \vec{0}$ and no polynomial in T of lower degree can annihilate v.

In the second example above, it was observed that the D-cyclic subspace of $C^\infty(\mathbf{R}, \mathbf{R})$ generated by the vector $v = \sin x$ under differentiation is the subspace $span\{\sin x, \cos x\}$. The characteristic (and minimal) polynomial of the matrix

$$\begin{pmatrix} 0 & -1 \\ 1 & 0 \end{pmatrix}$$

representing the restiction of differentiation to this subspace is $X^2 + 1$. In terms of our notation this says $M_{D, \sin x}(X) = X^2 + 1$. This shows that each element of $span\{\sin x, \cos x\}$ is a solution of the differential equation $f''(x) + f(x) = 0$, since this equation results from substituting D into the equation $X^2 + 1 = 0$.

If $T : V \to V$ has a cyclic vector, then we saw in Lemma 6.2.9 that there is a basis $\mathcal{B}$ of V for which $[T]_{\mathcal{B}}$ is a companion matrix. We use this to show that in the T-cyclic case, the minimal polynomial always coincides with the characteristic polynomial. Matrices for which the minimal polynomial equals the characteristic polynomial are called *nonderogatory*.

Lemma 6.6.3 *Suppose that $T : V \to V$ is a linear operator, V is a finite-dimensional vector space and $V = Z(T, v)$ for some $v \in V$. Then $M_{T,v}(X) = M_T(X) = C_T(X)$.*

Proof: Assume that V is n-dimensional. We know by Lemma 6.2.9 that setting $\mathcal{B} = \{v, T(v), T^2(v), \dots, T^{n-1}(v)\}$ gives an ordered basis for V where $[T]_{\mathcal{B}}$ is the companion matrix for $C_T(X)$. Suppose that $f(X) = X^m + a_{m-1}X^{m-1} + \cdots + a_1 X + a_0$ is a nonzero polynomial with $m < n = dim(V) = deg(C_T(X))$. Then $f(T)(v) = T^m(v) + a_{m-1}T^{m-1}(v) + \cdots + a_1 T(v) + a_0 v \neq \vec{0}$ since $T^m(v)$, $T^{m-1}(v), \dots$, $T(v), v$ are linearly independent. This shows that such $f(X) \notin An(T)$. It follows that $C_T(X)$ is the monic polynomial of least degree in $An(T)$; that is, $M_T(X) = C_T(X)$. But, $M_{T,v}(X) = M_T(X)$ by definition, since $V = Z(T, v)$. This proves the lemma. //

We next give a converse to the above lemma.

Lemma 6.6.4 *Suppose $T : V \to V$ is a linear operator.*
(i) *If $v \in V$ is such that $M_{T,v}(X) = C_T(X)$, then v is a cyclic vector for T.*
(ii) *If $M_T(X) = C_T(X)$, then V is T-cyclic.*

Proof: (i) Setting $W = Z(T, v) \subseteq V$, we see by Lemma 6.6.3 that $dim(W) = deg(C_{T|_W}(X)) = deg(M_{T,v}(X)) = deg(C_T(X)) = dim(V)$. Hence $V = W$, and v is a cyclic vector for V.

(ii)   We first treat the case where $C_T(X) = p(X)^r$, $p(X)$ a prime polynomial. A vector $v \in V$ will be a cyclic vector for T if and only if $p(T)^{r-1}(v) \neq \vec{0}$ [because then necessarily $M_{T,v}(X) = p(X)^r = C_T(X)$.] Thus, if V did not have a cyclic vector, we would have that $p(T)^{r-1}(v) = \vec{0}$ for all $v \in V$; that is, $p(X)^{r-1} \in An(T)$. This would contradict the assumption that $M_T(X) = C_T(X)$.

For the general case, let $V = V_1 \oplus V_2 \oplus \cdots \oplus V_s$ be the primary decomposition of V. The minimal polynomial for T is the product of the minimal polynomials of $T|_{V_i}$. Since $M_T(X)$ has degree $dim(V)$, it must happen that the minimal polynomials $M_{T|_{V_i}}(X)$ have degree $dim(V_i)$. By the prime power-case established above, this shows that each V_i is $T|_{V_i}$-cyclic. Let v_i be a cyclic vector for V_i. Then we claim that $v = v_1 + v_2 + \cdots + v_s$ is a cyclic vector for V. This is because if $g(X)$ is a polynomial, then by the T-invariance of each V_i, $g(T)(v) = g(T)(v_1) + \cdots + g(T)(v_s)$ where each $g(T)(v_i) \in V_i$. Thus, if $g(T)(v) = \vec{0}$, we must have (by directness) that $g(T)(v_i) = \vec{0}$ for all i. In particular, we find that $M_{T,v_i}(X) \mid M_{T,v}(X)$ for each i. This shows that $M_{T,v}(X) = C_T(X)$, which shows by part (i) that v is a T-cyclic vector. The lemma is proved. //

If $T : V \rightarrow V$ is an operator and V is T-cyclic, then there is a basis $\mathcal{B}$ for which $[T]_\mathcal{B}$ is a companion matrix. But what can be done if V is not T-cyclic? We shall see that it is always possible to express V as a direct sum of T-cyclic subspaces. After this has been done, we can find a basis $\mathcal{B}$ of V for which $[T]_\mathcal{B}$ is a block-diagonal matrix whose blocks are companion matrices. This result, known as the cyclic-decomposition theorem, is stated momentarily.

Example 6.6.5 Consider the 4×4 companion matrix

$$A = \begin{pmatrix} 0 & 0 & 0 & 1 \\ 1 & 0 & 0 & 0 \\ 0 & 1 & 0 & 0 \\ 0 & 0 & 1 & 0 \end{pmatrix}$$

for which $C_A(X) = X^4 - 1 = (X^2 - 1)(X^2 + 1)$. The vector e_1 is a cyclic vector for T_A. (We know it has to be, as A is a companion matrix.)

Set

$$v_1 = (T_A^2 + I)(e_1) = \begin{pmatrix} 1 \\ 0 \\ 1 \\ 0 \end{pmatrix} \quad \text{and} \quad v_2 = (T_A^2 - I)(e_1) = \begin{pmatrix} -1 \\ 0 \\ 1 \\ 0 \end{pmatrix}$$

Then,

$$T_A(v_1) = \begin{pmatrix} 0 \\ 1 \\ 0 \\ 1 \end{pmatrix} \quad \text{and} \quad T_A^2(v_1) = \begin{pmatrix} 1 \\ 0 \\ 1 \\ 0 \end{pmatrix}$$

so that $(T_A^2 - I)(v_1) = \vec{0}$; this shows that $M_{T,v_1} = X^2 - 1$. Similarly, $M_{T,v_2} = X^2 + 1$. We consider $W_1 = span\{v_1, T_A(v_1)\}$ and $W_2 = span\{v_2, T_A(v_2)\}$. Then W_1 and W_2 are two (T-invariant) T-cyclic subspaces of $\mathbf{F}^4$. Moreover, $\mathbf{F}^4 = W_1 \oplus W_2$. If $\mathcal{B} = \{v_1, T_A(v_1), v_2, T_A(v_2)\}$ is the corresponding ordered basis of $\mathbf{F}^4$, we find that

$$[T_A]_{\mathcal{B}} = \begin{pmatrix} 0 & 1 & 0 & 0 \\ 1 & 0 & 0 & 0 \\ 0 & 0 & 0 & -1 \\ 0 & 0 & 1 & 0 \end{pmatrix}$$

is a direct sum of two 2×2 blocks, each of which is a companion matrix.

The preceding example shows that given a linear operator, there may be several different block-diagonal matrices representing the operator, whose blocks are companion matrices. If we make some special assumptions about such a block decompositon, it is possible to guarantee that it is unique. (This will give the so-called rational form.) The cyclic-decomposition theorem (below) states that if $T : V \to V$ is a linear operator on a finite-dimensional vector space V, then V can be decomposed into a direct sum of T-invariant cyclic subspaces $V = Z_1 \oplus \cdots \oplus Z_s$. In this theorem, the order in which these subspaces are listed is crucial. The result requires that the minimal polynomial of the operator restricted to Z_1 is $M_T(X)$. In addition, the minimal polynomial of $T|_{Z_i}$ must divide the minimal polynomial of $T|_{Z_j}$ whenever $j < i$. Consequently, each of these polynomials is a factor of $M_T(X)$.

The key application of the cyclic-decomposition theorem is the existence and uniqueness of the rational form of an arbitrary matrix. This is discussed in the next section, which can be understood once the statement of the cyclic-decomposition theorem has been read (without reading the proof). We give two proofs of the cyclic-decomposition theorem, one in Sec. 6.8 and another in Sec. 6.9.

Theorem 6.6.6 (Cyclic-Decomposition Theorem) *Suppose that T : $V \to V$ is a linear operator on a finite-dimensional vector space V. Then there is a T-invariant direct-sum decomposition*

$$V = Z_1 \oplus Z_2 \oplus \cdots \oplus Z_s$$

which satisfies the following:
(i) $C_{T|Z_i}(X) = M_{T|Z_i}(X)$ *for all* $i = 1, 2, \ldots, s$; *that is, each Z_i is T-cyclic.*
(ii) $M_{T|Z_{i+1}}(X) \mid M_{T|Z_i}(X)$ *for all* $i = 1, 2, \ldots, s-1$.
(iii) $C_T(X) = M_{T|Z_1}(X) M_{T|Z_2}(X) \cdots M_{T|Z_s}(X)$, *and* $M_T(X) = M_{T|Z_1}(X)$. *Moreover, the number s and the sequence of polynomials*

$$M_{T|Z_1}(X), M_{T|Z_2}(X), \ldots, M_{T|Z_s}(X)$$

are uniquely determined.

Remark 6.6.7 The theorem does *not* assert that the subspaces Z_1, $Z_2, \ldots, Z_s$ are uniquely determined. Only the associated characteristic (minimal) polynomials of these T-cyclic subspaces are uniquely determined by the conditions (i), (ii), (iii). From this, it does follow that the dimensions of the Z_i are uniquely determined and that each Z_i is a T-cyclic subspace of V. The polynomials in the unique sequence $M_{T|Z_1}(X), \ldots, M_{T|Z_s}(X)$ are called the *invariant factors* of the linear operator T. The proof of the cyclic-decomposition theorem given in Sec. 6.8 will give an algorithm for determining the invariant factors of an arbitrary matrix.

We conclude this section by using the cyclic-decomposition theorem to prove that the minimal and characteristic polynomials alway have the same prime factors.

Corollary 6.6.8 *For any linear operator $T : V \to V$ on a finite-dimensional vector space V, the polynomials $M_T(X)$ and $C_T(X)$ have the same prime factors.*

Proof: Since $M_T(X) \mid C_T(X)$, every prime factor of $M_T(X)$ is also a prime factor of $C_T(X)$. Conversely, part (iii) of the cyclic-decomposition theorem shows that the characteristic polynomial of an operator is the product of polynomials, each of which divides the minimal polynomial. Hence every prime factor of $C_T(X)$ is a prime factor of $M_T(X)$. //

PROBLEMS

6.6.1. Find the T_A-cyclic subspace of $\mathbf{F}^3$ generated by the vector

$$\begin{pmatrix} 1 \\ 3 \\ 2 \end{pmatrix}$$

if A is

(a) $\begin{pmatrix} 1 & 2 & 3 \\ 0 & 2 & 0 \\ 1 & 2 & 3 \end{pmatrix}$ (b) $\begin{pmatrix} 2 & 0 & 0 \\ 0 & 1 & 0 \\ 0 & 0 & 3 \end{pmatrix}$ (c) $\begin{pmatrix} 4 & 0 & 0 \\ 2 & 2 & 2 \\ 3 & 1 & 1 \end{pmatrix}$

6.6.2. Suppose that $T : V \to V$ is a linear operator on a finite-dimensional vector space V. If T is diagonalizable and V is T-cyclic, prove that the eigenvalues of T are distinct.

6.6.3. For the following v, A find the minimal polynomial $M_{T,v}(X)$ of v under $T = T_A$.

(a) $v = \begin{pmatrix} 1 \\ 2 \\ 1 \end{pmatrix}$, $A = \begin{pmatrix} 2 & 1 & 2 \\ 2 & 4 & 1 \\ 4 & 1 & 3 \end{pmatrix}$

(b) $v = \begin{pmatrix} 1 \\ 1 \\ 1 \end{pmatrix}$, $A = \begin{pmatrix} 1 & 1 & 1 \\ 2 & 1 & 1 \\ 1 & 1 & 1 \end{pmatrix}$

(c) $v = \begin{pmatrix} 2 \\ 2 \\ 0 \end{pmatrix}$, $A = \begin{pmatrix} 1 & 2 & 1 \\ 2 & 1 & 0 \\ 1 & 1 & 0 \end{pmatrix}$

6.6.4. Give an example of a 4×4 real matrix A for which $\mathbf{F}^4$ has no nontrivial T_A-invariant subspaces.

6.6.5. Suppose that every nonzero vector in V is a cyclic vector for $T : V \to V$. If V is finite-dimensional, what can you say about $C_T(X)$? Prove your assertions.

6.6.6. Find a cyclic vector for the linear operator given by each of the following matrices. What companion matrix represents the operator in each case?

(a) $\begin{pmatrix} 1 & 0 & 0 \\ 0 & 2 & 0 \\ 0 & 0 & 3 \end{pmatrix}$
(b) $\begin{pmatrix} 1 & 0 & 2 \\ 0 & 4 & 0 \\ 2 & 0 & 1 \end{pmatrix}$
(c) $\begin{pmatrix} -1 & 0 & 2 \\ 3 & 2 & 1 \\ 0 & 0 & -1 \end{pmatrix}$

6.7   The Rational Form

Using the cyclic-decomposition theorem, one sees that every linear operator has a block-diagonal matrix representation, with each block a companion matrix. In this block representation of the linear operator, each block on the diagonal represents the operator acting upon a single summand in a cyclic decomposition of the vector space.

Definition 6.7.1 A square matrix R is said to be in rational form if R is a block-diagonal matrix:

$$R = \begin{pmatrix} Z_1 & 0 & \cdots & 0 & 0 \\ 0 & Z_2 & \cdots & 0 & 0 \\ \vdots & \vdots & \ddots & \vdots & \vdots \\ 0 & 0 & \cdots & Z_{s-1} & 0 \\ 0 & 0 & \cdots & 0 & Z_s \end{pmatrix}$$

where each Z_i is a companion matrix and $C_{Z_{i+1}}(X) \mid C_{Z_i}(X)$ for all $i = 1, 2, \ldots, s - 1$.

Of course, any companion matrix is already a rational-form matrix with a single block. Note that the order in which the blocks occur is important for the rational form. For example, consider the matrices

$$A = \begin{pmatrix} 0 & 0 & 1 & 0 & 0 \\ 1 & 0 & -1 & 0 & 0 \\ 0 & 1 & 1 & 0 & 0 \\ 0 & 0 & 0 & 0 & -1 \\ 0 & 0 & 0 & 1 & 0 \end{pmatrix} \quad \text{and} \quad B = \begin{pmatrix} 0 & -1 & 0 & 0 & 0 \\ 1 & 0 & 0 & 0 & 0 \\ 0 & 0 & 0 & 0 & 1 \\ 0 & 0 & 1 & 0 & -1 \\ 0 & 0 & 0 & 1 & 1 \end{pmatrix}$$

The matrix A is in rational form, while the matrix B is not. The first block of A is the companion matrix of the polynomial $X^3 - X^2 + X - 1 = (X^2 + 1)(X - 1)$ and the second block of A is the companion matrix of the factor $X^2 + 1$. The matrix B has these same two blocks, but they appear in the wrong order. Thus, B is not in rational form. Of course, modulo a change in the ordering of the basis, A and B represent the same operator. The point is that we define A to be the rational form in order to ensure that the associated rational form of an operator is uniquely determined.

Observe that for the operator $T_A : \mathbf{F}^5 \rightarrow \mathbf{F}^5$ determined by the matrix A just given, the cyclic decomposition of $\mathbf{F}^5$ arises by properly grouping the elements of the standard basis. In this case, setting $Z_1 = span\{e_1, e_2, e_3\}$ and $Z_2 = span\{e_4, e_5\}$ gives the desired cyclic decompositon $\mathbf{F}^5 = Z_1 \oplus Z_2$. Note that $Z_1 = Z(T_A, e_1)$ with $T_A \mid Z_1$ having minimal polynomial $X^3 - X^2 + X - 1$ and that $Z_2 = Z(T_A, e_4)$ with $T_A \mid Z_2$ having minimal polynomial $X^2 + 1$.

We explicitly record these observations in the next lemma.

Lemma 6.7.2 *If R is a matrix in rational form, then*
(i)   *The minimal polynomial $M_R(X) = C_{Z_1}(X)$.*
(ii) *The decomposition $\mathbf{F}^n = E_1 \oplus E_2 \oplus \cdots \oplus E_s$, corresponding to the given block decomposition of R, is a cyclic decomposition for $\mathbf{F}^n$ with respect to T_R.*

Proof: (i) Evidently, for any polynomial $f(X)$, one has that

$$
f(R) = \begin{pmatrix}
f(Z_1) & 0 & \cdots & 0 & 0 \\
0 & f(Z_2) & \cdots & 0 & 0 \\
\vdots & \vdots & \ddots & \vdots & \vdots \\
0 & 0 & \cdots & f(Z_{s-1}) & 0 \\
0 & 0 & \cdots & 0 & f(Z_s)
\end{pmatrix}
$$

Hence, the divisibility criteria $C_{Z_{i+1}}(X) \mid C_{Z_i}(X)$ gives that $0 = C_{Z_1}(Z_1) = C_{Z_1}(Z_2) = \cdots = C_{Z_1}(Z_s)$. It follows from Lemma 6.2.7 that $C_{Z_1}(X)$ is the minimal polynomial of $T \mid Z_1$. This gives statement (i).

(ii) Using Lemma 6.6.4 together with the assumption that Z_i is a companion matrix, it follows that the subspace corresponding to each block Z_i is T_{Z_i}-cyclic. So, Z_i is T-cyclic. As above, the minimal polynomial $M_{T_R \mid E_i}(X) = C_{Z_i}(X)$. The divisibility conditions guarantee that we have a cyclic decomposition of V with respect to T_R. This proves the lemma. //

Of course, a main reason for proving the cyclic-decomposition theorem was to show that every linear operator on a finite-dimensional vector space can be represented by a matrix in rational form. This is our next theorem.

Theorem 6.7.3 (i) *Suppose $T : V \to V$ is a linear operator on a finite-dimensional $\mathbf{F}$ vector space. Then there is an ordered basis $\mathcal{B}$ of V for which $[T]_{\mathcal{B}}$ is a matrix in rational form. Moreover, the matrix $[T]_{\mathcal{B}}$ is uniquely determined (although $\mathcal{B}$ need not be).*
(ii) *Every $n \times n$ matrix with coefficients in a field $\mathbf{F}$ is similar to a unique rational-form matrix.*

Proof: Both (i) and (ii) are consequences of the Cyclic-Decomposition Theorem 6.7.1, together with Lemma 6.7.2 and the fact that invariant direct-sum decompositions correspond to block-diagonal decompositions of matrix representations. //

Examples 6.7.4 (i) Consider the matrix

$$D = \begin{pmatrix} 0 & 3 & -5 \\ 2 & -5 & 10 \\ 1 & -3 & 6 \end{pmatrix}$$

Then one readily computes that the characteristic polynomial of D is $C_D(X) = X^3 - X^2 + X - 1 = (X+1)(X-1)^2$. However, as $D^2 = I_3$, we see that the minimal polynomial of D is $X^2 - 1$. Consequently, the sequence of polynomials arising in a cyclic decompositon of $\mathbf{F}^3$ for T_D must be $X^2 - 1$, $X - 1$. We conclude that the rational form of D is the matrix

$$\begin{pmatrix} 0 & 1 & 0 \\ 1 & 0 & 0 \\ 0 & 0 & 1 \end{pmatrix}$$

In order to find a basis $\mathcal{B}$ for which $[T_D]_{\mathcal{B}}$ is this rational-form matrix, we first compute the eigenspaces of T_D. We find that

$$E_1 = \ker \begin{pmatrix} -1 & 3 & -5 \\ 2 & -6 & 10 \\ 1 & -3 & 5 \end{pmatrix} = span \left\{ \begin{pmatrix} 3 \\ 1 \\ 0 \end{pmatrix}, \begin{pmatrix} 5 \\ 0 \\ -1 \end{pmatrix} \right\}$$

and

$$E_{-1} = \ker \begin{pmatrix} 1 & 3 & -5 \\ 2 & -4 & 10 \\ 1 & -3 & 7 \end{pmatrix} = span \left\{ \begin{pmatrix} -1 \\ 2 \\ 1 \end{pmatrix} \right\}$$

Any vector $v \in \mathbf{F}^3$ which does not lie in either of these subspaces will necessarily have $M_{T|_D,v}(X) = X^2 - 1 = M_{T|_D}(X)$. For example, e_1 is such a vector. We can take

$$Z_1 = Z(T_D, e_1) = span\left\{\begin{pmatrix} 1 \\ 0 \\ 0 \end{pmatrix}, \begin{pmatrix} 3 \\ -5 \\ -3 \end{pmatrix}\right\}$$

For Z_2 we can take the span of any vector in E_1 that does not already lie inside Z_1. For example

$$\begin{pmatrix} 3 \\ 1 \\ 0 \end{pmatrix}$$

is such a vector. We conclude the ordered basis

$$\mathcal{B} = \left\{\begin{pmatrix} 1 \\ 0 \\ 0 \end{pmatrix}, \begin{pmatrix} 3 \\ -5 \\ -3 \end{pmatrix}, \begin{pmatrix} 3 \\ 1 \\ 0 \end{pmatrix}\right\}$$

is one for which $[T_D]_{\mathcal{B}}$ is the rational form of D. Observe that the nonuniqueness of Z_1, Z_2 and $\mathcal{B}$ is illustrated by the variety of choices we had in selecting these elements.

(ii) Set

$$A = \begin{pmatrix} -47 & 71 & 4 & 10 \\ -30 & 43 & 2 & 6 \\ -6 & 10 & 0 & 1 \\ -2 & -4 & 3 & 0 \end{pmatrix}$$

Then it can be computed that $C_A(X) = (X^2+1)^2$, and since $A^2 = -I_4$, we find that $M_A(X) = X^2 + 1$. To find a cyclic decomposition for $T_A : \mathbf{F}^4 \to \mathbf{F}^4$, we arbitrarily choose

$$v_1 = \begin{pmatrix} 1 \\ 0 \\ 0 \\ 0 \end{pmatrix}$$

and set

$$v_2 = T_A(v_1) = \begin{pmatrix} -47 \\ -30 \\ -6 \\ -2 \end{pmatrix}$$

Then as $deg(M_A(X)) = 2$, we see that $V_1 = span\{v_1, v_2\}$ is a T-invariant T-cyclic subspace of $\mathbf{F}^4$ of maximal dimension. Set

$$v_3 = \begin{pmatrix} 0 \\ 0 \\ 0 \\ 1 \end{pmatrix} \quad \text{and} \quad v_4 = T_A(v_3) = \begin{pmatrix} 10 \\ 6 \\ 1 \\ 0 \end{pmatrix}$$

Then $\mathcal{B} = \{v_1, v_2, v_3, v_4\}$ is a basis for $\mathbf{F}^4$. Setting $V_2 = span\{v_3, v_4\}$, we obtain a cyclic decomposition $\mathbf{F}^4 = V_1 \oplus V_2$ for T_A. The rational form of A is

$$[T_A]_{\mathcal{B}} = \begin{pmatrix} 0 & -1 & 0 & 0 \\ 1 & 0 & 0 & 0 \\ 0 & 0 & 0 & -1 \\ 0 & 0 & 1 & 0 \end{pmatrix}$$

Examples 6.7.5 As with the Jordan form, the rational form of a matrix can often be found by investigating the properties of the minimal polynomial and the characteristic polynomial only. Here are some such examples.

(i) Whenever the characteristic and minimal polynomial are the same, the rational form is necessarily the companion matrix of this polynomial.

(ii) Suppose that G is a 4×4 matrix with characteristic polynomial $C_G(X) = (X^2 + 2)(X - 3)^2$. Then, there are two possibilities for the minimal polynomial $M_G(X)$: either $M_G(X) = C_G(X)$ or $M_G(X) = (X^2 + 2)(X - 3)$. In the first case, the rational form of G is necessarily the companion matrix

$$\begin{pmatrix} 0 & 0 & 0 & -18 \\ 1 & 0 & 0 & 12 \\ 0 & 1 & 0 & -11 \\ 0 & 0 & 1 & 6 \end{pmatrix}$$

and in the second case

$$\begin{pmatrix} 0 & 0 & 6 & 0 \\ 1 & 0 & -2 & 0 \\ 0 & 1 & 3 & 0 \\ 0 & 0 & 0 & 3 \end{pmatrix}$$

In the second case the sequence of invariant factors associated with any cyclic decompositon must be $(X^2 + 2)(X - 3) = X^3 - 3X^2 + 2X - 6$, $X - 3$.

(iii) Suppose that A is a 5×5 real matrix with characteristic polynomial $C_A(X) = (X^2 + 1)^2(X - 1)$ and minimal polynomial $M_A = (X^2 + 1)(X - 1)$. Since the product of the minimal polynomial on the cyclic factors must equal $C_A(X)$ and the first such minimal polynomial must be $M_A(X)$, it follows that necessarily $\mathbf{R}^5$ has a T_A-invariant decomposition $\mathbf{R}^5 = Z_1 \oplus Z_2$ where $M_{T_A|z_1}(X) = (X^2 + 1)(X - 1)$ and $M_{T_A|z_1}(X) = (X^2 + 1)$. Consequently, A is similar to the real matrix

$$\begin{pmatrix} 0 & 0 & 1 & 0 & 0 \\ 1 & 0 & -1 & 0 & 0 \\ 0 & 1 & 1 & 0 & 0 \\ 0 & 0 & 0 & 0 & -1 \\ 0 & 0 & 0 & 1 & 0 \end{pmatrix}$$

(iv) Suppose that B is a 6×6 matrix with characteristic polynomial $C_B(X) = (X - 1)^2(X + 1)^4$ and minimal polynomial $M_B(X) = (X - 1)(X + 1)^2$. Then there are two possible sequences of invariant factors in a cyclic decomposition:

$$(X - 1)(X + 1)^2, (X - 1)(X + 1)^2$$

and

$$(X - 1)(X + 1)^2, (X - 1)(X + 1), (X + 1)$$

Thus it follows that B is similar to one of the following two rational-form matrices:

$$\begin{pmatrix} 0 & 0 & 1 & 0 & 0 & 0 \\ 1 & 0 & 1 & 0 & 0 & 0 \\ 0 & 1 & -1 & 0 & 0 & 0 \\ 0 & 0 & 0 & 0 & 0 & 1 \\ 0 & 0 & 0 & 1 & 0 & 1 \\ 0 & 0 & 0 & 0 & 1 & -1 \end{pmatrix} \quad \text{or} \quad \begin{pmatrix} 0 & 0 & 1 & 0 & 0 & 0 \\ 1 & 0 & 1 & 0 & 0 & 0 \\ 0 & 1 & -1 & 0 & 0 & 0 \\ 0 & 0 & 0 & 0 & 1 & 0 \\ 0 & 0 & 0 & 1 & 0 & 0 \\ 0 & 0 & 0 & 0 & 0 & -1 \end{pmatrix}$$

The previous examples illustrate how rational forms arise, and we gave appropriate bases for their representation. Note that we did not

specify a specific algorithm for finding the rational form. Rather, we
determined the possibilities by using the characteristic and minimal
polynomials and then examined the operator to find the appropriate
basis. As illustrated, this procedure works well for reasonably small
matrices (assuming one can compute the characteristic polynomial!).
In the next section we shall give a computational algorithm for de-
terming the invariant factors of a matrix.

We conclude this section with an alternative proof of the existence
and uniqueness of the Jordan form (Theorem 6.4.7). This proof exploits the cyclic-decomposition theorem.

Proof of Theorem 6.4.7:   Let $T : V \rightarrow V$ be a linear operator on a
finite-dimensional vector space V. We assume that

$$M_T(X) = (X - a_1)^{r_1}(X - a_2)^{r_2} \cdots (X - a_s)^{r_s}$$

We apply the primary-decomposition theorem to obtain a T-invariant
direct-sum decompositon $V = V_1 \oplus V_2 \oplus \cdots \oplus V_s$ where $M_{T|_{V_i}}(X) = (X - a_i)^{r_i}$. It suffices to show that there is a basis for V_i in which $T|_{V_i}$
is represented by a Jordan matrix of eigenvalue a_i, because then T can
be represented by a block-diagonal matrix with these matrices as the
diagonal blocks.

We have reduced to the case where $M_T(X) = (X - a)^r$. Consider
the operator $N = T - aI$. We claim that $M_N(X) = X^r$, because clearly,
$N^r = (T - aI)^r = 0$ and since $M_T(X) = (X - a)^r$, no lower power of
X can annihilate $T - aI = N$. We now apply the cyclic-decomposition
theorem to N. The matrix of N with respect to a cyclic decomposition
of V is a block-diagonal matrix, with each block the companion matrix
for the polynomial X^s, $s \leq r$. Such a companion matrix looks like

$$\begin{pmatrix} 0 & 0 & \cdots & 0 & 0 \\ 1 & 0 & \cdots & 0 & 0 \\ 0 & 1 & \cdots & 0 & 0 \\ \vdots & \vdots & \ddots & \vdots & \vdots \\ 0 & 0 & \cdots & 1 & 0 \end{pmatrix}$$

and since the minimal polynomials of the lower blocks divide those of
the higher blocks, these blocks do not increase in size as they descend.
However, the matrix of the operator aI is the matrix aI_n [$n = dim(V)$]
in any basis. It follows that the matrix $[T]_\mathcal{B}$ of T in the basis $\mathcal{B}$ giving

the cyclic decomposition for $N = T - aI$ is the matrix $[N]_B + aI_n$. Such a matrix is a Jordan-form matrix of eigenvalue a. The uniqueness assertions can be derived from the uniqueness assertions in the cyclic-decomposition theorem. This proves the theorem. //

PROBLEMS

6.7.1. Find the cyclic decompositions and the rational forms of A if A is:

(a) $\begin{pmatrix} 0 & 2 & -1 \\ 2 & 3 & -2 \\ -1 & -2 & 0 \end{pmatrix}$ (b) $\begin{pmatrix} 3 & 1 & 0 \\ 0 & 3 & 9 \\ 0 & 0 & 5 \end{pmatrix}$ (c) $\begin{pmatrix} 3 & 0 & 0 \\ 0 & 3 & 9 \\ 0 & 0 & 5 \end{pmatrix}$

6.7.2. Find the cyclic decompositions and the rational forms of A if A is:

(a) $\begin{pmatrix} 1 & 3 & -3 \\ 3 & 1 & -3 \\ 3 & 3 & -5 \end{pmatrix}$ (b) $\begin{pmatrix} 3 & 0 & -1 \\ 0 & 2 & 0 \\ 1 & 0 & 1 \end{pmatrix}$ (c) $\begin{pmatrix} 1 & 2 & 2 & 2 \\ 0 & 1 & 0 & 2 \\ 0 & 0 & 1 & 2 \\ 0 & 0 & 0 & 1 \end{pmatrix}$

6.7.3. Give all possible real rational-form matrices R with $C_R(X) = (X^2+1)^3$. Over an algebraically closed field, how many possible Jordan forms are there?

6.7.4. Find the primary decompositions for each of the matrices in Probs. 6.7.1. and 6.7.2. Give the Jordan forms if they exist.

6.7.5. Suppose that $\mathbf{F}$ is a subfield of the field L. Prove that two $\mathbf{F}$ matrices A and B are similar over $\mathbf{F}$ if and only if they are similar over L.

6.7.6. Show that two 3×3 matrices with the same characteristic and minimal polynomials are similar. Is this true for 4×4 matrices?

6.7.7. For the 8×8 matrix

$$A = \begin{pmatrix} 0 & 0 & 0 & 0 \\ 1 & 0 & 0 & 0 \\ 0 & 0 & 0 & 0 \\ 0 & 0 & 1 & 0 \end{pmatrix} \oplus \begin{pmatrix} 1 & 0 & 0 & 0 \\ 1 & 1 & 0 & 0 \\ 0 & 0 & 1 & 0 \\ 0 & 0 & 0 & 1 \end{pmatrix}$$

find $M_A(X)$, $C_A(X)$, the eigenvalues (with multiplicity) of A, the invariant factors of A, and the rational form of A.

6.7.8. For the 7×7 matrix

$$A = \begin{pmatrix} 0 & 1 & 0 \\ 1 & 0 & 1 \\ 0 & 1 & 0 \end{pmatrix} \oplus \begin{pmatrix} 1 & 0 & 0 & 0 \\ 0 & 1 & 0 & 1 \\ 1 & 0 & 0 & 0 \\ 0 & 0 & 0 & 1 \end{pmatrix}$$

find $M_A(X)$, $C_A(X)$, the eigenvalues (with multiplicity) of A, the invariant factors of A, and the rational form of A.

6.8   Equivalence, Similarity, and the Smith Normal Form

In this section we prove the cyclic-decomposition theorem. The proof we give also gives an algorithm for determining the invariant factors of an arbitrary $n \times n$ matrix. For this proof it is necessary to work with matrices whose entries are polynomials. Recall that the collection of all polynomials in the variable X with coefficients in a field $\mathbf{F}$ is denoted by $\mathbf{F}[X]$. We emphasize that arbitrary quotients of polynomials *do not* lie in $\mathbf{F}[X]$; in fact the only elements of $\mathbf{F}[X]$ that have multiplicative inverses are the nonzero scalars in $\mathbf{F}$ (that is, the nonzero constant polynomials).

We denote by $M_{n \times n}(\mathbf{F}[X])$ the set of all $n \times n$ matrices with coefficients in $\mathbf{F}[X]$, and we denote the elements of $M_{n \times n}(\mathbf{F}[X])$ by $A(X)$, $B(X), \ldots \in M_{n \times n}(\mathbf{F}[X])$ in order to emphasize that polynomials occur as the entries of these matrices. Observe that any element $A(X) \in M_{n \times n}(\mathbf{F}[X])$ can be expressed uniquely as $A(X) = A_n X^n + A_{n-1}X^{n-1}+\cdots+A_1X+A_0$, where the matrices $A_n, A_{n-1}, \ldots, A_1, A_0 \in M_{n \times n}(\mathbf{F})$. We begin with the notion of matrix equivalence. Note that in this definition we allow *both* row and column operations.

Definition 6.8.1 Suppose that $A(X) \in M_{n \times n}(\mathbf{F}[X])$. An *elementary operation over* $\mathbf{F}[X]$ applied to $A(X)$ is one of the following three:

(i)   Interchange two rows or columns of $A(X)$.

(ii)   Multiply a row or column by a nonzero element of $\mathbf{F}$.

(iii) For $f(X) \in \mathbf{F}[X]$ replace a row by itself plus $f(x)$ times another row or replace a column by itself plus $f(x)$ times another column.

Two matrices $A(X)$, $B(X) \in M_{n \times n}(\mathbf{F}[X])$ are said to be *equivalent over* $\mathbf{F}[X]$, denoted $A(X) \sim B(X)$, if one can be obtained form the other by a (finite) sequence of elementary operations over $\mathbf{F}[X]$. A matrix in $M_{n \times n}(\mathbf{F}[X])$ obtained from the identity matrix I_n by a single elementary operation over $\mathbf{F}[X]$ will be called an *elementary matrix* (over $\mathbf{F}[X]$).

Remarks (i) Note that in the row operation (ii) above, it is required that the nonzero multiplication constant be a scalar in $\mathbf{F}$ (and *not* merely a polynomial in $\mathbf{F}[X]$). This is necessary to ensure that the operation can be reversed (for elements in $\mathbf{F}[X]$ which are not scalars in $\mathbf{F}$ fail to have multiplicative inverses).

(ii) Consider the equivalence of matrices in $M_{n \times n}(\mathbf{F})$. Two such matrices will be equivalent if one can be obtained from the other by applying a sequence of ususal row *and* column operations. We know that an $n \times n$ matrix has rank n if and only if it is row equivalent to I_n. More generally, an $n \times n$ matrix has rank r if and only if it is equivalent to the (pseudoidentity) matrix with r 1s on the diagonal and all other entries zero. (To see this, row-reduce the matrix to its reduced echelon form, and then apply column operations to bring all 1s to the diagonal and eliminate all the other entries.)

(iii) If $E(X)$ is an elementary matrix, then $E(X)$ has an elementary inverse $E(X)^{-1}$ in $M_{n \times n}(\mathbf{F}[X])$. [The reasons are the same as for the case of $M_{n \times n}(\mathbf{F})$.] In particular, products of elementary matrices in $M_{n \times n}(\mathbf{F}[X])$ have inverses in $M_{n \times n}(\mathbf{F}[X])$.

(iv) The determinant of any elementary matrix is an element of $\mathbf{F}$. Thus, elementary operations change the determinant of a matrix in $M_{n \times n}(\mathbf{F}[X])$ only by scalar factors.

This first lemma is analogous to Theorem 2.1.4, and its proof is omitted. The point is that row operations correspond to multiplication on the left by elementary matrices, while column operations correspond to multiplication on the right by elementary matrices.

Lemma 6.8.2 $A(X) \sim B(X)$ *if and only if there exist two matrices* $P(X)$, $Q(X) \in M_{n \times n}(\mathbf{F}[X])$, *each a product of elementary matrices, such that* $B(X) = P(X)A(X)Q(X)$.

The reason equivalence of matrices over $\mathbf{F}[X]$ is important is given in the next theorem. This result, together with the algorithm which follows, will enable us to prove the cyclic-decompositon theorem.

Theorem 6.8.3 *Let $A_0, B_0 \in M_{n \times n}(\mathbf{F})$. Then A_0 and B_0 are similar if and only if the characterisic matrices $(XI_n - A_0)$ and $(XI_n - B_0)$ are equivalent over $\mathbf{F}[X]$.*

In order to prove Theorem 6.8.3, it is convenient to give the following lemma first. This lemma is (a special case of) a division algorithm for elements of $M_{n \times n}(\mathbf{F}[X])$.

Lemma 6.8.4 *Let $P(X) \in M_{n \times n}(\mathbf{F}[X])$ and $A(X) = (XI_n - A_0)$ be the characteristic matrix of $A_0 \in M_{n \times n}(\mathbf{F})$. Then there exist matrices $N(X), M(X) \in M_{n \times n}(\mathbf{F}[X])$ and matrices $R, S \in M_{n \times n}(\mathbf{F})$ such that $P(X) = A(X)M(X) + R$ and $P(X) = N(X)A(X) + S$.*

Proof: Suppose that the maximal degree of all the polynomials in $P(X)$ is m and the coefficient of X^m in the ijth entry of $P(X)$ is a_{ij}. Set $P_{m-1}(X) = (a_{ij})X^{m-1}$. Then it is easily seen that the polynomial entries in $A(X)P_{m-1}(X) = (XI^n - A_0)(a_{ij})X^{m-1}$ all have degree at most m and the coefficient of X^m in the ijth entry is a_{ij}. Consequently, all the polynomial entries in $P(X) - A(X)P_{m-1}(X)$ have degree at most $m - 1$. This same argument can be applied to $P(X) - A(X)P_{m-1}(X)$ to produce a matrix $P_{m-2}(X)$ so that all the polynomial entries of $P(X) - A(X)P_{m-1}(X) - A(X)P_{m-2}(X)$ have degree at most $m - 2$. Continuing in this fashion and constructing the polynomial $M(X) = P_{m-1}(X) + P_{m-2}(X) + \cdots + P_0(X)$, we obtain the desired conclusion that the polynomial entries of $R = P(X) - A(X)M(X)$ have degree at most 0. The other assertion is proved by the same argument. The lemma follows. //

Using this lemma we can prove Theorem 6.8.3.

Proof of Theorem 6.8.3: Suppose that A_0 and B_0 are similar matrices in $M_{n \times n}(\mathbf{F})$. Then there exists some invertible matrix $P \in M_{n \times n}(\mathbf{F})$ with $B_0 = PA_0P^{-1}$. Consequently,

$$P(XI_n - A_0)P^{-1} = (PXI_nP^{-1} - PA_0P^{-1}) = (XI_n - B_0)$$

which shows that the characteristic matrices of A_0 and B_0 are equivalent.

Conversely, assume that $(XI_n - A_0)$ and $(XI_n - B_0)$ are equivalent over $\mathbf{F}[X]$. Then by Lemma 6.8.2 there is an expression $(XI_n - B_0) =$

$P(X)(XI_n - A_0)Q(X)$ for matrices $P(X), Q(X) \in M_{n \times n}(\mathbf{F}[X])$ which are products of elementary matrices. Setting $B(X) = (XI_n - B_0)$ and $A(X) = (XI_n - A_0)$ we have $B(X) = P(X)A(X)Q(X)$. Applying Lemma 6.8.4, we can express $P(X) = B(X)P'(X) + R$ and $Q(X) = Q'(X)B(X) + S$ for matrices $P'(X), Q'(X) \in M_{n \times n}(\mathbf{F}[X])$ and $R, S \in M_{n \times n}(\mathbf{F})$. Note that our original equation $B(X) = P(X)A(X)Q(X)$ gives $P(X)A(X) = B(X)Q(X)^{-1}$ and $A(X)Q(X) = P(X)^{-1}B(X)$. Substituting these expressions, we find:

$$
\begin{aligned}
B(X) \\
&= P(X)A(X)Q(X) \\
&= [B(X)P'(X) + R]A(X)Q(X) \\
&= B(X)P'(X)P(X)^{-1}B(X) + RA(X)Q(X) \\
&= B(X)P'(X)P(X)^{-1}B(X) + RA(X)[Q'(X)B(X) + S] \\
&= B(X)P'(X)P(X)^{-1}B(X) + RA(X)Q'(X)B(X) + RA(X)S
\end{aligned}
$$

Using $R = P(X) - B(X)P'(X)$ and expanding the middle summand of the last expression gives:

$$
\begin{aligned}
RA(X)Q'(X)B(X) \\
&= [P(X) - B(X)P'(X)]A(X)Q'(X)B(X) \\
&= [P(X)A(X)Q'(X) - B(X)P'(X)A(X)Q'(X)]B(X) \\
&= [B(X)Q(X)^{-1}Q'(X) - B(X)P'(X)A(X)Q'(X)]B(X) \\
&= B(X)[Q(X)^{-1}Q'(X) - P'(X)A(X)Q'(X)]B(X)
\end{aligned}
$$

Substituting this expression into the previous expression for $B(X)$ shows that $B(X) = B(X)T(X)B(X) + RA(X)S$ where $T(X) = P'(X)P(X)^{-1} + Q(X)^{-1}Q'(X) - P'(X)A(X)Q'(X)$.

To prove the theorem, we first show that $T(X) = 0$. Since the entries in $B(X)$ and $RA(X)S$ have degree at most one, the same must be true for the entries in $B(X)T(X)B(X)$. If $T(X) = T_m X^m + T_{m-1}X^{m-1} + \cdots + T_0$, then since $B(X) = (XI_n - B_0)$, we find that $B(X)T(X)B(X) = T_m X^{m+2} + T'(X)$ where the degrees of the entries of $T'(X)$ are each less than $m + 2$. Consequently, $T_m = 0$, and the conclusion that $T(X) = 0$ follows.

We now have that $B(X) = RA(X)S$. Expanding, we see that $XI_n = XRS$ and $B = RAS$. Consequently, $R = S^{-1}$ and $B = S^{-1}AS$. This proves the theorem. //

Our next task is to give an algorithm which enables us to tell if matrices in $M_{n \times n}(\mathbf{F}[X])$ are equivalent. The idea is somewhat like gaussian elimination, except that we perform elementary row and column operations (over $\mathbf{F}[X]$) to reduce the matrix to diagonal form. Two matrices will be equivalent if they have the same diagonal entries. As an example, we consider the characteristic matrix $(XI_3 - C)$ where C is the companion matrix of the polynomial $X^3 - 2X^2 - X - 1$. Applying elementary operations, we have:

$$
\begin{pmatrix} X & 0 & -1 \\ -1 & X & -1 \\ 0 & -1 & X-2 \end{pmatrix} \mapsto \begin{pmatrix} X & 0 & -1 \\ -1 & 0 & X(X-2)-1 \\ 0 & -1 & X-2 \end{pmatrix}
$$

$$
\mapsto \begin{pmatrix} 0 & 0 & X[X(X-2)-1]-1 \\ -1 & 0 & X(X-2)-1 \\ 0 & -1 & X-2 \end{pmatrix}
$$

$$
= \begin{pmatrix} 0 & 0 & X^3-2X^2-X-1 \\ -1 & 0 & X^2-2X-1 \\ 0 & -1 & X-2 \end{pmatrix}
$$

$$
\mapsto \begin{pmatrix} 0 & 0 & X^3-2X^2-X-1 \\ -1 & 0 & 0 \\ 0 & -1 & 0 \end{pmatrix}
$$

$$
\mapsto \begin{pmatrix} X^3-2X-X-1 & 0 & 0 \\ 0 & 1 & 0 \\ 0 & 0 & 1 \end{pmatrix}
$$

Observe that the resulting matrix has determinant $X^3 - 2X^2 - X - 1$, which is the characteristic polynomial of the original companion matrix C (as it should be).

The next result generalizes the preceding calculation to arbitrary companion matrices. We use the notation $Diag(f_1, f_2, \ldots, f_n)$ to denote the diagonal matrix with diagonal iith entry f_i.

Lemma 6.8.5 *The characteristic matrix of the companion matrix associated with the polynomial $f(X)$ is equivalent to the diagonal matrix $Diag(1, 1, \ldots, 1, f(X))$.*

Proof: The matrix in question is

$$\begin{pmatrix} X & 0 & \cdots & 0 & a_0 \\ -1 & X & \cdots & 0 & a_1 \\ \vdots & \vdots & \ddots & \vdots & \vdots \\ 0 & 0 & \cdots & X & a_{n-2} \\ 0 & 0 & \cdots & -1 & X - a_{n-1} \end{pmatrix}$$

Starting at the bottom, we add X times the bottom row to the next row, then add X times this second-from-bottom row to the next, and so forth. The result is the matrix

$$\begin{pmatrix} 0 & 0 & \cdots & 0 & X^n - a_{n-1}X^{n-1} - \cdots - a_1 X - a_0 \\ -1 & 0 & \cdots & 0 & X^{n-1} - a_{n-1}X^{n-2} - \cdots - a_1 \\ \vdots & \vdots & \ddots & \vdots & \vdots \\ 0 & 0 & \cdots & 0 & X^2 - a_{n-1}X - a_{n-2} \\ 0 & 0 & \cdots & -1 & X - a_{n-1} \end{pmatrix}$$

Applying column operations, we can eliminate all polynomials, except the upper right-hand polynomial, which is $f(X)$. Interchanging rows, multiplying by -1 where necessary, gives the desired conclusion. //

The matrix produced by Lemma 6.8.5 is a special case of an important class of matrices defined next.

Definition 6.8.6 A diagonal matrix D in $M_{n \times n}(\mathbf{F}[X])$ is said to be in *Smith normal form* if $D = Diag(f_1(X), f_2(X), \ldots, f_n(X))$, where each polynomial $f_i(X)$ is monic and divides $f_{i+1}(X)$ for $i = 1, 2, \ldots, n - 1$.

The next result will enable us to show that an arbitrary charactrisitic matrix is equivalent to the characteristic matrix of a direct sum of companion matrices.

Theorem 6.8.7 *Any matrix in $M_{n \times n}(\mathbf{F}[X])$ can be reduced by elementary operations to a Smith-normal-form matrix.*

Proof: Suppose that $A(X) \in M_{n \times n}(\mathbf{F}[X])$. Let $A'(X)$ be a matrix which is equivalent to $A(X)$ and is chosen to have an entry $f(X)$ which has the smallest possible degree among all entries of matrices equivalent to $A(X)$. By row and column interchanges, we can assume that $f(X)$ is the first-row, first-column entry of $A'(X)$. Suppose that $g(X)$ is some other first-column entry of $A'(X)$. By the division algorithm, we can express $g(X) = f(X)s(X) + r(X)$ for polynomials $s(X), g(X) \in \mathbf{F}[X]$ with $deg(r(X)) < deg(f(X))$. An elementary row operation can replace the entry $g(X)$ by $r(X)$. Our minimality assumption on the degree of $f(X)$ guarantees that $r(X) = 0$. Hence, $f(X)$ divides any other polynomial in the first column of $A'(X)$, and for the same reasons, $f(X)$ divides any entry in the first row of $A'(X)$.

It follows that $A'(X)$ can be reduced by elementary operations to a block-diagonal matrix of the form

$$\begin{pmatrix} f(X) & 0 & \cdots & 0 \\ \hline 0 & & & \\ \vdots & & A_1(X) & \\ 0 & & & \end{pmatrix}$$

where $A_1(X) \in M_{n-1 \times n-1}(\mathbf{F}[X])$. Moreover, an elementary column operation can move any entry of $A_1(X)$ into the first column of this matrix, without changing the first-row, first-column entry $f(X)$. The argument just given shows that $f(X)$ must divide this entry; in other words $f(X)$ must divide every entry of $A_1(X)$. By induction, the $(n-1) \times (n-1)$ matrix $A_1(X)$ can be reduced to a Smith-normal-form-matrix $S_1(X)$. As $f(X)$ divides each entry of $S_1(X)$, the matrix

$$\begin{pmatrix} f(X) & 0 & \cdots & 0 \\ \hline 0 & & & \\ \vdots & & S_1(X) & \\ 0 & & & \end{pmatrix}$$

is a Smith-normal-form matrix equivalent to $A(X)$. This proves the theorem. //

As a consequence of our work we obtain the existence of the Rational form.

Corollary 6.8.8 *Every matrix $A \in M_{n \times n}(\mathbf{F})$ is similar to a rational-form matrix.*

Proof: Suppose that the characteristic matrix $C(X) = (XI_n - A)$ is equivalent to the Smith-normal-form matrix

$$S(X) = Diag(f_1(X), f_2(X), \ldots, f_n(X))$$

Since $S(X) = P(X)C(X)Q(X)$ for $P(X), Q(X) \in M_{n \times n}(\mathbf{F}[X])$ which are products of elementary matrices, we conclude that $det(C(X)) = k \cdot det(S(X))$ for some $k \in \mathbf{F}$. Since $det(C(X))$ is the characteristic polynomial of C, we find that the sum of the degrees of the $f_i(X)$ is n.

Rearranging the rows and columns of $S(X)$, we see that the matrix $C(X)$ is equivalent to a direct sum of $d_i \times d_i$ matrices of the form $Diag(1, 1, \ldots, 1, f_i(X))$ where the degree of $f_i(X) = d_i$. Applying Lemma 6.8.5, each $d \times d$ block $Diag(1, 1, \ldots, 1, f_i(X))$ is equivalent to the characterisitic matrix of the companion matrix of $f_i(X)$. This shows that $C(X)$ is equivalent to the characteristic matrix of a rational-form matrix. Applying Theorem 6.8.3, we find that A is similar to a rational-form matrix. This proves the corollary. //

The last theorem of this section is the uniqueness of the Smith normal form. In order to prove this result, we need the following definition.

Definition 6.8.9 For any $A(X) \in M_{n \times n}(\mathbf{F}[X])$ we define $d_k(A(X))$ to be the monic polynomial which is the greatest common divisor of the determinants of all the $k \times k$ submatrices of $A(X)$.

This next lemma shows that the function d_k is invariant under elementary operations.

Lemma 6.8.10 *Suppose that $A(X)$ and $B(X)$ are equivalent matrices in $M_{n \times n}(\mathbf{F}[X])$. Then $d_k(A(X)) = d_k(B(X))$ for all k.*

Proof: It suffices to show that d_k does not change under elementary operations. So assume that $B(X)$ has been obtained from $A(X)$ by a single elementary operation. If rows (or columns) of $A(X)$ have been interchanged, then the collection of all $k \times k$ submatrices of the two matrices is the same, except for possible row (or column) interchanges. Since the determinant changes by only a sign upon such interchanges, the greatest common divisor of these determinants does not change. If a row (or column) of $A(X)$ is multiplied by a nonzero scalar $c \in \mathbf{F}$, the determinants of the $k \times k$ submatrices are either unchanged or multiplied by c. This does not change their greatest common divisor.

Finally, suppose that the ith row R_i of A has been replaced by $R_i + p(X)R_j$ where R_j is the jth row of $A(X)$. Let A_k denote some k $\times k$ submatrix of $A(X)$, and suppose that B_k denotes the corresponding

$k \times k$ submatrix of $B(X)$. In case the ith row of $A(X)$ does not occur in A_k, then $A_k = B_k$. If both rows i and j occur in the A_k submatrix, then B_k has been obtained from A_k by adding a multiple of one row to another, so $det(A_k) = det(B_k)$. Finally suppose that the ith row of $A(X)$ occurs in A_k, while the jth row does not. We denote by C_k the matrix obtained from A_k by replacing the ith row of $A(X)$ by the jth row. Note that C_k is, after possibly some row interchanges, a $k \times k$ submatrix of $A(X)$. Now, all rows of A_k, B_k, and C_k are identical except for ith. This row of B_k is the sum of the corresponding row of A_k with $p(X)$ times the corresponding row of C. Applying the n-linearity of det, we conclude that $det(B_k) = det(A_k) + p(X)det(C_k)$, in other words $det(B_k)$ is a sum of multiples of determinants of $k \times k$ submatrices of $A(X)$. From this it follows that the greatest common divisors of the determinants of the $k \times k$ submatrices of $A(X)$ and $B(X)$ are the same. $//$

We can now prove the uniqueness of the Smith normal form.

Theorem 6.8.11 *Any matrix in* $M_{n \times n}(\mathbf{F}[X])$ *can be reduced by elementary operations to a* unique *Smith-normal-form matrix.*

Proof: To verify the uniqueness assertion, it suffices to show that two Smith normal form matrices are equivalent if and only if they are equal. By Lemma 6.8.10 two equivalent Smith-normal-form matrices $S_1(X) = Diag(f_1, f_2, \ldots, f_n)$ and $S_2(X) = Diag(g_1, g_2, \ldots, g_n)$ have the same d_k values. Clearly $d_n(S_1(X)) = f_1 f_2 \cdots f_n$ and $d_n(S_2(X)) = g_1 g_2 \ldots g_n$, so the products of these polynomials must be the same. Next, by the divisibility condition in the definition of the Smith normal form, we see that $d_{n-1}(S_1(X)) = f_1 f_2 \cdots f_{n-1} = d_{n-1}(S_2(X)) = g_1 g_2 \cdots g_{n-1}$. Consequently, $f_n = g_n$. Continuing in this fashion, comparing d_{n-2}, d_{n-3}, and so forth, we conclude that $S_1(X) = S_2(X)$. $//$

As an immediate application of Theorem 6.8.11 we obtain the following.

Corollary 6.8.12 *Every matrix in* $M_{n \times n}(\mathbf{F})$ *is similar to a unique rational-form matrix.*

Whenever $A \in M_{n \times n}(\mathbf{F})$ and the Smith normal form of the characteristic matrix $(XI_n - A)$ is $Diag(f_1, f_2, \ldots, f_n)$, the nonconstant polynomials $f_n, f_{n-1}, \ldots, f_r$ $(r \leq n)$ are the invariant factors of A. (Recall that the invariant factors are the minimal polynomials of the blocks occurring in the rational form of A.)

Although the proof of Theorem 6.8.7 (giving the existence of the Smith normal form) is not entirely constructive, the ideas in the proof can be used to give an explicit algorithm for determining the Smith normal form. The basic idea is the following. In the proof of Theorem 6.8.7 we started with a matrix $A'(X)$ equivalent to A with first-row, first-column entry a polynomial $f(X)$ whose degree was the smallest possible. Such a polynomial entry can always be produced by elementary operations exploiting the division algorithm. This polynomial is the upper left-hand entry of the Smith normal form. The full Smith normal form can then be computed inductively as described in the proof of Theorem 6.8.7. We illustrate this process in the following example.

Example 6.8.13 Let

$$A = \begin{pmatrix} 2 & -4 & 1 & 3 \\ 2 & -3 & 0 & 2 \\ 0 & -1 & 1 & 2 \\ 1 & -1 & -1 & 0 \end{pmatrix}$$

Applying elementary operations to the matrix $XI_4 - A$, we have

$$\begin{pmatrix} X-2 & 4 & -1 & -3 \\ -2 & X+3 & 0 & -2 \\ 0 & 1 & X-1 & -2 \\ -1 & 1 & 1 & X \end{pmatrix}$$

$$\mapsto \begin{pmatrix} 1 & -1 & -1 & X \\ -2 & X+3 & 0 & -2 \\ 0 & 1 & X-1 & -2 \\ X-2 & 4 & -1 & -3 \end{pmatrix}$$

$$\mapsto \begin{pmatrix} 1 & -1 & -1 & -X \\ 0 & X+1 & -2 & -2X-2 \\ 0 & 1 & X-1 & -2 \\ 0 & X+2 & X-3 & X^2-2X-3 \end{pmatrix}$$

$$\longmapsto \begin{pmatrix} 1 & 0 & 0 & 0 \\ 0 & X+1 & -2 & -2X-2 \\ 0 & 1 & X-1 & -2 \\ 0 & X+2 & X-3 & X^2-2X-3 \end{pmatrix}$$

$$\longmapsto \begin{pmatrix} 1 & 0 & 0 & 0 \\ 0 & 1 & X-1 & -2 \\ 0 & X+1 & -2 & -2X-2 \\ 0 & X+2 & X-3 & X^2-2X-3 \end{pmatrix}$$

$$\longmapsto \begin{pmatrix} 1 & 0 & 0 & 0 \\ 0 & 1 & X-1 & X-2 \\ 0 & 0 & -X^2-1 & 0 \\ 0 & 0 & -X^2-2 & X^2+1 \end{pmatrix}$$

$$\longmapsto \begin{pmatrix} 1 & 0 & 0 & 0 \\ 0 & 1 & 0 & 0 \\ 0 & 0 & X^2+1 & 0 \\ 0 & 0 & X^2+1 & X^2+1 \end{pmatrix}$$

$$\longmapsto \begin{pmatrix} 1 & 0 & 0 & 0 \\ 0 & 1 & 0 & 0 \\ 0 & 0 & X^2+1 & 0 \\ 0 & 0 & 0 & X^2+1 \end{pmatrix}$$

We see that the characteristic polynomial of A is $C_A(X) = (X^2 + 1)^2$, the minimal polynomial of A is $M_A(X) = X^2 + 1$, and A has invariant factors $X^2 + 1$, $X^2 + 1$. In particular, the rational form of A is

$$\begin{pmatrix} 0 & -1 & 0 & 0 \\ 1 & 0 & 0 & 0 \\ 0 & 0 & 0 & -1 \\ 0 & 0 & 1 & 0 \end{pmatrix}$$

PROBLEMS

6.8.1. Find the Smith normal form for the matrix $XI - A$ for the following matrices A. Use the Smith normal form to determine the rational form of A.

(a) $\begin{pmatrix} 1 & 1 & 0 \\ 0 & 1 & 0 \\ 0 & 0 & 1 \end{pmatrix}$ (b) $\begin{pmatrix} 2 & 3 \\ 3 & 2 \end{pmatrix}$ (c) $\begin{pmatrix} 1 & 1 & 2 \\ 2 & 0 & 2 \\ 2 & 2 & 4 \end{pmatrix}$

6.8.2. Find the Smith normal form for the matrix $XI - A$ for the following matrices A. Use the Smith normal form to determine the rational form of A.

(a) $\begin{pmatrix} 1 & 0 & 0 \\ 2 & 3 & 6 \\ 5 & 6 & 12 \end{pmatrix}$ (b) $\begin{pmatrix} 2 & 2 & 0 & 0 \\ 1 & 3 & 0 & 0 \\ 1 & 0 & 0 & 1 \\ 0 & 1 & 1 & 0 \end{pmatrix}$ (c) $\begin{pmatrix} 2 & 0 & 0 \\ 1 & 2 & 0 \\ 1 & 1 & 2 \end{pmatrix}$

6.8.3. For which $n \times n$ matrices A does the Smith normal form of $A - XI_n$ have the form $Diag(f_1(X), f_2(X), \ldots, f_n(X))$ where each $f_i(X)$ is a nonscalar polynomial?

6.8.4. (a) Find the characteristic polynomial and invariant factors of

$$\begin{pmatrix} 0 & 1 & 1 & 0 & 0 & 1 \\ 1 & 0 & 0 & 0 & 1 & 1 \\ 0 & 0 & 1 & 0 & 0 & 0 \\ 0 & 0 & 0 & 0 & 1 & 1 \\ 1 & 0 & 1 & 0 & 0 & 0 \\ 0 & 0 & 0 & 0 & 0 & 1 \end{pmatrix}$$

(b) Find the characteristic polynomial and invariant factors of

$$\begin{pmatrix} 0 & 1 & 1 & 0 & 0 & 1 \\ 1 & 0 & 0 & 1 & 1 & 1 \\ 0 & 0 & 1 & 0 & 0 & 0 \\ 0 & 0 & 0 & 1 & 1 & 1 \\ 1 & 0 & 1 & 0 & 0 & 0 \\ 0 & 0 & 0 & 1 & 0 & 1 \end{pmatrix}$$

6.9 A Second Proof of the Cyclic-Decomposition Theorem

The cyclic-decomposition theorem can be deduced as a special case of a general result from abstract algebra known as the structure theorem for finitely generated modules over a principal ideal domain. (This latter result generalizes the structure theory of finitely generated abelian groups. For details the reader may consult the books on algebra listed in the Bibliography.) The proof given here for the cyclic-decomposition theorem is a version of the usual proof of this result, written in terms of the current setting of linear algebra. The reader familiar with this area of abstract algebra should be able to see that, after making the appropriate translations, the given proof also proves this more general result.

We emphasize however, that it is *not* necessary to have a background in abstract algebra (or to understand module theory) in order to understand this proof. What is crucial for reading this section is that the reader be comfortable with the viewpoint of polynomials acting upon linear operators. We begin with a few results about cyclic subspaces and polynomial ideals. This next theorem collects together a number of facts about T-cyclic subspaces. Recall (Definition 6.6.2) that whenever $T : V \to V$ and $v \in V$, $M_{T,v}(X)$ is the (unique) monic polynomial of least degree for which $M_{T,v}(T)(v) = \vec{0}$.

Theorem 6.9.1 *Suppose that $T : V \to V$ is a linear operator, $v \in V$, and $s = dim(Z(T, v))$. Then:*
(i)   *$\{v, T(v), T^2(v), \ldots, T^{s-1}(v)\}$ is a basis for $Z(T, v)$. In particular, every element $w \in Z(T, V)$ can be expressed uniquely as $w = g(T)(v)$ where $g(X)$ is a polynomial with $deg(g(X)) < s$.*
(ii)   *$deg(M_{T,v}(X)) = s$.*
(iii)   *$f(X) \in An(T \,|\, Z(T,v))$ if and only if $f(T)(v) = \vec{0}$.*
(iv)   *$M_{T,v}(X) \mid M_T(X)$, so $M_{T,v}(X) \mid C_T(X)$.*
(v)   *Every T-invariant subspace of $Z(T, v)$ is T-cyclic.*

Proof: (i) The first statement is proved in Lemma 6.2.9 (i), which says that $\mathcal{B} = \{v, T(v), \ldots, T^{s-1}(v)\}$ is a basis for cyclic subspace $Z(T, v)$. Since every element of $Z(T, v)$ can be written uniquely as a linear combination of $v, T(v), \ldots, T^{s-1}(v)$, the uniqueness of the polynomial $g(X)$ follows. This is the second part of (i).

(ii)   The linear independence of $v, T(v), \ldots, T^{s-1}(v)$ shows that whenever $deg(g(X)) < s$, then $g(T)(v) \neq \vec{0}$. Consequently, $deg(M_{T,v}(X)) \geq s$. But $dim(Z(T, v)) = s$, so $deg(M_{T,v}(X)) \leq s$ as well. This proves (ii).

(iii)   We define $An_{T,v}$ to be the ideal $\{f(X) \mid f(T)(v) = \vec{0}\}$. We claim that $An_{T,v} = An(T \mid Z(T,v))$. If $f(X) \in An(T \mid Z(T,v))$, then $f(T)(v) = \vec{0}$, so $f(X) \in An_{T,v}$. Conversely, if $f(T)(v) = \vec{0}$, then $\vec{0} = T(f(T)(v)) = f(T)(T(v))$, $\vec{0} = T^2(f(T)(v)) = f(T)(T^2(v)), \ldots$. Thus $f(T)$ annihilates a basis for $Z(T, v)$, so it follows that $f(T)(w) = \vec{0}$ for all $w \in Z(T, v)$. We have $An_{T,v} = An(T \mid (T,v))$ and (iii) follows.

(iv)   Evidently, the definitions give $M_T(X) \in An_{T,v}$. However, we just saw that $An_{T,v} = An(T|_{Z(T,v)}) = (M_{T,v}(X))$. Thus $M_{T,v}(X) \mid M_T(X)$ and (iv) follows.

(v) Suppose that $W \subseteq Z(T,v)$ is T-invariant. We may assume that $W \neq Z(T,v)$. We denote by $f(X) = M_{T,v}(X) = M_{T|_{Z(T,v)}}(X)$. Let $g(X) = M_{T|_W}(X)$. Then as $f(X) \in An(T|_W)$, we find that $g(X) \mid f(X)$. We express $f(X) = g(X)h(X)$ and set $u = deg(g(X))$, $s = deg(f(X))$, so we have $s - u = deg(h(X))$. To prove the result, we will show that $W = Z(T, h(T)(v))$.

Let $w \in ker(g(T))$. By part (i) we can express $w = r(T)(v)$, with $deg(r(X)) < s$. Since $\vec{0} = g(T)(w) = g(T)r(T)(v)$, we have that $f(X) \mid g(X)r(X)$. As $f(X) = g(X)h(X)$, we find that $h(X) \mid r(X)$, and we express $r(X) = h(X)r'(X)$. Since $deg(r(X)) < s$, we have $deg(r'(X)) < s - (s - u) = u$. It follows from this degree calculation that every element w of $ker(g(T))$ can be written as $r'(T)(h(T)(v))$ where $deg(r'(X)) < u$. Consequently, w must be a linear combination of $h(T)(v)$, $T(h(T)(v))$, $T^2(h(T)(v)), \ldots, T^{u-1}(h(T)(v))$. This shows that $ker(g(T))$ is a subspace of the u-dimensional vector space $Z(T, h(T)(v))$. Since $g(X) = M_{T|_W}(X)$, we have that $W \subseteq ker(g(T))$ and we know that $dim(W) \geq deg(g(X)) = u$. Consequently, $W = ker(g(T)) = Z(T, h(T)(v))$, and (v) is proved. //

For the next lemma we need to recall another definition from the theory of polynomials.

Definition 6.9.2 Two polynomials $f(X)$ and $g(X)$ are called *relatively prime* if $(f(X), g(X)) = \mathbf{F}[X]$.

Using this we obtain the following.

Lemma 6.9.3 *Suppose that $T : V \to V$ is a linear operator.*
(i) *If $W \subseteq V$ is a T-invariant subspace, $v \in V$, $v \notin W$, and $M_{T,v}(X)$ is prime, then $Z(T,v) \cap W = \{\vec{0}\}$.*
(ii) *If $v_1, v_2 \in V$ and the polynomials $M_{T,v_1}(X)$ and $M_{T,v_2}(X)$ are relatively prime, then $Z(T, v_1) \cap Z(T, v_2) = \{\vec{0}\}$.*

Proof: (i) Observe that $W \cap Z(T,v)$ is a T-invariant subspace of V. Let $w \in W \cap Z(T,v)$. If $w \neq \vec{0}$, then Theorem 6.9.1 (iv) applied to the subspace $Z(T,v)$ (which contains w) gives $M_{T,w}(X) \mid C_{T|_{Z(T,v)}}(X) = M_{T,v}(X)$. As $M_{T,v}(X)$ is prime, we find $M_{T,w}(X) = M_{T,v}(X)$. Hence,

$dim(Z(T,w)) = deg(M_{T,w}(X)) = deg(M_{T,v}(X)) = dim(Z(T,v))$. As $w \in W \cap Z(T,v)$, we have $Z(T,w) \subseteq W$ and $Z(T,w) \subseteq Z(T,v)$. Thus, $Z(T,v) = Z(T,w) \subseteq W$. But then $v \in W$, a contradiction. This proves (i).

(ii) By hypothesis there exist $f_1(X)$ and $f_2(X)$ so that

$$f_1(X)M_{T,v_1}(X) + f_2(X)M_{T,v_2}(X) = 1 \in \mathbf{F}[X]$$

Suppose that $v \in Z(T,v_1) \cap Z(T,v_2)$. Then since $\vec{0} = M_{T,v_1}(T)(v) = M_{T,v_2}(T)(v)$, we have

$$
\begin{aligned}
v &= I(v) = (f_1(T)M_{T,v_1}(T) + f_2(T)M_{T,v_2}(T))(v) \\
&= f_1(T)M_{T,v_1}(T)(v) + f_2(T)M_{T,v_2}(T)(v) \\
&= \vec{0} + \vec{0} = \vec{0}
\end{aligned}
$$

a contradiction. This proves the lemma. //

The proof we now give of the cyclic-decomposition theorem begins by first proving the result in the case where the minimal polynomial is a power of a prime polynomial (Theorem 6.9.5) and then proceeds by using the primary-decomposition theorem to "patch" together the general result. Before proving Theorem 6.9.5, we illustrate some of the ideas needed in its proof by examining the case where $M_T(X)$ is a prime polynomial. In this case a cyclic decomposition can be constructed by pushing some observations made earlier in Example 6.7.4 (ii) to their logical conclusion.

Suppose that $T : V \to V$ is a linear operator and $M_T(X)$ is a prime polynomial. We claim that the cyclic-decomposition theorem holds with $V_i = Z(T,v_i)$ where the v_i can be chosen arbitrarily subject only to the condition that $v_i \notin V_1 + V_2 + \cdots + V_{i-1}$. To see this, choose $v_1 \in V$ with $v_1 \neq \vec{0}$. By Theorem 6.9.1 (iv) $M_{T,v_1}(X) \mid M_T(X)$ which is prime, so $M_{T,v_1}(X) = M_T(X)$. Set $V_1 = Z(T,v_1)$. Choose $v_2 \in V$ with $v_2 \notin V_1$. Set $V_2 = Z(T,v_2)$. Then, as above, $M_{T,v_2}(X) = M_T(X)$ is prime, so by Lemma 6.9.3 (i) $V_1 \cap V_2 = \{\vec{0}\}$. Choose $v_3 \in V$ with $v_3 \notin V_1 + V_2$. Then again by Lemma 6.9.3 (i) $V_3 = Z(T,v_3)$ satisfies $V_3 \cap (V_1 + V_2) = \{\vec{0}\}$. Continuing this process yields a T-invariant direct-sum decomposition $V = V_1 \oplus V_2 \oplus \cdots \oplus V_s$ with each $V_i = Z(T,v_i)$ and $M_{T|V_i}(X) = M_T(X)$. The conditions required by the cyclic-decomposition theorem now follow in this case.

Unfortunately, the nice argument just given does *not* generalize from the prime case to the prime-power case. The difficulty is that the generating vectors $v_1, v_2, \ldots$ for the cyclic subspaces cannot be chosen in an arbitrary fashion. In fact, one must be extremely careful in choosing such vectors. For example, suppose that $T : V \to V$ is an operator with $M_T(X) = p(X)^2$ where $p(X)$ is a prime polynomial of degree n. Assume that $v_1 \in V$ where $p(T)(v_1) \neq \vec{0}$. Then as $p(T)^2(v_1) = \vec{0}$, we see that $M_{T,v_1}(X) = p(X)^2$, and thus $dim(Z(T, v_1)) = 2n$. Now suppose that $v_2 \notin Z(T, v_1)$ and $p(T)(v_2) = \vec{0}$. Then we also have that $v_1 + v_2 \notin Z(T, v_1)$. Note however that $p(T)(v_1 + v_2) = p(T)(v_1) + p(T)(v_2) = p(T)(v_1) \neq \vec{0} \in Z(T, v_1)$. This shows that $dim(Z(T, v_1 + v_2)) = 2n$ and $p(T)(v_1) \in Z(T, v_1) \cap Z(T, v_1 + v_2) \neq \{\vec{0}\}$. Consequently, we do not have a direct sum of the cyclic subspaces $Z(T, v_1)$ and $Z(T, v_1 + v_2)$ where the second vector (in this case $v_1 + v_2$) is chosen to satisfy the condition that it avoids $Z(T, v_1)$.

The difficulty just observed motivates the next lemma. Suppose that $T : V \to V$ has minimal polynomial $p(X)^r$, which is a power of a prime polynomial $p(X)$. If $v \in V$, then we know that $p(T)^r(v) = \vec{0}$, but possibly $p(T)^e(v) = \vec{0}$ for lower exponents e as well. Whenever $p(T)^e(v) = \vec{0}$ while $p(T)^{e-1}(v) \neq \vec{0}$, we shall say that v has *exact exponent* e (for the operator T). In other words, v has exact exponent e whenever $M_{T,v}(X) = p(X)^e$. Thus, if v has exact exponent e and $d = deg(p(X))$, then the subspace $Z(T, v)$ has dimension de.

We use quotient spaces (introduced in Sec. 6.3) in order to apply induction on the dimension of V in the proof of Theorem 6.9.5. Recall that whenever $W \subseteq V$ is a subspace, V/W (sometimes denoted $\overline{V}$) is the vector space whose elements are the cosets $v + W$ where $v \in V$. We denote by $\pi : V \to V/W$ the surjective linear transformation defined by $\pi(v) = v + W$. If, in addition, W is a T-invariant subspace, then we denote by $\overline{T} : V/W \to V/W$ the operator on V/W defined by $\overline{T}(v + W) = T(v) + W$. The point of the following lemma is that one can always "pull back" vectors (via π) from a quotient by a cyclic subspace in a way to match the exact exponents.

Lemma 6.9.4 *Suppose that $T : V \to V$ is a linear operator with $M_T(X) = p(X)^r$ where $p(X)$ is a monic prime polynomial. Let $v \in V$ have exact exponent r, set $W = Z(T, v) \subseteq V$ and $\overline{V} = V/W$. Suppose that $\overline{u} \in \overline{V}$ has exact exponent e (for the operator $\overline{T} : \overline{V} \to \overline{V}$). Then there is some $u' \in V$ of exact exponent e with $\pi(u') = \overline{u} \in V$.*

Proof: Choose any $u \in V$ with $\pi(u) = \bar{u}$. Then $p(T)^{e-1}(u) \neq \vec{0}$ as $\vec{0} \neq p(\bar{T})^{e-1}(\bar{u}) = p(T)^{e-1}(u) + W \in \bar{V}$. If $p(T)^e(u) = \vec{0}$, then we are done, so we assume that $p(T)^e(u) \neq \vec{0}$. It follows, using $p(\bar{T})^e(\bar{u}) = \vec{0} \in \bar{V}$, that $p(T)^e(u) = w$ for some $w \in W$. Suppose u has exact exponent s where $r \geq s > e$. Then, $\vec{0} = p(T)^s(u) = p(T)^{s-e}(p(T)^e(u)) = p(T)^{s-e}(w)$ and in addition $\vec{0} \neq p(T)^{s-1}(u) = p(T)^{s-1-e}(p(T)^e(u)) = p(T)^{s-1-e}(w)$. This shows that the exact exponent of w must be $s - e$.

We express $w \in Z(T, v)$ as $w = f(T)(v)$ for some polynomial $f(X)$. Since $p(T)^{s-e}(f(T)(v)) = \vec{0}$ and v has exact exponent r, we conclude that $p(X)^r \mid p(X)^{s-e}f(X)$. Moreover, as $p(T)^{s-1-e}(f(T)(v)) \neq \vec{0}$, we conclude that $p(X)^r$ does not divide $p(X)^{s-1-e}f(X)$. This shows that $p(X)^{r-(s-e)} \mid f(X)$. Also, as $r \geq s$, it follows that $r - (s - e) \geq e$. Thus $p(X)^e \mid f(X)$. Expressing $f(X) = p(X)^e g(X)$, we obtain that $w = f(T)(v) = p(T)^e(w')$ where $w' = g(T)(v) \in W$. Consider $u' = u - w'$. Observe that $\pi(u') = \pi(u - w') = \pi(u) - \pi(w') = \bar{u} - \vec{0} = \bar{u} \in \bar{V}$ and $p(T)^e(u') = p(T)^e(u - w') = p(T)^e(u) - p(T)^e(w') = w - w = \vec{0}$. This shows that u' has exact exponent e and proves the lemma. //

We are now ready to prove the existence assertion of the cyclic-decomposition theorem in the prime power case.

Theorem 6.9.5 *Suppose that $T : V \to V$ is a linear operator and $M_T(X) = p(X)^r$ is a power of a monic prime polynomial $p(X)$. Then the existence assertion of the cyclic-decomposition theorem holds for T.*

Proof: We proceed by induction on the dimension of V. The result is trivial if $dim(V) = 1$. Suppose that $dim(V) = n$. By induction we may assume the result is true for all vector spaces of dimension less than n. Since $M_T(X) = p(X)^r$, there must be some $v_1 \in V$ of exact exponent r for T; that is, $p(T)^{r-1}(v_1) \neq \vec{0}$. Set $V_1 = Z(T, v_1)$, $\bar{V} = V/V_1$, and consider the operator $\bar{T} : \bar{V} \to \bar{V}$. Since $V_1 \neq \{\vec{0}\}$, we see that $dim(\bar{V}) < n$. Also, as $p(T)^r = 0$, it must happen that $M_{\bar{T}}(X) \mid p(X)^r$. Thus, by our induction hypothesis there is a cyclic decomposition for the operator $\bar{T} : \bar{V} \to \bar{V}$.

We consider a $\bar{T}$-cyclic direct-sum decomposition $\bar{V} = \bar{V}_2 \oplus \bar{V}_3 \oplus \cdots \oplus \bar{V}_s$, where $\bar{V}_i = Z(T, \bar{v}_i)$ for $\bar{v}_i \in \bar{V}_i$. Assume that each $\bar{v}_i$ has exact exponent e_i. By Lemma 6.9.4, there exist $v_2, \ldots, v_s \in V$ with $v_i + V_1 = \bar{v}_i \in \bar{V}_i$ for each $i = 2, 3, \ldots, s$ where each v_i has exact exponent e_i. Set $V_i = Z(T, v_i) \subset V$ for these $v_2, \ldots, v_s$. Note that for each of these new V_i, since the exact exponents of v_i and $\overline{v}_i$ are both e_i, necessarily $dim(V_i) = dim(\overline{V}_i) = de_i$ where d is the degree of $p(X)$. We claim that $V = V_1 \oplus V_2 \oplus \cdots \oplus V_s$. Once established, this claim will complete the proof.

To check the independence of the V_i, consider $w_i \in V_i$ for $i = 1, 2, \ldots, s$ with $w_1 + w_2 + \cdots + w_s = \vec{0}$. As before, we denote by $\pi : V \to \overline{V} = V/V_1$ the projection linear transformation. Then, as $\pi(w_1) = \vec{0}$, $\vec{0} = \pi(w_1 + \cdots + w_s) = \pi(w_2) + \cdots + \pi(w_s)$. However, each $\pi(w_i) \in \overline{V}_i$, so by the independence of the $\overline{V}_i$, $i = 2, \ldots, s$, we conclude that $\pi(w_i) = \vec{0}$ for $i = 2, \ldots, s$. Observe that the restriction of π to each V_i, $\pi|_{V_i} : V_i \to \overline{V}$, has as image $\overline{V}_i$. Since $dim(V_i) = dim(\overline{V}_i)$, it follows from the rank plus nullity theorem that the restriction $\pi|_{V_i}$ is injective. Hence the previous conclusion that $\pi(w_i) = \vec{0}$ gives that $w_i = \vec{0}$ whenever $i = 2, 3, \ldots, s$. Necessarily $w_1 = \vec{0}$ as well; this gives the independence of $V_1, V_2, \ldots, V_s$.

To complete the proof, we see by Lemma 6.3.13 that $dim(V) = dim(V_1) + dim(\overline{V})$. Since $\overline{V} = \overline{V}_2 \oplus \cdots \oplus \overline{V}_s$, we have that $dim(\overline{V}) = dim(\overline{V}_2) + \cdots + dim(\overline{V}_s) = dim(V_2) + \cdots + dim(V_s)$. Altogether we have $dim(V) = dim(V_1) + dim(V_2) + \cdots + dim(V_s)$. We at last conclude by Corollary 6.3.5 (iii) that $V = V_1 \oplus V_2 \oplus \cdots \oplus V_s$. //

We are now able to combine the primary-decomposition theorem with Theorem 6.9.4 to prove the existence assertion of the cyclic-decomposition theorem in the general case.

Theorem 6.9.6 *The existence assertion of the cyclic-decomposition theorem is true for an arbitrary linear operator on a finite-dimensional vector space.*

Proof: Given $T : V \to V$, we first apply the Primary-Decomposition Theorem 6.5.7 to obtain a T-invariant direct-sum decomposition

$$V = W_1 \oplus \cdots \oplus W_m$$

where $M_{T|_{W_i}}(X) = p_i(X)^{r_i}$ with $M_T(X) = p_1(X)^{r_1} \cdots p_m(X)^{r_m}$ and each $p_i(X)$ prime. We next apply Theorem 6.9.5 to each $T \mid W_i : W_i \to W_i$. This gives for each i a T-cyclic T-invariant direct-sum decomposition $W_i = V_{i1} \oplus \cdots \oplus V_{is_i}$ where $s_i \geq 1$ depends upon i. Set $s = max\{s_1, \ldots, s_m\}$ and define $V_{ij} = \{\vec{0}\}$ for i, j with $s \geq j > s_i$. For each i, j we have $C_{T|_{V_{ij}}}(X) = M_{T|_{V_{ij}}}(X) = p_i(X)^{r_{ij}}$, where $r_i =$

$r_{i1} \geq r_{i2} \geq \cdots \geq r_{ij} \geq 0$ and $r_{ij} = 0$ in case $j > s_i$. Finally, for each $j = 1, \ldots, s$ we define $V_j = V_{1j} \oplus V_{2j} \oplus \cdots \oplus V_{mj}$. We claim that $V = V_1 \oplus V_2 \oplus \cdots \oplus V_s$ is the desired cyclic decomposition of V.

That each V_j is T-invariant follows since each V_{ij} is T-invariant. That V is the direct sum $V_1 \oplus \cdots \oplus V_s$ follows from the other directness assumptions. It is easy to see that the minimal polynomial $M_{T|V_j}(X) = p_1(X)^{r_{1j}} \cdots p_m(X)^{r_{mj}}$. The minimal polynomial on the summand V_{ij} is $p_i(X)^{r_{ij}}$, so the given product is the smallest-degree polynomial which can lie in $An(T \mid V_j)$. Since $dim(V_{ij}) = r_{ij}$, $dim(V_j) = deg(M_{T|V_j})$. Thus it follows from Lemma 6.6.4 that V_j is T-cyclic. Finally, since $r_{ij} \geq r_{ij+1}$, we obtain that $M_{T|V_{j+1}}(X) \mid M_{T|V_j}(X)$ for all j. This gives the existence of the cyclic decomposition. $//$

Only the uniqueness assertion of the cyclic-decomposition theorem remains. In order to prove this, we need a technical lemma that describes how T-cyclic subspaces intersect other T-invariant subspaces. This lemma involves checking numerous technical facts; the key point to remember is that the T-cyclic subspace generated by a vector v consists of all vectors of the form $g(T)(v)$ where $g(X)$ is a polynomial [see Theorem 6.9.1 (i)].

Lemma 6.9.7 *Suppose that $T : V \to V$ is a linear operator and $V = Z_1 \oplus Z_2 \oplus \cdots \oplus Z_s$ is a T-cyclic decomposition of V where $Z_i = Z(T, v_i)$. Assume that $p^e(X) \mid M_T(X)$ where $p(X)$ is an irreducible polynomial of degree d. Then:*
(i) $p(T)^e(V) \cap Z_i = Z(T, p(T)^e(v_i)) = p(T)^e(Z(T, v_i))$.
(ii) $[p(T)^e(V) \cap Z_i]/[p(T)^{e+1}(V) \cap Z_i]$ *is a d-dimensional vector space if and only if $p(X)^{e+1} \mid M_{T,v_i}(X)$. When this fails, the quotient is zero-dimensional; that is, $p(T)^e(V) \cap Z_i = p(T)^{e+1}(V) \cap Z_i$.*
(iii) $p(T)^e(V)/p(T)^{e+1}(T)(V)$ *is isomorphic to the direct sum*

$$\oplus_{i=1}^{s}[p(T)^e(V) \cap Z_i]/[p(T)^{e+1}(T)(V) \cap Z_i]$$

Proof: (i) The inclusion $Z(T, p(T)^e(v_i)) \subseteq p(T)^e(V) \cap Z_i$ is clear since we have $p(T)^e(v_i) \in Z(T, v_i) = Z_i$. Conversely, consider some element $p(T)^e(u) \in V \cap Z_i$ where we express $u = w_1 + w_2 + \cdots + w_s$ with $w_i \in Z_i$. We have that $p(T)^e(u) = p(T)^e(w_1) + \cdots + p(T)^e(w_s) \in Z_i$. By the directness of our T-cyclic decomposition of V, since each $p(T)^e(w_j) \in Z_j$, we conclude that $p(T)^e(w_j) = \vec{0}$ for all $j \neq i$. Hence,

$p(T)^e(u) = p(T)^e(w_i)$. However, as $w_i \in Z_i$, there exists some polynomial $g(X)$ with $g(T)(v_i) = w_i$. Consequently, $p(T)^e(u) = p(T)^e(w_i) = p(T)^e(g(T)(v_i)) = g(T)(p(T)^e(v_i)) \in Z(T, p(T)^e(v_i))$, as desired.

The subspace $Z(T, p(T)^e(v_i))$ consists of all vectors in V of the form $g(T)(p(T)^e(v_i))$ for $g(X)$ a polynomial. Since $g(T)(p(T)^e(v_i)) = p(T)^e(g(T)(v_i))$, this subspace must be the image under the operator $p(T)^e$ of the T-cyclic subspace generated by v_i. In other words, $Z(T, p(T)^e(v_i)) = p(T)^e(Z(T, v_i))$, as required. This proves (i).

(ii) By part (i) we have that

$$[p(T)^e(V) \cap Z_i]/[p(T)^{e+1}(V) \cap Z_i] = Z(T, p(T)^e(v_i))/Z(T, p(T)^{e+1}(v_i))$$

We let $W_1 = Z(T, p(T)^e(v_i))$ and $W_2 = Z(T, p(T)^{e+1}(v_i))$. Since both W_1 and W_2 are T-invariant subspaces of V, the operator T induces an operator $\overline{T} : W_1/W_2 \to W_1/W_2$ by setting $\overline{T}(w + W_2) = T(w) + W_2$ for all $w \in W_1$. Evidently, W_1/W_2 is $\overline{T}$-cyclic generated by $u = p(T)^e(v_i) + W_2$; that is, $W_1/W_2 = Z(\overline{T}, u)$. Since $p(\overline{T})(u) = p(T)p(T)^e(v_i) + W_2 = \vec{0} \in W_1/W_2$, we conclude [since $p(X)$ is prime] that the minimal polynomial of $\overline{T}$ is either $p(X)$ or 1. Since W_1/W_2 is $\overline{T}$-cyclic, this means that by Lemma 6.6.3 $dim(W_1/W_2) = d$ in the first case and 0 in the second. In particular, $dim(W_1/W_2) = d$ if and only if $p(T)^e(v_i) \notin W_2$.

Suppose that $M_{T,v_i}(X) = p(X)^{e+1}q(X)$ for some polynomial $q(X)$. If $p(T)^e(v_i) \in W_2$, then we could write $p(T)^e(v_i) = s(T)(p(T)^{e+1}(v_i))$ for some polynomial $s(X) \in \mathbf{F}[X]$. This would give $q(T)p(T)^e(v_i) = q(T)s(T)p(T)^{e+1}(v_i) = \vec{0}$ since $q(X)p(X)^{e+1} = M_{T,v_i}(X)$. As the degree of $q(X)p(X)^e$ is less than the degree of $M_{T,v_i}(X)$, there is a contradiction. Thus, $p(T)^e(v_i) \notin W_2$ in this case.

Conversely, assume $M_{T,v_i}(X) = p(X)^e q(X)$ for some polynomial $q(X)$ with $p(X)$ and $q(X)$ relatively prime. Suppose that $W_1/W_2 \neq \{\vec{0}\}$. Then, as $\overline{T}$ has minimal polynomial the prime $p(X)$, it follows that for any nonzero $w \in W_1/W_2$ the minimal polynomial $M_{\overline{T},w} = p(X)$ also. In particular, as $p(X)$ does not divide $q(X)$, we see that $q(T)(p(T)^e(v_i) + W_2) \neq \vec{0} \in W_1/W_2$. However, $q(T)p(T)^e(v_i) = \vec{0} \in V$; this is a contradiction. It follows that $W_1/W_2 = \{\vec{0}\}$ in this case, and this proves (ii).

(iii) Consider the linear transformation

$$G : p(T)^e(V) \to p(T)^e(Z_1) \oplus \cdots \oplus p(T)^e(Z_s)$$

defined by $G(p(T)^e(w_1 + \cdots + w_s)) = (p(T)^e(w_1), \ldots, p(T)^e(w_s))$ and the linear transformation

$$H : \oplus_{i=1}^s p(T)^e(Z_i) \to \oplus_{i=1}^s p(T)^e(Z_i)/p(T)^{e+1}(Z_i)$$

defined on each summand by the usual projection. Evidently, the composition $H \circ G : p(T)^e(V) \to \oplus_{i=1}^s p(T)^e(Z_i)/p(T)^{e+1}(Z_i)$ is a surjective linear transformation, as G and H are. Now, $H(G(w_1 + \cdots + w_s)) = \vec{0}$ if and only if $w_i \in p(T)^{e+1}(Z_i)$ for each $i = 1, 2, \ldots, s$. By part (i) $p(T)^{e+1}(Z_i) = p(T)^{e+1}(V) \cap Z_i$, so we find that $ker(H \circ G) = p(T)^{e+1}(V)$. Corollary 6.3.13 (ii) says that $p(T)^e(V)/ker(H \circ G)$ is isomorphic to $\oplus_{i=1}^s p(T)^e(Z_i)/p(T)^{e+1}(Z_i)$. This completes (iii). //

We can finally conclude the proof of Theorem 6.9.1.

Theorem 6.9.8   *The sequence of minimal polynomials given in the cyclic-decomposition theorem is uniquely determined.*

Proof: We first consider the case where $M_T(X) = p(X)^r$ is a power of a prime polynomial $p(X)$ of degree d. Suppose that $V = Z_1 \oplus \cdots \oplus Z_s$ is a cyclic decomposition of V with respect to T. We must show that the number s and the minimal polynomials $M_{T|Z_i}(X)$ are uniquely determined. By Lemma 6.9.7 (iii) we see that the quotient $V/p(T)(V)$ is isomorphic to the direct sum $\oplus_{i=1}^s Z_i/p(T)(Z_i)$. Since each summand in this direct sum is a T-cyclic subspace with minimal polynomial $p(X)$, each summand has dimension d. It follows that the dimension of $V/p(T)(V)$ is ds. This shows that the number s must be the same in any T-cyclic decompositon of V since the dimension of $V/p(T)(V)$ depends only upon T and V.

We now apply Lemma 6.9.6 (i) and (iii) to see that for each exponent e with $1 \le e \le r$, $p(T)^e(V)/p(T)^{e+1}(V)$ is isomorphic to $\oplus_{i=1}^s p(T)^e(Z_i)/p(T)^{e+1}(Z_i)$. Lemma 6.9.6 (ii) shows that the dimension of this direct sum is dg where g is the number of subspaces Z_i for which $p(X)^{e+1} \mid M_{T,v_i}(X)$. It now follows that the number of minimal polynomials divisible by $p(X)^{e+1}$ in any T-cyclic decomposition of V is g, a number which depends only upon $p(T)^e(V)/p(T)^{e+1}(V)$. The uniqueness of the minimal polynomials in the cyclic-decomposition theorem follows, because these polynomials are ordered (in the unique manner) so that the powers of $p(X)$ occurring do not increase.

It remains to consider the case where the polynomial $M_T(X)$ is not a power of a prime. Assume that $M_T(X) = p_1(X)^{d_1} \cdots p_r(X)^{d_r}$ where each $p_i(X)$ is a prime polynomial. Let $V = W_1 \oplus W_2 \oplus \cdots \oplus W_r$ be the primary decompositon of V, and suppose that $V = Z_1 \oplus Z_2 \oplus \cdots \oplus Z_s$ is a cyclic decompositon of V. For each i, j let $Z_{i,j} := Z_i \cap W_j$. Recall that $W_j = ker(p_j(T)^{d_j})$. As the Z_i are T-invariant, whenever $z_i \in Z_i$, $z_1 + z_2 + \cdots + z_s \in ker(p_j(T)^{d_j})$ if and only if $z_i \in ker(p_j(T)^{d_j})$ for each i. Thus, $W_j = Z_{1,j} \oplus Z_{2,j} \oplus \cdots \oplus Z_{s,j}$ is a T-cyclic decompositon of W_j. Moreover, if $p(X)$ is the prime corresponding to W_j and the minimal polynomial of T restricted to $Z_{i,j}$ is $p(X)^e$, then the exact power of $p(X)$ dividing $M_{T|_{Z_i}}(X)$ is $p(X)^e$. The uniqueness assertion of the cyclic-decompostion theorem now follows from the prime-power case. //

PROBLEMS

6.9.1.   Let $T : V \to V$ be a linear operator on a finite-dimensional vector space V with minimal polynomial $M_T(X) = f(X)$. Prove there is some $v \in V$ with $M_{T,v}(X) = f(X)$.

6.9.2.   If $T : V \to V$ is a linear operator and W is a T-invariant subspace of V, W is said to have a *complementary T-invariant subspace* W' if $W \oplus W' = V$ is a T-invariant direct-sum decompositon of V.
(a) If T is diagonalizable, show that every T-invariant subspace of V has a complementary T-invariant subspace.
(b) If $M_T(X) = C_T(X) = p(X)^2$ for a prime polynomial $p(X)$, show that there is a nontrivial T-invariant subspace of V which does not have a complementary T-invariant subspace.
(c) Suppose that V is finite-dimensional and $im(T) \cap ker(T) = \{\vec{0}\}$. Prove that $im(T)$ has a unique complementary T-invariant subspace.

6.9.3. Give an example of a linear operator $T : \mathbf{F}^4 \to \mathbf{F}^4$ for which $\mathbf{F}^4$ is T-cyclic and which also has a T-invariant direct-sum decomposition $\mathbf{F}^4 = V_1 \oplus V_2$ with $dim(V_1) = dim(V_2) = 2$ and with each V_i T-cyclic. Explain why such an example does not contradict the cyclic-decomposition theorem.

6.9.4.   Suppose that $T : V \to V$ is a linear operator on a finite-dimensional vector space V and there is no decomposition $V = V_1 \oplus V_2$ where V_1 and V_2 are T-invariant subspaces of V. Show that $M_T(X) = C_T(X) = p(X)^e$ for some prime polynomial $p(X)$.

6.9.5. Show for any $n \times n$ matrix A that A and A^t are similar.

Chapter 7

Some Applications to Infinite-Dimensional Vector spaces

7.1 The Wronskian

This chapter studies infinite-dimensional vector spaces. The discussion is barely an introduction to a vast subject, and we make no attempt to be complete. It is *not* our goal to present the theory. Rather, we state the results we need, and instead of proving them, we indicate how these results can be understood using what we know about the finite-dimensional case. In the process we will illustrate some of the similarities and differences between the finite- and infinite-dimensional cases. Unless specified otherwise, all vector spaces will be infinite-dimensional real vector spaces (although most everything described for the real case applies equally well to the complex case). At the end of Sec. 7.2 we utilize the Jordan form (developed in Secs. 6.2 and 6.4). Aside from this, the only material necessary for reading this chapter can be found in Chaps. 0 through 4.

Infinite-dimensional vector spaces usually arise as vector spaces of functions. We begin by recalling some notation. For any subset $\mathcal{U} \subseteq \mathbf{R}^p$ we denote by $C^s(\mathcal{U}, \mathbf{R})$ the real vector space of all functions $f : \mathcal{U} \to \mathbf{R}$ for which all sth order (partial) derivatives are defined and are continuous. $C^\infty(\mathcal{U}, \mathbf{R})$ denotes the vector space of all real-valued infinitely differentiable functions defined on $\mathcal{U}$. Hence, for any $f \in$

$C^\infty(\mathcal{U}, \mathbf{R})$, all possible (partial) derivatives are defined and continuous.

The first difficulty encountered in studying function spaces is that of determining if a collection of vectors (which are functions) are linearly independent. The definition of linear independence is clumsy to apply when studying functions. For example, consider the functions $\sin x$, $\sin 2x$, $\sin 3x$ and $\cos x$ in $C(\mathbf{R}, \mathbf{R})$. To see they are linearly independent, one must verify that whenever $a\sin x + b\sin 2x + c\sin 3x + d\cos x$ is the zero function, then the scalars a, b, c, and d are all zero. To check this, one usually finds four values of x with the property that the system of four homogeneous equations in the variables a, b, c, and d, resulting from substituting these x values, has only the trivial solution. But it is not always clear how to find such values for x. Alternatively, one might hope to apply some sort of gaussian elimination to the vectors. Unfortunately, unlike the finite-dimensional case where most vector spaces have a natural basis (and consequently coordinates), one cannot represent a vector with an n-tuple and apply gaussian elimination.

To resolve these difficulties, we introduce the Wronskian, one use of which is to test for the linear independence of vectors in real function spaces. The Wronskian is particularly useful in the theory of differential equations, where it is usually first encountered by students.

Definition 7.1.1 Suppose $\mathcal{U} \subseteq \mathbf{R}$ is an open interval and f_1, $f_2, \ldots,$ $f_n \in C^{n-1}(\mathcal{U}, \mathbf{R})$. We define the *Wronskian* of f_1, $f_2, \ldots,$ f_n by

$$W(f_1, f_2, \ldots, f_n) = \det \begin{pmatrix} f_1 & f_2 & \cdots & f_n \\ f_1' & f_2' & \cdots & f_n' \\ f_1'' & f_2'' & \cdots & f_n'' \\ \vdots & \vdots & \ddots & \vdots \\ f_1^{(n-1)} & f_2^{(n-1)} & \cdots & f_n^{(n-1)} \end{pmatrix}$$

Observe that $W(f_1, f_2, \ldots, f_n) \in C(\mathcal{U}, \mathbf{R})$.

Our main use of the Wronskian is the following.

Lemma 7.1.2 *Suppose that* f_1, $f_2, \ldots,$ $f_n \in C^{n-1}(\mathcal{U}, \mathbf{R})$ *for some open interval* $\mathcal{U} \subseteq \mathbf{R}$. *If the function* $W(f_1, f_2, \ldots, f_n) \neq \vec{0} \in C(\mathcal{U}, \mathbf{R})$, *then* f_1, $f_2, \ldots,$ f_n *are linearly independent vectors.*

Proof: Suppose that $f_1, f_2, \ldots, f_n$ were linearly dependent. Then there would exist an expression $c_1 f_1 + c_2 f_2 + \cdots + c_n f_n = \vec{0}$ for some real numbers c_1, $c_2, \ldots$, c_n not all 0. Differentiating we find that $c_1 f_1' + c_2 f_2' + \cdots + c_n f_n' = \vec{0}$, $c_1 f_1'' + c_2 f_2'' + \cdots + c_n f_n'' = \vec{0}, \ldots$, and $c_1 f_1^{(n-1)} + c_2 f_2^{(n-1)} + \cdots + c_n f_n^{(n-1)} = \vec{0}$. It follows that for any $r \in \mathcal{U}$, $(c_1, c_2, \ldots, c_n)$ is a nontrivial solution to the homogeneous matrix equation

$$
\begin{pmatrix}
f_1(r) & f_2(r) & \cdots & f_n(r) \\
f_1'(r) & f_2'(r) & \cdots & f_n'(r) \\
f_1''(r) & f_2''(r) & \cdots & f_n''(r) \\
\vdots & \vdots & \ddots & \vdots \\
f_1^{(n-1)}(r) & f_2^{(n-1)}(r) & \cdots & f_n^{(n-1)}(r)
\end{pmatrix}
\begin{pmatrix}
X_1 \\
X_2 \\
\vdots \\
X_n
\end{pmatrix}
=
\begin{pmatrix}
0 \\
0 \\
0 \\
\vdots \\
0
\end{pmatrix}
$$

Consequently, the $n \times n$ matrix above must be singular for all $r \in \mathcal{U}$. Since the determinant of a singular matrix is 0, we see for all $r \in \mathcal{U}$ that $W(f_1, f_2, \ldots, f_n)(r) = 0$. This proves the lemma. //

Example 7.1.3 Consider the functions $f_1 = 1$, $f_2 = x$, $f_3 = \sin x$, $f_4 = \cos x \in \mathcal{C}^\infty([0, 2\pi], \mathbf{R})$. We compute the Wronskian:

$$
W(f_1, f_2, f_3, f_4) = \det
\begin{pmatrix}
1 & x & \sin x & \cos x \\
0 & 1 & \cos x & -\sin x \\
0 & 0 & -\sin x & -\cos x \\
0 & 0 & -\cos x & \sin x
\end{pmatrix}
$$

$$
= \det
\begin{pmatrix}
-\sin x & -\cos x \\
-\cos x & \sin x
\end{pmatrix}
= -1
$$

It follows that the four vectors f_1, f_2, f_3, and f_4 are linearly independent in $\mathcal{C}^\infty([0, 2\pi], \mathbf{R})$.

Unfortunately, the converse of Lemma 7.1.2. is not true in general. There are linearly independent functions whose Wronskian is zero. For example, the functions f_1, $f_2 : \mathbf{R} \to \mathbf{R}$ defined by $f_1(x) = x^3$ and $f_2(x) = |x^3|$ are both differentiable. Moreover, they are linearly independent since they are both nonzero and one is not a multiple of another. However, we can compute that $f_1'(x) = 3x^2$ and that $f_2'(x) = 3x^2$ for $x \geq 0$ while $f_2'(x) = -3x^2$ for $x \leq 0$. This shows that $W(f_1, f_2) = x^3 3x^2 - x^3 3x^2 = 0$ for $x \geq 0$ and $W(f_1, f_2) = x^3(-3x^2) - (-x^3)3x^2 = 0$ for $x \leq 0$. Consequently, the converse to Lemma 7.1.2 fails.

There is however an important situation in which the converse to Lemma 7.1.2 does hold. The converse holds if the functions considered are all solutions to the same nth-order homogeneous linear differential equation. This result is proved by using the uniqueness of solutions to such equations which satisfy n initial conditions. This is explained in the proof.

Theorem 7.1.4 *Suppose that $f_1, f_2, \ldots, f_n$ are linearly independent solutions to the mth-order homogeneous linear differential equation*

$$y^{(n)} + p_1(x)y^{(n-1)} + \cdots + p_{n-1}(x)y' + p_n(x)y = 0 \qquad (*)$$

*where $p_1(x), p_2(x), \ldots, p_n(x)$ are continuous on the open interval $\mathcal{U} \subseteq$ **R**. Then $W(f_1, f_2, \ldots, f_n)(r) \neq 0$ for all $r \in \mathcal{U}$.*

Proof: Suppose that $f_1, f_2, \ldots, f_n$ are linearly independent solutions to $(*)$. Let $r \in \mathcal{U}$ and suppose that $W(f_1, f_2, \ldots, f_n)(r) = 0$. It follows that there exist $c_1, c_2, \ldots, c_n \in$ **R**, not all zero, such that

$$\begin{pmatrix} f_1(r) & f_2(r) & \cdots & f_n(r) \\ f_1'(r) & f_2'(r) & \cdots & f_n'(r) \\ f_1''(r) & f_2''(r) & \cdots & f_n''(r) \\ \vdots & \vdots & \ddots & \vdots \\ f_1^{(n-1)}(r) & f_2^{(n-1)}(r) & \cdots & f_n^{(n-1)}(r) \end{pmatrix} \begin{pmatrix} c_1 \\ c_2 \\ \vdots \\ c_n \end{pmatrix} = \begin{pmatrix} 0 \\ 0 \\ 0 \\ \vdots \\ 0 \end{pmatrix}$$

Set $y = c_1 f_1 + c_2 f_2 + \cdots c_n f_n$. Then y is a solution to $(*)$, $y \neq \vec{0}$, and the matrix of equations shows that $y(r) = 0$, $y'(r) = 0$, $\ldots$, $y^{(n-2)}(r) = 0$, and $y^{(n-1)}(r) = 0$. However, the zero function is also a solution to $(*)$ which satisfies these same initial conditions at r. This is a contradiction to the theorem which says that for any $r \in \mathcal{U}$ and any sequence of real numbers $k_0, k_1, \ldots, k_{n-1}$, there exists a *unique* solution y for $(*)$ which satisfies $y(r) = k_0$, $y' = k_1$, $y''(r) = k_2$, $\ldots$, $y^{(n-1)}(r) = k_{r-1}$. For more discussion of this theorem (which we do not prove), we refer the reader to Boyce and DiPrima [Chap. 5] and Coddington [Chap. 6]. //

For example, each of the functions e^t, te^t, $t^2 e^t$ are a solution to the homogeneous linear differential equation $f'''(t) - 3f''(t) + 3f'(t) - f(t) = 0$. Theorem 7.1.4. says the computation of the Wronskian determinant $W(e^t, te^t, t^2 e^t)$ will determine with certainty whether these three functions are linearly independent (which the reader may check they are).

In earlier chapters we considered the infinite-dimensional vector space P^∞ of all polynomials with coefficients in a field $\mathbf{F}$. This vector space is infinite-dimensional since the vectors 1, X, X^2, X^3,... are all linearly independent. (By definition the polynomial $a_n X^n + a_{n-1} X^{n-1} + \cdots + a_0 \cdot 1$ is zero only if $a_n = a_{n-1} = \cdots = a_0 = 0$.) However, it does *not* follow from this that the vector space of *polynomial functions* on a field is infinite-dimensional. In fact, over a finite field, the vector space of functions is finite-dimensional since there are only finitely many functions. (For example, over the field with two elements, $\mathbf{Z}/2\mathbf{Z}$, the nonzero polynomial $X^2 - X$ defines the zero function.) Using the Wronskian, we can show that whenever $\mathcal{U} \subseteq \mathbf{R}$ is an open interval, the vector space $C^\infty(\mathcal{U}, \mathbf{R})$ is infinite-dimensional. To do this, we show that the functions defined by the polynomials 1, X, X^2, X^3,... are linearly independent in this vector space.

Theorem 7.1.5   *If $\mathcal{U} \subseteq \mathbf{R}$ is any open interval, then vector space $C^\infty(\mathcal{U}, \mathbf{R})$ is infinite-dimensional.*

Proof: Consider the functions $f_1 = 1$, $f_2 = x$, $f_3 = x^2,\ldots$, $f_n = x^{n-1}$. The Wronskian $W(f_1, f_2, \ldots, f_n)$ is the determinant of a upper triangular matrix, in fact $W(f_1, f_2, \ldots, f_n) = 1(2)(6) \cdots ((n-1)!) \neq 0$. Thus f_1, $f_2, \ldots$, f_n are linearly independent by Lemma 7.1.2. Since n can be arbitrarily large, $C^\infty(\mathcal{U}, \mathbf{R})$ must be infinite-dimensional. //

PROBLEMS

7.1.1.   Use the Wronskian to show that the following collections of functions are linearly independent in $C^\infty(\mathbf{R}, \mathbf{R})$.
(a) $1, x, e^x$
(b) $x, x^2, \sin x, x\sin x$
(c) $\sin x, \cos x, e^x$
(d) $1, e^x, e^{2x}, e^{3x}$
(e) $\sin x, \sin 2x, \sin 3x$

7.1.2.   Use Theorem 7.1.4 to determine which of the following collections of functions in $C^\infty(\mathbf{R}, \mathbf{R})$ are linearly independent. In each case identify a linear differential equation for which each function is a solution.

(a) $\sin x, x\sin x, \cos x$
(b) $e^t, e^{-t}, \sinh t$
(c) $\cos x, -\sin x, \cos x + \pi \sin x$

7.1.3. Let V be the vector space of real-valued functions on the three points $\{0, 1, -1\}$. Which of the following functions are linearly independent in V? Why does Lemma 7.1.2 fail to apply in this case?
(a) $1, X, X^2$
(b) $1, X, X^2, X^3$
(c) $1, X^2, X^4$

7.1.4. What is the dimension of the $\mathbf{Z}/p\mathbf{Z}$ vector space of all functions $f : (\mathbf{Z}/p\mathbf{Z})^n \to \mathbf{Z}/p\mathbf{Z}$?

7.2 Systems of Linear Differential Equations

In this section we see how the techniques of linear algebra can be used to study systems of linear differential equations. In what follows, we set $V = (\mathcal{C}^\infty(\mathbf{R}, \mathbf{R}))^n$. Thus, an element of V is an n-tuple

$$F(t) = \begin{pmatrix} f_1(t) \\ f_2(t) \\ \vdots \\ f_n(t) \end{pmatrix}$$

of functions, with each $f_i(t) \in \mathcal{C}^\infty(\mathbf{R}, \mathbf{R})$. For such an n-tuple, we denote by $F'(t)$ the n-tuple

$$F'(t) = \begin{pmatrix} f_1'(t) \\ f_2'(t) \\ \vdots \\ f_n'(t) \end{pmatrix}$$

of derivatives of each component function.

Let A denote any $n \times n$ real matrix. Consider the equation $F'(t) = AF(t)$. This equation represents a system of first order linear differential equations. For example, suppose that

$$A = \begin{pmatrix} 1 & 2 \\ 2 & 1 \end{pmatrix}$$

Then $F'(t) = AF(t)$ is the system of differential equations

$$
\begin{aligned}
f_1'(t) &= f_1(t) + 2f_2(t) \\
f_2'(t) &= 2f_1(t) + f_2(t)
\end{aligned}
$$

Before describing how to solve such a system of differential equations, we state without proof the theorem describing the solutions to such systems. The existence part of this theorem will, in fact, be a consequence of our later calculations. For further discussion we refer the reader to Boyce and DiPrima [Chap. 7].

Theorem 7.2.1 *For any $n \times n$ real matrix A, the system of differential equations $F'(t) = AF(t)$ has an n-dimensional vector subspace of solutions in $V = (C^\infty(\mathbf{R}, \mathbf{R}))^n$.*

In order to study the problem of determining the solutions to such a system of differential equations, we first recall the situation when $n = 1$. In this case we have a 1×1 matrix $A = (a)$, and the system of equations reduces to $f'(t) = af(t)$. It is well known that all the solutions to this equation have the form $f(t) = Ce^{at}$ where $C \in \mathbf{R}$ is an arbitrary constant. In particular, the set of solutions is the one-dimensional subspace of $V = (C^\infty(\mathbf{R}, \mathbf{R}))^1$ spanned by $\{e^{at}\}$.

Proceeding by analogy from the $n = 1$ case, one expects that the solutions to $F'(t) = AF(t)$ should come from the functions e^{At}. But what kind of nonsense is this? Actually it turns out not to be nonsense, but it requires some work to interpret the matrix exponential function e^{At} properly. To begin, one first defines the matrix exponential e^A for any $n \times n$ real matrix A.

Definition 7.2.2 Let A be an $n \times n$ real matrix. We define e^A to be the matrix whose entries are the limits of the (finite) sums that arise as the entries in the infinite sum

$$
e^A = I_n + A + (1/2)A^2 + (1/3!)A^3 + (1/4!)A^4 + \cdots
$$

Note that e^A is defined by precisely the same power series that gives e^r for any real number r. The next lemma shows that this definition of e^A really makes sense, that is, that the definition specifies a well-defined $n \times n$ matrix e^A for any real matrix A. The key fact needed to prove the lemma is that the power series for e^r converges absolutely for all real numbers r.

Lemma 7.2.3 *For any $n \times n$ real matrix A, the infinite sum in each entry of the series for e^A above converges absolutely, so the definition of e^A specifies a unique $n \times n$ real matrix e^A.*

Proof: For any matrix $A = (a_{ij})$, we define the norm of A by $\| A \| = max\{|a_{ij}| \mid 1 \leq i, j \leq n\}$. Since each entry of A^2 is a sum of n products of pairs of entries of A, we find that $\| A^2 \| \leq n \| A \|^2$. Similarly, $\| A^3 \| \leq n \| A^2 \| \cdot \| A \| \leq n^2 \| A \|^3$, $\| A^4 \| \leq n^3 \| A \|^4$, and so forth.

We now consider the infinite sum of $n \times n$ matrices

$$e^A = I_n + A + (1/2)A^2 + (1/3!)A^3 + (1/4!)A^4 + \cdots$$

We can bound the norm of the $(m+1)$th summand of this expression as

$$\| (1/m!)A^m \| \leq (n^{m-1}/m!) \| A \|^m < (n \| A \|)^m/(m!)$$

Recall that for any $x \in \mathbf{R}$, the sum $\sum_{m=0}^{\infty} x^m/m!$ converges absolutely to e^x. For any i and j, the ijth entry of the mth summand in the infinite sum defining e^A has absolute value bounded by $(n \| A \|)^m/(m!)$. It now follows from the comparison test that the sum in the ijth entry of the series for e^A converges absolutely. Since the sums for each entry converge, the limiting matrix e^A makes sense. This proves the lemma. //

Let t be a real variable. For any matrix of functions $B(t) = (f_{ij}(t))$, we define $B'(t)$ to be the matrix of derivatives; that is, $B'(t) = (f'_{ij}(t))$. For a fixed real $n \times n$ matrix A, At is an $n \times n$ matrix of functions, and using the matrix exponential, we obtain an $n \times n$ matrix of functions e^{At}. What happens if we differentiate each entry function in the matrix e^{At}? We show in the next lemma that Ae^{At} is the derivative of the matrix exponential e^{At}. This generalizes the familiar 1×1 case, where we know that the derivative of e^{at} is ae^{at}.

Lemma 7.2.4 *Suppose that A is an $n \times n$ real matrix. If $B(t)$ is the matrix of functions $B(t) = e^{At}$, then the matrix of derivatives is given by $B'(t) = A \cdot (B(t))$.*

Proof: Recall that if $f(x) = a_0 + a_1 x + a_2 x^2 + \cdots$ is a power series that converges absolutely for $x \in (r, s) \subset \mathbf{R}$, then the derivative of

$f(x)$ is given by the power series $f'(x) = a_1 + 2a_2 x + 3a_3 x^2 + \cdots$ which also converges absolutely for all $x \in (r, s)$. We apply this result to the power series that occur as entries of the matrix exponential. By definition

$$e^{At} = I + At + (1/2)A^2 t^2 + (1/3!)A^3 t^3 + \cdots$$

Thus, the coefficient of t^m in some entry of e^{At} is the corresponding (scalar) entry of $(1/m!)A^m$. From this, by computing the derivative of each (power series) entry of e^{At}, it follows that

$$
\begin{aligned}
(e^{At})' &= A + 2(1/2)A^2 t + 3(1/3!)A^3 t^2 + \cdots \\
&= A(I + At + (1/2)A^2 t^2 + (1/3!)A^3 t^3 + \cdots) \\
&= A(e^{At})
\end{aligned}
$$

This gives the lemma. //

Using Lemma 7.2.4, we can now describe all solutions to a first-order system of linear differential equations.

Theorem 7.2.5 *For any $n \times n$ real matrix A, all solutions to the system of linear differential equations $F'(t) = AF(t)$ are given by $F(t) = e^{At}\vec{C}$, where $\vec{C}$ is a column vector of constants from $\mathbf{R}^n$.*

When $n = 1$, Theorem 7.2.5 gives the well-known solution to $f'(t) = af(t)$. Note that the n constants in the vector $\vec{C}$ (which can be chosen arbitrarily) give the freedom necessary to generate the n-dimensional vector space of solutions to $F'(t) = AF(t)$ that is guaranteed by Theorem 7.2.1.

Proof: We know by Lemma 7.2.4 that if $B(t) = e^{At}$, then $B'(t) = A \cdot B(t)$. Consider any

$$\vec{C} = \begin{pmatrix} c_1 \\ c_2 \\ \vdots \\ c_n \end{pmatrix} \in \mathbf{R}^n$$

Since the entries of $B(t)\vec{C}$ are linear combinations of the entries of $B(t)$ and differentiation is linear, we have that $(B(t)\vec{C})' = B'(t)\vec{C}$. As $B'(t)\vec{C} = A \cdot B(t)\vec{C}$, $B(t)\vec{C}$ is a solution to $F'(t) = AF(t)$.

To see that this process gives all solutions, we show that the solutions given by $B(t)e_1, B(t)e_2, \ldots, B(t)e_n$ are all linearly independent in $V = (C^\infty(\mathbf{R}, \mathbf{R}))^n$. The result will then follow by Theorem 7.2.1. The solutions listed are the columns of $B(t)$. If they were linearly dependent, then such a dependency would give a nonzero $\vec{C} \in \mathbf{R}^n$ with $B(t)\vec{C} = \vec{0} \in V$. Substituting the value $t = 0$ into this expression gives that $B(0)\vec{C} = \vec{0}$. But $B(0) = e^{A\vec{0}} = I_n + 0 + 0 + \cdots = I_n$; that is, $I\vec{C} = \vec{0}$. Consequently, $\vec{C} = \vec{0}$, a contradiction. This proves the theorem. //

Example 7.2.6 Let

$$A = \begin{pmatrix} 2 & 1 \\ 0 & -1 \end{pmatrix}$$

Then

$$A^2 = \begin{pmatrix} 4 & 1 \\ 0 & 1 \end{pmatrix}, \quad A^3 = \begin{pmatrix} 8 & 3 \\ 0 & -1 \end{pmatrix}, \quad A^4 = \begin{pmatrix} 16 & 5 \\ 0 & 1 \end{pmatrix}, \quad \ldots$$

These calculations show that

$$
e^{At} = \begin{pmatrix} 1 + 2t + (4/2!)t^2 + \cdots & 0 + t + (1/2)t^2 + (3/3!)t^3 + \cdots \\ 0 & 1 + -t + (1/2)t^2 - (1/3!)t^3 - \cdots \end{pmatrix}
$$
$$
= \begin{pmatrix} e^{2t} & ? \\ 0 & e^{-t} \end{pmatrix}
$$

where it is not quite clear what ? is. To compute ?, we see that the series

$$
0 + t + (1/2)t^2 + (3/3!)t^3 + \cdots
$$
$$
= (1/3)[1 + 2t + (4/2)t^2 + (8/3!)t^3 + \cdots] +
$$
$$
(-1/3)[1 - t + (1/2)t^2 - (1/3!)t^3 + \cdots]
$$
$$
= (1/3)e^{2t} + (-1/3)e^{-t}
$$

(Where did this come from? See Example 7.2.8.) This shows that the solutions to the system of differential equations $F'(t) = AF(t)$ are given by

$$
F(t) = e^{At}\vec{C} = \begin{pmatrix} e^{2t} & (1/3)e^{2t} + (-1/3)e^{-t} \\ 0 & e^{-t} \end{pmatrix} \begin{pmatrix} c_1 \\ c_2 \end{pmatrix}
$$
$$
= \begin{pmatrix} (c_1 + c_2/3)e^{2t} - (c_2/3)e^{-t} \\ c_2 e^{-t} \end{pmatrix} \quad \text{where } c_1, c_2 \in \mathbf{R}
$$

As is evident from the above example, the calculation of the exponential matrix e^{At} is not always easy. However, the situation is simplified greatly if we change basis. Before illustrating this, we need the following lemma.

Lemma 7.2.7 *Suppose that A and P are $n \times n$ real matrices, where P is invertible. Then $e^{(P^{-1}AP)} = P^{-1}e^A P$. In particular, $e^{(P^{-1}APt)} = P^{-1}e^{(At)}P$.*

Proof: According to the definition,

$$
\begin{aligned}
e^{(P^{-1}AP)} &= I + P^{-1}AP + (1/2)(P^{-1}AP)^2 + \cdots \\
&= I + P^{-1}AP + (1/2)P^{-1}A^2P + (1/3!)P^{-1}A^3P + \cdots
\end{aligned}
$$

The ijth entry in this matrix is the limit of partial sums of the ijth entries in the matrix partial sums $I + P^{-1}AP + \cdots + (1/m!)P^{-1}A^m P$. This matrix partial sum is the same as $P^{-1}(I + A + \cdots + (1/m!)A^m)P$. In this latter expression, the ijth entry is a linear combination of the entries of the sum $I + A + \cdots + (1/m!)A^m$. Moreover, the coefficients of this linear combination depend only upon P and not upon m. In calculus it is proved that a linear combination of absolutely convergent series converges to the (same) linear combination of the limits of the individual series. In particular, this shows that the ijth entry of the partial sums of $I + P^{-1}AP + (1/2)P^{-1}A^2P + \cdots$ converges to the ijth entry of the partial sums of $P^{-1}(I + A + (1/2)A^2 + \cdots)P = P^{-1}e^A P$. This proves the lemma. $//$

Example 7.2.8 We again solve the problem of Example 7.2.6, this time by diagonalizing our matrix

$$
A = \begin{pmatrix} 2 & 1 \\ 0 & -1 \end{pmatrix}
$$

We have that $C_A(X) = (X - 2)(X + 1)$, so the eigenvalues of A are 2 and -1. The eigenspaces are

$$
E_2 = span\left\{ \begin{pmatrix} 1 \\ 0 \end{pmatrix} \right\} \quad \text{and} \quad E_{-1} = span\left\{ \begin{pmatrix} -1 \\ 3 \end{pmatrix} \right\}
$$

Setting

$$
P = \begin{pmatrix} 1 & -1 \\ 0 & 3 \end{pmatrix}
$$

we find that
$$P^{-1} = \frac{1}{3}\begin{pmatrix} 3 & 1 \\ 0 & 1 \end{pmatrix}$$

and consequently
$$D = P^{-1}AP = \begin{pmatrix} 2 & 0 \\ 0 & -1 \end{pmatrix}$$

As
$$D^m = \begin{pmatrix} 2^m & 0 \\ 0 & (-1)^m \end{pmatrix}$$

for all m, it is easily checked that
$$e^{(Dt)} = \begin{pmatrix} e^{2t} & 0 \\ 0 & e^{-t} \end{pmatrix}$$

Now applying Lemma 7.2.7, we find that
$$e^{At} = P(e^{Dt})P^{-1} = \begin{pmatrix} e^{2t} & (1/3)(e^{2t} - e^{-t}) \\ 0 & e^{-t} \end{pmatrix}$$

as obtained previously. This diagonalization procedure illustrates why the unknown function ? in Example 7.2.6 turned out to be a linear combination of the two exponential functions e^{2t} and e^{-t}.

The example above shows how to solve the system of differential equations $F'(t) = AF(t)$ whenever A is an $n \times n$ diagonalizable matrix. What happens when A fails to be diagonalizable? When A has a Jordan form, the problem can be solved by using the same ideas as in the diagonalizable case. (This can always be done if A is viewed as a complex matrix.) First, one finds a matrix P so that $P^{-1}AP$ is the Jordan form of A. The solutions to the differential equation can then be computed by the same method as in Example 7.2.8 once the exponential of a Jordan block matrix has been computed. This comes from the following calculation: Fix

$$J = \begin{pmatrix} c & 0 & 0 & \cdots & 0 & 0 \\ 1 & c & 0 & \cdots & 0 & 0 \\ 0 & 1 & c & \cdots & 0 & 0 \\ \vdots & \vdots & \vdots & \ddots & \vdots & \vdots \\ 0 & 0 & 0 & \cdots & c & 0 \\ 0 & 0 & 0 & \cdots & 1 & c \end{pmatrix}$$

to be an $n \times n$ Jordan block matrix of eigenvalue c. Then we can write $J = cI_n + N$ where

$$N = \begin{pmatrix} 0 & 0 & 0 & \cdots & 0 & 0 \\ 1 & 0 & 0 & \cdots & 0 & 0 \\ 0 & 1 & 0 & \cdots & 0 & 0 \\ \vdots & \vdots & \vdots & \ddots & \vdots & \vdots \\ 0 & 0 & 0 & \cdots & 0 & 0 \\ 0 & 0 & 0 & \cdots & 1 & 0 \end{pmatrix}$$

Observe that the matrix N satisfies

$$N^2 = \begin{pmatrix} 0 & \cdots & 0 & 0 & 0 \\ 0 & \cdots & 0 & 0 & 0 \\ 1 & \cdots & 0 & 0 & 0 \\ \vdots & \ddots & \vdots & \vdots & \vdots \\ 0 & \cdots & 1 & 0 & 0 \end{pmatrix}, \ldots, \quad N^{n-1} = \begin{pmatrix} 0 & 0 & \cdots & 0 & 0 \\ 0 & 0 & \cdots & 0 & 0 \\ 0 & 0 & \cdots & 0 & 0 \\ \vdots & \vdots & \ddots & \vdots & \vdots \\ 1 & 0 & \cdots & 0 & 0 \end{pmatrix}$$

and $N^n = 0$. The powers of J can be expanded as

$$\begin{aligned} J^m &= (cI_n + N)^m \\ &= (c^m I_n + mc^{m-1}N + (m(m-1)/2)c^{m-2}N^2 + \cdots + N^m) \end{aligned}$$

Observe that in this latter sum there are at most n summands, since $N^m = 0$ for $m \geq n$. Using this expression, we can now compute the matrix of functions $e^{(Jt)} = (f_{ij}(t))$.

Whenever $j > i$, the ijth entries of I_n and N^s are always 0. From this we see that $f_{ij}(t) = 0$ whenever $j > i$. When $i = j$, we see that the iith entries of N^s are always 0. Hence, the power series that results in the iith entries of $e^{(Jt)}$ is precisely $1+ct+(1/2)c^2t^2+(1/3!)c^3t^3+\cdots$. In other words $f_{ii}(t) = e^{ct}$. Next we compute $f_{i(i-1)}(t)$. The coefficients in this series must arise from the coefficients of N in the expansions of J^m above. Hence we obtain that

$$\begin{aligned} f_{i(i-1)}(t) &= 0 + 1t + (1/2)2ct^2 + (1/3!)3c^2t^3 + \cdots \\ &= t(1 + ct + (1/2)c^2t^2 + \cdots) = te^{ct} \end{aligned}$$

Analogously, we obtain that

$$\begin{aligned} f_{i(i-2)}(t) &= 0 + 0 + (1/2)t^2 + (1/3!)3ct^3 + (1/4!)(4 \cdot 3/2)c^2t^4 + \cdots \\ &= (1/2)t^2(1 + ct + (1/2)c^2t^2 + \cdots) = (1/2)t^2 e^{ct} \end{aligned}$$

In general, $f_{i(i-s)}(t) = (1/s!)t^s e^{ct}$. This shows that $e^{(Jt)}$ is

$$e^{ct} \begin{pmatrix} 1 & 0 & 0 & \cdots & 0 & 0 \\ t & 1 & 0 & \cdots & 0 & 0 \\ (1/2)t^2 & t & 1 & \cdots & 0 & 0 \\ \vdots & \vdots & \vdots & \ddots & \vdots & \vdots \\ (1/n!)t^n & (1/(n-1)!)t^{n-1} & (1/(n-2)!)t^{n-2} & \cdots & t & 1 \end{pmatrix}$$

We apply the preceding calculation in this final example.

Example 7.2.9 Consider the matrix

$$A = \begin{pmatrix} 23 & 0 & -60 \\ 2 & 3 & -5 \\ 8 & 0 & -21 \end{pmatrix}$$

Computing $e^{(At)}$ by the definition would be a nightmare! A direct calculation gives that $C_A(X) = M_A(X) = (X-3)^2(X+1)$. One then computes

$$ker(A - 3I_3) = ker \begin{pmatrix} 20 & 0 & -60 \\ 2 & 0 & -5 \\ 8 & 0 & -24 \end{pmatrix} = span \left\{ \begin{pmatrix} 0 \\ 1 \\ 0 \end{pmatrix} \right\}$$

$$ker(A - 3I_3)^2 = ker \begin{pmatrix} -80 & 0 & 240 \\ 0 & 0 & 0 \\ -32 & 0 & 96 \end{pmatrix} = span \left\{ \begin{pmatrix} 3 \\ 0 \\ 1 \end{pmatrix}, \begin{pmatrix} 0 \\ 1 \\ 0 \end{pmatrix} \right\}$$

$$ker(A + I_3) = ker \begin{pmatrix} 24 & 0 & -60 \\ 2 & 4 & -5 \\ 8 & 0 & -20 \end{pmatrix} = span \left\{ \begin{pmatrix} 5 \\ 0 \\ 2 \end{pmatrix} \right\}$$

This means that if we set

$$B = \left\{ \begin{pmatrix} 3 \\ 0 \\ 1 \end{pmatrix}, \begin{pmatrix} 0 \\ 1 \\ 0 \end{pmatrix}, \begin{pmatrix} 5 \\ 0 \\ 2 \end{pmatrix} \right\}$$

the matrix $[T_A]_B$ is the Jordan-form matrix

$$J = \begin{pmatrix} 3 & 0 & 0 \\ 1 & 3 & 0 \\ 0 & 0 & -1 \end{pmatrix}$$

We find from the above computation that

$$e^{(Jt)} = \begin{pmatrix} e^{3t} & 0 & 0 \\ te^{3t} & e^{3t} & 0 \\ 0 & 0 & e^{-t} \end{pmatrix}$$

Setting

$$P = \begin{pmatrix} 3 & 0 & 5 \\ 0 & 1 & 0 \\ 1 & 0 & 2 \end{pmatrix} \qquad P^{-1} = \begin{pmatrix} 2 & 0 & -5 \\ 0 & 1 & 0 \\ -1 & 0 & 3 \end{pmatrix}$$

we apply Lemma 7.2.7 and obtain that

$$
\begin{aligned}
e^{At} &= e^{PJP^{-1}} = Pe^{Jt}P^{-1} \\
&= \begin{pmatrix} 3 & 0 & 5 \\ 0 & 1 & 0 \\ 1 & 0 & 2 \end{pmatrix} \begin{pmatrix} e^{3t} & 0 & 0 \\ te^{3t} & e^{3t} & 0 \\ 0 & 0 & e^{-t} \end{pmatrix} \begin{pmatrix} 2 & 0 & -5 \\ 0 & 1 & 0 \\ -1 & 0 & 3 \end{pmatrix} \\
&= \begin{pmatrix} 6e^{3t} - e^{-t} & 0 & 15e^{3t} + 3e^{-t} \\ 2te^{3t} & e^{3t} & 5te^{3t} \\ 2e^{3t} - 2e^{-t} & 0 & 5e^{3t} + 6e^{-t} \end{pmatrix}
\end{aligned}
$$

Altogether this shows that the solutions to the system of differential equations

$$
\begin{aligned}
f_1'(t) &= 23f_1(t) & & & - & 60f_3(t) \\
f_2'(t) &= -2f_1(t) & + & 3f_2(t) & - & 5f_3(t) \\
f_3'(t) &= 8f_1(t) & & & - & 21f_3(t)
\end{aligned}
$$

are all of the form

$$
\begin{pmatrix} f_1(t) \\ f_2(t) \\ f_3(t) \end{pmatrix} = \begin{pmatrix} 6e^{3t} - e^{-t} & 0 & 15e^{3t} + 3e^{-t} \\ 2te^{3t} & e^{3t} & 5te^{3t} \\ 2e^{3t} - 2e^{-t} & 0 & 5e^{3t} + 6e^{-t} \end{pmatrix} \begin{pmatrix} c_1 \\ c_2 \\ c_3 \end{pmatrix}
$$

for arbitrary real numbers c_1, c_2, and c_3. In other words the solutions are given by

$$
\begin{aligned}
f_1(t) &= (6c_1 + 5c_3)e^{3t} + (-c_1 + 3c_3)e^{-t} \\
f_2(t) &= (2c_1t + 5c_3t + c_2)e^{3t} \\
f_3(t) &= (-2c_1 + 5c_3)e^{3t} + (2c_1 + 6c_3)e^{-t}
\end{aligned}
$$

PROBLEMS

7.2.1. Compute the matrix e^A where

(a) $A = \begin{pmatrix} 3 & 0 \\ 0 & 4 \end{pmatrix}$ (b) $A = \begin{pmatrix} 0 & 1 \\ 0 & 0 \end{pmatrix}$ (c) $A = \begin{pmatrix} 1 & 2 \\ 0 & 1 \end{pmatrix}$

7.2.2. Compute the matrix e^A where

(a) $A = \begin{pmatrix} 2 & 0 & 0 \\ 1 & 2 & 0 \\ 0 & 1 & 2 \end{pmatrix}$ (b) $A = \begin{pmatrix} 2 & 0 & 0 \\ 1 & 2 & 0 \\ 0 & 1 & 3 \end{pmatrix}$ (c) $A = \begin{pmatrix} 1 & 0 & 1 \\ 0 & 1 & 0 \\ 1 & 0 & 1 \end{pmatrix}$

7.2.3. For the following matrices A, find e^{At} and all solutions to the system of first-order linear differential equations $F'(t) = AF(t)$.

(a) $A = \begin{pmatrix} 1 & 3 \\ 3 & 1 \end{pmatrix}$ (b) $A = \begin{pmatrix} 0 & -1 \\ 1 & 2 \end{pmatrix}$

7.2.4. For the following matrices A, find $e^{(At)}$ and all solutions to the system of first-order linear differential equations $F'(t) = AF(t)$.

(a) $A = \begin{pmatrix} 0 & 0 & -1 \\ 1 & 0 & 3 \\ 0 & 1 & 3 \end{pmatrix}$ (b) $A = \begin{pmatrix} 3 & 1 & -1 \\ 2 & 2 & -1 \\ 2 & 2 & 0 \end{pmatrix}$

7.2.5. Show that there exist 2×2 matrices A, B for which $e^A e^B \neq e^{(A+B)}$. Why do matrices behave differently from real numbers in this respect?

7.2.6. Show that the solutions to the nth-order homogeneous linear differential equation $y^{(n)} + a_{n-1} y^{(n-1)} + \cdots + a_0 y = 0$ are precisely the first entries of solutions to $F'(t) = AF(t)$ where

$$A = \begin{pmatrix} 0 & 1 & 0 & \cdots & 0 \\ 0 & 0 & 1 & \cdots & 0 \\ \vdots & \vdots & \vdots & \ddots & \vdots \\ 0 & 0 & 0 & \cdots & 1 \\ -a_0 & -a_1 & -a_2 & \cdots & -a_{n-1} \end{pmatrix}$$

Deduce from Theorem 7.2.1 the basic fact that such an nth-order homogeneous linear differential equation has an n-dimensional vector space of solutions and a unique solution satisfying any set of initial conditions $y(b) = c_0$, $y'(b) = c_1, \ldots, y^{(n-1)}(b) = c_{n-1}$.

7.2.7. Use the preceding problem to find all solutions to the following homogeneous linear differential equations. In each case use the Wronskian determinant test from Theorem 7.1.4 to check that your basis of solutions is linearly independent.

(a) $y'' + 4y' + 4y = 0.$
(b) $y'' + y' - 6y = 0.$
(c) $y''' - y'' + y' - y = 0.$

7.3   Orthogonal Polynomials

In this section we consider several different inner products on the function space $V = \mathcal{C}([a, b], \mathbf{R})$, $[a, b] \subseteq \mathbf{R}$. For these inner products we give a sequence of mutually orthogonal vectors, each of which is a polynomial. The importance of these polynomials is the following: Given an element of V, the projection onto the span of the first n of these polynomials is a good (nth-order) approximation to the original function. This approximation, which is a polynomial, is a least-squares approximation, where now "least squares" means an approximation in terms of the norm associated with the given inner product. In each case the sequence of polynomials will be complete in the sense that any element of V is the limit of a Cauchy sequence of linear combinations of the polynomials. This completeness concept is described in more detail in the section which follows. The existence of such polynomials has both applied and theoretical uses.

We begin with the definition of the type of polynomials we are looking for.

Definition 7.3.1 Suppose that $< , >$ is an inner product on the real vector space P^∞. A sequence of *orthogonal polynomials* with respect to $< , >$ is an orthogonal sequence of elements of P^∞, $f_0, f_1, f_2, \ldots$ with the degree of f_i equal to i and chosen so that the leading coefficient of f_i is positive.

We already have an algorithm for obtaining a sequence of orthogonal polynomials. Starting with the linearly independent sequence 1, X, $X^2, \ldots$, we can apply the Gram-Schmidt process. (It is not hard to see that the degree of the ith polynomial will be i and the positivity condition can be realized by multiplying by -1 if necessary.) However, in many cases it is possible to describe the entire class of polynomials directly, using properties of the inner product in question. Then, using a recurrence relation (Theorem 7.3.2), we can generate the specific orthogonal polynomials with much greater efficiency than we can using the Gram-Schmidt process.

As a first example, we consider the inner product on $\mathcal{C}([-1,1], \mathbf{R})$ defined by

$$<f(X), g(X)> = \int_{-1}^{1} f(X)g(X) \, dX$$

For $n = 0, 1, 2, \ldots$ we define the polynomial $f_n(X)$ by

$$f_n = \frac{d^n}{dX^n}(X^2 - 1)^n$$

$f_n(X)$ is a polynomial of degree n since it is the nth derivative of a polynomial of degree $2n$. For $n \geq 1$ we compute

$$<f_n(X), 1> = \int_{-1}^{1} \frac{d^n}{dX^n}(X^2 - 1)^n = \frac{d^{n-1}}{dX^{n-1}}(X^2 - 1)^n \Big|_{-1}^{1} = 0$$

the last equality following because the polynomial $(X^2 - 1)$ is a factor of $d^{n-1}/dX^{n-1}(X^2 - 1)^n$. Moreover, for $m \geq n$ integration by parts gives

$$<f_n(X), f_m(X)>$$

$$= \int_{-1}^{1} \frac{d^n}{dX^n}(X^2 - 1)^n \frac{d^m}{dX^m}(X^2 - 1)^m \, dX$$

$$= \frac{d^n}{dX^n}(X^2 - 1)^n \frac{d^{m-1}}{dX^{m-1}}(X^2 - 1)^m \Big|_{-1}^{1}$$

$$\quad - \int_{-1}^{1} \frac{d^{n+1}}{dX^{n+1}}(X^2 - 1)^n \frac{d^{m-1}}{dX^{m-1}}(X^2 - 1)^m \, dX$$

$$= 0 - \frac{d^{n+1}}{dX^{n+1}}(X^2 - 1)^n \frac{d^{m-2}}{dX^{m-2}}(X^2 - 2)^m \Big|_{-1}^{1}$$

$$\quad + \int_{-1}^{1} \frac{d^{n+2}}{dX^{n+2}}(X^2 - 1)^n \frac{d^{m-2}}{dX^{m-2}}(X^2 - 1)^m \, dX$$

$$= \cdots$$

$$= (-1)^n \int_{-1}^{1} \frac{d^{2n}}{dX^{2n}}(X^2 - 1)^n \frac{d^{m-n}}{dX^{m-n}}(X^2 - 1)^m \, dX$$

$$= (-1)^n(2n)! \int_{-1}^{1} \frac{d^{m-n}}{dX^{m-n}}(X^2 - 1)^m \, dX$$

$$= \begin{cases} 0 & \text{if } m > n \\ (-1)^n(2n)! \int_{-1}^{1}(X^2 - 1)^m dX & \text{if } m = n \end{cases}$$

This shows that the polynomials 1, $f_1(X)$, $f_2(X)$,... are orthogonal in P^∞, and moreover, since $deg(f_i(X)) = i$, they are a sequence of orthogonal polynomials. Setting $L_n(X) = 1/2^n n! f_n(X)$ [this definition for $L_n(X)$ is known as *Rodrigues' formula*] gives the sequence of orthogonal polynomials for V known as the *Legendre polynomials*. We shall obtain a recurrence relation for the Legendre polynomials in Example 7.3.3.

In order to compute orthogonal polynomials effectively, we derive a recurrence relation which calculates $f_{n+1}(X)$ in terms of $f_n(X)$ and $f_{n-1}(X)$. We consider the general situation of a sequence $f_0(X)$, $f_1(X)$, $f_2(X)$,... of orthogonal polynomials with respect to some inner product $< , >$ on P^∞. Observe that the first $n + 1$ polynomials $f_0, f_1, \ldots, f_n$ form an orthogonal basis for the subspace P^n of P^∞. Consequently, f_{n+1} lies in the orthogonal complement of P^n in P^{n+1}; that is, $f_{n+1} \in (P^n)^\perp$. We need this observation to prove the next result.

The type of inner products we shall be interested in all have the form $< f(X), g(X) > = \int_a^b f(X)g(X)K(X)\, dX$, where $K(X)$ is some fixed positive continuous function defined on the interval $[a, b]$. Because of this assumption, our inner products have the additional property that $<h(X)f(X), g(X)> = <f(X), h(X)g(X)>$. This property is critical for our next result.

Theorem 7.3.2 *Suppose that f_0, f_1, f_2,... is a sequence of orthogonal polynomials inside P^∞. Assume that $< , >$ is an inner product on P^∞ which satisfies $<Xf(X), g(X)> = <f(X), Xg(X)>$ for all $f(X)$, $g(X) \in P^\infty$. Suppose that a_i is the coefficient of X^i in f_i and set*

$$r_n = \frac{a_{n-1}}{a_n} \;,\quad s_n = \frac{<f_n, Xf_n>}{<f_n, f_n>} \;,\quad \text{and} \;\; t_n = \frac{<f_n, f_n>}{<f_{n-1}, f_{n-1}>}$$

Then the f_i satisfy the recurrence relation

$$r_{n+1} f_{n+1} = (X - s_n) f_n - r_n t_n f_{n-1}$$

This relation remains valid for $n = 0$ if we interpret $f_{-1}(X) = 0$.

Proof: We set

$$d_i = \frac{<Xf_n, f_i>}{<f_i, f_i>} = \frac{<f_n, Xf_i>}{<f_i, f_i>}$$

The polynomial $Xf_n \in P^{n+1}$. Thus as $f_0, \ldots, f_{n+1}$ form an orthogonal basis for P^{n+1}, we obtain that

$$Xf_n = Proj_{P^{n+1}}(Xf_n) = \sum_{i=0}^{n+1} d_i f_i$$

Note that if $i < n - 1$, then $Xf_i \in P^{n-1}$ so that as $f_n \in (P^{n-1})^\perp$, we find $<f_n, Xf_i> = 0$; that is, $d_i = 0$ for such i. Our sum becomes

$$Xf_n = d_{n-1}f_{n-1} + d_n f_n + d_{n+1}f_{n+1}$$

which can be expressed as

$$d_{n+1}f_{n+1} = (X - d_n)f_n - d_{n-1}f_{n-1}$$

This last relation is precisely what we seek; we must check only that the coefficients d_{n+1}, d_n, d_{n-1} are correct. That $d_{n+1} = a_n/a_{n+1} = r_{n+1}$ follows by comparing the coefficients of X^{n+1} on both sides of this relation. d_n is what it should be by definition. Finally,

$$d_{n-1} = \frac{<f_n, Xf_{n-1}>}{<f_{n-1}, f_{n-1}>} = \frac{<f_n, Xf_{n-1}>}{<f_n, f_n>} \cdot \frac{<f_n, f_n>}{<f_{n-1}, f_{n-1}>} = r_n t_n$$

by applying the definition of t_n and the fact just established that $d_n = r_n$. This gives the recurrence relation. The validity of the relation where $n = 0$ with $f_{-1}(X) = 0$ follows by the same argument applied to this special case. //

We now give several examples of how this relation can generate sequences of orthogonal polynomials for various inner products on P^∞.

Example 7.3.3 (Legendre Polynomials) In this example we apply Theorem 7.3.2 to the Legendre polynomials

$$L_n(X) = \frac{1}{2^n n!} \frac{d^n}{dX^n} (X^2 - 1)^n$$

defined above. Using integration by parts, a calculation analogous to the earlier computation of $<f_n(X), f_m(X)>$, shows that

$$<f_{n+1}(X), Xf_n(X)> = (-1)^{n+1}(2n)!(n + 1) \int_{-1}^{1} (X^2 - 1)^{n+1} dX$$

Thus,

$$<L_{n+1}(X), XL_n(X)>$$

$$= \frac{1}{2^{2n+1}(n+1)!n!}(-1)^{n+1}(2n)!(n+1)\int_{-1}^{1}(X^2-1)^{n+1}\,dX$$

and using our previous computation of $<f_n(X), f_n(X)>$, we have

$$<L_{n+1}(X), L_{n+1}(X)>$$

$$= \frac{1}{2^{2n+2}(n+1)!^2}(-1)^{n+1}(2n+2)!\int_{-1}^{1}(X^2-1)^{n+1}\,dX$$

Using the notation from the proof of Theorem 7.3.2, we find that

$$d_{n+1} = \frac{<L_{n+1}(X), XL_n(X)>}{<L_{n+1}(X), L_{n+1}(X)>} = \frac{n+1}{2n+1}$$

Since the product $L_n(X)XL_n(X)$ has only odd-degree monomials, it follows that $<L_n(X), XL_n(X)> = 0$. Consequently, $d_n = 0$. Finally, it can be shown that $<L_n(X), L_n(X)> = 2/(2n+1)$, so we obtain

$$d_{n-1} = \frac{<L_n(X), XL_{n-1}(X)>}{<L_n(X), L_n(X)>} \cdot \frac{<L_n(X), L_n(X)>}{<L_{n-1}(X), L_{n-1}(X)>}$$

$$= \frac{n}{2n-1} \cdot \frac{2/(2n+1)}{2/(2n-1)} = \frac{n}{2n+1}$$

Multiplying by $2n+1$ gives the following recurrence relation for the Legendre polynomials:

$$(n+1)L_{n+1}(X) = (2n+1)XL_n(X) - nL_{n-1}(X)$$

Using this relation and starting with the initial polynomials $L_0(X) = 1$, $L_1(X) = X$ we find that $2L_2(X) = 3X^2 - 1$; that is, $L_2(X) = (1/2)(3X^2 - 1)$. For the next term, $3L_3(X) = 5(1/2)(3X^2 - 1)X - 2X$; that is, $L_3(X) = (1/2)(5X^3 - 3X)$, and so forth.

The Legendre polynomial $L_n(X)$ turns out to be a solution to the second-order differential equation $(1 - X^2)y'' - 2Xy' + n(n+1)y = 0$ (known as the *Legendre equation*). This equation can be rewritten as $-[(1 - X^2)y']' = n(n+1)y$. From this point of view, $L_n(X)$ is an eigenvector (eigenfunction) for the linear operator defined by

$T(y) = -[(1 - X_2)y']'$. This observation is important in the theory of differential equations.

Example 7.3.4 (Hermite Polynomials) In this example we take for our inner product

$$<f(X), g(X)> = \int_{-\infty}^{\infty} f(X)g(X)e^{-X^2}\, dX$$

A system of orthogonal polynomials H_n for this inner product is defined by the equation

$$e^{-X^2} H_n(X) = (-1)^n \frac{d^n}{dX^n} e^{-X^2}$$

It turns out that the $H_n(X)$ satisfy the recurrence relation

$$H_n(X) = 2XH_{n-1}(X) - 2(n-1)H_{n-2}(X)$$

Thus, starting with the polynomials $H_0(X) = 1$ and $H_1(X) = 2X$, we apply the recurrence relation to find that $H_2 = 4X^2 - 2$, $H_3 = 8X^3 - 12X$, and so on. The Hermite polynomial $H_n(X)$ is a solution to the differential equation $y'' - 2Xy' + 2ny = 0$. Readers who have studied differential equations will recall that the solutions to this equation can be described as linear combinations of two power series. When $n \geq 0$, one of these power series can be taken to be the polynomial $H_n(X)$. See Prob. 7.3.7 for more details.

Example 7.3.5 (Tchebycheff Polynomials) In this example we take for our inner product

$$<f(X), g(X)> = \int_{-1}^{1} f(X)g(X)(1 - X^2)^{-1/2}\, dX$$

The Tchebycheff polynomials T_n are defined by the recurrence relation

$$T_{n+1}(X) = 2XT_n(X) - T_{n-1}(X)$$

starting with $T_0(X) = 1$ and $T_1(X) = X$. Tchebycheff polynomials turn out to be useful in problems of electrical engineering, as well as applied mathematics. One reason is their relationship to Fourier series, which we briefly indicate.

Suppose that $\cos\theta$ is substituted for X in the Tchebycheff polynomials. We have $T_0(\cos\theta) = 1$, $T_1(\cos\theta) = \cos\theta$, and using the recurrence relation, we see inductively that

$$T_{n+1}(\cos\theta) = 2\cos\theta\cos(n\theta) - \cos[(n-1)\theta] = \cos[(n+1)\theta]$$

Consider a polynomial in $\cos\theta$, say $f(\cos\theta)$. If

$$f(X) = a_n T_n(X) + a_{n-1}T_{n-1}(X) + \cdots + a_0 T_0(X)$$

is the Tchybecheff series for this polynomial [that is, $f(X)$ expressed as a linear combination of $T_0(X), T_1(X), \ldots, T_n(X)$ where $deg(f(X)) = n$]. Substituting $\cos\theta$ for X in this expression, we obtain that

$$f(\cos\theta) = a_n\cos(n\theta) + a_{n-1}\cos[(n-1)\theta] + \cdots + a_0$$

This latter sum is the cosine Fourier series for $f(\cos\theta)$. What we have seen, shows that the substitution $X = \cos\theta$ gives an isometry between P^∞ with the Tchebycheff inner product and the vector space of polynomials in $\cos\theta$ with the inner product given by

$$<f(X), g(X)> = \int_0^\pi f(X)g(X)\,dX$$

We conclude this section by showing how orthogonal polynomials can be used to give least-squares approximations of functions. The vector space P^n can be viewed as an $(n+1)$-dimensional subspace of functions in V. The orthogonal projection from V to P^n can be computed as the sum of the projections onto the orthogonal polynomials $f_0(X), f_1(X), \ldots, f_n(X)$ which span P^n. This projection is the least-squares degree n polynomial approximation to functions in V; that is, it is the degree n polynomial which is closest to the original function in the norm associated with $< , >$. This type of approximation is important because it reflects the geometry of the inner product under consideration.

As an example we return to the Legendre polynomials and the inner product given by

$$<f(X), g(X)> = \int_{-1}^1 f(X)g(X)\,dX$$

The second-degree polynomial least-squares approximation to $f(X) = e^X$ can be computed as follows.

$$Proj_{L_0}(e^X) = \frac{<e^X, 1>}{<1, 1>} \cdot 1 = \frac{e - e^{-1}}{2}$$

$$Proj_{L_1}(e^X) = \frac{<e^X, X>}{<X, X>} \cdot X = \frac{2e^{-1}}{2/3} \cdot X$$

and

$$Proj_{L_2}(e^X) = (5/2)[e - 7e^{-1}](1/2)(3X^2 - 1)$$

Thus this second-order least-squares approximation to the function e^X is

$$(1/4)(-3e + 33e^{-1}) + 3e^{-1}X + (15/4)(e - 7e^{-1})X^2$$

which is approximately $0.996 + 1.104X + 0.537X^2$. The reader will observe that this expression is close to the initial three summands of the usual Taylor series for e^X [that is, it is close to $1 + X + (1/2)X^2$]. This occurs because as one considers higher-order such least-squares approximations, one will obtain a sequence of polynomials which converge to the Taylor series for e^X. Further discussion of the use of such approximations is left to a another course.

PROBLEMS

7.3.1. Use the recursion formula given in Example 7.3.4 to compute H_4 and H_5. Verify by a direct calculation that H_0, H_1, and H_2 are orthogonal.

7.3.2. Use the recursion formula given in Example 7.3.5 to compute T_2 and T_3. Verify by a direct calculation that T_0, T_1, and T_2 are orthogonal.

7.3.3. Compute the best quadratic least squares approximation to the following functions in $C([-1, 1], \mathbf{R})$ with respect to the inner product

$$<f(X), g(X)> = \int_{-1}^{1} f(X)g(X)\,dX$$

by projecting onto the Legendre polynomials:
(a) X^3
(b) $\sin(\pi - X)$
(c) $1 + X + X^2$

7.3.4. Compute the best quadratic least-squares approximation to the following functions in $C(\mathbf{R}, \mathbf{R})$ with respect to the inner product

$$<f(X), g(X)> = \int_{-\infty}^{\infty} f(X)g(X)e^{-X^2}\, dX$$

by projecting onto the Hermite polynomials:
(a) X^3
(b) e^X
(c) $1 + X + X^2$

7.3.5.   Verify that the Legendre polynomials L_n where $n = 0, 1, 2$ satisfy the Legendre equation $(1 - X^2)y'' - 2Xy + n(n+1)y = 0$.

7.3.6.   Verify that the Tchebycheff polynomials T_n where $n = 0, 1, 2$ satisfy the Tchebycheff equation $(1 - X^2)y'' - Xy' + n^2 y = 0$.

7.3.7.   Solve the Hermite equation $y'' - 2Xy' + 2ny = 0$ by formally substituting the power series $y = a_0 + a_1 X + a_2 X^2 + \cdots$ into the equation. Use the fact that $y' = a_1 X + 2a_2 X + 3a_3 X^2 + \cdots$ and $y'' = 2a_2 + 6a_3 X + 12a_4 X^2 + \cdots$ wherever the original series for y converges. By collecting like powers of X after this substitution, obtain the recurrence relation

$$a_{i+2} = \frac{(2i - n)a_i}{(i+2)(i+1)}$$

This shows that a solution to the Hermite equation is determined as a power series once the coefficients a_0 and a_1 are known. For n with $0 \le n \le 4$ show that the Hermite polynomial $H_n(X)$ emerges as a solution.

7.3.8.   The *Laguerre polynomials* $L_n(X)$ are defined by

$$L_n(X) = \sum_{k=0}^{n} \frac{(-1)^k \binom{n}{k}}{k!} X^k$$

(a) Check that $L_0(X) = 1$, $L_1(X) = X$, $L_2(X) = 1 - 2X + X^2/2!$, and compute $L_n(X)$ for $n = 3, 4$.
(b) Show that

$$L_n(X) = \frac{e^X}{n!} \frac{d^n}{dX^n}(X^n e^{-X})$$

(c) Use (b) and integration by parts to verify that the Laguerre polynomials are orthonormal with respect to the inner product

$$<f(X), g(X)> = \int_0^\infty f(X)g(X)e^{-X} dX$$

(d) Show that $L_n(X)$ is a solution to the differential equation

$$Xy'' + (1 - X)y' + ny = 0$$

7.4 Fourier Series and Hilbert Space

In this section we continue to investigate infinite-dimensional inner product spaces. We will show how to understand some important infinite-dimensional results as analogues of these same results from the finite-dimensional theory. In the last section we studied orthogonal polynomials, and we used them to find polynomial approximations of functions. These approximations were obtained by computing the projection of the function onto P^n (taken with respect to the specified inner product). Often it is desirable to approximate a function as a sum of other standard functions, which are not necessarily polynomials. The same ideas utilized in the case of orthogonal polynomials apply, and they lead naturally to the concept of a *Fourier series*.

As an example, we consider $V = \mathcal{C}([0, 2\pi], \mathbf{R})$, the vector space of continuous functions on the interval $[0, 2\pi]$. It is customary to think of V as a vector space of periodic functions on the real line (with period 2π). For this reason it is natural to try to decompose elements of V as linear combinations of the familiar periodic functions, rather than as linear combinations of polynomials (which, unless constant, are never periodic).

We consider the inner product

$$<f(x), g(x)> = \frac{1}{\pi} \int_0^{2\pi} f(x)g(x) dx$$

The trigonometric functions are familiar periodic functions, so we consider the collection $\{1/\sqrt{2}, \cos x, \sin x, \cos(2x), \sin(2x), \ldots\}$. This collection of functions is orthogonal in V [which can be checked using the facts that $\int_0^{2\pi} \sin(nx)\cos(mx) dx = 0$ for all integers $m, n \geq 0$ and $\int_0^{2\pi} \sin(nx)\sin(mx) dx = 0 = \int_0^{2\pi} \cos(nx)\cos(mx) dx$ for all integers

$m \neq n \geq 0$.] Moreover, each function in this collection has length 1, so the collection is orthonormal.

Let W^{2n+1} denote the $(2n+1)$-dimensional subspace of V spanned by the orthonormal vectors $\{1/\sqrt{2}, \cos x, \sin x, \ldots, \cos(nx), \sin(nx)\}$. For any $f(x) \in V$ we compute the projection $Proj_{W^{2n+1}}(f(x))$ according to the formula given in Theorem 4.3.3. We find:

$$Proj_{W^{2n+1}}(f(x))$$
$$= \ <f(x), 1/\sqrt{2}>(1/\sqrt{2}) + \ <f(x), \cos x>\cos x$$
$$+ <f(x), \sin x>\sin x + \cdots + <f(x), \cos(nx)>\cos(nx)$$
$$+ \qquad <f(x), \sin(nx)> \sin(nx)$$
$$= \ \frac{a_0}{2} + \sum_{k=1}^{n} [a_k \cos(kx) + b_k \sin(kx)]$$

where

$$a_0 = \frac{1}{\pi} \int_0^{2\pi} f(x)\, dx,$$

$$a_k = \frac{1}{\pi} \int_0^{2\pi} f(x) \cos(kx)\, dx$$

$$b_k = \frac{1}{\pi} \int_0^{2\pi} f(x) \sin(kx)\, dx$$

The vector $Proj_{W^{2n+1}}(f(x))$ is an appropriate projection of $f(x)$ onto W^{2n+1} since it can be shown (see Rudin [Theorem 4.16]) that it is the vector in W^{2n+1} whose distance from $f(x)$ in V is minimal. (We proved this in the finite-dimensional case in Lemma 4.4.4.) Thus, like those computed in the last section, this projection is a least squares approximation to $f(x)$ inside the vector space W^{2n+1} with respect to the specified inner product. In other words, we have a best possible approximation for $f(x)$ in terms of the $2n+1$ familiar periodic functions $\{1/\sqrt{2}, \cos x, \sin x, \ldots, \cos(nx), \sin(nx)\}$.

The projection $Proj_{W^{2n+1}}(f(x))$ gives the first $2n + 1$ terms in what is called the (trigonometric) Fourier series for $f(x)$. It can be shown that $lim_{n\to\infty} \| f(x) - Proj_{W^{2n+1}}(f(x)) \| = 0$. As a consequence the function $f(x)$ is the limit of the approximating functions $Proj_{W^{2n+1}}(f(x))$. In other words we can view $f(x)$ as

$$lim_{n\to\infty} \left(\frac{a_0}{2} + \sum_{k=1}^{n} (a_k \cos(kx) + b_k \sin(kx)] \right)$$

which is the *Fourier series* of $f(x)$.

The reader may wonder in what sense the function $f(x)$ is the limit of its Fourier series. It would be nicest if at every point $r \in [0, 2\pi]$ the value $f(r)$ was the limit of the values of the approximating functions $Proj_{W^{2n+1}}(f(r))$. This turns out to be true for almost all r whenever $f(x)$ is continuous, but unfortunately need not be true for every r. Thus, pointwise convergence is not the right point of view to adopt. The correct point of view is to consider the convergence of the Fourier series as the convergence *of functions* inside an appropriate inner product space. This is discussed further in Example 7.4.9.

The convergence of the functions given in the above Fourier series is expressed in terms of the distance function associated with the norm on V. Thus, instead of being able to study only finite sums of vectors (as is usual in linear algebra), the inner product on V permits us to consider infinite sums. This enables us to give a different notion of a basis, which uses a generalized definition of "span," allowing infinite sums.

In order to develop this theme, we digress at this point and next discuss some definitions from metric space analysis. The purpose of this discussion is to give the reader some feeling of how the basic ideas from linear algebra interact with these concepts. The discussion is *extremely* incomplete, and the reader who finds this material interesting should pursue his or her interest by studying an advanced text devoted to the topic (see the Bibliography). The following definitions are given in most advanced calculus courses for the case of $\mathbf{R}^n$. We need them for our study of infinite-dimensional inner product spaces.

Definition 7.4.1 (i) A *metric space* is a set V together with a distance function $d(\ ,\ ) : V \times V \to \mathbf{R}^+ \cup \{0\}$ which satisfies
(a) $d(v, w) = 0$ if and only if $v = w$.
(b) $d(v, w) = d(w, v)$ for all $v, w \in V$.
(c) $d(v, w) \geq d(v, u) + d(u, w)$ for all $u, v, w \in V$.

(ii) A *Cauchy sequence* $v_1, v_2, v_3, \ldots$ in a metric space V is a sequence of elements in V with the property that for every real $\varepsilon > 0$ there exists a natural number N for which $d(v_i, v_j) < \varepsilon$ whenever $i, j > N$.

(iii) A Cauchy sequence $v_1, v_2, v_3, \ldots$ *converges* to an element $v \in V$ if for every real $\varepsilon > 0$ there exists a natural number N for which $d(v, v_i) < \varepsilon$ whenever $i > N$.

(iv) A metric space is called *complete* if every Cauchy sequence converges to an element of the space.

The most familiar examples of metric spaces are the real numbers and $\mathbf{R}^n$. In $\mathbf{R}$, as is proved in advanced calculus, the Cauchy sequences are precisely all the sequences of real numbers that converge. Since any Cauchy sequence of real numbers converges to a (unique) real number, the reals are a complete metric space. In contrast, the subset $\mathbf{Q}$ of all rational numbers is a metric subspace of $\mathbf{R}$ that is not complete, because there are Cauchy sequences of rationals that do not converge to a rational number.

The next lemma asserts that whenever one has a norm on a real vector space, the distance function defined above gives a metric.

Lemma 7.4.2 *If $\| \ \|: V \to \mathbf{R}$ is a norm, then $d(\ ,\ )$ defined by $d(v, w) = \| v - w \|$ is a metric on V.*

Proof: This result follows directly from the definition of a norm. (In fact norms were defined so that this would be true.) The connection is as follows: That $d(v, w) = 0$ if and only if $v = w$ follows from the positive-definite condition on a norm; that is, $\| u \| = 0$ if and only if $u = 0$. That $d(v, w) = d(w, v)$ follows as $\| v - w \| = |-1| \cdot \| w - v \| = \| w - v \|$. Finally to see that $d(v, w) \leq d(v, u) + d(u, w)$, we apply the triangle inequality (as stated for norms):

$$\| v - w \| = \| (v - u) + (u - w) \| \leq \| v - u \| + \| u - w \|$$

This concludes the proof. //

We next give two important definitions.

Definition 7.4.3 If a normed vector space V is complete in the metric induced by its norm, then V is called a *Banach space*. If $< , >: V \times V \to \mathbf{R}$ is an inner product and V is a complete metric space with respect to the induced norm $\| \ \|$, then V is called a *Hilbert space*.

In the remainder of this section we show how to apply the inner product concepts developed in Chap. 4 to Hilbert spaces. It is not hard to check that any finite-dimensional inner product space (normed space) is a Hilbert space (Banach space). This follows (essentially) from the completeness of the real numbers. But it is *not* true that infinite-dimensional inner product spaces are complete.

For example, $C([1, -1], \mathbf{R})$ is an inner product space with inner product $<f(x), g(x)> = \int_{-1}^{1} f(x)g(x)\, dx$. However, $C([-1, 1], \mathbf{R})$ is not complete in the associated norm. To see this, note that the step function $s(x)$, defined by $s(x) = 1$ if $x \in [0, 1]$ and $s(x) = 0$ otherwise, does not lie in $C([-1, 1], \mathbf{R})$, yet $s(x)$ is a limit of functions in $C([-1, 1], \mathbf{R})$. In order to rectify this problem of noncompleteness, we must consider a larger class of functions on $[-1, 1]$, namely those that are integrable (or more precisely, the *Lebesgue measurable functions*). For details we mention Rudin [Chap. 2]. Our discussion for the remainder of this section will be informal, and the reader unfamiliar with the definition of Lebesgue measure can still absorb the basic ideas of this discussion by viewing these integrable functions as a generalization of Riemann integrable functions.

By expanding to the measurable functions, denoted $\mathcal{M}([-1, 1], \mathbf{R})$, we will obtain the desired completeness property for our space (note that the step function above is integrable, even though noncontinuous). However, at the same time we have introduced a new difficulty. The inner product defined by $< f(x), g(x) > = \int_{-1}^{1} f(x)g(x)\, d\mu$ (μ stands for Lebesgue measure) fails to be positive definite. Note, for example, that $<p(x), p(x)> = 0$ where $p(x)$ is the nonzero function defined by $p(0) = 1$ and $p(x) = 0$ for all $x \neq 0$. This second difficulty is resolved by identifying functions which differ only on a set of measure zero. In other words, we call the functions $f(x), g(x) \in \mathcal{M}([-1, 1], \mathbf{R})$ equivalent if $\int_{-1}^{1} (f(x) - g(x))^2\, d\mu = 0$. So, the function $p(x)$ just defined is equivalent to the zero function. If, instead of using the measurable functions on $[-1, 1]$, we take the equivalence classes of such functions as elements of our vector space, we obtain a positive-definite inner product which is also complete. In formal algebraic language, we define the vector space $L^2([-1, 1], \mathbf{R})$ to be the quotient space $\mathcal{M}([-1, 1], \mathbf{R})/Z$ where Z is the subspace $Z = \{p(x) \in \mathcal{M}([-1, 1], \mathbf{R}) \mid \int_{-1}^{1} p(x)^2\, d\mu = 0\}$. We synopsize this discussion in the next example.

Example 7.4.4 Let $X \subseteq \mathbf{R}^n$. We define $L^2(X)$ to be the real vector space of equivalence classes of all functions $f : X \to \mathbf{R}$ for which the integral $\int_X f^2\, d\mu$ exists. Here, two functions $f, g : X \to \mathbf{R}$ are called equivalent if $\int_X (f - g)^2\, d\mu = 0$ (that is, they differ on a set of measure zero). Then, exactly as above, we define an inner product on $L^2(X)$ by

$<f, g> = \int_X fg\, d\mu$. Our example asserts that $L^2(X)$ is a Hilbert space. [The completeness of the metric in $L^2(X)$ is known as the *Riesz-Fischer theorem.* See Rudin [Theorem 4.1.7].] We remark that the subspace of $L^2(X)$ of (equivalence classes of) continuous functions is *not* a Hilbert space since it fails to be complete.

Example 7.4.5 We denote by $\ell^2(\mathbf{N})$ the collection of all sequences of real numbers $(r_1, r_2, \ldots)$ with the property that r_1^2, $r_1^2 + r_2^2$, $r_1^2 + r_2^2 + r_3^2, \ldots$ is a Cauchy sequence of real numbers. $\ell^2(\mathbf{N})$ is a real vector space with the obvious componentwise addition and scalar multiplication. We define the inner product $<v, w>$ of two vectors $v = (r_1, r_2, r_3, \ldots)$ and $w = (q_1, q_2, q_3, \ldots)$ by $<v, w> = lim_{n \to \infty} s_n$ where $s_n = r_1 q_1 + r_2 q_2 + \cdots + r_n q_n$. With this definition it is not difficult to verify that $\ell^2(\mathbf{N})$ is a Hilbert space.

In $\ell^2(\mathbf{N})$ consider the vectors $e_1 = (1, 0, 0, \ldots)$, $e_2 = (0, 1, 0, 0, \ldots)$, $e_3 = (0, 0, 1, 0, 0, 0, \ldots)$, etc. Evidently $\mathcal{B} = \{e_1, e_2, e_3, \ldots\}$ is a linearly independent subset of $\ell^2(\mathbf{N})$. However, $\mathcal{B}$ is not a basis for $\ell^2(\mathbf{N})$. Consider the element $v = (1, 1/2, 1/3, 1/4, \ldots) \in \ell^2(\mathbf{N})$. [Recall that the series $\sum_{n=1}^{\infty}(1/n^2)$ converges.] Clearly v cannot be expressed as a finite linear combination of elements of $\mathcal{B}$. However, v can be expressed as the limit of a Cauchy sequence $v_1, v_2, \ldots$ where v_i is defined as $e_1 + (1/2)e_2 + \cdots + (1/i)e_i$. In particular, we can write $v = \sum_{i=1}^{\infty}(1/i)e_i$. Clearly, similar expressions exist for all elements of $\ell^2(\mathbf{N})$. In fact, more is true. Note that for $v = (r_1, r_2, \ldots) \in \ell^2(\mathbf{N})$, we have $<v, e_i> = r_i$. From this we obtain the expression $v = \sum_{i=1}^{\infty} <v, e_i> e_i$. The similarity of this observation with Theorem 4.2.2 motivates the following.

Definition 7.4.6 (i) Let V be a Hilbert space (or a Banach space). A subset $S \subseteq V$ is called *complete* in V if every element of V is a limit of a Cauchy sequence of (finite) linear combinations of elements of S.

(ii) A subset $\mathcal{B}$ of a Hilbert space V is called a *complete orthonormal set* for V if $\| v \| = 1$ for each $v \in \mathcal{B}$, $<v, w> = 0$ for each $v, w \in \mathcal{B}$ with $v \neq w$, and $\mathcal{B}$ is complete in V.

The discussion above shows that $\mathcal{B} = \{e_1, e_2, e_3, \ldots\}$ is a complete orthonormal set for $\ell^2(\mathbf{N})$. The following theorem (which we do not prove) shows that complete orthonomal sets can be used in Hilbert spaces in much the same way an orthonormal basis is used in finite-dimensional inner product spaces (for example see Theorem 4.2.2).

This result shows why the Fourier series considered earlier in this section converged. For more details see Rudin [Chap. 4].

Theorem 7.4.7 *Suppose that $\mathcal{B} = \{u_1, u_2, u_3, \ldots\}$ is a complete orthonormal set for a Hilbert space V ($\mathcal{B}$ is indexed by the natural numbers 1, 2, ...). Then every $v \in V$ can be expressed uniquely as a convergent infinite sum*

$$v = \sum_{i=1}^{\infty} a_i u_i$$

where $a_i \in \mathbf{R}$. Moreover, the coefficients are determined by the expression $a_i = <v, u_i>$.

Remark 7.4.8 That the above theorem holds for $\ell^2(\mathbf{N})$ with the complete orthonormal set $\mathcal{B} = \{e_1, e_2, \ldots\}$ was illustrated above. Note that in order for the series to converge, (that is, in order for the sequence of finite sums to be Cauchy) it must happen that $|<v, u_i>| \to 0$ as $i \to \infty$. In analogy with the terminology from Fourier series, the coefficients $<v, u_i>$ are called the *Fourier coefficients of v with respect to* $\{u_1, u_2, \ldots\}$.

Example 7.4.9 Set $\hat{V} = L^2([0, 2\pi], \mathbf{R})$. [$\hat{V}$ contains the vector space $V = \mathcal{C}([0, 2\pi], \mathbf{R})$ which was considered at the beginning of this section as a subspace and can also be viewed as a space of periodic functions on $\mathbf{R}$ with period 2π.] We consider the same inner product as before, namely $<f(x), g(x)> = (1/\pi) \int_0^{2\pi} f(x)g(x)\, dx$. It turns out that the collection of vectors $\{1/\sqrt{2}, \cos x, \sin x, \ldots\}$ forms a *complete* orthonormal set for $\hat{V}$. This is why the Fourier series for $f(x)$

$$lim_{n\to\infty} \left(\frac{a_0}{2} + \sum_{k=1}^{n} [a_k \cos(kx) + b_k \sin(kx)] \right)$$

converges. We emphasize however, that this convergence is convergence in $\hat{V}$. As remarked earlier in this section, the Fourier series of a function $f(x)$ need not converge (pointwise) to the exact function $f(x)$.

Another particularly nice consequence of Theorem 7.4.7 is that the linear transformation $C : L^2([0, 2\pi], \mathbf{R}) \to \ell^2(\mathbf{N})$ defined by

$$C(f) = (<f, 1/\sqrt{2}>, <f, \cos x>, <f, \sin x>, \ldots)$$

is an isometry of Hilbert spaces. In other words one can interpret $\hat{V} = L^2([0, 2\pi])$ as the subspace $\ell^2(\mathbf{N})$ of $\mathbf{R}^\infty$ consisting of those sequences whose sums of squares converge.

In this section we have shown how to think of Fourier series from a purely algebraic point of view. Usually when students first encounter Fourier series, they are left in the dark as to where the mysterious Fourier coefficients (which appear to be just more ugly integrals) come from. The point is that the coefficients $<f(x), f_i(x)>$ are simply the lengths of the vectors which are $f(x)$ projected upon the unit vectors $f_i(x)$. In $\mathbf{R}^n$, the formula $v = <v, e_1>e_1 + <v, e_2>e_2 + \cdots + <v, e_n>e_n$ is easily understood, both algebraically and geometrically. The Fourier series is simply an infinite-dimensional version of this fact, which, while impossible to comprehend geometrically, is reasonable from the perspective of linear algebra.

PROBLEMS

7.4.1. Find the Fourier series for the following functions in the vector space $L^2([0, 2\pi], \mathbf{R})$:
(a) $f(x) = x$.
(b) $f(x) = \cos^2 x$.
(c) $f(x) = x^2$.

7.4.2. Verify that Example 7.4.5 really is a Hilbert space.

7.4.3. Show that the inner product space P^∞ of all polynomials with the inner product $<f(X), g(X)> = \int_0^1 f(X)g(X)\, dX$ is *not* a complete metric space.

7.5   Bases for Arbitrary Vector Spaces

In this section we study further the question of what "basis" means for an infinite-dimensional vector space. Unlike the last section, where we used metric space ideas to modify the notion of "span" (specifically replacing it with the notion of a complete set), we shall work directly with the usual notion of "span". We prove that every vector space has a basis. However, it is important to note that the proof is *not* constructive (as it is in the finite-dimensional case). In fact it is known that this proof *cannot* be made constructive.

We begin by restating the usual definition of a basis for a vector space, making sure that it is clear what is meant in the infinite-dimensional case.

Definition 7.5.1 Let V be a vector space, which is possibly infinite-dimensional. A subset $S = \{v_i \mid i \in I\} \subseteq V$ indexed by a set I is said to be *linearly independent* if each finite subset of S is linearly independent. S is said to *span* V if every vector in V can be written as a finite linear combination of elements of S. S is called a *basis* for V if S is linearly independent and spans V.

Corollary 1.4.9 (i) says that every finitely spanned vector space has a basis. The theorem we prove next says that *every* vector space has a basis. The proof uses (and needs) the axiom of choice. The axiom of choice is an axiom of abstract set theory which cannot be proved from more elementary axioms. A version of the axiom of choice, known as Zorn's lemma, is stated in the proof. As you will see, the proof tells you nothing about how to find a basis; it just shows bases exist.

Theorem 7.5.2 (Uses Axiom of Choice) *Every vector space V over any field $\mathbf{F}$ has a basis.*

Proof: We use the axiom of choice in the form of Zorn's cemma. Let S be any set. We call a collection $\mathcal{C}$ of subsets of S a *chain* if $\mathcal{C}$ is linearly ordered by inclusion (that is, whenever U_1, $U_2 \in \mathcal{C}$, then $U_1 \subseteq U_2$ or $U_2 \subseteq U_1$). Zorn's lemma says the following:

> Suppose that S is a set and T is a collection of subsets of S with the property that every chain $\mathcal{C}$ of elements of T has an upper bound in T (that is, there is an element $M \in T$ with $U \subseteq M$ for all $U \in \mathcal{C}$). Then T has a maximal element.

To prove the theorem, let V be a vector space. We let T be the collection of *all* linearly independent subsets of V. Suppose that $\mathcal{C}$ is a chain of elements of T. We claim that the union $M = \bigcup_{U \in \mathcal{C}} U \in T$. To prove this, we must show that M is linearly independent. Pick any finite collection $v_1, v_2, \ldots, v_m \in M$. By the definition of M each $v_i \in U_i$ for some $U_i \in \mathcal{C}$. Since $\mathcal{C}$ is linearly ordered by inclusion, there is some j with $U_i \subseteq U_j$ for $i = 1, 2, \ldots, m$. We find that $v_1, v_2, \ldots, v_m \in U_j$.

By hypothesis U_j is a linearly independent subset of V, and this shows that $v_1, v_2, \ldots, v_m$ are linearly independent. Since $v_1, v_2, \ldots, v_m \in M$ were arbitrary, the linear independence of M follows. Clearly $M \in T$ is an upper bound for the chain C. This shows that the hypothesis of Zorn's lemma is true for T.

Zorn's lemma now says that T has a maximal element B. We claim that B is a basis for V. Since $B \in T$, B is linearly independent. We must show that B spans V. Let $v \in V$, and suppose that $v \notin span(B)$. We claim that $B \cup \{v\}$ is linearly independent. Consider any finite subset of $B \cup \{v\}$, which is necessarily of the form $\{v_1, v_2, \ldots, v_n\}$ or $\{v_1, v_2, \ldots, v_n\} \cup \{v\}$ where $v_1, v_2, \ldots, v_n \in B$. The first subset lies in B, hence is linearly independent. Since $v \notin span\{v_1, v_2, \ldots, v_n\}$ by hypothesis, we conclude that $\{v_1, v_2, \ldots, v_n\} \cup \{v\}$ is linearly independent also. Thus, $B \cup \{v\}$ is linearly independent, so $B \cup \{v\} \in T$. This contradicts the maximality of B in T. From this contradiction we conclude that any $v \in V$ lies in the span of B. Thus B is a basis of V, and the theorem is proved. //

Note that the last paragraph of the proof of Theorem 7.5.2 is the same as the proof of Lemma 1.4.8 (i) where it was shown that a maximally linearly independent subset of a finitely spanned vector space is a basis. The difference between the proof of Theorem 7.5.2 and the proof of Lemma 1.4.8 (i) is that in the infinite-dimensional case one needs Zorn's lemma to obtain a maximally linearly independent subset, a fact which comes easily in the finitely spanned case.

Theorem 7.5.2 has some interesting applications. Consider the real numbers $\mathbf{R}$ as a vector space over the field $\mathbf{Q}$ of rational numbers. Viewed as such, $\mathbf{R}$ is infinite-dimensional. Theorem 7.5.2 says that $\mathbf{R}$ has a basis B over the field $\mathbf{Q}$. In other words there exists a subset $B \subset \mathbf{R}$ with the property that every real number can be expressed uniquely as a finite linear combination $q_1 b_1 + q_2 b_2 + \cdots + q_n b_n$ for $q_1, q_2, \ldots, q_n \in \mathbf{Q}$ and $b_1, b_2, \ldots, b_n \in B$. Such a basis B must be infinite (in fact it is uncountably infinite). Thus there is no way to explicitly write down such a set. So, although we do not know explicitly what they look like, if one believes in Zorn's lemma, then one knows (many) such B exist.

Chapter 8

Multilinear Algebra

8.1 Multilinear Functions and Duality

As with the preceding chapter, this chapter is not intended to be a comprehensive treatment of the subject. The first two sections discuss multilinear functions, tensors, and forms. The last section is an informal discussion of how tensors and forms are used in multivariable calculus. Tensors are often first encountered by students in physics and engineering courses, before (if at all) in any mathematics course. It is customary in such courses to teach students how to manipulate tensors, as is necessary to solve certain problems. But usually there is insufficient time to develop the foundations of the subject. This chapter will develop these foundations (in abbreviated form). Most of the results in Secs. 8.1 and 8.2 will be true for arbitrary vector spaces, while the applications studied in Sec. 8.3 will use real (and possibly complex) vector spaces. The prerequisite for Secs. 8.1 and 8.2 is the material of Chap. 3 together with some knowledge of the inner product (dot product) on $\mathbf{R}^n$. Section 8.3 requires knowledge of multivariable calculus.

We begin with the topic of multilinear functions. Recall that a bilinear form on a vector space V is a function $(\ ,\ ) : V \times V \to \mathbf{F}$ which is linear in each variable; that is, for every $v \in V$ the functions $R_v(u) = (v, u)$ and $L_v(u) = (u, v)$ are both linear (see Lemma 5.1.3 for details). Multilinear functions generalize this. A multilinear function will be a function on a product of vector spaces which gives a linear transformation in one variable whenever all the other variables are fixed. This is made precise in the next definition. Note that we allow different vector spaces as factors of the domain of an arbitrary multilinear function, and note as well that the function need not be scalar-valued (as it was for bilinear forms).

Definition 8.1.1 Suppose that $V_1, V_2, \ldots, V_m, W$ are $\mathbf{F}$ vector spaces. A function $f : V_1 \times V_2 \times \cdots \times V_m \to W$ is called *multilinear* if whenever $v_1 \in V_1$, $v_2 \in V_2$, $\ldots$, $v_m \in V_m$ and $i \in \{1, 2, \ldots, m\}$, then the function $T_i : V_i \to W$ defined by $T(u) = f(v_1, \ldots, v_{i-1}, u, v_{i+1}, \ldots, v_m)$ is a linear transformation. When $m = 2$, f is called *bilinear*; when $m = 3$, f is called *trilinear*, and so forth.

In addition to bilinear forms, which are the scalar-valued bilinear functions on $V \times V$, we have studied another type of multilinear function, namely the determinant function (when interpreted appropriately). The point is this. Instead of using the usual viewpoint where $det : M_{n \times n}(\mathbf{F}) \to \mathbf{F}$ is a function of $n \times n$ matrices, we view

$$det : \underbrace{\mathbf{F}^n \times \mathbf{F}^n \times \cdots \times \mathbf{F}^n}_{n \ times} \longrightarrow \mathbf{F}$$

as a function of the n individual rows of a matrix (each of which is a "copy" of $\mathbf{F}^n$). It follows from Theorem 2.1.4 that *det* is a multilinear function when viewed in this context. (Similarly, *det* is a multilinear function of the columns of a matrix.)

In this chapter we study a number of special types of multilinear functions and develop their properties. In each case we will consider the vector space consisting of all the multilinear functions of a given type. Thus, before treating these examples, we give the following lemma (the straightforward proof of which is omitted).

Lemma 8.1.2 *The set of all multilinear functions*

$$f : V_1 \times V_2 \times \cdots \times V_m \to W$$

together with the usual operations of function addition and scalar multiplication, is a vector space.

The dual space is an extremely important tool in the study of general multilinear functions.

Definition 8.1.3 Let V be a **F** vector space. The vector space of all linear transformations from V to **F** $\{T \mid T : V \to \mathbf{F}\}$ is denoted by V^*. V^* is called the *dual space* of V and its elements are called *linear functionals*.

The dual space of a vector space is closely related to the original vector space. Indeed, we shall see shortly that whenever V is finite-dimensional, the vector space V^* is isomorphic to V. (The identification of V and V^* plays a big role in the applications of multilinear algebra.) The infinite-dimensional case is also important. For example, suppose that $V = \mathcal{C}([0, 1], \mathbf{R})$. Then the linear functional $Int : V \to \mathbf{R}$ defined by $Int(f) = \int_0^1 f(x)\, dx$ is an element of V^*. So are the functionals which give the various Fourier coefficients of f. When one studies Fourier series, in some sense, one is using the vector space V^* to analyze the original vector space V.

Let $f \in V^*$ and suppose that $v \in V$. Since $f : V \to \mathbf{F}$ is a linear transformation, $f(v)$ is defined and is an element of **F**. It is important to view this pairing of f and v as a function $< , >: V^* \times V \to \mathbf{F}$. This function is bilinear, as the next lemma shows.

Lemma 8.1.4 *Let V be a **F** vector space. Then $< , >: V^* \times V \to \mathbf{F}$ defined by $<f, v> = f(v)$ is bilinear.*

Proof: To see that $< , >$ is bilinear, we first note that for fixed f, the function $v \mapsto <f, v> = f(v)$ is linear as $f \in V^*$. For fixed v, $<f_1 + f_2, v> = (f_1 + f_2)(v) = f_1(v) + f_2(v) = <f_1, v> + <f_2, v>$ by the definiton of functional addition. Similarly $<kf, v> = kf(v) = k<f, v>$, which shows that $< , >$ is linear on the left as well. This gives the lemma. //

Whenever V is finite-dimensional, it is possible to create a basis for V^* out of any basis for V by using the above pairing.

Definition 8.1.5 Suppose that V is a finite-dimensional vector space with basis $\mathcal{B} = \{v_1, v_2, \ldots, v_n\}$. We define the *dual basis* $\mathcal{B}^*$ for V^* by setting $\mathcal{B} = \{f_1, f_2, \ldots, f_n\}$ where $f_i : V \to \mathbf{F}$ is the (unique) linear transformation which satisfies $f_i(v_i) = 1$ and $f_i(v_j) = 0$ whenever $j \neq i$.

In addition to showing that the dual basis really is a basis for V^*, the next theorem details the relationship between the vector spaces V and V^* by making use of the pairing $< , >: V^* \times V \to \mathbf{F}$. In this theorem, $V^{**} = (V^*)^*$ is the so-called *double dual* of V. In other words, V^{**} is the vector space of all linear functionals on V^*.

Theorem 8.1.6 *Let V be a $\mathbf{F}$ vector space. Then:*
(i)   *The linear transformation $Q : V \to V^{**}$ defined by $v \mapsto < , v>$ is injective.*
(ii) *If V is finite-dimensional, then for any basis $\mathcal{B}$ of V the dual basis $\mathcal{B}^*$ is a basis for V^*. In particular, $dim(V) = dim(V^*)$. Consequently, in the finite-dimensional case $Q : V \to V^{**}$ is an isomorphism.*

Proof: (i) The linear transformation $Q : V \to V^{**}$ is given as follows: For each $v \in V$, $Q(v)$ is the linear functional $< , v>$ on V^* given by $(Q(v))(f) = < f, v> \;[= f(v)] \in \mathbf{F}$ for each functional $f \in V^*$. Note that the bilinearity of $< , >$ shows that $Q(v)$ is a linear functional on V^* (this is linearity on the left), so that in fact $Q(v) \in V^{**}$. The bilinearity of $< , >$ also shows that Q is a linear transformation from V to V^{**}.

We must check that Q is injective. Suppose that $Q(v) = \vec{0} \in V^{**}$. This means that $<f, v> = 0 \in \mathbf{F}$ for all $f \in V^*$; that is, $f(v) = \vec{0}$ for all $v \in V^*$. The Basis-Extension Theorem 1.4.11 together with Lemma 3.1.4 shows (in the finite-dimensional case) that if $v \neq \vec{0} \in V$, then there is some $f \in V^*$ with $f(v) \neq 0$. This shows that $v = \vec{0}$, and the injectivity of Q follows.

(ii) Suppose that $\mathcal{B} = \{v_1, \ldots, v_n\}$ is a basis for V. We claim that the dual basis $\mathcal{B}^* = \{f_1, f_2, \ldots, f_n\}$ is a basis for V^*. Let $f \in V^*$ and suppose that $f(v_i) = a_i \in \mathbf{F}$. Then by Lemma 3.1.4 $f = a_1 f_1 + a_2 f_2 + \cdots + a_n f_n$ since $(a_1 f_1 + a_2 f_2 + \cdots + a_n f_n)(v_i)$ is also a_i. This shows that $\mathcal{B}^*$ spans V^*. This also gives the linear independence of $f_1, f_2, \ldots, f_n$, because if $\vec{0} = a_1 f_1 + a_2 f_2 + \cdots + a_n f_n$, then we must have $0 = (a_1 f_1 + a_2 f_2 + \cdots + a_n f_n)(v_i) = a_i$ for all i. Hence $\mathcal{B}^*$ is a basis of V^*.

The final observation in (ii) follows from the injectivity of $Q : V \to V^{**}$ together with the fact that $dim(V) = dim(V^*) = dim(V^{**})$. This proves the theorem. //

The linear transformation $Q : V \to V^{**}$ is an example of what mathematicians call a *natural map*. The point is this. Whenever V is a finite-dimensional vector space, Theorem 8.1.6 (ii) shows the vector spaces V and V^* are isomorphic. One explicit isomorphism is to start with a basis $\mathcal{B} = \{v_1, v_2, \ldots, v_n\}$, form $\mathcal{B}^* = \{f_1, f_2, \ldots, f_n\}$, and define an isomorphism $S : V \to V^*$ by $S(v_i) = f_i$. But note that this isomorphism depends upon the choice of the basis $\mathcal{B}$ for V and a different isomorphism will result if a different basis for V is chosen. In contrast, the isomorphism $Q : V \to V^{**}$ defined by $v \mapsto < \, , v >$ does not depend upon any choice of a basis. Thus this second isomorphism is natural. The naturality of Q will surface again in Sec. 8.3. For more discussion of the term "natural" (and a definition) see Prob. 8.1.2.

The concept of the dual space enables us to give a new perspective to the concept of a transpose of a matrix. Let $T : V \to W$ be a linear transformation. We construct a linear transformation $T^* : W^* \to V^*$, called the *transpose* of T, as follows: For $f \in W^*$ we define $T^*(f)$ to be the linear functional on V given by $T^*(f)(v) = f(T(v))$ for all $v \in V$. Note that as $T^*(f)$ is the composition of two linear transformations, it is linear. Thus, $T^*(f)$ is a linear functional in V^*. It is easy to verify that $T^* : W^* \to V^*$ is a linear transformation. The next result shows why T^* is called the transpose of T.

Lemma 8.1.7 *Suppose that V and W are finite-dimensional vector spaces with ordered bases $\mathcal{B}$ and $\mathcal{C}$ respectively. We denote by $\mathcal{B}^*$ and $\mathcal{C}^*$ the dual (ordered) bases for V^* and W^* respectively. If $T : V \to W$ is a linear transformation, then the matrix $[T^*]_{\mathcal{C}^*,\mathcal{B}^*} = ([T]_{\mathcal{B},\mathcal{C}})^t$.*

Proof: Suppose that $\mathcal{B} = \{v_1, v_2 \ldots, v_n\}$ and $\mathcal{C} = \{w_1, w_2, \ldots, w_m\}$. We denote $\mathcal{B}^* = \{f_1, f_2, \ldots, f_n\}$ and $\mathcal{C}^* = \{g_1, g_2, \ldots, g_m\}$. Assume that $[T]_{\mathcal{B},\mathcal{C}} = (a_{ij})$. By definition, $T(v_i) = a_{1i}w_1 + a_{2i}w_2 + \cdots + a_{mi}w_m$ for all $i = 1, 2, \ldots, n$. We compute $T^*(g_j)$. By definition $T^*(g_j)(v_i) = g_j(T(v_i)) = g_j(a_{1i}w_1 + a_{2i}w_2 + \cdots + a_{mi}w_m) = a_{ji}$. It follows that $T^*(g_j)$ is the linear functional $a_{j1}f_1 + a_{j2}f_2 + \cdots + a_{jn}f_n$. In particular, $[T^*]_{\mathcal{C}^*,\mathcal{B}^*} = (a_{ji}) = ([T]_{\mathcal{B},\mathcal{C}})^t$, as required. //

We saw in Theorem 8.1.6 that in the finite-dimensional case, the bilinear pairing $< \, , >: V^* \times V \to \mathbf{F}$ gave a natural identification of V with V^{**}. The idea of using bilinear functions to identify vector spaces with their dual spaces can be extended to a more general situation.

Suppose $P : V \times W \to \mathbf{F}$ is bilinear. We shall say that P is *nondegenerate* if for any nonzero $v \in V$ and any nonzero $w \in W$, the elements $<v, \ > \in W^*$ and $<\ , w > \in V^*$, are nonzero. (In other words, for any $v \in W$ there is some $z \in W$ with $<v, z> \neq 0$ and for any $w \in W$ there is some $u \in V$ with $<u, w> \neq 0$.) The important feature of nondegenerate pairings is given in the following lemma.

Lemma 8.1.8 (Duality) *Suppose that $P : V \times W \to \mathbf{F}$ is a nondegenerate bilinear pairing of finite-dimensional vector spaces. Then V is isomorphic to W^* (via P) and W is isomorphic to V^* (via P).*

Proof: We define a linear transformation $R : V \to W^*$ as follows: For $v \in V$ set $R(v) = <v, \ > \in W^*$. That R is linear follows from the bilinearity of $<\ , \ >$. Since $<\ , \ >$ is nondegenerate, we know that whenever $v \neq \vec{0}$, $<v, \ >$ is nonzero, and consequently R is injective. This shows that $dim(V) \leq dim(W^*)$. To see that R is surjective, we consider $L : W \to V^*$ defined by $L(w) = <\ , w >$. L is an injective linear transformation for the same reasons as R is, and thus $dim(W) \leq dim(V^*)$. Applying Theorem 8.1.6, we obtain that $dim(V^*) = dim(V) \leq dim(W^*) = dim(W) \leq dim(V^*)$, so equality holds throughout. Consequently, both R and L are isomorphsims. This proves the result. //

The special case of Lemma 8.1.7 where $V = W$ is extremely important. It says that if $<\ , \ >$ is a nondegenerate bilinear form on a finite-dimensional vector space V, then V and V^* are isomorphic. Of course, we already know that V and V^* are (abstractly) isomorphic by Lemma 8.1.5. However, rather than using some (arbitrarily chosen) basis to construct the isomorphism, the proof of Lemma 8.1.7 gives an isomorphism that is intimately associated to the pairing $<\ , \ >$. (This is the isomorphism $V \to V^*$ given by $v \mapsto <v, \ >$.) This identification is particularly important for inner product spaces, because then it becomes natural (in the presence of an inner product) to identify V and V^*. This identification is used heavily in the development of tensor analysis in the case where $V = \mathbf{R}^n$.

PROBLEMS

8.1.1. (a) Find the basis $\mathcal{B}^* = \{f_1, f_2, f_3\}$ for $\mathbf{F}^{3*}$ which is dual to the basis $\mathcal{B} = \{v_1, v_2, v_3\}$ where $v_1 = (1, 1, 0), v_2 = (1, 0, 1)$, and $v_3(0, 1, 0)$.

(b) Show that the isomorphism $T_{\mathcal{B}} : \mathbf{F}^3 \to (\mathbf{F}^3)^*$ given by $T_{\mathcal{B}}(v_i) = f_i$ is *not* the same as the analogous isomorphism given by the standard basis.

8.1.2. (Naturality) (a) Show that for any vector space V, the injective linear transformation $Q_V : V \to V^{**}$ is natural in the following sense: Whenever $T : V \to W$ is a linear transformation, then $Q_W \circ T = T^{**} \circ Q_V$. This is symbolized by saying that the following diagram commutes:

$$
\begin{array}{ccc}
V & \xrightarrow{\;T\;} & W \\
Q_V \downarrow & & \downarrow Q_W \\
V^{**} & \xrightarrow{\;T^{**}\;} & W^{**}
\end{array}
$$

(b) Show that if V is a finite-dimensional inner product space, then V and V^* are naturally isomorphic in the sense outlined in part (a).

8.1.3.   Let V be a finite-dimensional vector space and V^* its dual space. For a subspace $W \subseteq V$, define $W^0 \subseteq V^*$ to be the subspace $W^0 = \{f \in V^* \mid f(w) = \vec{0} \text{ for all } w \in W\}$. W^0 is called the *annihilator* of W.
(a) If $W_1 \subseteq W_2 \subseteq V$, show that $W_2^0 \subseteq W_1^0$.
(b) Show that $dim(W^0) = dim(V) - dim(W)$.
(c) If W_1 and W_2 are subspaces of V, show that $(W_1 \cap W_2)^0 = W_1^0 + W_2^0$ and $(W_1 + W_2)^0 = W_1^0 \cap W_2^0$.
(d) Show that $W_1 = W_2 (\subset V)$ if and only if $W_1^0 = W_2^0$.

8.1.4. If V is an arbitrary vector space and $v \in V$, show that there is some $f \in V^*$ for which $f(v) \neq 0$. [This was used (and sketched) in the proof of Theorem 8.1.6 (i).]

8.1.5. If V is finite-dimensional, show that any basis of V^* is the dual basis for some basis of V.

8.1.6.   Give direct proofs of the following properties of the transposes of linear mappings. (These results are well known for the transposes of matrices.)
(a) $(S + T)^* = S^* + T^*$.
(b) $(S \circ T)^* = T^* \circ S^*$.
(c) If $S : V \to W$ is an isomorphism, then S^* is an isomorphsim and $(S^*)^{-1} = (S^{-1})^*$.
(d) What does $(T^*)^* = T$ mean in the context of the transpose of a linear transformation T?

8.2 Tensors and Tensor Products

In this section, all vector spaces considered will be finite-dimensional. Suppose that $V_1, V_2, \ldots V_r$ are $\mathbf{F}$ vector spaces. Then Lemma 8.1.2 says that the set of all multilinear functions $f : V_1^* \times V_2^* \times \cdots \times V_r^* \to \mathbf{F}$ is a $\mathbf{F}$ vector space. This important vector space is called the *tensor product* of $V_1, V_2, \ldots, V_k$. An important special case of the tensor product occurs where all the V_i's are the same vector space V and is called a *tensor power* of V. The tensor powers of $\mathbf{R}^n$ and $(\mathbf{R}^n)^*$ are extremely important in multivariable calculus and differential geometry. We will survey their use in the next section.

In order to study tensor products, we first construct multilinear functions using linear functionals.

Definition 8.2.1 Whenever $V_1, V_2, \ldots, V_r, W$ are $\mathbf{F}$ vector spaces, the vector space of all multilinear functions $f : V_1 \times V_2 \times \cdots \times V_r \to W$ is denoted $ML(V_1, V_2, \ldots, V_r; W)$. Suppose that $V_1, V_2, \ldots, V_r$ are $\mathbf{F}$ vector spaces and $t_1 \in V_1^*$, $t_2 \in V_2^*, \ldots, t_r \in V_r^*$ are linear functionals. Define the multilinear function $t_1 \otimes t_2 \otimes \cdots \otimes t_r : V_1 \times V_2 \times \cdots \times V_r \to \mathbf{F}$ by setting $t_1 \otimes t_2 \otimes \cdots \otimes t_r(v_1, v_2, \ldots, v_r) = t_1(v_1)t_2(v_2) \ldots t_r(v_r)$. (The multiplication in the right side of this expression is the usual field multiplication in $\mathbf{F}$.) This multilinear function is called the *tensor product* of the linear functionals $t_1, t_2, \ldots, t_r$.

In the next lemma we construct a basis for $ML(V_1, V_2, \ldots, V_r; \mathbf{F})$ using the tensor product of bases for V_i^*.

Lemma 8.2.2 *Suppose that $V_1, V_2, \ldots, V_r$ are finite-dimensional $\mathbf{F}$ vector spaces and that $\{f_{i1}, f_{i2}, \ldots, f_{id_i}\}$ is a basis for V_i^*. Then the collection of tensors*

$$f_{1j_1} \otimes f_{2j_2} \otimes \cdots \otimes f_{rj_r} \quad \text{where} \quad 1 \le j_i \le d_i$$

forms a basis for $ML(V_1, V_2, \ldots, V_r; \mathbf{F})$. Consequently, if each V_i is d_i-dimensional, then $ML(V_1, V_2, \ldots, V_r; \mathbf{F})$ is $d_1 d_2 \cdots d_r$-dimensional.

Proof: We treat the case where $d = 2$, the general case being similar (and following by induction from this case). We denote $V_1 = V$ and $V_2 = W$. Let $\{f_1, f_2, \ldots, f_s\}$ be the basis of V^* dual to the basis $\{v_1, v_2, \ldots, v_s\}$ of V and $\{g_1, g_2, \ldots, g_t\}$ be the basis of W^* dual to the basis $\{w_1, w_2, \ldots, w_t\}$ of W. According to the defintions, $f_i \otimes g_j(v_k, w_m) = 1$ if $i = k$ and $j = m$, and $f_i \otimes g_j(v_k, w_m) = 0$ otherwise. Consider a linear combination $\sum a_{ij}(f_i \otimes g_j) \in ML(V, W; \mathbf{F})$. We find that $\sum a_{ij}(f_i \otimes g_j)(v_k, w_m) = a_{km}$. From this we see that $\sum a_{ij}(f_i \otimes g_j) = 0 \in ML(V, W; \mathbf{F})$ if and only if $a_{km} = 0$ for all k, m. This proves the linear independence of the $f_i \otimes g_j$. To see that they span $ML(V, W; \mathbf{F})$, we check that for any $B(\, , \,) \in ML(V, W; \mathbf{F})$, if $B(v_i, w_j) = a_{ij}$, then $B(\, , \,) = \sum a_{ij}(f_i \otimes g_j)$. This is proved exactly as in the proof of Theorem 5.1.4 (i) (the uniqueness of the matrix representing a bilinear form). The details are omitted. $//$

We next give the definition of the tensor product of vector spaces. An alternative formulation is given in Remark 8.2.5.

Definition 8.2.3 Suppose that $V_1, V_2, \ldots, V_r$ are finite-dimensional $\mathbf{F}$ vector spaces. Then $V_1 \otimes V_2 \otimes \cdots \otimes V_r = ML(V_1^*, V_2^*, \ldots, V_r^*; \mathbf{F})$ is called the *tensor product* of $V_1, V_2, \ldots, V_r$. We denote by $\otimes^r V = V \otimes V \otimes \cdots \otimes V$ (r times). The elements of $\otimes^r V$ are called *contravariant r-tensors* on V. The elements of $\otimes^r V^*$ are called *covariant r-tensors* on V. An element of the tensor product $T_q^p(V) = \otimes^p V \otimes \otimes^q V^*$ is called a *tensor of type* $\begin{pmatrix} p \\ q \end{pmatrix}$ on V.

Observe that by definition, $\otimes^1 V$ is simply V^{**}, because the multilinear functions from V^* to $\mathbf{F}$ are simply the linear functionals on V^*. In particular (as V is finite-dimensional) Theorem 8.1.6 shows that $\otimes^1 V$ can be naturally identified with V. When working with such tensors, one customarily identifies a vector $v \in V$ as a linear functional on V^*. Specifically, if $v \in V$, then v is identified with the element v^{**} of V^{**} specified by $v^{**}(f) = f(v)$ for all $f \in V^*$. This point of view will dominate the next section.

In the notation of Definition 8.2.3 a vector in V is a tensor type $\begin{pmatrix} 1 \\ 0 \end{pmatrix}$. Similarly, a vector in V^* is a tensor of type $\begin{pmatrix} 0 \\ 1 \end{pmatrix}$. Sometimes a tensor of type $\begin{pmatrix} 0 \\ 1 \end{pmatrix}$ is called a *covector*. A bilinear form on a vector space V is a tensor of type $\begin{pmatrix} 0 \\ 2 \end{pmatrix}$ a trilinear form is a tensor of type $\begin{pmatrix} 0 \\ 3 \end{pmatrix}$ and so forth. Suppose that f is a tensor of type $\begin{pmatrix} p \\ q \end{pmatrix}$ and g is a tensor

of type $\binom{r}{s}$. It can be shown that $(\otimes^p V \otimes \otimes^q V^*) \otimes (\otimes^r V \otimes \otimes^s V^*)$
is isomorphic to $\otimes^{p+r} V \otimes \otimes^{q+s} V^*$. Consequently $f \otimes g$ can be viewed
as a tensor of type $\binom{p+r}{q+s}$. (See Prob. 8.2.6.)

As an immediate corollary of Lemma 8.2.2 we obtain the following
result.

Corollary 8.2.4 *Suppose that $V_1, V_2, \ldots, V_r$ are finite-dimensional* **F**
vector spaces and $\{v_{i1}, v_{i2}, \ldots, v_{id_i}\}$ is a basis for V_i. If we identify
$V_i = V_i^{**}$, *then the tensors $v_{1j_1} \otimes v_{2j_2} \otimes \cdots \otimes v_{rj_r}$ where $1 \le j_i \le d_i$*
form a basis for $V_1 \otimes V_2 \otimes \cdots \otimes V_r$. In particular, if V_i is d_i-dimensional,
then $V_1 \otimes V_2 \otimes \cdots \otimes V_r$ is $d_1 d_2 \cdots d_r$-dimensional.

Remark 8.2.5 Often, the tensor product of vector spaces is described
by the following (more abstract) construction. Given two vector spaces
V and W, we take $V \otimes W$ to be the quotient vector space G/R. Here G
is the (presumably infinite-dimensional) vector space with basis given
by all symbols $v \otimes w$ where $v \in V$ and $w \in W$. R is the subspace of G
spanned by the following elements of G: $\vec{0} \otimes w$, $v \otimes \vec{0}$, $av \otimes w - a(v \otimes w)$,
$v \otimes aw - a(v \otimes w)$, $(v_1 + v_2) \otimes w - v_1 \otimes w - v_2 \otimes w$, and
$v \otimes (w_1 + w_2) - v \otimes w_1 - v \otimes w_2$. The vectors in R are usually called
relations since they represent elements of G which become zero in G/R.
One can check that G/R is isomorphic to $V \otimes W$ by the identification of
$v \otimes w \in V \otimes W$ with $(v \otimes w) + R \in G/R$. The point is that the relations
generating the subspace R correspond to the defining conditions of a
bilinear function, and these elements (which are vectors in G) become
zero in the quotient G/R.

The next result further illustrates the relationship between the tensor product and multilinear functions. The result shows that the tensor
product is universal for bilinear functions. (This characterization could
be used to give a third definition of the tensor product.) Since we do
not use this result in the sequel, the proof (which is not difficult) is
omitted. Further discussion of this approach can be found in the text
by Lang cited in the Bibliography.

Lemma 8.2.6 *The function $F : V \times W \to V \otimes W$ defined by $F(v, w) =$*
$v \otimes w$ *is bilinear. Moreover, for any bilinear function $G : V \times W \to U$,*
there is a unique linear transformation $T : V \otimes W \to U$ for which
$T \circ F = G$.

One of the more important uses of tensors in multivariable calculus is in defining differential forms. This is developed in the next section. In order to treat this material, we need to study alternating tensors. Recall that if $f \in \bigotimes^r V^*$, then by definition $f \in ML(V^{**}, \ldots, V^{**}; \mathbf{F})$. Identifying $V = V^{**}$, such an f is r-linear on V.

Definition 8.2.7 An r-tensor $f \in \bigotimes^r V^*$ is called *alternating* if whenever $v_1, v_2, \ldots, v_r \in V$ and $i, j \in \{1, 2, \ldots, r\}$, then

$$f(v_1, \ldots, v_i, \ldots, v_j, \ldots, v_r) = -f(v_1, \ldots, v_j, \ldots, v_i, \ldots, v_r)$$

(that is, interchanging v_i and v_j changes the sign of f by -1). The subspace of $\bigotimes^r V^*$ of all such alternating r-tensors is denoted $\bigwedge^r V^*$, and its elements are called *r-forms*.

The elements of $\bigwedge^r V^{**}$ $(= \bigwedge^r V)$ are called *r-vectors*. All the constructions given in the sequel apply to r-vectors as well as r-forms. The details are omitted.

When $V = \mathbf{F}^r$, then we are already familiar with one r-form on V, namely the determinant function. When $r = 1$, the definition becomes vacuous; that is, every element of V^* is a one-form on V. In order to give more examples of r-forms on a finite-dimensional vector space V we need the wedge product. For convenience, we shall henceforth assume that $\mathbf{F}$ has characteristic 0 (which is true if $\mathbf{F}$ is $\mathbf{R}$ or $\mathbf{C}$). This guarantees that $n! \neq 0$, for all natural numbers n.

Definition 8.2.8 Suppose that $f \in \bigotimes^r V^*$. For any permutation $\sigma \in S_r$ we define $\sigma(f) \in \bigotimes^r V^*$ by the formula

$$\sigma(f)(v_1, v_2, \ldots, v_r) = f(v_{\sigma_1}, v_{\sigma_2}, \ldots, v_{\sigma_r})$$

We then define

$$\mathbf{A}f := \frac{1}{r!} \sum_{\sigma \in S_r} sg(\sigma)\sigma(f)$$

called the *alternation* of f. Whenever $f \in \bigwedge^p V$ and $g \in \bigwedge^q V$, the *exterior* or *wedge product* is defined by

$$f \wedge g = \frac{(p+q)!}{p!q!} \mathbf{A}(f \otimes g) = \frac{1}{p!q!} \sum_{\sigma \in S_n} sg(\sigma)\sigma(f \otimes g)$$

The reader will surely note the similarity of the definition of the alternation of a tensor with the permutation definition of the determinant. Suppose that $\{f_1, f_2, \ldots, f_n\}$ is the dual basis to the standard basis of $\mathbf{F}^n$. For any matrix $A \in M_{n \times n}(\mathbf{F})$ we can view A as an n-tuple of column matrices $A = (A_1, A_2, \ldots, A_n)$. Then it is not hard to check that $n! \mathbf{A}(f_1 \otimes f_2 \otimes \cdots \otimes f_n)(A_1, A_2, \ldots, A_n) = det(A)$. (In fact $f_1 \wedge f_2 \wedge \ldots \wedge f_n$ *is det* when viewed in this context.) In particular we have $\mathbf{A}(f_1 \otimes f_2 \otimes \cdots \otimes f_n) \in \bigwedge^n \mathbf{F}^n$. The next lemma generalizes this observation.

Lemma 8.2.9 (i) *If* $f \in \bigotimes^r V^*$, *then* $\mathbf{A}(f) \in \bigwedge^r V^*$.
(ii) *For every* $f \in \bigwedge^r V^*$, $\mathbf{A}(f) = f$.

Proof: (i) Since f is r-linear, it is clear that for each $\sigma \in S_n$, $\sigma(f)$ is r-linear. As $\mathbf{A}(f)$ is a linear combination of r-linear functions, it must be r-linear. This shows that $\mathbf{A}(f) \in \bigotimes^r V^*$. We must show that $\mathbf{A}(f)$ is alternating. For this we recall from Lemma 2.2.7 that whenever the permutation σ' is obtained from a permutation σ by interchanging two entries, then $sg(\sigma') = -sg(\sigma)$. The alternating property of $\mathbf{A}(f)$ now follows from the definition, because the sums $\mathbf{A}(f)(v_1, \ldots, v_i, \ldots, v_j, \ldots, v_r)$ and $\mathbf{A}(f)(v_1, \ldots, v_j, \ldots, v_i, \ldots, v_r)$ are identical except for a change in sign at each summand.

(ii) In case $f \in \bigwedge^r V^*$, then for all $\sigma \in S_r$ it is not hard to check that $\sigma(f) = sg(\sigma)f$. Since there are $r!$ elements of S_r, the result follows from the definition. This gives the lemma. //

As a consequence of the lemma, we obtain that whenever $f \in \bigwedge^p V^*$ and $g \in \bigwedge^q V^*$, then $f \wedge g \in \bigwedge^{p+q} V^*$. Thus we have a method of obtaining new forms from old forms. It is easy to check that the wedge product is bilinear; that is, $(kf_1 + f_2) \wedge g = k(f_1 \wedge g) + f_2 \wedge g$ and similarly on the other side. (To show this, apply the definition and use the corresponding properties for $\otimes$.) Two other important properties of the wedge product are its associativity and anticommutativity. These are given next.

Lemma 8.2.10 (i) $\wedge$ *is associative; that is, for any three forms* f, g, *and* h, $(f \wedge g) \wedge h = f \wedge (g \wedge h)$.
(ii) *Whenever* $f \in \bigwedge^p V^*$ *and* $g \in \bigwedge^q V^*$, *then* $f \wedge g = (-1)^{pq} g \wedge f$.

Proof: (i) Suppose that $f \in \bigwedge^p V^*$, $g \in \bigwedge^q V^*$, and $h \in \bigwedge^r V^*$. Then by definition

$$(f \wedge g) \wedge h = \frac{1}{(p+q)!r!} \sum_{\sigma \in S_{p+q+r}} sg(\sigma)\sigma((f \wedge g) \otimes h)$$

$$= \frac{1}{(p+q)!r!} \sum_{\sigma \in S_{p+q+r}} sg(\sigma)\sigma(\frac{1}{p!q!} \sum sg(\tau)\tau(f \otimes g)) \otimes h)$$

$$= \frac{1}{(p+q)!r!p!q!} \sum_{\sigma \in S_{p+q+r}, \tau \in S_{p+q}} sg(\sigma)\sigma((sg(\tau)\tau(f \otimes g)) \otimes h)$$

Using the definitions, one can check that for each $\tau \in S_{p+q}$ and each $\sigma \in S_{p+q+r}$ there is a unique $\sigma' \in S_{p+q+r}$ for which $\sigma(\tau(f \otimes g) \otimes h) = \sigma'((f \otimes g) \otimes h)$. Moreover, for fixed τ the σ's corresponding to distinct σ are distinct, and one can check that $sg(\sigma)sg(\tau) = sg(\sigma')$ (see Prob. 8.2.4). Consequently, for each permutation $\tau \in S_{p+q}$ one has that

$$\sum_{\sigma \in S_{p+q+r}} sg(\sigma)\sigma((sg(\tau)\tau(f \otimes g)) \otimes h) = \sum_{\sigma' \in S_{p+q+r}} sg(\sigma')\sigma'((f \otimes g) \otimes h)$$

Thus, in the latter expression above for $(f \wedge g) \wedge h$, there are $(p+q)!$ sums identical to $\sum_{\sigma \in S_{p+q+r}} sg(\sigma)\sigma((sg(\tau)\tau(f \otimes g)) \otimes h)$. It follows that

$$(f \wedge g) \wedge h = \frac{1}{p!q!r!} \sum_{\sigma' \in S_{p+q+r}} sg(\sigma')\sigma'((f \otimes g)) \otimes h)$$

An argument identical to this gives the same result for $f \wedge (g \wedge h)$. This proves (i).

(ii) Consider $\sigma = (i_1, \ldots, i_p, i_{p+1}, \ldots, i_{p+q}) \in S_{p+q}$. For such σ we define $\sigma^* = (i_{p+1}, \ldots, i_{p+q}, i_1, \ldots, i_p)$. Note that there are pq pairs of the form (i_j, i_k) where $j \in \{1, 2, \ldots, p\}$ and $k \in \{p+1, p+2, \ldots, p+q\}$. For each such pair, the number of inversions going from σ to σ^* changes by 1 or -1. This shows that $sg(\sigma) = (-1)^{pq}sg(\sigma^*)$.

Using the definitions, it follows that for any $f \in \bigwedge^p V$ and any $g \in \bigwedge^q V$, $\sigma(f \otimes g) = \sigma^*(g \otimes f)$. Consequently,

$$f \wedge g = \frac{1}{p!q!} \sum_{\sigma \in S_{p+q}} sg(\sigma)\sigma(f \otimes g)$$

$$= \frac{1}{p!q!} \sum_{\sigma^* \in S_{p+q}} (-1)^{pq} sg(\sigma^*)\sigma^*(g \otimes f)$$

$$= (-1)^{pq} g \wedge f$$

This proves the lemma. //

Since the wedge product is associative, we shall omit the parentheses and write $f_1 \wedge f_2 \wedge \cdots \wedge f_r$ whenever taking an r-fold wedge product. But be careful: the order in which the f_i's are written is important. We remarked earlier that the wedge product was bilinear. For the same reasons, the r-fold wedge product is multilinear (or r-linear).

Suppose that $f_1, f_2, \ldots, f_r \in V^*$ are all nonzero. It is important to know when the r-fold wedge product is nonzero. Unlike the tensor product (where the r-fold tensor product of the f_i's would necessarily be nonzero) we have the following result.

Theorem 8.2.11 *Suppose that $f_1, f_2, \ldots, f_r \in V^*$. Then the wedge product $f_1 \wedge f_2 \wedge \cdots \wedge f_r \neq 0 \in \bigwedge^r V^*$ if and only if $f_1, f_2, \ldots, f_r$ are linearly independent in V^*.*

Proof: We first note that for any $f \in V^*$, $f \wedge f = (1/2)(f \otimes f - f \otimes f) = 0$. Using this, together with the multilinear properties of $\wedge$, one sees that if $f_i = a_1 f_1 + a_2 f_2 + \cdots + a_{i-1} f_{i-1}$, then $f_1 \wedge f_2 \wedge \cdots \wedge f_i = 0$. Hence, whenever $f_1, f_2, \ldots, f_r$ are linearly dependent, $f_1 \wedge f_2 \wedge \cdots \wedge f_r = 0$. Conversely, suppose that $f_1, f_2, \ldots, f_r$ are linearly independent. Then there exist $v_1, v_2, \ldots, v_r \in V$ for which $f_i(v_j) = 1$ for $i = j$ and is 0 otherwise. Now see that $(f_1 \otimes f_2 \otimes \cdots \otimes f_r)(v_1, v_2, \ldots, v_r) = 1$ while $\sigma(f_1 \otimes f_2 \otimes \cdots \otimes f_r)(v_1, v_2, \ldots, v_r) = 0$ for any $\sigma \in S_r$ different from $(1, 2, \ldots, r)$. It follows from the definition that $(f_1 \wedge f_2 \wedge \cdots \wedge f_r)(v_1, v_2, \ldots, v_r) = 1/r! \neq 0$. Consequently $f_1 \wedge f_2 \wedge \cdots \wedge f_r \neq 0$, and the theorem is proved. //

We are at last ready to produce a basis for $\bigwedge^r V$ where V is any finite-dimensional vector space (over a field of characteristic 0).

Theorem 8.2.12 *Suppose that $\{f_1, f_2, \ldots, f_n\}$ is a basis for the vector space V^*. Then the collection of r-fold wedge products*

$$\{f_{i_1} \wedge f_{i_2} \wedge \cdots \wedge f_{i_r} \mid 1 \leq i_1 < i_2 < \cdots < i_r \leq n\}$$

is a basis for $\bigwedge^r V^*$. In particular,

$$dim(\overset{r}{\bigwedge} V) = \binom{n}{r} = \frac{n!}{r!(n-r)!}$$

If $r > n = dim(V)$, then $\bigwedge^r V^*$ is zero.

Proof: Let $f \in \bigwedge^r V^*$. According to Lemma 8.2.2 f is a linear combination of the tensors $f_{i_1} \otimes f_{i_2} \otimes \cdots \otimes f_{i_r}$ for $i_1, i_2, \ldots, i_r \in \{1, 2, \ldots, n\}$. Since $\mathbf{A} : \bigotimes^r V^* \to \bigwedge^r V^*$ is linear, we see that $f = \mathbf{A}(f)$ is a linear combination of the $\mathbf{A}(f_{i_1} \otimes f_{i_2} \otimes \cdots \otimes f_{i_r}) = f_{i_1} \wedge f_{i_2} \wedge \cdots \wedge f_{i_r}$. Using Lemma 8.2.10 (ii) together with Theorem 8.2.11, we see that $\bigwedge^r V$ is generated by the elements $f_{i_1} \wedge f_{i_2} \wedge \cdots \wedge f_{i_r}$ where $1 \leq i_1 < i_2 < \cdots < i_r \leq n$. This shows the desired elements span $\bigwedge^r V^*$.

To see that the listed elements are linearly independent, suppose that $\{f_1, f_2, \ldots, f_n\}$ is the dual basis to the basis $\{v_1, v_2, \ldots, v_n\}$ of V. Then $f_{i_1} \otimes f_{i_2} \otimes \cdots \otimes f_{i_r}(v_{j_1}, v_{j_2}, \ldots, v_{j_r}) = 1$ if $i_1 = j_1, i_2 = j_2, \ldots, i_r = j_r$ and is 0 otherwise. Consequently, whenever $1 \leq j_1 < j_2 < \cdots < j_r \leq n$, we have that $f_{i_1} \wedge f_{i_2} \wedge \cdots \wedge f_{i_r}(v_{j_1}, v_{j_2}, \ldots, v_{j_r}) = 1$ for $i_1 = j_1, i_2 = j_2, \ldots, i_r = j_r$ and is zero otherwise. Now consider any linear combination $f = \sum a_{i_1, i_2, \ldots, i_r} f_{i_1} \wedge f_{i_2} \wedge \cdots \wedge f_{i_r}$. We have that $f(v_{i_1}, v_{i_2}, \ldots, v_{i_r}) = a_{i_1, i_2, \ldots, i_r}$. It follows that $f = 0$ only in case each $a_{i_1, i_2, \ldots, i_r} = 0$. This proves the theorem. //

In the special case $r = n$ the theorem says that $dim(\bigwedge^n (\mathbf{F}^n)^*) = 1$ and $\bigwedge^n (\mathbf{F}^n)^*$ has as a basis the nonzero element $f_1 \wedge f_2 \wedge \cdots \wedge f_n$ where $\{f_1, f_2, \ldots, f_n\}$ is a basis of $(\mathbf{F}^n)^*$. This may seem surprising at first, but in fact we previously proved this result. The elements of $\bigwedge^n (\mathbf{F}^n)^*$ are n-linear, alternating functions on $\mathbf{F}^n$, all of which by Theorem 2.1.8 must be multiples of the determinant function. In case $\{f_1, f_2, \ldots, f_n\}$ is dual to the standard basis of $\mathbf{F}^n$, then viewed as a function on the n columns of a matrix, $f_1 \wedge f_2 \wedge \cdots \wedge f_n = det$. This follows by Theorem 2.1.8 since $f_1 \wedge f_2 \wedge \cdots \wedge f_n(e_1, e_2, \ldots, e_n) = 1 = det(I_n)$.

PROBLEMS

8.2.1. If V and W are finite-dimensional vector spaces, show that $V \otimes W$ is isomorphic to $(ML(V, W; \mathbf{F}))^*$.

8.2.2. Prove Lemma 8.2.6.

8.2.3. Find the coordinates of $(1, 1) \otimes (-1, 2)$ with respect to the standard ordered basis of $V \otimes V$.

8.2.4. Suppose $\tau = (i_1, i_2, \ldots, i_p) \in S_p$ and $\sigma = (j_1, j_2, \ldots, j_{p+r}) \in S_{p+r}$. Set $\sigma' := (j_{i_1}, j_{i_2}, \ldots, j_{i_p}, j_{p+1}, j_{p+2}, \ldots, j_{p+r})$. Show that $\sigma' \in S_{p+r}$ and $sg(\sigma') = sg(\tau)sg(\sigma)$. If one views $S_p \subset S_{p+r}$ via the function $(i_1, i_2, \ldots, i_p) \mapsto (i_1, i_2, \ldots, i_p, p + 1, p + 2, \ldots, p + r)$, then σ' is the composite function $\sigma \circ \tau$ (that is, the product in the group S_{p+q}). This result was used in the proof of Lemma 8.2.10.

8.2.5. (a) If $\{f_1, f_2\}$ is the dual basis of the standard basis $\{e_1, e_2\}$ of $\mathbf{F}^2$, show explicitly (using the defining formulas) that $f_1 \wedge f_2$ is given by the determinant on $(\mathbf{F}^2)^2$.
(b) Extend the result of part (a) to $\mathbf{F}^n$.

8.2.6. Suppose that U, V, and W are $\mathbf{F}$ vector spaces.
(a) Show that $U \otimes V$ is isomorphic to $V \otimes U$.
(b) Show that $(U \otimes V) \otimes W$ is isomorphic to $U \otimes (V \otimes W)$.
(c) Use (a) and (b) to see that $(\bigotimes^p V \otimes \bigotimes^q V^*) \otimes (\bigotimes^r V \otimes \bigotimes^s V^*)$ is isomorphic to $\bigotimes^{p+r} V \otimes \bigotimes^{q+s} V^*$.

8.3   Differential Forms

In this section we restate the basic definitions from multivariable calculus, using the notation of tensors and forms. No results will be proved. The point is to show how these tools can clarify what the definitions mean. For example, students of advanced calculus are often perplexed by the fact that sometimes they have $dx\, dy = dy\, dx$, while at other times they find $dx\, dy = -dy\, dx$. This section will illuminate why (and under what circumstances) both expressions are correct. Another mystery for many students of advanced calculus is the difference between Green's theorem and Stokes' theorem. Specifically, students are not certain why *curl* (F) is a vector field on $\mathbf{R}^3$, whereas *curl* (F) is a scalar-valued function on $\mathbf{R}^2$.

It is necessary to assume that the reader of this section already has some background in multivariable calculus. For simplicity we shall assume that all functions considered are C^∞; that is, they can be differentiated as often as necessary. The interested reader may consult a text on vector calculus to see precisely what continuity conditions are necessary to give the definitions and prove the results.

We begin by recalling the notion of the line integral. Suppose that $\gamma : [a, b] \to \mathbf{R}^n$ is a curve in $\mathbf{R}^n$. If $\gamma(t) = (\gamma_1(t), \gamma_2(t), \ldots, \gamma_n(t))$, then $\gamma' : \mathbf{R} \to \mathbf{R}^n$ denotes the function $\gamma'(t) = (\gamma_1'(t), \gamma_2'(t), \ldots, \gamma_n'(t))$. Suppose that $F = (F_1, F_2, \ldots, F_n) : \mathbf{R}^n \to \mathbf{R}^n$ is a continuous function (that is, a vector field). Then the *line integral* of F along the curve γ is defined to be the (real) integral $\int_a^b <F(\gamma(t)), \gamma'(t)> dt$ where $<, >$ denotes the dot product on $\mathbf{R}^n$. This integral is usually denoted by $\int_\gamma F \cdot dx$, which is shorthand for $\int_\gamma F_1 \, dx_1 + F_2 \, dx_2 + \cdots + F_n \, dx_n$. [This latter expression recaptures the dot product if one views $F = (F_1, F_2, \ldots, F_n)$ and $\gamma'(t)$ as $(dx_1, dx_2, \ldots, dx_n)$.]

The intuitive idea behind the definition of the line integral is that if the vector field F represents a force field, then the work required to push a particle through such a force field is, at any given point, proportional to the dot product of the direction of motion with the force vector. The integral then represents the total work required to move the particle from $\gamma(a)$ to $\gamma(b)$ along the curve specified by γ.

We shall return to line integration shortly. Next we recall the definition of surface integrals. For convenience we consider the situation of a surface embedded in $\mathbf{R}^3$. Suppose that $A \subseteq \mathbf{R}^2$ and $g : A \to \mathbf{R}^3$ smoothly parameterizes a piece of surface $S \subset \mathbf{R}^3$. We fix a vector field $F : \mathbf{R}^3 \to \mathbf{R}^3$. The *surface integral* of F over S is defined to be the integral

$$\int_S F \cdot dS = \int_A \left< F(g(u,v)), \left(\frac{\partial g(u,v)}{\partial u} \times \frac{\partial g(u,v)}{\partial v} \right) \right> du \, dv$$

Here, both $\partial g(u,v)/\partial u$ and $\partial g(u,v)/\partial v$ are the (vector-valued) partial derivatives of the parameterizing function g, and the cross product is a vector normal to the tangent plane of the surface at a given point. Observe the actual integral computed is that of a real-valued function over a subset (namely A) of $\mathbf{R}^2$.

The physical interpretation of the surface integral just defined is that of flux. The idea is that if the vector field F represents the speed and direction of a flow (say of a fluid for example), then the surface integral will represent the rate of flow across the piece of surface parameterized by g. Returning to the definition of the surface integral, we see the cross product

$$\frac{\partial g(u,v)}{\partial u} \times \frac{\partial g(u,v)}{\partial v} = \left(\frac{\partial(y,z)}{\partial(u,v)}, \frac{\partial(z,x)}{\partial(u,v)}, \frac{\partial(x,y)}{\partial(u,v)} \right)$$

where $g(u,v) = (x(u,v), y(u,v), z(u,v))$ and where

$$\frac{\partial(x,y)}{\partial(u,v)} \quad \text{means} \quad det \begin{pmatrix} \partial x/\partial u & \partial y/\partial u \\ \partial x/\partial v & \partial y/\partial v \end{pmatrix}$$

Using this notation, one abbreviates the surface integral of F over S as

$$\int_S F \cdot dS = \int F_1 \frac{\partial(y,z)}{\partial(u,v)} + F_2 \frac{\partial(z,x)}{\partial(u,v)} + F_3 \frac{\partial(x,z)}{\partial(u,v)}$$

or in more compact form as $\int F_1\,dy\,dz + F_2\,dz\,dx + F_3\,dx\,dy$. It is this use of the symbols $dz\,dx$ that causes the confusion alluded to earlier, because one cannot replace $dz\,dx$ with $dx\,dz$ in this latter expression (as can be done with the ordinary Riemann integral).

Consider the following general questions. Suppose that $A \subseteq \mathbf{R}^k$ and $g : A \to \mathbf{R}^n$ parameterizes a k-dimensional surface S in $\mathbf{R}^n$ (here $n > k$). One would like to integrate over this surface. But for what kind of a function is such an integral meaningful, and given such a function, how is the integral defined? There are two features we require of our answer. First, the integral should depend upon the k-dimensional tangent planes of the surface S. Second, the actual integral computed should, in the end, be an ordinary Riemann integral in $\mathbf{R}^k$ over the subset A.

In order to answer our two questions, we must digress and define the tangent bundle TS to the (k-dimensional) surface $S \subset \mathbf{R}^n$. Usually, one views a tangent vector to the surface S at a point $p \in S$ as a vector $v \in \mathbf{R}^n$ with starting point p instead of the origin. (In this setting, such a v as in the accompanying figure is sometimes called a *geometric vector*.)

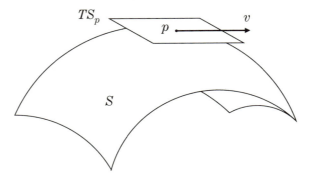

We formalize this idea with the following construction. Suppose S is parameterized by $g = (g_1, g_2, \ldots, g_k) : A \to \mathbf{R}^n$ (where $A \subseteq \mathbf{R}^k$). As part of our smoothness conditititions we assume that at each $a \in A$, the rank of the Jacobian $g'(a)$ is k. For each point $p = g(a_1, a_2, \ldots, a_k) \in S$ the *tangent space* TS_p of S at $p \in S$ is defined to be the k-dimensional vector space of all pairs $(p, v) \in \mathbf{R}^{2n}$ where $v \in \mathbf{R}^n$ is a tangent vector to S at p. The vector operation in TS_p comes from the vector operations on the second component; that is, $(p, v_1) + (p, v_2) = (p, v_1 + v_2)$. Consider the k partial derivatives

$\partial g(a)/\partial x_1$, $\partial g(a)/\partial x_2,\ldots,$ $\partial g(a)/\partial x_k$ where $g(a) = p$. In view of our assumption that $rk(g'(a)) = k$, these partial derivatives give linearly independent tangent vectors to S at $p = g(a)$. Thus they give a basis for the k-dimensional vector space TS_p. We denote

$$\left(p, \frac{\partial g}{\partial x_i}(a)\right) \in TS_p \quad \text{by} \quad \left.\frac{\partial}{\partial x_i}\right|_p$$

and with this notation our basis for TS_p becomes

$$\left\{ \left.\frac{\partial}{\partial x_1}\right|_p, \left.\frac{\partial}{\partial x_2}\right|_p, \ldots, \left.\frac{\partial}{\partial x_k}\right|_p \right\}$$

The *tangent bundle* TS is obtained by "gluing together" all the individual tangent spaces, so $TS = \bigcup_{p \in S} TS_p$. For example, according to this definition, the tangent bundle to $\mathbf{R}^n$ (inside $\mathbf{R}^n$) is nothing other than $\mathbf{R}^{2n}$, where for each $p \in \mathbf{R}^n$, the subset $T(\mathbf{R}^n)_p = \{(p, v) \mid v \in \mathbf{R}^n\}$ is viewed as a vector space (isomorphic to $\mathbf{R}^n$.) For any k-dimensional surface $S \subset \mathbf{R}^n$ we may then view $TS \subset T\mathbf{R}^n$, where for any point $p \in S$ the vector space TS_p is a k-dimensional subspace of the n-dimensional vector space $T(\mathbf{R}^n)_p$. Using these notions, we define vector fields and differential k-forms on $\mathbf{R}^n$.

Definition 8.3.1 (i) A *vector field* on $\mathbf{R}^n$ is a function F on $\mathbf{R}^n$ which assigns for each $p \in \mathbf{R}^n$ a vector $F(p) \in T(\mathbf{R}^n)_p$.

(ii) For $0 \le k \le n$ a *k-form* (or a *differential k-form*) ω on $\mathbf{R}^n$ is a C^∞ function ω defined on $\mathbf{R}^n$ which for each $p \in \mathbf{R}^n$ assigns $\omega(p) \in \bigwedge^k(T(\mathbf{R}^n)_p)^*$.

(iii) A tensor of type $\binom{r}{q}$ on $\mathbf{R}^n$ is a function ω on $\mathbf{R}^n$ which assigns to each $p \in \mathbf{R}^n$ an element $\omega(p) \in T_q^r(T(\mathbf{R}^n)_p)$.

Literally, a vector field (in the above sense) is a function $F : \mathbf{R}^n \to \mathbf{R}^{2n} = \mathbf{R}^n \times \mathbf{R}^n$, with $F(p) = (p, F_2(p))$ for all $p \in \mathbf{R}^n$. (The second component function $F_2 : \mathbf{R}^n \to \mathbf{R}^n$ is what was previously called the vector field.) In what follows we describe how to integrate k-forms over k-dimensional surfaces. The advantage to working with k-forms (as opposed to vector fields) is that they operate naturally upon k-tuples of vectors in $T(\mathbf{R}^n)_p$. For example, suppose that $S \subseteq \mathbf{R}^n$ is a k-dimensional smooth surface parameterized by a function

$g(X_1, X_2, \ldots, X_k) : A \to \mathbf{R}^n$. Let ω be a k-form on $\mathbf{R}^n$. Then for $p \in S$, we can evaluate

$$\omega\left(\left.\frac{\partial g}{\partial x_1}\right|_p, \left.\frac{\partial g}{\partial x_2}\right|_p, \ldots, \left.\frac{\partial g}{\partial x_k}\right|_p\right) \in \mathbf{R}$$

so that ω determines a real-valued function on S. Composing this function with the parameterization $g : A \to \mathbf{R}^n$ of S, we obtain a scalar-valued function on A, which can be integrated over A (inside $\mathbf{R}^k$). Explicitly we have the following.

Definition 8.3.2 If $g(X_1, X_2, \ldots, X_k) : A \to \mathbf{R}^n$ smoothly parameterizes a k-dimensional surface and ω is a k-form on $\mathbf{R}^n$, then we define

$$\int_S \omega = \int_A \omega\left(\left.\frac{\partial g}{\partial x_1}\right|_p, \left.\frac{\partial g}{\partial x_2}\right|_p, \ldots, \left.\frac{\partial g}{\partial x_k}\right|_p\right) dx_1\, dx_2 \cdots dx_k$$

We next illustrate how one integrates one-forms over curves in $\mathbf{R}^n$, and how to integrate two-forms over surfaces in $\mathbf{R}^3$. The following notation proves to be convenient. We denote by $\{dx^1|_p, dx^2|_p, \ldots, dx^n|_p\}$ the ordered basis for $(T(\mathbf{R}^n)_p)^*$, dual to the standard basis $\{\partial/\partial x_1|_p, \partial/\partial x_2|_p, \ldots, \partial/\partial x_k|_p\}$. For $i = 1, 2, \ldots, n$ we denote by dx^i the one-form defined by $dx^i(p) = dx^i|_p$ for all $p \in \mathbf{R}^n$. Thus, for any vector $v = (p, v_p) \in T(\mathbf{R}^n)_p$, $dx^i(v)$ is the ith coordinate of v_p. Let $\gamma : [a, b] \to \mathbf{R}^n$ be a curve. Since $\{dx^1|_p, dx^2|_p, \ldots, dx^n|_p\}$ is a basis of $\bigwedge^1(T(\mathbf{R}^n)_p)^*$, a one-form on $\mathbf{R}^n$ can be expressed as $\omega = F_1\, dx^1 + F_2\, dx^2 + \cdots + F_n\, dx^n$, where each of the functions $F_i : \mathbf{R}^n \to \mathbf{R}$. By definition the integral

$$\begin{aligned}
\int_\gamma \omega &= \int_a^b \omega(\gamma(t))\left(\left.\frac{d\gamma}{dt}(t)\right|_{\gamma(t)}\right) dt \\
&= \int_a^b \left[F_1(\gamma(t))\frac{d\gamma_1}{dt}(t) + F_2(\gamma(t))\frac{d\gamma_2}{dt}(t)\right. \\
&\qquad \left. + \cdots + F_n(\gamma(t))\frac{d\gamma_n}{dt}(t)\right] dt
\end{aligned}$$

This latter expression is none other than the line integral $\int_\gamma F \cdot dx$ as defined earlier, where $F = (F_1, F_2, \ldots, F_n)$ is a vector field and $\gamma = (\gamma_1, \gamma_2, \ldots, \gamma_n)$.

In effect, what happened is that in defining the line integral, we identified the vector field $F = (F_1, F_2, \ldots, F_n)$ with the one-form $\omega = F_1 \, dx^1 + F_2 \, dx^2 + \cdots + F_n \, dx^n$. This may seem peculiar, because vector fields are vector-valued functions on $\mathbf{R}^n$, whereas one-forms are covector-valued. But there is no inconsistency. The point is that when the vector field was used in the definition of the line integral, it was used *as if* it was a one-form; the inner product $< \, , \, >$ in the defining expression of $\int_\gamma F \cdot dx$ changed the vector $F(\gamma(t))$ into the functional $< F(g(t)), \, >$. Observe that this exploits the natural identification of $\mathbf{R}^n$ with $\mathbf{R}^{n*}$, arising from the inner product.

We next analyze surface integrals in $\mathbf{R}^3$. In this case it turns out that the vector field on $\mathbf{R}^3$ is identified with a two-form. Consider $A \subseteq \mathbf{R}^2$ and suppose that $g : A \to \mathbf{R}^3$ parameterizes a piece of smooth surface S. According to Theorem 8.2.12, $\bigwedge^2(T\mathbf{R}_p^3)^*$ is the three-dimensional vector space with basis $\{dx^1 \, |_p \wedge dx^2 \, |_p, \; dx^1 |_p \wedge dx^3 |_p, \; dx^2 |_p \wedge dx^3 |_p\}$. A two-form on $\mathbf{R}^3$ is therefore a function of the form $\omega = F_1 \, dx^2 \wedge dx^3 + F_2 \, dx^3 \wedge dx^1 + F_3 \, dx^1 \wedge dx^2$, where $F_i : \mathbf{R}^3 \to \mathbf{R}$. (The reason for using $dx^3 \wedge dx^1$ instead of $dx^1 \wedge dx^3$ in this particular representation will become apparant momentarily.)

Applying the definitions, we have

$$\int_S \omega = \int_A \Big[F_1(g(u, v)) \, dx^2 \wedge dx^3(\partial g(u, v))$$
$$+ \; F_2(g(u, v)) \, dx^3 \wedge dx^1(\partial g(u, v))$$
$$+ \; F_3(g(u, v)) \, dx^1 \wedge dx^2(\partial g(u, v)) \Big] \, du \, dv$$

where $\partial g(u, v)$ denotes the pair of tangent vectors

$$\left(\frac{\partial g}{\partial u}(u, v), \frac{\partial g}{\partial v}(u, v) \right) \in TS_p \times TS_p$$

We compute that

$$dx^2 \wedge dx^3(\partial g(u, v)) = dx^2 \wedge dx^3 \left((\frac{\partial x}{\partial u}, \frac{\partial y}{\partial u}, \frac{\partial z}{\partial u}), (\frac{\partial x}{\partial v}, \frac{\partial y}{\partial v}, \frac{\partial z}{\partial v}) \right)$$
$$= \frac{\partial y}{\partial u} \frac{\partial z}{\partial v} - \frac{\partial z}{\partial u} \frac{\partial y}{\partial v}$$

which was denoted earlier by $\partial(y, z)$. Together with the analogous results for $dx^3 \wedge dx^1(\partial g(u, v))$ and $dx^1 \wedge dx^2(\partial g(u, v))$, we see that the integral $\int_S \omega$ is precisely the surface integral denoted previously by $\int_S F_1 \, dx_2 \, dx_3 + F_2 \, dx_3 \, dx_1 + F_3 \, dx_1 \, dx_2$. In particular, the two-form $F_1 \, dx^2 \wedge dx^3 + F_2 \, dx^3 \wedge dx^1 + F_3 \, dx^1 \wedge dx^2$ and the vector field $F = (F_1, F_2, F_3)$ are identified when one views the surface integral as that of a two-form. In multivariable calculus, the expression $dx_1 \, dx_2$ is an abbreviation for the two-form $dx^1 \wedge dx^2$. Since $dx^1 \wedge dx^2 = -dx^2 \wedge dx^1$ we have uncovered the origin of the mysterious equation $dx_1 \, dx_2 = -dx_2 \, dx_1$.

This section concludes with a discussion of the exterior derivative of a differential form. Using the exterior derivative, one can interpret the various generalizations of the fundamental theorem of calculus (namely Green's and Stokes' theorems) each as a piece of one grand scheme.

According to convention, $\bigwedge^0 V^* = \mathbf{R}$ for all real vector spaces V. Thus, a zero-form on $\mathbf{R}^n$ is (essentially) a function $f : \mathbf{R}^n \to \mathbf{R}$. The *differential*, df, of a zero-form is the one-form defined by

$$df = \frac{\partial f}{\partial x_1} \, dx^1 + \frac{\partial f}{\partial x_2} \, dx^2 + \cdots + \frac{\partial f}{\partial x_n} \, dx^n$$

If $x^i : \mathbf{R}^n \to \mathbf{R}$ is the ith-coordinate function, the definition gives that $d(x^i) = dx^i$. In other words, the differential of the ith-coordinate function is the one-form which at each tangent space is the ith-coordinate functional.

Now consider a k-form

$$\omega = \sum w_{i_1 i_2 \ldots i_k} dx^{i_1} \wedge dx^{i_2} \wedge \cdots \wedge dx^{i_k}$$

(where the $w_{i_1 i_2 \ldots i_k} : \mathbf{R}^n \to \mathbf{R}$ are scalar valued, that is, zero-forms). We define

$$d\omega := \sum (dw_{i_1 i_2 \ldots i_k}) \, dx^{i_1} \wedge dx^{i_2} \wedge \cdots \wedge dx^{i_k}$$

a $(k{+}1)$-form. $d\omega$ is called the *differential* of ω or sometime the *exterior derivative* of ω. The following theorem summarizes some of the main properties of the exterior derivative. The proof of this theorem is not hard and is omitted.

Theorem 8.3.3 *For differential forms ω and η on $\mathbf{R}^n$*
(i) $d(\omega + \eta) = d\omega + d\eta$.
(ii) $d(\omega \wedge \eta) = d\omega \wedge \eta + (-1)^k \omega \wedge d\eta$ *if ω is a k-form.*
(iii) $d(d\omega) = 0$.

In case $f : \mathbf{R} \to \mathbf{R}$ is a real-valued function on $\mathbf{R}$, then the differential df is the one-form $f'(x)\,dx$. Note that the Riemann integral $\int_a^b f'(x)\,dx$ is precisely the same as the integral of the one-form df over the subset $[a, b] \subset \mathbf{R}$. For any interval $[a, b] \subset \mathbf{R}$, the fundamental theorem of calculus tells us that $\int_a^b f'(x)\,dx = f(b) - f(a)$. In other words the value of this integral is determined by the value of the function f at the boundary of the interval.

Using the language of differential forms, we can formulate a generalization of the fundamental theorem of calculus to higher dimensions. Suppose that $S \subset \mathbf{R}^n$ is a smooth piece of k-dimensional surface. The *boundary* of S is denoted by ∂S and is assumed to be a finite union of smooth $(k - 1)$-dimensional surfaces. We will not formally define the concept of the boundary, assuming the reader has an intuitive idea of what it should be. For example, the boundry of the disk in $\mathbf{R}^2$ defined by $x^2 + y^2 \le 1$ is the circle given by $x^2 + y^2 = 1$, while the sphere in $\mathbf{R}^3$ defined by $x^2 + y^2 + z^2 = 1$ has no boundry (that is, it has $\emptyset$ as boundry). With this notation we can state Stokes' theorem.

Stokes' Theorem *Suppose that S is a smooth k-dimensional surface in $\mathbf{R}^n$ with boundary ∂S. Assume that ω is a $(k - 1)$-form on $\mathbf{R}^n$. Then*

$$\int_S d\omega = \int_{\partial S} \omega$$

We will not even attempt to describe the proof of this fundamental result here, but we do recommend Michael Spivak's book *Calculus on Manifolds* mentioned in the Bibliography for a careful treatment of this subject. The two-dimensional version of this result should be familiar to most advanced calculus students. Known as *Green's theorem* it says the following.

Green's Theorem *Suppose that D is a plane region with boundary ∂D given by a piecewise smooth curve γ parameterized so that it is traced once with D on the left. Suppose that F_1 and F_2 are smooth real-valued functions defined on $D \cup \partial D$. Then*

$$\int_D \left(\frac{\partial F_2}{\partial x_1} - \frac{\partial F_1}{\partial x_2} \right) dx_1\,dx_2 = \int_\gamma F_1\,dx_1 + F_2\,dx_2$$

To see that Green's theorem is indeed a special case of our formulation of Stokes' theorem, one identifies the vector field $F = (F_1, F_2)$ on $\mathbf{R}^2$ with the one-form $\omega = F_1\, dx^1 + F_2\, dx^2$. Then, by definition

$$
\begin{aligned}
d\omega &= \left(\frac{\partial F_1}{\partial x_1}\, dx^1 + \frac{\partial F_1}{\partial x_2}\, dx^2\right) \wedge dx^1 + \left(\frac{\partial F_2}{\partial x_1}\, dx^1 + \frac{\partial F_2}{\partial x_2}\, dx^2\right) \wedge dx^2 \\
&= \frac{\partial F_1}{\partial x_2}\, dx^2 \wedge dx^1 + \frac{\partial F_2}{\partial x_1}\, dx^1 \wedge dx^2 \\
&= \left(\frac{\partial F_2}{\partial x_1} - \frac{\partial F_1}{\partial x_2}\right) dx^1 \wedge dx^2
\end{aligned}
$$

Thus the integral

$$
\int_D d\omega = \int_D \left(\frac{\partial F_2}{\partial x_1} - \frac{\partial F_1}{\partial x_2}\right) dx_1\, dx_2
$$

as an ordinary Riemann integral. This shows that in this case the conclusion of Stokes theorem, $\int_D d\omega = \int_{\partial D} d\omega$, is Green's theorem. The two-form

$$
d\omega = \left(\frac{\partial F_2}{\partial x_1} - \frac{\partial F_1}{\partial x_2}\right) dx^1 \wedge dx^2
$$

is usually denoted $curl\,(F)$. Since $dim(\wedge^2(\mathbf{R}^2)^*) = 1$, $curl\,(F)$ is actually a real-valued function of $\mathbf{R}^2$.

For a two-dimensional surface in $\mathbf{R}^3$, Stokes' theorem takes the following form (also called Stokes' theorem). We continue to identify vector fields on $\mathbf{R}^3$ with both one-forms and two-forms (under appropriate circumstances). Suppose that the surface S has boundary ∂S, which is parameterized by a curve $\gamma : [a, b] \to \mathbf{R}^3$ which traces ∂S once with S on the left. Let $F = (F_1, F_2, F_3)$ be a vector field on $\mathbf{R}^3$, and as the case with Green's theorem, we really integrate the one-form $\omega = F_1\, dx^1 + F_2\, dx^2 + F_3\, dx^3$. Using the facts that $dx^1 \wedge dx^1 = 0$ and $dx^i \wedge dx^j = -dx^j \wedge dx^i$ we compute

$$
\begin{aligned}
d\omega &= \left(\frac{\partial F_3}{\partial x_2} - \frac{\partial F_2}{\partial x_3}\right) dx^2 \wedge dx^3 + \left(\frac{\partial F_1}{\partial x_3} - \frac{\partial F_3}{\partial x_1}\right) dx^3 \wedge dx^1 \\
&\quad + \left(\frac{\partial F_2}{\partial x_1} - \frac{\partial F_1}{\partial x_2}\right) dx^1 \wedge dx^2
\end{aligned}
$$

The vector field

$$
\left(\left(\frac{\partial F_3}{\partial x_2} - \frac{\partial F_2}{\partial x_3}\right), \left(\frac{\partial F_1}{\partial x_3} - \frac{\partial F_3}{\partial x_1}\right), \left(\frac{\partial F_2}{\partial x_1} - \frac{\partial F_1}{\partial x_2}\right)\right)
$$

associated with the two-form $d\omega$ is known as $curl(F)$, so in this case Stokes' theorem says $\int_S curl\,(F)dS = \int_{\partial S} F \cdot dx$. Note that in this version of Stokes' theorem, one is integrating a two-form on the left-hand side and a 1-form on the right, *even though* both are commonly interpreted as vector fields. Of course, from the point of view of multivariable calculus (without differential forms), the left-hand integral is a surface integral, while the right-hand integral is a line integral. The ability to identify vector fields with both one- and two-forms is a luxury that occurs only in $\mathbf{R}^3$. (We emphasize that this does not occur in higher dimensions.)

Stokes' theorem also shows why vector fields of the form $grad(F)$ are conservative. If $F : \mathbf{R}^n \to \mathbf{R}$ is a C^∞ function, then by definition $grad(F)$ is the vector field

$$grad(F) = \left(\frac{\partial F}{\partial x_1}, \frac{\partial F}{\partial x_2}, \ldots, \frac{\partial F}{\partial x_n} \right)$$

which is the vector field usually identified with the one-form

$$dF = \frac{\partial F}{\partial x_1}\,dx^1 + \frac{\partial F}{\partial x_2}\,dx^2 + \cdots + \frac{\partial F}{\partial x_n}\,dx^n$$

Using the fact that $d(dF) = 0$ [known as $curl(grad(F)) = 0$ in multivariable calculus], together with Stokes' theorem, one obtains that line integrals of the vector field $grad(F)$ over closed curves are zero. This is because whenever $\gamma : [a, b] \to \mathbf{R}^n$ is a closed curve [that is, $\gamma(a) = \gamma(b)$], $\int_\gamma grad(F)\,dx = \int_S curl(grad(F)) = 0$ for any smooth surface S with boundary γ. This latter condition is equivalent to a vector field being conservative.

Suppose one applies Stokes' theorem to a three-dimensional solid in $\mathbf{R}^3$ with (piecewise) smooth two-dimensional boundary. Then one obtains what is known as Gauss' theorem. If $F = (F_1, F_2, F_3)$ is a vector field, then the real-valued function obtained by computing the three-form $d\omega$ (where ω is the two-form associated with F) is known as the divergence $div(F)$. Explicitly, if $\omega = F_1\,dx^2 \wedge dx^3 + F_2\,dx^3 \wedge dx^1 + F_3\,dx^1 \wedge dx^2$ one obtains that

$$d\omega = \left(\frac{\partial F_1}{\partial x_1} + \frac{\partial F_2}{\partial x_2} + \frac{\partial F_3}{\partial x_3} \right) dx^1 \wedge dx^2 \wedge dx^3$$

or as is commonly written

$$div(F) = \left(\frac{\partial F_1}{\partial x_1} + \frac{\partial F_2}{\partial x_2} + \frac{\partial F_3}{\partial x_3} \right) dx_1\,dx_2\,dx_3$$

Using this notation, Stokes' theorem gives $\int_B div(F)\,dV = \int_{\partial B} F \cdot dS$ for any solid reigon $B \subseteq \mathbf{R}^3$ with smooth boundary ∂B ("positively oriented"). Whenever using this theorem, one should keep in mind that $div(F)$ is really a three-form on $\mathbf{R}^3$ (so that dV abbreviates $dx^1 \wedge dx^2 \wedge dx^3$) and the vector field F is interpreted as a two-form.

The beauty of Stokes' theorem is its validity for k-dimensional surfaces in $\mathbf{R}^n$ where $k < n$ are arbitrary. For $n > 3$ we cannot understand the meaning of this result using vector fields alone. The full force of the theory of differential forms becomes necessary. We do not pursue the details here.

Appendix: Notation from Set Theory, Functions, and Complex Numbers

This book assumes that the reader is familiar with the language of sets, functions, and complex numbers. In this Appendix we informally recall the basic definitions and notation for reference. If you are not familiar with the contents of this Appendix, have someone help you figure it out, because the book will not make any sense if you do not understand the language.

Notation from Set Theory

A.1 Usually capital letters denote sets, which are collections of mathematical objects. The objects that constitute the set are called the *elements* of the set. The symbol $\in$ is used to denote the *element of* relation. For example, in this book $\mathbf{R}$ denotes the set (collection) of all real numbers. Then $0 \in \mathbf{R}$, $\pi \in \mathbf{R}$, and $-1 \in \mathbf{R}$. However $i = \sqrt{-1}$ is not a real number, so we write $i \notin \mathbf{R}$ to denote this. ($\notin$ means *not an element of*.)

A.2 We use the braces $\{\ ,\ \}$ to denote sets by listing the elements of the set between the braces. For example $A = \{3, \pi\}$ means that A is the set with two elements, the real numbers 3 and π.

A.3 To describe more complicated sets, often one needs to use English and/or mathematical sentences. There is a standard method of doing this. One writes what is necessary inside the brackets $\{\ ,\ \}$ and the symbol $|$ is used in place of the English phrase "such that." As examples, the set of positive real numbers $\mathbf{R}^+$ can be written as

495

$\mathbf{R}^+ = \{x \in \mathbf{R} \mid x > 0\}$ and the even integers can be described as $\{n \mid n$ is an integer and $n = 2m$ for some other integer $m\}$.

A.4 For any set S we denote by $S \times S$ the set of ordered pairs of elements of S. Thus, $S \times S = \{(s_1, s_2) \mid s_1 \in S$ and $s_2 \in S\}$. When using parentheses in ordered pairs, the order is important; that is, $(s_1, s_2) = (s_3, s_4)$ if and only if $s_1 = s_3$ and $s_2 = s_4$. More generally S^n is the set of ordered n-tuples of elements in S, i.e. $S^n = \{(s_1, s_2, \ldots, s_n) \mid s_1, s_2, \ldots, s_n \in S\}$.

A.5 If S and T are sets, we define the *intersection* of S and T, denoted $S \cap T$, by $S \cap T = \{x \mid x \in S$ and $x \in T\}$. We define the *union* of the sets S and T, denoted $S \cup T$, by $S \cup T = \{x \mid x \in S$ or $x \in T\}$.

A.6 If A and B are sets, we say that A is a *subset* of B if whenever $x \in A$, then $x \in B$. This is denoted by $A \subseteq B$. The bar underneath means that possibly $A = B$, although not necessarily. Observe that for any two sets A and B, $A = B$ if and only if both $A \subseteq B$ and $B \subseteq A$. Usually, to prove that two sets are equal, we prove both of these inclusions.

A.7 It is easily checked by using the definitions of A.5 and A.6 that for any two sets A and B, $A \cap B \subseteq A$, $A \cap B \subseteq B$, $A \subseteq A \cup B$, and $B \subseteq A \cup B$.

Functions

A.8 A *function* from a set A to a set B is, loosely speaking, a rule which assigns to each element of A a unique element of B. For example the familiar function $\sin x$ is a function between the set of real numbers $\mathbf{R}$ and the closed interval $[-1, 1]$. If f is a function between two sets A and B, we use the notation $f : A \to B$ to denote that f is a function which assigns to each element of A a unique value in B. The set A is called the *domain* of the function f, and the set B is called the *range* of the function f.

Formally, a function $f : A \to B$ is a collection of ordered pairs $F \subseteq A \times B$ [so each element of F is a pair of the form (a, b) where $a \in A$ and $b \in B$] which satisfies the condition that for each $a \in A$ there is a unique $b \in B$ with $(a, b) \in F$. However, for the purposes of this book, such formality is not necessary and the reader may view a function simply as a rule which assigns to each element of A a unique element of B.

A.9 A function $f : A \rightarrow B$ is called *one-one* or *injective* if whenever $f(a_1) = f(a_2) \in B$, then necessarily $a_1 = a_2$. In other words f is one-one if any two distinct elements $a_1, a_2 \in A$ have distinct images $f(a_1), f(a_2) \in B$. The function $f : A \rightarrow B$ is called *onto* or *surjective* if for every $b \in B$ there exists some $a \in A$ with $f(a) = b$. F is surjective precisely if every element of B is the image of some element of A. If $f : A \rightarrow B$ is both one-one and onto, then f is called *bijective*. Bijective functions are precisely those functions $f : A \rightarrow B$ for which there is a well-defined (and unique) *inverse function* $f^{-1} : B \rightarrow A$ which is defined by $f^{-1}(b) = a$ if and only if $f(a) = b$. According to the definition of f^{-1}, one sees that $f^{-1}(f(a)) = a$ for all $a \in A$ and $f(f^{-1}(b)) = b$ for all $b \in B$.

A.10 A *binary operation* on a set S is a function $f : S \times S \rightarrow S$. Thus, a binary operation combines two elements of a set and produces a third. Familiar examples of binary operations are addition and multiplication of real numbers.

Complex Numbers

A.11 The *field of complex numbers* **C** is defined to be

$$\mathbf{C} = \{a + bi \mid a, b \in \mathbf{R}\}$$

The *addition of complex numbers* is defined according to

$$(a + bi) + (c + di) = (a + c) + (b + d)i$$

and *multiplication* by

$$(a + bi)(c + di) = (ac - bd) + (ad + bc)i$$

Observe that with this rule for multiplication $i \cdot i = -1$. (Here, we write i for $0 + 1i$.) Note also that the real numbers are contained naturally in the complex numbers (as a subfield): Every real number $r \in \mathbf{R}$ can be viewed as the complex number $r + 0i \in \mathbf{C}$.

It is not difficult to verify that the complex numbers form a field, and we omit the details. The *zero element* of **C** is $0 = 0 + 0i$, and the *multiplicative identity* is the real number $1 = 1 + 0i$. Perhaps the only tricky fact to remember is the formula which gives the *multiplicative inverse*:

$$(a + bi)^{-1} = (a^2 + b^2)^{-1}(a - bi)$$

[Here, $(a^2 + b^2)^{-1}$ is the inverse of the real number $a^2 + b^2$.]

A.12 It is customary to identify the complex number $a + bi$ with the point $(a, b) \in \mathbf{R}^2$. With this geometric point of view, the real numbers correspond to the x axis, and the pure imaginary complex numbers (those of the form $0 + bi$) correspond to the y axis. The *absolute value* (or modulus) of a complex number $a + bi$ is defined by

$$|a + bi| = \sqrt{a^2 + b^2}$$

Note that the absolute value of $a + bi$ is the length of the line segment in $\mathbf{R}^2$ with endpoints $(0, 0)$ and (a, b).

A.13 We observe that the product $(a+bi)(a-bi) = a^2 + b^2 = |a+bi|^2$ is a real number. The relationship between the complex numbers $a+bi$ and $a - bi$ is important for a variety of reasons. We call $a - bi$ the *complex conjugate* of the number $a + bi$, which is usually denoted by an overbar as in $\overline{a + bi} = a - bi$.

The following basic facts about the complex absolute value and conjugation are given without proof.

Facts *For all* $z, w \in \mathbf{C}$:
(i) $|z + w| \leq |z| + |w|$.
(ii) $|zw| = |z| \cdot |w|$.
(iii) $\overline{z + w} = \overline{z} + \overline{w}$.
(iv) $\overline{zw} = \overline{z}\,\overline{w}$.
(v) $z\overline{z} = |z|^2$.

A.14 We noted above that the real number -1 has a square root in the complex numbers. In fact, much more is true. The complex numbers are what is known as an *algebraically closed field*. This is formulated in the following theorem.

Fundamental Theorem of Algebra *Every complex polynomial*

$$f(X) = a_n X^n + a_{n-1} X^{n-1} + \cdots + a_1 X + a_0$$

(where $a_1, a_2, \ldots, a_n \in \mathbf{C}$*) has a root in the field of complex numbers. Consequently one can factor* $f(X)$ *as*

$$f(X) = a_n(X - b_1)(X - b_2) \cdots (X - b_n)$$

for some $b_1, b_2, \ldots, b_n \in \mathbf{C}$.

Answers to Most
Odd-Numbered Problems

SECTION 0.2

0.2.1. (a) $X = (-1/2)t + 5/2, Y = (-5/2)t + 5/2, Z = t$ where t is a parameter.

(b) $X = 0, Y = 0, Z = 0$.

(c) $X_1 = t + 12, X_2 = -t - 6, X_3 = -2t - 17, X_4 = t$ where t is a parameter.

(d) $X_1 = -t + 3/2, X_2 = t + 1/2, X_3 = t, X_4 = -9/2$ where t is a parameter.

(e) $(7 - t - u - v, t, u, v)$ where t, u, v are parameters.

(f) $(3 + t, 2 - u, t, u)$ where t and u are parameters.

(g) $t(-5, 1, 2)$ where t is a parameter.

(h) $\emptyset$, the empty set (that is, there are no solutions).

0.2.3. (a) The reduced row-echelon form is:

$$
\begin{array}{rcrcl}
X & + & & (1/2)Z & = & 5/2 \\
& & Y + & (5/2)Z & = & 5/2
\end{array}
$$

All possible echelon forms are:

$$
\begin{array}{rcrcccl}
X & + & kY & + & ((5k+1)/2)Z & = & (5k+5)/2 \\
& & Y & + & (5/2)Z & = & 5/2
\end{array}
$$

where k is an arbitrary scalar.

0.2.5. Applying Gauss-Jordan elimination, we can assume that this system is in reduced echelon form. (Row operations do not change the solution set.) If the system is inconsistent, it has no solutions, so in particular it cannot have a unique solution. Thus we can assume the system is consistent. Since there are at most three equations in this reduced echelon form, there are at most three determined variables. In particular there must be at least one free variable (since there are four variables). Using the process of back-substitution, it follows that there is a solution to the system with this free variable taking any scalar value. This shows the system has more than one solution, and this proves the result.

0.2.7. This problem is a special case of Theorem 0.5.2, so you can see Sec. 0.5 for a proof.

0.2.9. To show that these two systems are row equivalent, first show that each system can be row reduced to the reduced echelon form system

$$X \quad = \quad 0$$
$$Y \quad = \quad 0$$

Thus both systems are row equivalent to the same system. Since row operations can always be reversed, it follows that the two original systems must be row equivalent.

0.2.11. Since these systems are consistent, they can be reduced to one of the following types of reduced echelon form systems:

$$X \quad = \quad a \qquad X \; + \; aY \quad = \quad b \qquad Y \quad = \quad a$$
$$Y \quad = \quad b \qquad\qquad 0Y \quad = \quad 0 \qquad 0 \quad = \quad 0$$

In the first case, the system has a unique solution, completely determined by the echelon form. In the second case, the solution set is precisely $\{(b - at, t) \mid t \in \mathbf{F}\}$, and in the third case, the solution set is $\{(t, a) \mid t \in \mathbf{F}\}$. In each case, the set of solutions is completely determined by the echelon form, and moreover, if the echelon forms are different, the solution sets are different. Thus, two consistent systems which have precisely the same solution set, must have the same reduced echelon form. It follows that they are row equivalent.

0.2.13.
$$X_1 \quad - \quad 2X_2 \quad + \qquad\qquad X_4 \qquad\qquad = \quad 3$$
$$X_3 \; + \; X_4 \qquad\qquad = \quad 2$$
$$X_5 \quad = \quad 2$$

SECTION 0.3

0.3.1. (a) $\begin{pmatrix} 3 & 6 & 9 \\ 3 & 4 & 5 \\ 6 & 6 & 6 \end{pmatrix}$ (b) Nonsense (c) $\begin{pmatrix} 21 & 20 & 19 \\ 21 & 20 & 19 \\ 21 & 20 & 19 \end{pmatrix}$

(d) $\begin{pmatrix} 12 & 24 & 36 \\ 6 & 12 & 18 \\ 12 & 24 & 36 \end{pmatrix}$ (e) $\begin{pmatrix} 5 & 9 \\ 5 & 9 \\ 5 & 9 \end{pmatrix}$ (f) Nonsense

(g) $\begin{pmatrix} 5 & 9 \\ 6 & 18 \end{pmatrix}$ (h) $\begin{pmatrix} 19 & 2 & 25 \\ 20 & 4 & 29 \\ 0 & 0 & 0 \end{pmatrix}$

0.3.3. Assume that

$$A = \begin{pmatrix} x & y \\ z & w \end{pmatrix} \quad \text{and} \quad B = \begin{pmatrix} s & t \\ u & v \end{pmatrix}$$

Then by the definition of matrix multiplication we have

$$AB = \begin{pmatrix} sx + uy & * \\ * & tu + vw \end{pmatrix} \quad \text{and} \quad BA = \begin{pmatrix} sx + tz & * \\ * & uy + vw \end{pmatrix}$$

where the values in the $*$ places are omitted. Subtracting, we find that

$$AB - BA = \begin{pmatrix} uy - tz & * \\ * & tz - uy \end{pmatrix} = \begin{pmatrix} a & b \\ c & d \end{pmatrix}$$

Hence, we find $a + d = (uy - tz) + (tz - uy) = 0$.

0.3.5. Suppose that A is an $m \times n$ matrix for which $AB = 0$ whenever B is any matrix. Let $B_k = (b_{ij})$ be the $n \times 1$ (column) matrix defined by $b_{k1} = 1$ and $b_{i1} = 0$ whenever $k \neq i$. Then, according to the definition of matrix multiplication, the $m \times 1$ matrix AB_i is precisely the ith row of A. Since AB_i must be a zero matrix, it follows that the ith row of A is zero. This is true for all i; therefore $A = 0$, as required.

0.3.7. Four such matrices are given by

$$\begin{pmatrix} \pm 1 & 0 \\ 0 & \pm 1 \end{pmatrix}$$

Two more are

$$\begin{pmatrix} 0 & 1 \\ 1 & 0 \end{pmatrix} \quad \text{and} \quad \begin{pmatrix} 0 & -1 \\ -1 & 0 \end{pmatrix}$$

0.3.9. $M = \begin{pmatrix} t & u & v \\ -t & 2-u & -v \end{pmatrix}$ where $t, u, v \in \mathbf{F}$.

0.3.11. Suppose that $T = (t_{ij})$ is an upper-triangular $n \times n$ matrix and $T^n = 0$. The definition of matrix multiplication shows that $T^n(i, i) = t_{ii}^n$ for each i. Hence $t_{ii}^n = 0$; this shows that $t_{ii} = 0$ for each i. Yes, the converse is also true. Hint: Show that $T^k(i, j) = 0$ whenever $i \leq j + k$ by induction on k. The case of $k = n$ gives $T^n = 0$.

0.3.13. $B^2 = \begin{pmatrix} 7 & 9 & 0 \\ 6 & 10 & 0 \\ \hline 0 & 0 & 9 \end{pmatrix}$ and $B^3 = \begin{pmatrix} 25 & 39 & 0 \\ 26 & 38 & 0 \\ \hline 0 & 0 & 27 \end{pmatrix}$

SECTION 0.4

0.4.1. (a) $\begin{pmatrix} 3 & -1 & -1 \\ 2 & 0 & 1 \end{pmatrix} \begin{pmatrix} X \\ Y \\ Z \end{pmatrix} = \begin{pmatrix} 5 \\ 5 \end{pmatrix}$

(c) $\begin{pmatrix} 1 & -1 & 1 & 0 \\ 1 & 2 & 0 & 1 \\ 1 & 1 & 0 & 0 \end{pmatrix} \begin{pmatrix} X_1 \\ X_2 \\ X_3 \\ X_4 \end{pmatrix} = \begin{pmatrix} 1 \\ 0 \\ 6 \end{pmatrix}$

0.4.3. The solution sets for these systems are:
(a) $\{((23 - 5t)/3, (8 + t)/3, 5 - t, t) \mid t \in \mathbf{R}\}$
(b) $\{(1, -t, t) \mid t \in \mathbf{R}\}$
(c) $\{(u - w, 5v + w, u, v, w) \mid u, v, w \in \mathbf{R}\}$
(d) $\emptyset$, that is, the system is inconsistent

0.4.5. All solutions are

$$\vec{X} = \begin{pmatrix} 2u + 2v - 7 \\ 2u - v - 4 \\ u \\ v \end{pmatrix}$$

where $u, v \in \mathbf{F}$.

0.4.7. (a) $\begin{pmatrix} 1 & 2 & 3 & 4 \\ 0 & 0 & 0 & 0 \end{pmatrix}$, rank 1   (b) $\begin{pmatrix} 1 & 0 & 0 & -1 \\ 0 & 1 & 0 & -1/2 \\ 0 & 0 & 1 & 5 \end{pmatrix}$, rank 3

(c) $\begin{pmatrix} 1 \\ 0 \\ 0 \\ 0 \end{pmatrix}$, rank 1   (d) $\begin{pmatrix} 1 & 0 & 0 & 0 \\ 0 & 1 & 0 & 0 \\ 0 & 0 & 0 & 0 \\ 0 & 0 & 0 & 0 \end{pmatrix}$, rank 2

0.4.9. The reduced row-echelon form of the first matrix is always

$$\begin{pmatrix} 1 & 0 & 0 \\ 0 & 1 & 0 \\ 0 & 0 & 1 \end{pmatrix}$$

while the reduced row-echelon form of the second matrix is

$$\begin{pmatrix} 1 & 0 & 1/2 \\ 0 & 1 & 3/2 \\ 0 & 0 & 0 \end{pmatrix}$$

Call the first matrix A and the second matrix B. The systems of equations $A\vec{X} = \vec{0}$ and $B\vec{X} = \vec{0}$ have different solutions. Therefore A and B cannot be row equivalent. It follows that the original two matrices cannot be row equivalent.

0.4.11.   If a matrix has rank 1, then it is row equivalent to a matrix that looks like

$$\begin{pmatrix} r_1 & r_2 & \cdots & r_n \\ 0 & 0 & \cdots & 0 \\ \vdots & \vdots & \ddots & \vdots \\ 0 & 0 & \cdots & 0 \end{pmatrix}$$

Since row operations can be reversed, one can apply row operations to this matrix and recover the original matrix. Any row operation applied to this matrix will give a matrix whose rows are multiples of the top row. Any further row operations will preserve this property. Thus all nonzero rows of the original matrix are multiples of one another.

SECTION 0.5

0.5.1 (a) The homogeneous solutions are

$$t \cdot \begin{pmatrix} 1/2 \\ -1 \\ 0 \\ 1 \end{pmatrix}$$

where $t \in \mathbf{F}$. A specific solution is

$$\begin{pmatrix} 0 \\ 1 \\ 0 \\ 0 \end{pmatrix}$$

so the general solutions are

$$\begin{pmatrix} 0 \\ 1 \\ 0 \\ 0 \end{pmatrix} + t \cdot \begin{pmatrix} 1/2 \\ -1 \\ 0 \\ 1 \end{pmatrix} \quad \text{where } t \in \mathbf{F}$$

(b) The only homogeneous solution is $X = Y = 0$. The only nonhomogeneous solution is $X = 1, Y = 0$.

(c) The homogeneous solutions are of the form

$$t \cdot \begin{pmatrix} -2 \\ 0 \\ 1 \end{pmatrix} + u \cdot \begin{pmatrix} 0 \\ -4 \\ 3 \end{pmatrix}$$

and the nonhomogeneous solutions are

$$\begin{pmatrix} 1 \\ 0 \\ 0 \end{pmatrix} + t \cdot \begin{pmatrix} -2 \\ 0 \\ 1 \end{pmatrix} + u \cdot \begin{pmatrix} 0 \\ -4 \\ 3 \end{pmatrix}$$

(d) The homogeneous solutions are

$$t \cdot \begin{pmatrix} -1 \\ -1/3 \\ 1 \end{pmatrix}$$

and there are no nonhomogeneous solutions.

0.5.3. This system has a solution if and only if $2a - b + c = 0$.

0.5.5. If A is 3×4, B is 3×2, and $AM = B$, then M is 4×2. Suppose that M_1 and M_2 are the columns of M, which we view as columns of variables. In order to solve the equation for M, it suffices to solve the equations $AM_1 = B_1$ and $AM_2 = B_2$ where B_1 and B_2 are the columns of B. Since the rank of A is 3, we necessarily have that $rk(A) = rk(A : B_1) = 3$ and $rk(A) = rk(A : B_2) = 3$. By Theorem 0.5.7 (i) the two equations $AM_1 = B_1$ and $AM_2 = B_2$ can be solved for M_1 and M_2. It follows that the equation $AM = B$ can be solved.

0.5.7. Note that $(a_1, b_1), (a_2, b_2), \ldots, (a_n, b_n)$ are collinear in $\mathbf{R}^2$ if and only if they all are solutions to a single equation $r + sX + tY = 0$ for $r, s, t \in \mathbf{R}$ not all zero. The existence of such r, s, t (not all zero) is equivalent to the existence of a nontrivial solution to the matrix equation

$$\begin{pmatrix} 1 & a_1 & b_1 \\ 1 & a_2 & b_2 \\ \vdots & \vdots & \vdots \\ 1 & a_n & b_n \end{pmatrix} \begin{pmatrix} R \\ S \\ T \end{pmatrix} = \begin{pmatrix} 0 \\ 0 \\ \vdots \\ 0 \end{pmatrix}$$

Such a matrix equation has a nontrivial solution if and only if the coefficient matrix has rank at most 2 by Corollary 0.5.8. This proves what was required.

SECTION 0.6

0.6.1 (a) This matrix has no right or left inverses since it has rank 2.

(b) This matrix has no right inverses. The general form of all left inverses is

$$\begin{pmatrix} (X_3 - 2)/10 & (4 - 7X_3)/10 & X_3 \\ (3 + 2Y_3)/20 & (-1 - 14Y_3)/20 & Y_3 \end{pmatrix}$$

where X_3 and Y_3 are arbitrary scalars.

(c) This matrix has inverse

$$\begin{pmatrix} -3/4 & 4/5 & 3/20 \\ 1/2 & -1/5 & -1/10 \\ -1/4 & 0 & 1/4 \end{pmatrix}$$

(g) This matrix has inverse

$$\begin{pmatrix} 1 & -1 & -4 & -3 \\ -2 & 3 & 3 & 4 \\ 0 & 0 & -3 & -2 \\ 0 & 0 & 2 & 1 \end{pmatrix}$$

0.6.3. To see that $(A^{-1})^{-1} = A$, we note that A^{-1} is invertible with inverse B in case there is some matrix B for which $(A^{-1})B = I_n$ and $B(A^{-1}) = I_n$. But since A^{-1} is the inverse of A, by definition we have that $(A^{-1})A = I_n$ and $A(A^{-1}) = I_n$. Thus A is the desired matrix B.

0.6.5. First verify that for any matrix A, $S = A + A^t$ is always symmetric and $U = A - A^t$ is always skew symmetric. Then observe that $A = (1/2)S + (1/2)U$.

0.6.7. If B is invertible, we claim that A is invertible and $A^{-1} = CB^{-1}C^{-1}$. This is because, using $A = CBC^{-1}$, we find that

$$(CB^{-1}C^{-1})A = (CB^{-1}C^{-1})(CBC^{-1}) = CB^{-1}(C^{-1}C)B^{-1}C^{-1}$$
$$= CB^{-1}BC^{-1} = CC^{-1} = I_n$$

and similarly, $A(CB^{-1}C^{-1}) = (CBC^{-1})(CB^{-1}C^{-1}) = I_n$. Thus, $A^{-1} = CB^{-1}C^{-1}$. Now note that as $A = CBC^{-1}$, then $C^{-1}AC = C^{-1}CBC^{-1}C = B$. Thus, by the argument just given, if A is invertible, then so is B with $B^{-1} = C^{-1}A^{-1}C$. This gives the result.

0.6.9. We proceed by contradiction. Assume that the matrix A has a left inverse, say L. Then $LA = I$. Multiplying the equation $AC = 0$ gives $L(AC) = L0 = 0$. But $L(AC) = (LA)C = C$, so $C = 0$, a contradiction. This shows that A does not have a left inverse.

0.6.11. (a) According to Theorem 0.6.6, the product of invertible matrices is always invertible. If a nilpotent matrix N is invertible, then so is $NN = N^2$, $N^3, \ldots, N^k$ for any k. But $N^k = 0$ for some k. This is a contradiction since the zero matrix cannot be invertible.

(b) We have been told what the inverse of $I_n - N$ should be, so to prove invertibility, we must only multiply these matrices together and see if we get I_n. We assume $N^k = 0$. Then $(I_n - N)(I_n + N + N^2 + \cdots + N^{k-1}) = (I_n + N + N^2 + \cdots + N^{k-1}) - (N + N^2 + \cdots + N^{k-1} + N^k) = I_n - N^k = I_n$, as desired. Similarly, $(I_n + N + N^2 + \cdots + N^{k-1})(I_n - N) = I_n$; this gives the result.

(c) In view of the result proved in (b) above, it suffices to show that $-AN$ is nilpotent, because then $(I_n + AN) = (I_n - (-AN))$ will be invertible. It is easily checked that $-AN$ is nilpotent if and only if AN is nilpotent. Assume that $N^k = 0$. Then $(AN)^k = (AN)(AN) \cdots (AN) = A(NA)N(AN) \cdots (AN) = A(AN)N \cdots (AN) = A^2N^2(AN) \cdots (AN) = \cdots = A^kN^k = A^k0 = 0$; this gives the result.

0.6.13. (a) If $U, S \in O_{\mathbf{F}}(2)$, then $UU^t = I_2$ and $SS^t = I_2$. We compute that $(US)(US)^t = (US)(S^tU^t) = U(SS^t)U^t = UU^t = I_2$, as required. Thus $US \in O_{\mathbf{F}}(2)$.

(b) If $U \in O_{\mathbf{F}}(2)$, then $UU^t = I_2$. Since U is square, the right inverse U^t for U is necessarily the inverse of U; that is, $U^t = U^{-1}$. Consequently, $U^{-1}(U^{-1})^t = U^t(U^t)^t = U^tU = I_2$; this shows that $U^{-1} \in O_{\mathbf{F}}(2)$.

0.6.15. Assume that A and B are invertible and $AB = BA$. Then, AB is invertible, and $(AB)^{-1} = B^{-1}A^{-1}$ by Theorem 0.6.6. By the same reason, $B^{-1}A^{-1} = (AB)^{-1} = (BA)^{-1} = A^{-1}B^{-1}$, so A^{-1} and B^{-1} commute.

0.6.17. (a) If A is $m \times n$ and $A = A^t$, A^t is $n \times m$, and we must have that $m = n$; that is, A is square.
(b) Note that $(AA^t)^t = (A^t)^t A^t = AA^t$ by Theorem 0.6.15, so AA^t is symmetric. Easily, $(A + A^t) = (A^t + A) = A + A^t$, so $A + A^t$ is symmetric.
(c) Assume that both A and B are symmetric. If AB is symmetric, then $AB = (AB)^t = B^t A^t = BA$ so B and A commute. Conversely, if A and B commute, then $(AB)^t = B^t A^t = BA = AB$, so AB is symmetric. This proves the result.

0.6.19. There are many possible solutions to these problems. Some possible solutions are:

(a) $\begin{pmatrix} 1 & 4 \\ 3 & 2 \end{pmatrix} = \begin{pmatrix} 1 & 0 \\ 3 & 1 \end{pmatrix} \begin{pmatrix} 1 & 0 \\ 0 & -10 \end{pmatrix} \begin{pmatrix} 1 & 4 \\ 0 & 1 \end{pmatrix}$

(b) $\begin{pmatrix} 0 & 0 & 1 \\ 0 & 1 & 0 \\ 1 & 0 & 0 \end{pmatrix} = \begin{pmatrix} 1 & 0 & -1 \\ 0 & 1 & 0 \\ 0 & 0 & 1 \end{pmatrix} \begin{pmatrix} 1 & 0 & 0 \\ 0 & 1 & 0 \\ 1 & 0 & 1 \end{pmatrix}$
$\begin{pmatrix} 1 & 0 & -1 \\ 0 & 1 & 0 \\ 0 & 0 & 1 \end{pmatrix} \begin{pmatrix} 1 & 0 & 0 \\ 0 & 1 & 0 \\ 0 & 0 & -1 \end{pmatrix}$

0.6.21. Both A and B are invertible and consequently can be written as a product of elementary matrices. Since $A(BR) = (AB)R$, ABR is obtained from R by multiplying R on the left by a sequence of elementary matrices. Thus ABR can be obtained form R by performing a sequence of elementary row operations. This shows that R and ABR are row equivalent.

SECTION 1.1

1.1.1. (a) $\begin{pmatrix} 2 \\ 1 \\ 3 \\ 2 \end{pmatrix} = 2 \begin{pmatrix} 1 \\ 0 \\ 0 \\ 1 \end{pmatrix} + 2 \begin{pmatrix} 0 \\ 1 \\ 2 \\ 0 \end{pmatrix} + (-1) \begin{pmatrix} 0 \\ 1 \\ 1 \\ 0 \end{pmatrix}$

(b) Impossible!

(c) $\begin{pmatrix} 2 \\ 1 \\ 3 \\ 2 \end{pmatrix} = 2 \begin{pmatrix} 1 \\ 0 \\ 0 \\ 1 \end{pmatrix} + 3 \begin{pmatrix} 1 \\ 0 \\ 1 \\ 0 \end{pmatrix} + 1 \begin{pmatrix} 1 \\ 1 \\ 0 \\ 0 \end{pmatrix} + (-4) \begin{pmatrix} 1 \\ 0 \\ 0 \\ 0 \end{pmatrix}$

1.1.3. (a) $\left\{ \begin{pmatrix} 2r \\ 2r \\ r \end{pmatrix} \mid r \in \mathbf{F} \right\}$

(b) $\mathbf{F}^3$

(c) $\left\{\begin{pmatrix} r \\ 2r + 3s \\ r + 5s \\ s \end{pmatrix} \mid r, s \in \mathbf{F}\right\}$

(d) $\mathbf{F}^2$

1.1.5. (a) Yes (b) No (c) No (d) Yes (e) No (f) Yes (g) No (h) No (i) Yes

1.1.7. We set $V = span\{v_1, v_2, \ldots, v_n\}$. Easily, $V \subseteq span\{v, v_1, v_2, \ldots, v_n\}$, since any linear combination of $v_1, v_2, \ldots, v_n$ is also a linear combination of $v, v_1, v_2, \ldots, v_n$. We must show the reverse inclusion, that $span\{v, v_1, v_2, \ldots, v_n\} \subseteq V$. We have that $v = a_1 v_1 + a_2 v_2 + \cdots + a_n v_n$ for appropriate scalars $a_1, a_2, \ldots, a_n$ since v is a linear combination of $v_1, v_2, \ldots, v_n$. Consider $z \in span\{v, v_1, v_2, \ldots, v_n\}$. If $z = cv + b_1 v_1 + b_2 v_2 + \cdots + b_n v_n$, then $z = c(a_1 v_1 + a_2 v_2 + \cdots + a_n v_n) + b_1 v_1 + b_2 v_2 + \cdots + b_n v_n = (ca_1 + b_1)v_1 + (ca_2 + b_2)v_2 + \cdots + (ca_n + b_n)v_n \in span\{v_1, v_2, \ldots, v_n\} = V$. This concludes the proof.

1.1.9. $8X_1 + 10X_3 - 29X_4 = 0, X_2 = 0$.

1.1.11. We suppose that A is $m \times n$ and B is $n \times p$, so that C is $m \times p$. We assume that $A = (a_{ij})$, $B = (b_{jk})$, and $C = (c_{ik})$. We expand both A and C into columns $A = (A_1 \quad A_2 \quad \cdots \quad A_n)$ and $C = (C_1 \quad C_2 \quad \cdots \quad C_p)$. The definition of matrix multiplication gives that $c_{ik} = \sum_{j=1}^{n} a_{ij} b_{jk}$. Since c_{ik} is the ith entry of the kth column of C, that is, the ith entry of C_k, and similarly as a_{ij} is the ith entry of A_j, it follows that the column $C_k = \sum_{j=1}^{n} b_{jk} A_j$. This shows that the columns of C are linear combinations of the columns of A.

SECTION 1.2

1.2.1. (a) $(1 \quad 0 \quad 0 \quad 3 \quad 2)$ and $(3 \quad 8 \quad 9 \quad 4 \quad 1)$

(b) $\begin{pmatrix} 1 \\ 3 \\ 3 \end{pmatrix}$ and $\begin{pmatrix} 0 \\ 1 \\ 2 \end{pmatrix}$ (c) $\begin{pmatrix} -3 \\ 5/8 \\ 0 \\ 1 \end{pmatrix}$ and $\begin{pmatrix} 0 \\ -9/8 \\ 1 \\ 0 \end{pmatrix}$

1.2.3. (a) The matrices

$$\begin{pmatrix} 1 & 1 \\ 1 & 1 \end{pmatrix} \quad \text{and} \quad \begin{pmatrix} 1 & 0 \\ 1 & 0 \end{pmatrix}$$

have the same column space, namely

$$span\left\{\begin{pmatrix} 1 \\ 1 \end{pmatrix}\right\}$$

Clearly, however, their row spaces are different.
(b) Take the transposes of the matrices in (a).

1.2.5. (a) Is not a vector space; is not closed under scalar multiplication.
(b) Is not a vector space; is not closed under scalar multiplication.
(c) Is a vector space; same as $\mathbf{R}^n$.
(d) Is a vector space; is the set of solutions to a homogeneous system.
(e) Is not a vector space.
(f) Is a vector space.

1.2.7. (a) No   (b) Yes   (contains only the zero matrix) (c) No   (d) Yes

1.2.9. (a) Yes   (b) Yes   (c) No   (d) Yes   (e) Yes   (f) No   (g) Yes
(h) No

1.2.11. Yes. Suppose that $f(x), g(x) \in Fun_0(S, \mathbf{F})$, $f(x)$ is zero for all $x \in S$ except for $x \in S_1$ which is finite, and $g(x)$ is zero for all $x \in S$ except for $x \in S_2$ which also is finite. Then the union $S_1 \cup S_2$ is finite, and $(f + g)(x)$ is zero for all $x \in S$ except possibly those $x \in S_1 \cup S_2$. This shows that $(f + g)(x) \in Fun_0(S, \mathbf{F})$. Let $k \in \mathbf{F}$ be a scalar. Then $kf(x)$ is zero for all $x \in S$ except possibly those $x \in S_1$. So, $kf(x) \in Fun_0(S, \mathbf{F})$. This shows that $Fun_0(S, \mathbf{F})$ is a vector subspace of $Fun(S, \mathbf{F})$.

1.2.13. (a) Suppose that $v_1, v_2 \in V \cap W$ and $k \in \mathbf{F}$ is a scalar. Then, using the fact that both V and W are subspaces, we see that $v_1 + v_2 \in V$, $v_1 + v_2 \in W$, $kv_1 \in V$, and $kv_2 \in W$. Consequently, $v_1 + v_2, kv_1 \in V \cap W$; this shows that $V \cap W$ is a subspace.
(b) Suppose $v_1, v_2 \in V$, $w_1, w_2 \in W$, and $k \in \mathbf{F}$. Then $v_1 + v_2, kv_1 \in V$ and $w_1 + w_2, kw_1 \in W$. Consequently, $(v_1 + w_1) + (v_2 + w_2) = (v_1 + v_2) + (w_1 + w_2) \in V + W$ and $k(v_1 + w_1) = kv_1 + kw_1 \in V + W$. This shows that $V + W$ is a subspace.
(c) Suppose that $u = v_1 + w_1 = v_2 + w_2$ where $v_1, v_2 \in V$ and $w_1, w_2 \in W$. Then $\vec{0} = u - u = (v_1 - v_2) + (w_1 - w_2)$ so $(v_1 - v_2) = -(w_1 - w_2) \in V \cap W$. Since $V \cap W = \{\vec{0}\}$, we obtain that $(v_1 - v_2) = -(w_1 - w_2) = \vec{0}$; that is, $v_1 = v_2$ and $w_1 = w_2$. This gives the desired uniqueness of the representation of the vector u.

SECTION 1.3

1.3.1. (a),(b),(d),(e),(f) are linearly independent; (c) is linearly dependent.

1.3.3. Note: There is more than one solution to each of these problems. Possible answers are:

(a) $\begin{pmatrix} 2 \\ 3 \\ 1 \end{pmatrix}, \begin{pmatrix} 2 \\ 5 \\ 2 \end{pmatrix}$ (b) $\begin{pmatrix} 1 \\ 1 \\ 3 \end{pmatrix}, \begin{pmatrix} 1 \\ 3 \\ 3 \end{pmatrix}, \begin{pmatrix} 2 \\ 4 \\ 3 \end{pmatrix}$

(c) $\begin{pmatrix} 1 \\ 1 \\ 1 \end{pmatrix}, \begin{pmatrix} 3 \\ 2 \\ 1 \end{pmatrix}, \begin{pmatrix} 5 \\ 1 \\ 1 \end{pmatrix}$ (d) $\begin{pmatrix} 0 \\ 2 \\ 4 \\ 1 \end{pmatrix}, \begin{pmatrix} 1 \\ 3 \\ 6 \\ 2 \end{pmatrix}$

1.3.5.   To show that the vectors $v_1, v_2, \ldots, v_{n-1}$ are linearly independent, we must show that the requirements of the definition of linear independence are met for these vectors. Note that the definition of linear independence is an "if, then" statement. Hence in such linear independence proofs, we will assume the "if" part, and *prove* the "then" part.

We have that $v_1, v_2, \ldots, v_n$ are linearly independent. To show that $v_1, v_2, \ldots, v_{n-1}$ are linarly independent, we assume $a_1 v_1 + a_2 v_2 + \cdots + a_{n-1} v_{n-1} = \vec{0}$ where $a_1, a_2, \ldots, a_{n-1} \in \mathbf{F}$. It follows that $a_1 v_1 + a_2 v_2 + \cdots + a_{n-1} v_{n-1} + 0 v_n = \vec{0}$. The linear independence of $v_1, v_2, \ldots, v_n$ gives that $a_1 = 0, a_2 = 0, \ldots, a_{n-1} = 0$ (and $0 = 0!$). Consequently, $v_1, v_2, \ldots, v_{n-1}$ are linearly independent.

1.3.7.   We proceed by induction on n. In case $n = 1$, the result is clear, so we may assume that the result is true for $n - 1$. Thus, $v_1, v_2, \ldots, v_{n-1}$ are linearly independent. Now suppose that $a_1 v_1 + a_2 v_2 + \cdots + a_{n-1} v_{n-1} + a_n v_n = \vec{0}$ where the $a_i \in \mathbf{F}$. If $a_n \neq 0$, we can write $v_n = (-1/a_n)(a_1 v_1 + a_2 v_2 + \cdots + a_{n-1} v_{n-1})$. But this contradicts the assumption that $v_n \notin span\{v_1, v_2, \ldots, v_{n-1}\}$. Consequently, $a_n = 0$. But now as $a_1 v_1 + a_2 v_2 + \cdots + a_{n-1} v_{n-1} = \vec{0}$, the linear independence of $v_1, v_2, \ldots, v_{n-1}$ gives that $a_1 = 0, a_2 = 0, \ldots, a_{n-1} = 0$. This shows that $v_1, v_2, \ldots, v_{n-1}, v_n$ are linearly indpendent.

1.3.9.   Assume the contrary, that $w_1, w_2, \ldots, w_m$ are linearly independent. Since $V = span\{v_1, v_2, \ldots, v_n\}$, we know that each of $w_1, w_2, \ldots, w_m$ is a linear combnation of $v_1, v_2, \ldots, v_n$. Theorem 1.3.9 shows that $m \leq n$, a contradiction. Thus $w_1, w_2, \ldots, w_m$ are linearly dependent.

1.3.11.   Assume that $a_1 v_1 + a_2(v_1 + v_2) + a_3(v_1 + v_2 + v_3) + \cdots + a_n(v_1 + \cdots + v_n) = \vec{0}$. We must show that $a_1 = 0, a_2 = 0, \ldots, a_n = 0$. Using the distributive and commutative laws for vector addition, our equation can be rewritten as

$$(a_1 + a_2 + \cdots + a_n)v_1 + (a_2 + a_3 + \cdots + a_n)v_2 + \cdots + (a_{n-1} + a_n)v_{n-1} + a_n v_n = \vec{0}$$

Since $v_1, v_2, \ldots, v_n$ are linearly independent, we see that the coefficients of each of these in the preceding equation vanish; that is, $a_1 + a_2 + \cdots + a_n = 0$, $a_2 + a_3 + \cdots + a_n = 0, \ldots, a_{n-1} + a_n = 0$, $a_n = 0$. Back-substituting, starting from the last equation $a_n = 0$, we have $a_{n-1} = 0$, so $a_{n-2} = 0, \ldots,$ so $a_1 = 0$. This proves the result.

1.3.13.   Assume that $v_1, v_2, \ldots, v_n$ are linearly independent and A is invertible. Suppose that $a_1 A v_1 + a_2 A v_2 + \cdots + a_n A v_n = \vec{0}$ where $a_i \in \mathbf{F}$. Then $\vec{0} = A^{-1}\vec{0} = A^{-1}(a_1 A v_1 + \cdots + a_n A v_n) = a_1 v_1 + \cdots + a_n v_n$. The linear independence of $v_1, v_2, \ldots, v_n$ gives that $a_1 = 0, a_2 = 0, \ldots, a_n = 0$. This proves that $A v_1, A v_2, \ldots, A v_n$ are linearly independent.

1.3.15. $c + 4b - 2a \neq 0$.

1.3.17. $\{1, X, X^2, X^3, \ldots\}$.

SECTION 1.4

1.4.1. (a) $\left\{ \begin{pmatrix} 4/3 \\ 1 \\ 0 \end{pmatrix}, \begin{pmatrix} -2/3 \\ 0 \\ 1 \end{pmatrix} \right\}$ is an answer (others possible, bases are not unique).

(b) $\left\{ \begin{pmatrix} 1 \\ 0 \\ 0 \\ 1 \end{pmatrix}, \begin{pmatrix} 0 \\ 1 \\ 0 \\ 0 \end{pmatrix}, \begin{pmatrix} 0 \\ 0 \\ 1 \\ 0 \end{pmatrix} \right\}$ is an answer (others possible).

(c) $\left\{ \begin{pmatrix} -4 \\ -48/33 \\ 20/11 \\ 0 \\ 1 \end{pmatrix}, \begin{pmatrix} 0 \\ -28/33 \\ 8/11 \\ 1 \\ 0 \end{pmatrix} \right\}$ is an answer (others possible).

1.4.3. (a) The REF of

$$\begin{pmatrix} 1 & 1 & 3 & 0 & 2 \\ 0 & 1 & 2 & 0 & 1 \\ 2 & 1 & 4 & 1 & 3 \\ 1 & 0 & 1 & 2 & 1 \end{pmatrix} \quad \text{is} \quad \begin{pmatrix} 1 & 0 & 1 & 0 & 1 \\ 0 & 1 & 2 & 0 & 1 \\ 0 & 0 & 0 & 1 & 0 \\ 0 & 0 & 0 & 0 & 0 \end{pmatrix}$$

(b) From the REF we see that the third and fifth columns of A are linear combinations of the first and second columns (as this is clearly true in the REF). Thus the dimension of the span of the first, second, third, and fifth columns of A is 2.

(c) A basis for the row space of A will be the three nonzero rows of the REF, that is, $\{(1,0,1,0,1),(0,1,2,0,1),(0,0,0,1,0)\}$.

(d) $\left\{ \begin{pmatrix} -1 \\ -1 \\ 0 \\ 0 \\ 1 \end{pmatrix}, \begin{pmatrix} -1 \\ -2 \\ 1 \\ 0 \\ 0 \end{pmatrix} \right\}$

1.4.5. (a) $\left\{ \begin{pmatrix} 1 \\ 2 \\ 3 \\ 4 \end{pmatrix}, \begin{pmatrix} 4 \\ 3 \\ 2 \\ 1 \end{pmatrix} \right\}$ (b) $\left\{ \begin{pmatrix} 1 \\ 0 \\ 1 \end{pmatrix}, \begin{pmatrix} 0 \\ 1 \\ 2 \end{pmatrix} \right\}$

1.4.7. Suppose that $dim(V) = dim(W) = n$ and $v_1, v_2, \ldots, v_n$ is a basis of V. By Theorem 1.4.5 (i), since $v_1, v_2, \ldots, v_n$ are linearly independent vectors inside W and W is n-dimensional, $v_1, v_2, \ldots, v_n$ must be a basis of W. In particular, since $v_1, v_2, \ldots, v_n$ span W, every element of W is a linear combination of them. But $span\{v_1, v_2, \ldots, v_n\} = V$, so W is a subspace of V. Hence $V = W$.

1.4.9. Since V is n-dimensional, and the collection $v_1 + v_2, \ldots, v_n + v_1$ has n elements, it suffices to show they are linearly independent if and only if n

is odd. First we look at the case where n is even. Observe then that

$$(v_1 + v_2) - (v_2 + v_3) + (v_3 + v_4) - \cdots + (v_{n-1} + v_n) - (v_n + v_1) = \vec{0}$$

where the signs at the end turn out the way they do because n is even. Thus when n is even, the vectors fail to be linearly independent. Now suppose that n is odd. Assume that

$$a_1(v_1 + v_2) + a_2(v_2 + v_3) + \cdots + a_n(v_n + v_1) = \vec{0}$$

Then rewriting after using the distributive and commutative laws gives

$$(a_1 + a_n)v_1 + (a_1 + a_2)v_2 + \cdots + (a_{n-1} + a_n)v_n = \vec{0}$$

so by the linear independence of $v_1, v_2, \ldots, v_n$ we find that $a_1 + a_n = 0, a_1 + a_2 = 0, \ldots, a_{n-1} + a_n = 0$. Row-reducing the matrix of the homogeneous equation in the a_i's yields

$$\begin{pmatrix} 1 & 0 & 0 & \cdots & 0 & 1 \\ 1 & 1 & 0 & \cdots & 0 & 0 \\ 0 & 1 & 1 & \cdots & 0 & 0 \\ \vdots & \vdots & \vdots & \ddots & \vdots & \vdots \\ 0 & 0 & 0 & \cdots & 1 & 1 \end{pmatrix} \mapsto \begin{pmatrix} 1 & 0 & 0 & \cdots & 0 & 1 \\ 0 & 1 & 0 & \cdots & 0 & -1 \\ 0 & 0 & 1 & \cdots & 0 & 1 \\ \vdots & \vdots & \vdots & \ddots & \vdots & \vdots \\ 0 & 0 & 0 & \cdots & 0 & -2 \end{pmatrix}$$

where the nonzero entry in the last row occurs as n is odd. This shows that the matrix has rank n, and $a_1 = 0, a_2 = 0, \ldots, a_n = 0$ follow.

1.4.11. For example (there are many)

$$\left\{ \begin{pmatrix} 1 \\ 1 \\ 0 \\ 1 \end{pmatrix}, \begin{pmatrix} 2 \\ 2 \\ 3 \\ 1 \end{pmatrix}, \begin{pmatrix} 1 \\ 0 \\ 0 \\ 0 \end{pmatrix}, \begin{pmatrix} 0 \\ 1 \\ 0 \\ 0 \end{pmatrix} \right\}$$

is a basis of $\mathbf{R}^4$.

1.4.13. Suppose that $\{v_1, \ldots, v_n\}$ form a basis for $V \cap W$. Applying the basis-extension theorem, let $\{v_1, \ldots, v_n, v_{n+1}, \ldots, v_m\}$ be a basis for V and $\{v_1, \ldots, v_n, w_{m+1}, \ldots, w_p\}$ be a basis for W. To prove the result, it suffices to verify that $\{v_1, \ldots, v_n, v_{n+1}, \ldots, v_m, w_{n+1}, \ldots, w_p\}$ is a basis for $V + W$. That these vectors span $V + W$ follows since they contain spanning sets for both V and W. Now assume that

$$a_1 v_1 + \cdots + a_n v_n + b_{n+1} v_{n+1} + \cdots + b_m v_m + c_{n+1} w_{n+1} + \cdots + c_p w_p = \vec{0}$$

We find that

$$\begin{aligned} u &= a_1 v_1 + \cdots + a_n v_n + b_{n+1} v_{n+1} + \cdots + b_m v_m \\ &= -(c_{n+1} w_{n+1} + \cdots + c_p w_p) \in V \cap W = \text{span}\{v_1, \ldots, v_n\} \end{aligned}$$

Expressing u as a linear combination of $v_1, \ldots, v_n$, we see that by the linear independence of $v_1, \ldots, v_n, w_{n+1}, \ldots, w_p$, $c_{n+1} = 0, \ldots, c_p = 0$. Finally, as $u = \vec{0}$, the linear independence of $v_1, \ldots, v_n, v_{n+1}, \ldots, v_m$ shows that $a_1 = 0, \ldots, a_n = 0, b_{m+1} = 0, \ldots, b_m = 0$, and the result is proved.

1.4.15. Note that

$$\begin{pmatrix} 1 & 0 \\ 0 & 0 \end{pmatrix} \quad \text{and} \quad \begin{pmatrix} 1 & 1 \\ 0 & 0 \end{pmatrix}$$

are both rank 1, 2×2, matrices. Since they are both reduced echelon form matrices and are different, they cannot be row equivalent. (Alternatively, to see they are not row equivalent, observe that their row spaces are different.)

1.4.17 (a) The columns of AB are linear combinations of the columns of A by Prob. 1.1.11. But A has a single column, and consequently, the columns of AB must be linearly dependent. This shows that AB cannot be invertible. (b) If A is $n \times m$, B is $m \times n$, and $m < n$, then AB is not invertible. To see this note, that the columns of AB must be linear combinations of the m columns of A. In particular, they cannnot span the n-dimensional vector space $\mathbf{F}^n$. Consequently the matrix AB cannot be invertible.

1.4.19. $\{1, X, X^2, X^3, \ldots\}$ is a basis.

1.4.21. According to Prob. 1.1.11, the columns of AB are linear combinations of the columns of A. It follows from this that the column space of AB is a subspace of the column space of A. Thus the dimension of the column space of AB is at most the dimension of the column space of A. As these dimensions are the rank, we find that $rk(AB) \leq rk(A)$. Now, $rk(AB) = rk((AB)^t) = rk(B^t A^t) \leq rk(B^t) = rk(B)$, where the inequality comes from the result just established. Thus $rk(AB) \leq rk(A), rk(B)$, as desired.

SECTION 1.5

1.5.1. $\begin{pmatrix} -2 \\ 3 \end{pmatrix}$

1.5.3. $(1/3) \begin{pmatrix} a - c \\ a + c \\ 2b - c - a \end{pmatrix}$

1.5.5. $A = \begin{pmatrix} 8 & -5 \\ -3 & 2 \end{pmatrix}$

1.5.7. $\begin{pmatrix} 1 & 1 & 1 & 1 \\ 0 & 1 & 2 & 3 \\ 0 & 0 & 1 & 3 \\ 0 & 0 & 0 & 1 \end{pmatrix}$

1.5.9. R_θ has inverse

$$\begin{pmatrix} \cos\theta & -\sin\theta \\ \sin\theta & \cos\theta \end{pmatrix}$$

The basis is

$$B = \left\{ \begin{pmatrix} \cos\theta \\ \sin\theta \end{pmatrix}, \begin{pmatrix} -\sin\theta \\ \cos\theta \end{pmatrix} \right\}$$

1.5.11. Apply Theorem 1.5.5 repeatedly.

SECTION 1.6

1.6.1. The collections (b) and (c) are linearly independent.

1.6.3.

$$\begin{pmatrix} 1 \\ 1 \\ 0 \\ 0 \\ 1 \\ 0 \\ 1 \end{pmatrix}, \begin{pmatrix} 0 \\ 0 \\ 1 \\ 0 \\ 0 \\ 1 \\ 0 \end{pmatrix}, \begin{pmatrix} 0 \\ 0 \\ 0 \\ 0 \\ 1 \\ 1 \\ 0 \end{pmatrix}$$

are not codewords. Their corrections are

$$\begin{pmatrix} 0 \\ 1 \\ 0 \\ 0 \\ 1 \\ 0 \\ 1 \end{pmatrix}, \begin{pmatrix} 0 \\ 0 \\ 1 \\ 0 \\ 1 \\ 1 \\ 0 \end{pmatrix}, \begin{pmatrix} 0 \\ 0 \\ 1 \\ 0 \\ 1 \\ 1 \\ 0 \end{pmatrix}$$

SECTION 2.1

2.1.1. (a) -6 (b) 0 (c) 3 (d) 7 (e) 7 (f) 0

2.1.3.   Assume that $N^k = 0$ but $det(N) = d \neq 0$. Then $det(N^k) = det(NN \cdots N) = det(N)det(N) \cdots det(N) = det(N)^k = d^k \neq 0$. However, $det(N^k) = det(0) = 0$, a contradiction. Hence whenever N is nilpotent, $det(N) = 0$.

2.1.5. Since $AB = -BA$ and each are $n \times n$, $det(AB) = (-1)^n det(BA)$. If n is odd, necessarily, $-det(BA) = det(AB) = det(A)det(B) = det(B)det(A) = det(BA)$. Thus $det(B)det(A) = 0$. It follows that one of $det(A)$ or $det(B) = 0$; that is, one of A or B fails to be invertible.

For $n = 2$ setting

$$A = \begin{pmatrix} 0 & 1 \\ 1 & 0 \end{pmatrix} \quad \text{and} \quad B = \begin{pmatrix} 0 & 1 \\ -1 & 0 \end{pmatrix}$$

gives an example of anticommuting invertible matrices.

2.1.7. $det(AB) = det(A)det(B) = det(B)det(A) = det(BA)$ by Theorem 2.2.13.

2.1.9. $det(B^t A^{-1} BA) = det(B^t)det(A^{-1})det(B)det(A) = det(B)det(A)^{-1} \cdot det(B)det(A) = det(B)^2$. But $det(B) = 6$, so the answer is 36.

2.1.11. (a) 2.

(b) If $det(A) = 1$, then A is invertible, so $rk(A) = 4$. Thus the dimension of the rowspace of A is 4 so the rowspace of A must be $\mathbf{F}^4$.

2.1.13. 2

SECTION 2.2

2.2.1. (a) 1 (10 inversions)

(b) −1 (9 inversions)

(c) −1 (1 inversion)

2.2.3.   For such an $n \times n$ matrix A, there are less than n nonzero entries. Hence any product of n entries of A is necessarily zero. By Theorem 2.2.9, $det(A)$ must be zero since it is a sum of (signed) products of n entries from A.

2.2.5.   If A is orthogonal, then $AA^t = I$. Thus, $1 = det(I) = det(AA^t) = det(A)det(A^t) = det(A)det(A) = det(A)^2$. Since $det(A) \in \mathbf{R}$, $det(A) = \pm 1$.

2.2.7.   Suppose that n is odd. Since the matrix $-A$ is obtained from A by multiplying each of the rows of A by -1 and there are an odd number of rows, $det(-A) = (-1)^n det(A) = -det(A)$. If A is skew symmetric, we find that $det(A) = det(A^t) = det(-A) = -det(A)$. From this we must have $det(A) = 0$.

2.2.9.   Half the permutations in S_n have the form $(i_1, i_2, i_3, \ldots, i_n)$ where $i_1 < i_2$. The permutations $(i_1, i_2, i_3, \ldots, i_n)$ and $(i_2, i_1, i_3, \ldots, i_n)$ (which are the same except that the first two entries are reversed) have opposite signs since the number of transpositions differs by 1. Pairing off all the permutations in S_n in this fashion shows that half the permutations have sign 1 and the other half sign -1.

SECTION 2.3

2.3.1. (a) −9   (b) −4   (c) −5   (d) −4   (e) 1   (f) 0

2.3.3. $(1/5) \begin{pmatrix} -1 & 1 & 0 \\ 3 & 2 & 5 \\ 3 & 2 & 0 \end{pmatrix}$

2.3.5. $X = 1, Y = -1, Z = 1$.

2.3.7. (a) 84   (b) 0   (c) −8   (d) −3

2.3.9.   If A is not invertible, apply the previous problem to see both determinants are 0. If A is invertible, then $adj(A) = det(A)A^{-1}$. Since A is $n \times n$, we find that $det(adj(A)) = (det(A))^n det(A^{-1}) = (det(A))^n (det(A))^{-1} = (det(A))^{n-1}$.

SECTION 3.1

3.1.1. (a) Not linear
(b) Linear, matrix is

$$\begin{pmatrix} 1 & -1 \\ 0 & 0 \end{pmatrix}$$

(c) Not linear
(d) Linear, matrix is

$$\begin{pmatrix} 0 & 0 & 0 \\ 0 & 0 & 0 \end{pmatrix}$$

(e) Linear, matrix is

$$\begin{pmatrix} 1 & 0 & 0 \\ 0 & 1 & 0 \end{pmatrix}$$

(f) Not linear
(g) Linear, matrix is

$$\begin{pmatrix} 1 & 0 & 0 \\ 1 & 0 & 0 \\ 0 & 1 & 0 \\ 1 & 0 & \pi \end{pmatrix}$$

(h) Not linear

3.1.3. (a) Yes, apply Lemma 3.1.4.
(b) No, such T would have to satisfy

$$T\begin{pmatrix} 1 \\ 0 \\ 0 \end{pmatrix} + 2T\begin{pmatrix} 0 \\ 1 \\ 0 \end{pmatrix} = \begin{pmatrix} 1 \\ 2 \\ 3 \end{pmatrix} + 2\begin{pmatrix} 3 \\ 2 \\ 1 \end{pmatrix} \neq \begin{pmatrix} 4 \\ 4 \\ 4 \end{pmatrix} = T\begin{pmatrix} 1 \\ 2 \\ 0 \end{pmatrix}$$

3.1.5. (a) $\begin{pmatrix} -1 & 0 \\ 0 & 1 \end{pmatrix}$ (b) $\begin{pmatrix} 0 & 1 \\ 1 & 0 \end{pmatrix}$ (c) $\begin{pmatrix} 1 & 0 \\ 0 & 0 \end{pmatrix}$

3.1.7. To prove that $v_1, v_2, \ldots, v_n$ are linearly independent, we assume that $a_1 v_1 + a_2 v_2 + \cdots + a_n v_n = \vec{0}$. Applying T, we find that $\vec{0} = T(a_1 v_1 + a_2 v_2 + \cdots + a_n v_n) = a_1 T(v_1) + a_2 T(v_2) + \cdots + a_n T(v_n)$. Since $T(v_1), T(v_2), \ldots, T(v_n)$ are linearly independent, we conclude that $a_1 = 0$, $a_2 = 0, \ldots, a_n = 0$. This shows that $v_1, v_2, \ldots, v_n$ are linearly independent.

3.1.9. (a) Yes (b) No

3.1.11. Suppose that $v_1, v_2 \in V$ and $k \in \mathbf{F}$. As $T(v_1), T(v_2) \in U$, we see that $T(v_1 + v_2) = T(v_1) + T(v_2), T(kv_1) = kT(v_1) \in U$ since $U \subseteq W$ is a subspace. This shows that $v_1 + v_2, kv_1 \in S$, so S is a subspace of V.

3.1.13. (a) If $T_1, T_2 : \mathbf{F}^n \to \mathbf{F}^m$ are linear transformations, then by definition $(T_1 + T_2)(v) = T_1(v) + T_2(v)$ and $(kT_1)(v) = k(T_1(v))$. All the vector space axioms can now be seen to be consequences of the corresponding vector space laws for $\mathbf{F}^m$.

(b) We know that the ith column of the matrix $[T]$ is given by $T(e_i)$. Consequently, the ith column of $[T_1 + T_2]$ is given by $(T_1 + T_2)(e_i) = T_1(e_i) + T_2(e_i)$.
This shows that $[T_1 + T_2] = [T_1] + [T_2]$. Analogously, $[kT_1] = k[T_1]$ for any
$k \in \mathbf{F}$. Altogether we have shown that the function $Mat : Hom(\mathbf{F}^n, \mathbf{F}^m) \rightarrow$
$M_{m \times n}(\mathbf{F})$ given by $Mat(T) = [T]$ is a linear transformation.

SECTION 3.2

3.2.1. (a) $\begin{pmatrix} 1 & 0 & -1 \\ 1 & 0 & 1 \\ 2 & 0 & 2 \\ 0 & 0 & 1 \end{pmatrix}$ has rank 2 and has nullity 1.

(b) $\begin{pmatrix} 1 & 1 & -1 & -1 \\ 1 & 0 & 1 & 0 \end{pmatrix}$ has rank 2 and has nullity 2.

(c) $(0 \ \ 0 \ \ 0 \ \ 0)$ has rank 0 and has nullity 4.

(d) (3) has rank 1 and has nullity 0.

(e) $\begin{pmatrix} -1 & 0 & 0 & 0 & 0 \\ 0 & -1 & 0 & 0 & 0 \\ 0 & 0 & -1 & 0 & 0 \\ 0 & 0 & 0 & -1 & 0 \\ 0 & 0 & 0 & 0 & -1 \end{pmatrix}$ has rank 5 and has nullity 0.

3.2.3. There are many possible answers to these. Some solutions are:
(a) Let T be T_A where

$$A = \begin{pmatrix} 1 & 1 & 1 \\ 1 & 1 & 2 \\ 0 & 0 & 9 \end{pmatrix}$$

(b) Let T be T_A where

$$A = \begin{pmatrix} 1 & -1 & 1/9 \\ 0 & 0 & 0 \\ 0 & 0 & 0 \end{pmatrix}$$

3.2.5. No. To see this, note that the image of the standard basis is not a
basis for $\mathbf{R}^3$.

3.2.7. An example is T_A where A is

$$\begin{pmatrix} 0 & 0 & 1 \\ 0 & 0 & 0 \\ 0 & 0 & 0 \end{pmatrix}$$

for $A \neq 0$, but $A^2 = 0$. The rank of any such example must be 1.

3.2.9. Let $W = im(T) \subseteq V$. Define $T' : W \to V$ by $T'(v) = T(v)$ for all $v \in W$. (T' is called the *restriction* of T to W.) It is easily checked that $im(T \circ T) = im(T')$. The condition that $rk(T \circ T) = rk(T)$ thus shows that in fact $im(T') = im(T) = W$. As $rk(T') = dim(W)$, the rank plus nullity theorem shows that $null(T') = 0$. Hence $ker(T') = \{\vec{0}\}$. However, by the definition of T', $ker(T') = ker(T) \cap im(T)$. This proves the result.

3.2.11. Given $T : V \to V$ with $ker(T) = im(T)$, we necessarily have that $rk(T) = null(T)$. But $dim(V) = rk(T) + null(T)$, so that $dim(V)$ must be even.

3.2.13. Since T is one-one, we know that $null(T) = 0$. As $rank(T) + null(T) = dim(V) = m$, we find that $rk(T) = m$. As $im(T) = ker(S)$, we find that $null(S) = rk(T) = m$. Since S is onto, $im(S) = U$, so as $dim(U) = n$, we have $rk(S) = n$. Finally, $dim(W) = rk(S) + null(S) = n + m$.

3.2.15. It is easy to see that $im(T^i)$ is spanned by $v_{i+1}, v_{i+2}, \ldots, v_n$ since these are the images of the basis $v_1, v_2, \ldots, v_n$ under T^i. But, $v_{i+1}, v_{i+2}, \ldots, v_n$ are linearly independent, hence a basis for $im(T^i)$. From this $rk(T^i) = n - i$ follows.

3.2.17. Note that the image $im(T_B)$ is the column space of B. It is readily checked that $rk(B) = 3$, so $im(T_B)$ is a three-dimensional subspace of $\mathbf{R}^4$. The vector space $V = \{(a, b, c, d) \in \mathbf{R}^4 \mid 3a + 2b + c + d = 0\}$ is also a three-dimensional subspace of $\mathbf{R}^4$, since it is the set of solutions to a single nontrivial homogeneous equation. Since each column of B solves the equation $3a + 2b + c + d = 0$, we find that $im(T_B) \subseteq V$ and consequently $im(T_B) = V$.

3.2.19. Let $v \in V$. We find that $T(v - T(v)) = T(v) - T(T(v)) = T(v) - T(v) = \vec{0}$. Consequently, $v - T(v) \in ker(T)$ for all $v \in V$. Since $T(v) \in im(T)$, we find that $v = (v - T(v)) + T(v) \in ker(T) + im(T)$, as required.

3.2.21. (a) Suppose that $v \in im(T)$. Then for some $w \in \mathbf{F}^n, v = T_A(w)$; that is, $v = Aw$. We find that $T_A(v) = Av = A(Aw) = A^2 w = \vec{0}$. Hence $v \in ker(T_A)$, as was needed.
(b) Part (a) shows that $rk(T_A) \leq null(T_A)$. Since $rk(T_A) + null(T_A) = n$, necessarily $rk(T_A) \leq n/2$ follows.

3.2.23. (a) $A = \begin{pmatrix} -1 & 2 & 0 \\ 0 & 0 & 1 \\ 1 & 0 & 1 \\ 0 & 1 & 0 \end{pmatrix}$

(b) $rk(L) = rk(A) = 3$, and $null(L) = null(A) = 0$.
(c) Since $null(L) = 0$, L is one-one, but L is not onto.
(d) The columns of A give a basis for $im(L)$.

3.2.25. Since $Aw = \vec{0}$ if and only if $w = \vec{0}$, we see that $ABv = \vec{0}$ if and only if $Bv = \vec{0}$. This shows that AB and B have the same nullspace. The rows of AB are linear combinations of the rows of B, so the row space of AB is a subspace of the row space of B. For the same reason, the row space of $A^{-1}AB = B$ is a subspace of the row space of AB. Thus they are equal.

3.2.27. $rk(T) = 2$, so $null(T) = 4 - 2 = 2$.

SECTION 3.3

3.3.1. (a) $\begin{pmatrix} 1 & 2 \\ 2 & 1 \end{pmatrix}$ (b) $\begin{pmatrix} 3 & -1 \\ 5 & -2 \end{pmatrix}$

(c) $\begin{pmatrix} 0 & -1 & 0 \\ 1 & 0 & 0 \\ -3 & 2 & 1 \end{pmatrix}$ (d) $\begin{pmatrix} 5 & -6 & 3 \\ 0 & 4 & -1 \\ 3 & -1 & 2 \end{pmatrix}$

3.3.3. If $[T]_{\mathcal{B},\mathcal{B}}$ is zero, then for $v \in V$, $(T(v))_{\mathcal{B}} = [T]_{\mathcal{B},\mathcal{B}}(v)_{\mathcal{B}} = 0$ for all $v \in V$. This shows that $T(v) = \vec{0}$ for all $v \in V$. Conversely, if $T(v) = \vec{0}$ for all $v \in V$, then for any ordered basis $\mathcal{B} = \{v_1, v_2, \ldots, v_n\}$, $(T(v_i))_{\mathcal{B}} = (\vec{0})_{\mathcal{B}} = \vec{0}$. But $(T(v_i))_{\mathcal{B}}$ is the ith column of $[T]_{\mathcal{B},\mathcal{B}}$, so $[T]_{\mathcal{B},\mathcal{B}}$ is zero.

3.3.5. $\begin{pmatrix} 0 & 0 & 0 & 0 \\ 1 & 0 & 0 & 0 \\ 0 & 1/2 & 0 & 0 \\ 0 & 0 & 1/3 & 0 \\ 0 & 0 & 0 & 1/4 \end{pmatrix}$

3.3.7. (a) $\begin{pmatrix} 0 & 0 & 1 \\ 1 & 0 & -1 \\ 1 & 1 & 0 \end{pmatrix}$ (b) $\begin{pmatrix} -1 & 0 & 1 \\ 1 & 1 & 0 \\ -1 & 0 & 0 \end{pmatrix}$

3.3.9. $\begin{pmatrix} 1 & 0 \\ 0 & -1 \end{pmatrix}$

3.3.11. (a) $\begin{pmatrix} -1 & -2 & -2 \\ 5 & 6 & 5 \end{pmatrix}$ (b) $\begin{pmatrix} 1 & 2 & -2 \\ 0 & 0 & 1 \end{pmatrix}$

3.3.13. The proof of the Rank Plus Nullity Theorem 3.2.6 shows that there is a basis $\mathcal{B} = \{v_1, \ldots, v_s, v_{s+1}, \ldots, v_n\}$ for V such that $\{v_1, \ldots, v_s\}$ is a basis for $ker(T)$ and $\{T(v_{s+1}), \ldots, T(v_n)\}$ is a basis for $im(T)$. In particular, $T(v_{s+1}), \ldots, T(v_n)$ are linearly independent in W. Applying the basis-extension theorem, we obtain a basis $\mathcal{C} = \{w_1, \ldots, w_s, T(v_{s+1}), \ldots, T(v_n)\}$ for W. Since $T(v_i) = \vec{0}$ for $i \le s$ and $T(v_i) = T(v_i) \in \mathcal{C}$ for $i > s$, it follows that $[T]_{\mathcal{B},\mathcal{C}}$ is a diagonal matrix.

SECTION 3.4

3.4.1. (a) $X^2 - 4X - 12$, $E_{-2} = span\left\{\begin{pmatrix} -1 \\ 2 \end{pmatrix}\right\}$, $E_6 = span\left\{\begin{pmatrix} 3 \\ 2 \end{pmatrix}\right\}$

(b) $X^2 - 1$, $E_1 = span\left\{\begin{pmatrix} 1 \\ 1 \end{pmatrix}\right\}$, $E_{-1} = span\left\{\begin{pmatrix} -1 \\ 1 \end{pmatrix}\right\}$

(c) $X^2 + 1$; there are no real eigenvalues; the complex eigenvalues are $\pm i$, the complex eigenspaces are

$$E_i = span\left\{\begin{pmatrix} 1 \\ i \end{pmatrix}\right\}, E_{-i} = span\left\{\begin{pmatrix} -1 \\ i \end{pmatrix}\right\}$$

(d) $X^3 + 2X^2 - 2X - 2$, the eigenspaces are

$$E_1 = span\left\{\begin{pmatrix} 1 \\ 1 \\ 1 \end{pmatrix}\right\}, E_{-1} = span\left\{\begin{pmatrix} 1 \\ -1 \\ 1 \end{pmatrix}\right\}, E_{-2} = span\left\{\begin{pmatrix} 1 \\ -2 \\ 4 \end{pmatrix}\right\}$$

3.4.3 $\begin{pmatrix} 0 & 2 \\ -1 & 3 \end{pmatrix}$

3.4.5. Suppose that N^k is zero and v is an eigenvector for N with eigenvalue $r \in \mathbf{F}$. Then $\vec{0} = N^k v = N^{k-1} r v = N^{k-2} r^2 v = \cdots = r^k v$. As $v \neq \vec{0}$, necessarily $r = 0$.

3.4.7. (a) If A is an $n \times n$ matrix with n distinct eigenvalues $c_1, c_2, \ldots, c_n$, then necessarily the characteristic polynomial

$$C_A(X) = (X - c_1)(X - c_2) \cdots (X - c_n)$$

Substituting 0 for X in $det(XI_n - A) = C_A(X)$, we find that $det(-A) = (-1)^n c_1 c_2 \cdots c_n$. As A is $n \times n$, we find that $det(A) = c_1 c_2 \cdots c_n$.
(b) Observe that the sum of the diagonal entries of a matrix A is precisely the negative of the coefficient of X^{n-1} in $C_A(X) = det(XI_n - A)$. To see this, consider the permutation expansion of $det(XI_n - A)$. All summands in this expansion are polynomials of degree less than $n - 1$ *except* the product of the diagonal entries.] Since $C_A(X) = (X - c_1)(X - c_2) \cdots (X - c_n)$, we see that $tr(A) = c_1 + c_2 + \cdots + c_n$.

3.4.9. $det(XI_n - A) = det((XI_n - A)^t) = det(XI_n^t - A^t) = det(XI_n - A^t)$, as required.

3.4.11. Observe that

$$\begin{pmatrix} 1 \\ 1 \\ \vdots \\ 1 \end{pmatrix}$$

is an eigenvector of A with eigenvalue r.

3.4.13. Substituting 0 for X in $C_A(X)$ gives

$$C_A(0) = det(0I_n - A) = det(-A) = (-1)^n det(A)$$

But if $C_A(X) = X^n + a_{n-1}X^{n-1} + \cdots + a_1 X + a_0$, then $C_A(0) = a_0$. Hence, $det(A) = (-1)^n a_0$. This gives the first assertion. If $T : V \to V$ is a linear operator, then T is invertible if and only if $[T]_\mathcal{B}$ is invertible for any basis $\mathcal{B}$ of V. But $[T]_\mathcal{B}$ is invertible if and only if $det([T]_\mathcal{B}) \neq 0$; the second statement follows.

SECTION 3.5

3.5.1. (a) For

$$A = \begin{pmatrix} 0 & 1 & 1 \\ 1 & 0 & 1 \\ 1 & 1 & 0 \end{pmatrix}$$

one finds $C_A(X) = X^3 - 3X - 2 = (X - 2)(X + 1)^2$. The eigenvalues of T_A are 2 and -1.

$$E_2 = span\left\{ \begin{pmatrix} 1 \\ 1 \\ 1 \end{pmatrix} \right\} \quad \text{and} \quad E_{-1} = span\left\{ \begin{pmatrix} -1 \\ 0 \\ 1 \end{pmatrix}, \begin{pmatrix} 0 \\ -1 \\ 1 \end{pmatrix} \right\}$$

It follows that T_A is diagonalizable and

$$P^{-1}AP = \begin{pmatrix} 2 & 0 & 0 \\ 0 & -1 & 0 \\ 0 & 0 & -1 \end{pmatrix}$$

where P is the matrix

$$\begin{pmatrix} 1 & -1 & 0 \\ 1 & 0 & -1 \\ 1 & 1 & 1 \end{pmatrix}$$

(b) For

$$A = \begin{pmatrix} 0 & 0 & 1 \\ 1 & 0 & -3 \\ 0 & 1 & 3 \end{pmatrix}$$

we find that $C_A(X) = X^3 - 3X^2 + 3X - 1 = (X - 1)^3$. Hence T_A has the single eigenvalue 1. The eigenspace associated with 1 is one-dimensional and has as a basis

$$\left\{ \begin{pmatrix} 1 \\ -2 \\ 1 \end{pmatrix} \right\}$$

It follows that T_A is not diagonalizable.
(c) The characteristic polynomial of

$$\begin{pmatrix} 0 & -1 \\ 1 & -1 \end{pmatrix}$$

is $X^2 + X + 1$, and thus this matrix has no real eigenvalues.

3.5.3. The hypothesis that $T_A \circ T_A = 0$ guarantees that A is not invertible. Since the nonzero matrix A is 2×2, we find that $rk(A) = 1$. Choose $v_1 \in \mathbf{R}^2$ with $Av_1 = v_2 \neq \vec{0}$. We have that $Av_2 = \vec{0}$; it follows that v_1, v_2 must be linearly independent. Setting $B = \{v_1, v_2\}$ to be an ordered basis of $\mathbf{R}^2$, we find that

$$[T_A]_B = \begin{pmatrix} 0 & 0 \\ 1 & 0 \end{pmatrix}$$

From this the existence of P follows (P is the transition matrix from the standard basis to B).

3.5.5. If a triangular matrix has distinct diagonal entries $a_1, a_2, \ldots, a_n$, then its characteristic polynomial is $(X - a_1)(X - a_2) \cdots (X - a_n)$. Thus it has n distinct eigenvalues, and hence is diagonalizable. Consider

$$A = \begin{pmatrix} 1 & 1 \\ 0 & 1 \end{pmatrix}$$

The only eigenvalue of A is 1. However, the eigenspace of A associated with 1 is one-dimensional. Hence A is not diagonalizable.

3.5.7. Since A has n distinct eigenvalues, A is diagonalizable. Choose P invertible so that PAP^{-1} is diagonal. Setting $C = PAP^{-1}$ and $D = PBP^{-1}$, we find that $PBP^{-1}PAP^{-1} = PBAP^{-1} = PABP^{-1} = PAP^{-1}PBP^{-1}$; that is, $CD = DC$. Now, C is a diagonal matrix with distinct entries on the diagonal. The matrix CD is obtained from D by multiplying the ith row of D by the ith diagonal entry of C. Similarly DC is the matrix obtained from D by multiplying the ith column of D by the ith diagonal entry of C. If c_i denotes the ith diagonal entry in C and $D = (d_{ij})$, this means that the ijth entry of CD is $c_i d_{ij}$ and the ijth entry of DC is $c_j d_{ij}$. For $i \neq j$ we have that $c_i \neq c_j$, so as $c_i d_{ij} = c_j d_{ij}$, we see that $d_{ij} = 0$ whenever $i \neq j$. Thus D is diagonal. As $D = PBP^{-1}$, we find that B is diagonalizable.

3.5.9. (a) No (b) No (c) Yes

3.5.11. Suppose that $w \in W$. Then w can be written as $w = v_1 + v_2 + \cdots + v_m$ where $v_1, v_2, \ldots, v_m$ are eigenvectors. If the eigenvalue of v_i is r_i, then $T(w) = r_1 v_1 + r_2 v_2 + \cdots + r_m v_m$, which lies in W since it is a sum of eigenvectors. Thus W is invariant.

3.5.13. Note that $\ker(T)$ is the eigenspace E_0 of T with associated eigenvalue 0. Since T has a nonzero eigenvalue k, there is a second eigenspace E_k of T. Since $\dim(E_0) = n - 1$ and $\dim(E_k) \geq 1$, it follows that the sum of the geometric multiplicities of the eigenspaces of T is n. Theorem 3.5.8 shows that T is diagonalizable.

3.5.15. Suppose that v is an eigenvector of a linear operator T with associated eigenvalue k. Then, as $T(v) = kv$, we find that $T^2(v) = T(T(v)) = T(kv) = k^2 v$ and similarly $T^n(v) = k^n v$ for all n. Thus v is an eigenvector for T^n. Now suppose that $T : V \to V$ is diagonalizable. Let $\{v_1, v_2, \ldots, v_n\}$ be an eigenbasis for T. What we have just seen shows that $\{v_1, v_2, \ldots, v_n\}$ is an eigenbasis for T^n. Hence T^n is diagonalizable.

SECTION 3.6

3.6.1. (a) Not regular   (b) Regular   (c) Regular

3.6.3. $\begin{pmatrix} 1/2 & 1/2 \\ 1/2 & 1/2 \end{pmatrix}$

3.6.5. $\begin{pmatrix} 1/4 \\ 1/2 \\ 1/4 \end{pmatrix}$

3.6.7. (a) The Gershgorin circles given by the rows are the circles of radius 0 with center 1, radius 3 with center 6, and radius 3 with center 4. This shows that the matrix (which must have a real eigenvalue since it is 3×3) has a positive real eigenvalue e with $1 \le e \le 9$. The Gershgorin circles coming from the columns are the circles of radius 3 around 1, radius 2 around 6, and radius 1 around 4. This additional information shows that $1 \le e \le 8$.
(b) This matrix has a real eigenvalue e with $-3 \le e \le 3$.
(c) Gerschgorin's theorem shows that all eigenvalues $z \in \mathbf{C}$ of the matrix satisfy $|z| \le 3$.

3.6.9. The matrix A of this problem is *not* stochastic, however A^t is. Thus Theorem 3.6.4 can be applied to compute $\lim_{n \to \infty} (A^t)^n$. The answer, which is the transpose of this limit is the matrix

$$\frac{1}{20} \begin{pmatrix} 5 & 7 & 8 \\ 5 & 7 & 8 \\ 5 & 7 & 8 \end{pmatrix}$$

SECTION 4.1

4.1.1. (a) For any $v \in V$ one has $0 = 0 < v, v > \; = \; <0v, v> \; = \; <\vec{0}, v>$.
(b) If $<T(v), w> \; = \; 0$ for all $w \in V$, then in particular, $<T(v), T(v)> \; = 0$. By the positive-definite property, $T(v) = \vec{0}$.

4.1.3. (a) Bilinearity is straightforward. Note also that $< (x, y), (x, y) > \; = \; 2x^2 + 3y^2$, which is zero if and only if both $x = 0$ and $y = 0$. This gives the positive-definite property.
(b) $< (1, 1), (-1, 1) > \; = \; -2 + 3 = 1$ and $\| (2, 7) \| = \sqrt{<(2, 7), (2, 7)>} = \sqrt{2 \cdot 2 \cdot 2 + 3 \cdot 7 \cdot 7} = \sqrt{155}$.
(c) $span\{(-3, 2)\}$.

4.1.5. (a) $<v, w>' \; = \; < Av, Aw > \; = \; < Aw, Av > \; = \; <w, v>'$; this gives symmetry.
(b) $<kv, w>' = <Akv, Aw> = <kAv, Aw> = k<Av, Aw> = k<v, w>'$.
(c) $<v_1 + v_2, u>' \; = \; <A(v_1 + v_2), Au> = <Av_1 + Av_2, Au> = <Av_1, Au> + <Av_2, Au> = <v_1, u>' + <v_2, u>'$.
(d) $<v, v>' \; = \; < Av, Av > \; \ge 0$. If $<v, v>' \; = 0$, then $< Av, Av> = 0$, so $Av = \vec{0}$. But A is invertible, so $v = \vec{0}$.

4.1.7. Consider the linear transformation $T : V \to \mathbf{R}^n$ defined by

$$T(w) = \begin{pmatrix} <v_1, w> \\ <v_2, w> \\ \vdots \\ <v_n, w> \end{pmatrix}$$

(Linearity follows from the linearity of $< , >$.) Suppose $T(w) = \vec{0}$. We claim that $w = \vec{0}$. Write $w = \sum a_i v_i$. Then

$$<w, w> = <\sum a_i v_i, w> = \sum a_i <v_i, w> = \sum a_i \cdot 0 = 0$$

so by the positive-definite property of $< , >$ we conclude $w = \vec{0}$. This shows that T is one-one, and as $dim(V) = n$, T is also onto. The existence of w with $<v_i, w>$ taking specified values follows.

4.1.9. It is clear from the definition that $tr(A + B) = tr(A) + tr(B)$ for $A, B \in M_{n \times n}(\mathbf{R})$ and $tr(kA) = k \, tr(A)$ for $k \in \mathbf{R}$ and $tr(A) = tr(A^t)$. From this one sees that $< A, B >= tr(AB^t) = tr((AB^t)^t) = tr(BA^t) =< B, A >$, $< kA, B >= tr(kAB^t) = k \cdot tr(AB^t) = k < A, B >$, and $< A_1 + A_2, B >= tr(A_1 B^t + A_2 B^t) = tr(A_1 B^t) + tr(A_2 B^t) =< A_1, B> + < A_2, B>$. Finally, if $A = (a_{ij})$, then $< A, A >= tr(AA^t) = \sum a_{ij}^2$. But $\sum a_{ij}^2 = 0$ if and only if $a_{ij} = 0$ for all i, j. This shows that $tr(AB^t)$ defines an inner product on $M_{n \times n}(\mathbf{R})$.

4.1.11. First note that

$$
\begin{aligned}
\| kT \| &= max\{\| kT(v) \| / \| v \|\} \\
&= max\{| k ||| T(v) \| / \| v \|\} \\
&= | k | max\{\| T(v) \| / \| v \|\} = | k | \cdot \| T \|
\end{aligned}
$$

Next,

$$
\begin{aligned}
\| T + U \| &= max\{\| (T + U)(v) \| / \| v \|\} \\
&= max\{\| T(v) + U(v) \| / \| v \|\} \\
&\le max\{(\| T(v) \| + \| U(v) \| / \| v \|\} \\
&\le max\{\| T(v) \| / \| v \|\} + max\{\| U(v) \| / \| v \|\} \\
&= \| T \| + \| U \|
\end{aligned}
$$

as required.

4.1.13. Yes. No. Positive definiteness requires $b^2 - 4ac < 0$ and $a > 0$.

4.1.15. Suppose that $T(v) = \vec{0}$. Then as $|T(v)| = |v|$, the positive-definite property of the norm gives that $v = \vec{0}$. Hence T is injective and consequently T is an isomorphism.

SECTION 4.2

4.2.1. $\{(0, 3/5, 4/5), (1, 0, 0), (0, 4/5, -3/5)\}$

4.2.3. $\{(1, 0), (1, -2)\}$ (many possible)

4.2.5. $\{1, X, 3X^2 - 1\}$ (many possible)

4.2.7. False, for example take $u = (1, 0, 0), v = (0, 1, 0), w = (1, 0, 1) \in \mathbf{R}^3$

4.2.9. (a) $\begin{pmatrix} 1 & 0 \\ 0 & 1 \end{pmatrix} \begin{pmatrix} 1 & 1 \\ 0 & 1 \end{pmatrix}$ (b) $\begin{pmatrix} 1/\sqrt{2} & -1/\sqrt{6} \\ 1/\sqrt{2} & 1/\sqrt{6} \\ 0 & 2/\sqrt{6} \end{pmatrix} \begin{pmatrix} \sqrt{2} & 1/\sqrt{2} \\ 0 & \sqrt{3}/\sqrt{2} \end{pmatrix}$

SECTION 4.3

4.3.1. (a) $(1/18) \begin{pmatrix} 2 & 4 & 4 \\ 4 & 17 & -1 \\ 4 & -1 & 17 \end{pmatrix}$ (b) $(1/14) \begin{pmatrix} 13 & 2 & -3 \\ 2 & 10 & 6 \\ -3 & 6 & 5 \end{pmatrix}$

(c) $(1/21) \begin{pmatrix} 1 & 2 & -4 \\ 2 & 4 & -8 \\ -4 & -8 & 16 \end{pmatrix}$

4.3.3. (a) $span\{(-4, 1, 1)\}$ (b) $span\{(1, -2, 3)\}$
(c) $span\{(-2, 1, 0), (4, 0, 1)\}$

4.3.5. (a) If $w \in W$, then for all $u \in W^\perp$ $<w, u> \ge 0$. This shows that $w \in (W^\perp)^\perp$. If $dim(V) = n$ and $dim(W) = m \le n$, then $dim(W^\perp) = n - m$, and consequently $dim((W^\perp)^\perp) = n - (n - m) = m$. As $W \subseteq (W^\perp)^\perp$, we have $W = (W^\perp)^\perp$.
(c) Suppose that $v \in (U + W)^\perp$. Since $U, W \subseteq (U + W)$, we have for all $u \in U, w \in W$ that $<v, u> = 0 = <v, w>$ so that $v \in U^\perp \cap W^\perp$. Conversely suppose that $v \in U^\perp \cap W^\perp$. For $u \in U$ and $w \in W$ we have $0 = <v, u> = <v, w>$ so that $0 = <v, u + w>$ for all $u = w \in U + W$. This shows $v \in (U + W)^\perp$, as required.

4.3.7. $O_{dd}^\perp$ is the even functions; that is, those $g(x) \in \mathcal{C}([-1, 1], \mathbf{R})$ for which $g(-x) = g(x)$ whenever $x \in [-1, 1]$. [Note that if $f(x)$ is odd and $g(x)$ is even, then $f(x)g(x)$ is odd so that $0 = \int_{-1}^1 f(x)g(x)\, dx = <f(x), g(x)>$.]

4.3.9. Define a linear transformation $T : Hom(V, W) \to Hom(V_1, W) \oplus Hom(V_2, W)$ by $T(f) = (f_1, f_2)$ where $f_1 : V_1 \to W$ is the restriction of f to V_1 and $f_2 : V_2 \to W$ is the restriction of f to V_2. T is one-one, because if both f_1 and f_2 are zero, then so is f since it is zero on a spanning set for V. T is onto, because given $(f_1, f_2) \in Hom(V_1, W) \oplus Hom(V_2, W)$, if one defines $f(v_1 + v_2) = f_1(v_1) + f_2(v_2)$ where $v_1 \in V_1$ and $v_2 \in v_2$, then $T(f) = (f_1, f_2)$.

4.3.11. (a) Suppose that $T : V \to V$ is an orthogonal projection. Then for any $v \in V$, $T(v)$ and $v - T(v)$ are orthogonal. Hence,

$$\| T(v) \|^2 + \| v - T(v) \|^2 = \| v \|^2$$

and consequently $\| T(v) \| \leq \| v \|$. [Note that $\| T(v) \| < \| v \|$ whenever $v \neq T(v)$.]

(b) We now assume that $T : V \to V$ satisfies $T^2 = T$ and $\| T(v) \| \leq \| v \|$ for all $v \in V$. (V is finite-dimensional.) Set $W = im(T)$. To prove that T is an orthogonal projection, we must show that $ker(T) = W^\perp$. Assume the contrary and set $U = ker(T)$. Then, $U \neq W^\perp$ and consequently $U^\perp \neq W$. We choose $w = T(v') \in W = im(T)$ with $w \notin U^\perp$. If $v = v' - Proj_U(v')$, then since $U = ker(T)$, $v \in U^\perp$ satisfies $T(v) = T(v') = w$.

Claim: $Proj_{U^\perp}(T(v)) = v$.

To see this, since $T^2 = T$, we have that $T(v) - v \in ker(T) = U$. Thus we have $Proj_{U^\perp}(T(v) - v) = \vec{0}$ and $Proj_{U^\perp}(T(v)) = Proj_{U^\perp}(v) = v$ as, required.

Now, $T(v) = w \notin U^\perp$, and consequently what was noted in (a) above shows that $\| T(v) \| > \| Proj_{U^\perp}(T(v)) \| = \| v \|$. This contradicts our hypothesis that $\| T(v) \| \leq \| v \|$ for all $v \in V$. This shows that $U = ker(T) = W^\perp$, as required.

SECTION 4.4

4.4.1. (a) The normal equation is

$$\begin{pmatrix} 3 & 7 \\ 7 & 21 \end{pmatrix} \begin{pmatrix} X \\ Y \end{pmatrix} = \begin{pmatrix} 10 \\ 31 \end{pmatrix}$$

The solution to the original least-squares problem is $X = -1/2, Y = 23/14$.

4.4.3. (a) The projection operator is

$$(1/30) \begin{pmatrix} 5 & 5 & 10 \\ 5 & 29 & -2 \\ 10 & -2 & 26 \end{pmatrix}$$

and the solution is $X = 58/15, Y = 4/5$.

4.4.5. (a) $\begin{pmatrix} 1/2 & 1/2 & 0 & 0 \\ 1/2 & 1/2 & 0 & 0 \\ 0 & 0 & 1/2 & 1/2 \\ 0 & 0 & 1/2 & 1/2 \end{pmatrix}$ (b) $(1/30) \begin{pmatrix} 4 & 8 & 6 & 2 \\ 8 & 16 & 12 & 4 \\ 6 & 12 & 9 & 3 \\ 2 & 4 & 3 & 1 \end{pmatrix}$

SECTION 4.5

4.5.1. (a), (d) and (e) only.

4.5.3. Suppose that $Au = ku$ for $k \in \mathbf{R}$ and $u \neq \vec{0}$. Then, as A is orthogonal, $< u, u > = < Au, Au > = < ku, ku > = k^2 < u, u >$. Thus $k^2 = 1$ so that $k = \pm 1$ follows.

4.5.5. If $(I + S)v = \vec{0}$, then $0 = v^t(I + S)v = v^t v + v^t Sv$ and $0 = (v^t(I + S)v)^t = v^t(I - S)v = v^t v - v^t Sv$. Adding these equations, we find that

$0 = 2v^t v = 2 <v, v>$. This shows that $v = \vec{0}$ and consequently, $(I + S)$ must be invertible. We now can compute $((I - S)(I + S)^{-1})^t (I - S)(I + S)^{-1} = (I + S^t)^{-1}(I - S^t)(I - S)(I + S)^{-1} = (I - S)^{-1}(I + S)(I - S)(I + S)^{-1} = (I - S)^{-1}(I - S^2)(I + S)^{-1} = (I - S)^{-1}(I - S)(I + S)(I + S)^{-1} = I$. This shows that $(I - S)(I + S)^{-1}$ is orthogonal.

4.5.7. If A is self-adjoint, then $C_A(X) = det(XI_n - A) = det(XI_n - \overline{A}^t) = det(\overline{(XI_n - A)}^t) = det(\overline{XI_n - A}) = \overline{C_A(X)}$. Consequently, all the coefficients of $C_A(X)$ must be real.

4.5.9. (a) The eigenvalues are 5, -1 and the spectral resolution is

$$\begin{pmatrix} 2 & 3 \\ 3 & 2 \end{pmatrix} = 5(1/2)\begin{pmatrix} 1 & 1 \\ 1 & 1 \end{pmatrix} + (-1)(1/2)\begin{pmatrix} 1 & -1 \\ -1 & 1 \end{pmatrix}$$

(b) The eigenvalues are 1, 2, 3 and the spectral resolution is

$$\begin{pmatrix} 2 & 0 & -1 \\ 0 & 2 & 0 \\ -1 & 0 & 2 \end{pmatrix} = 1(1/2)\begin{pmatrix} 1 & 0 & 1 \\ 0 & 0 & 0 \\ 1 & 0 & 1 \end{pmatrix} +$$

$$2\begin{pmatrix} 0 & 0 & 0 \\ 0 & 1 & 0 \\ 0 & 0 & 0 \end{pmatrix} + 3(1/2)\begin{pmatrix} 1 & 0 & -1 \\ 0 & 0 & 0 \\ -1 & 0 & 1 \end{pmatrix}$$

4.5.11. The key point is that hermitian matrices are diagonalizable. The diagonal entries of a diagonalization of such matrices are completely determined by the (factors of) the characteristic polynomial. The result follows from this.

4.5.13. If $O^{-1}AO$ is diagonal, then $O^{-1}AO = (O^{-1}AO)^t = O^t A^t (O^{-1})^t = O^{-1}A^t O$. Multiplication on the left by O and on the right by O^{-1} gives that $A = A^t$, as required.

4.5.15. Suppose that $T : V \rightarrow V$ is the orthogonal projection onto $W \subseteq V$. Then $I - T : V \rightarrow V$ is the orthogonal projection onto $W^\perp$. Let $u, v \in V$. Since $T(u) \in W$ and $(I - T)(v) \in W^\perp$, we have that $<T(u), (I - T)(v)> = 0$. Consequently,

$$\begin{aligned} <T(u), v> &= <T(u), T(v) + (I - T)(v)> \\ &= <T(u), T(v)> + <T(u), (I - T)(v)> \\ &= <T(u), T(v)> \\ &= <T(u), T(v)> + <(I - T)(u), T(v)> \\ &= <T(u) + (I - T)(u), T(v)> = <u, T(v)> \end{aligned}$$

This shows that T is self-adjoint, as required.

SECTION 4.6

4.6.1. (a) $\begin{pmatrix} 1 & 1 \\ 0 & 0 \end{pmatrix} = \begin{pmatrix} 1 & 0 \\ 0 & 1 \end{pmatrix} \begin{pmatrix} \sqrt{2} & 0 \\ 0 & 0 \end{pmatrix} \begin{pmatrix} 1/\sqrt{2} & 1/\sqrt{2} \\ 1/\sqrt{2} & -1/\sqrt{2} \end{pmatrix}$

is singular-value decomposition, and

$$\begin{pmatrix} 1/2 & 0 \\ 1/2 & 0 \end{pmatrix}$$

is the pseudoinverse.

4.6.3. If A is pseudodiagonal, the result is clear. Suppose that $A^t = U_2 D^t U_1^t$ and $A^\dagger = U_2 D^\dagger U_1^t$. Then the column spaces of $D^t U_2^t$, D^t, $D^\dagger$, $D^\dagger U_2^t$ are each the same. (This is because multiplication on the right by invertible matrices does not change the column space.) For any matrix M the column space of $U_2 M$ is $T_{U_2}(col(M))$. Consequently the column spaces of A^t and $A^\dagger$ are the same.

To see that $T_{A^\dagger A}$ is the orthogonal projection onto $col(A^t)$, first check that $A^\dagger A A^t = A^t$. Hence, whenever C is a column of A^t, $A^\dagger A C = C$, and therefore $T_{A^\dagger A}$ is the identity on $col(A^t)$. If $v \in col(A^t)^\perp = row(A)^\perp$, then $Av = \vec{0}$. Thus $T_{A^\dagger A}(v) = \vec{0}$; this shows that $T_{A^\dagger A}$ is the orthogonal projection onto $col(A^t)$.

4.6.5. (a) The pseudoinverse is

$$\begin{pmatrix} 1/6 & 1/6 & 1/6 \\ 1/6 & 1/6 & 1/6 \end{pmatrix}$$

and the optimal solution is $X = 3/4$, $Y = 3/4$.

4.6.7. (a) $\begin{pmatrix} -1 & -9 \\ 7 & 13 \end{pmatrix} = \begin{pmatrix} 3/5 & -4/5 \\ 4/5 & 3/5 \end{pmatrix} \begin{pmatrix} 5 & 5 \\ 5 & 15 \end{pmatrix}$

(b) $\begin{pmatrix} 1 & 0 & 2 \\ 2 & 3 & 0 \\ 4 & 2 & 1 \end{pmatrix} = \begin{pmatrix} 0 & 0 & 1 \\ 0 & 1 & 0 \\ 1 & 0 & 0 \end{pmatrix} \begin{pmatrix} 4 & 2 & 1 \\ 2 & 3 & 0 \\ 1 & 0 & 2 \end{pmatrix}$

SECTION 5.1

5.1.1. (a) $\begin{pmatrix} 0 & 1 \\ -2 & 1 \end{pmatrix}$ (b) $\begin{pmatrix} 1 & 0 & 1 \\ 0 & -1 & 0 \\ 0 & -1 & 0 \end{pmatrix}$ (c) $\begin{pmatrix} 1 & -1 & 0 \\ -1 & 1 & 2 \\ 0 & 2 & 1 \end{pmatrix}$

5.1.3. (a) $\begin{pmatrix} 1 & 0 & 0 \\ 0 & 1 & 1 \\ 1 & 0 & 1 \end{pmatrix} \begin{pmatrix} 0 & 1 & 0 \\ 1 & 0 & 2 \\ 0 & 2 & 1 \end{pmatrix} \begin{pmatrix} 1 & 0 & 1 \\ 0 & 1 & 0 \\ 0 & 1 & 1 \end{pmatrix} = \begin{pmatrix} 0 & 1 & 0 \\ 1 & 5 & 4 \\ 0 & 4 & 1 \end{pmatrix}$

5.1.5. We assume that $(\ ,\ )$ is represented by a matrix A in some basis for V. If $rk(A) < n = dim(V)$, then there is some nonzero $v \in \mathbf{F}^n$ with $Av = \vec{0}$. In particular, for every $w \in \mathbf{F}^n$, $w^t A v = 0$. Thus if condition (iii) fails, so must condition (ii). Conversely, if there is some $\vec{0} \neq v \in \mathbf{F}^n$ with $w^t Av = 0$ for all $w \in \mathbf{F}^n$, then necessarily $Av = \vec{0}$. This shows that $rk(A) < n$, and this establishes the equivalence of (ii) and (iii). The equivalence of (i) and (iii) follows analogously.

5.1.7. These bilinear forms correspond to skew-symmetric 3×3 matrices (those satisfying $A^t = -A$). (The collection of such bilinear forms is a three-dimensional $\mathbf{F}$ vector space.)

5.1.9. All such bilinear forms have the form $(v, u) = L_1(v)L_2(u)$ where $L_1, L_2 : V \to \mathbf{F}$ are linear transformations.

SECTION 5.2

5.2.1. There are many possible solutions. The different diagonalizations can differ by squares on the diagonal and the order in which the diagonal entries are listed. Possible solutions:

(a) $\begin{pmatrix} 1 & 0 \\ -3 & 2 \end{pmatrix} \begin{pmatrix} 2 & 3 \\ 3 & 1 \end{pmatrix} \begin{pmatrix} 1 & -3 \\ 0 & 2 \end{pmatrix} = \begin{pmatrix} 2 & 0 \\ 0 & -14 \end{pmatrix}$

(b) $\begin{pmatrix} 1 & 1 & 0 \\ 1 & -1 & 0 \\ 1 & 0 & -1 \end{pmatrix} \begin{pmatrix} 0 & 2 & 0 \\ 2 & 0 & 2 \\ 0 & 2 & 0 \end{pmatrix} \begin{pmatrix} 1 & 1 & 1 \\ 1 & -1 & 0 \\ 0 & 0 & -1 \end{pmatrix} = \begin{pmatrix} 4 & 0 & 0 \\ 0 & -4 & 0 \\ 0 & 0 & 0 \end{pmatrix}$

5.2.3. If $A = P^t S P$ where S is symmetric, then $A^t = (P^t S P)^t = P^t S^t P = P^t S P = A$, as required.

5.2.5. If the 2×2 matrix A has two orthogonal eigenvectors, let $\mathcal{B}$ be a basis of orthonormal eigenvectors for T_A. If P is the transition matrix from the standard basis to $\mathcal{B}$, then P is orthogonal; that is, $P^{-1} = P^t$. But the matrix $P^{-1} A P = P^t A P$ is the diagonalization of A, and consequently $P^t A^t P = (P^t A P)^t = P^t A P$. Multiplication on the left by P and on the right by P^t gives $A^t = A$, as required.

SECTION 5.3

5.3.1. There are many possible solutions.
(a) $R^2 + (3/4)S^2 - (13/12)T^2$ since $X^2 - XY + Y^2 + XZ - Z^2 = (X - (1/2)Y + (1/2)Z)^2 + (3/4)(Y + (1/3)Z)^2 - (13/12)Z^2$
(c) $R^2 - S^2 + T^2 - U^2$

5.3.3. (a) $\begin{pmatrix} 1 & -1/2 & 1/2 \\ -1/2 & 1 & 0 \\ 1/2 & 0 & -1 \end{pmatrix}$ (c) $\begin{pmatrix} 0 & 1/2 & 0 & 0 \\ 1/2 & 0 & 0 & 0 \\ 0 & 0 & 0 & 1/2 \\ 0 & 0 & 1/2 & 0 \end{pmatrix}$

5.3.5. (a) $2X^2 + 8XY + Y^2$
(b) $X^2 + 4XY + 2XZ + 6YZ + 2Z^2$
(c) $X^2 + 6XY + 8XZ + 2YZ + 2Z^2$

5.3.7. (a) $((a, b), (c, d)) = ac$.
(b) $((a, b), (c, d)) = (1/2)(ad + bc)$.
(c) $((a, b), (c, d)) = ac - (1/2)(ad + bc)$.
(d) $((a, b), (c, d)) = ac + 3bd$.

SECTION 5.4

5.4.1. (a) 0 (b) 3 (c) 1 (d) 0

5.4.3. If $ac - b^2 \leq 0$, then the quadratic equation $aX^2 + 2bX + c = 0$ has a real solution r. Then $q((r, 1)) = 0$ so that q fails to be positive definite. Conversely, suppose $q((r, s)) = 0$ where $(r, s) \neq (0, 0)$. Then in case $s \neq 0$, we find that r/s is a solution to $aX^2 + 2bX + c = 0$. In case $s = 0$, necessarily $a = 0$. In either case $ac - b^2 \leq 0$. If q is negative definite, necessarily $a < 0$. This gives the result.

5.4.5. The key point is that the bilinear form associated with S is positive definite. If S is orthogonally diagonalized, then its diagonal entries are its eigenvalues, and thus they are all positive. Consequently, the signature of S is n, and the positive definite property holds.

5.4.7.

$$C = \begin{pmatrix} 1 & -2 & 1 & 2 \\ 0 & 1 & -2 & -3 \\ 0 & 0 & 1 & 0 \\ 0 & 0 & 0 & 1 \end{pmatrix}$$

will work. The rank is 2 and the signature is 0.

SECTION 6.1

6.1.1. $\begin{pmatrix} 0 & 0 & 0 \\ 0 & 0 & 0 \\ 0 & 0 & 0 \end{pmatrix}$, $\begin{pmatrix} 0 & 0 & 1 \\ 0 & 0 & 0 \\ 0 & 0 & 0 \end{pmatrix}$, $\begin{pmatrix} 0 & 1 & 0 \\ 0 & 0 & 1 \\ 0 & 0 & 0 \end{pmatrix}$

6.1.3. (a) $C_A(X) = M_A(X) = X^3 - cX^2 - bX - a$.
(b) $C_A(X) = (X - 1)^2(X + 1) = X^3 - X^2 - X + 1$, $M_A(X) = X^2 - 1$.
(c) $C_A(X) = X^4 - 4X^2$, $M_A(X) = X^3 - 4X$.

6.1.5. Check that for any polynomial $f(X)$ and any such A, $f(A^t) = (f(A))^t$. Consequently, $f(A) = 0$ if and only if $f(A^t) = 0$. From this the result follows.

6.1.7. X^{n+1}.

6.1.9. $C_{P_W}(X) = X^{n-m}(X - 1)^m$ and $M_{P_W}(X) = X(X - 1)$.

6.1.11. Suppose that c is an eigenvalue of T and $T(v) = cv$ for some $v \neq \vec{0}$. By induction on n, we find $T^n(v) = T^{n-1}(T(v)) = T^{n-1}(cv) = cT^{n-1}(v) = c \cdot c^{n-1}v = c^n v$. Suppose that $f(X) = a_n X^n + a_{n-1} X^{n-1} + \cdots + a_1 X + a_0$. Then $f(T)(v) = a_n T^n(v) + a_{n-1} T^{n-1}(v) + \cdots + a_1 T(v) + a_0 v = a_n c^n v + a_{n-1} c^{n-1} v + \cdots + a_1 cv + a_0 v$. Thus, $f(T)(v) = f(c)v$, as required.

SECTION 6.2

6.2.1.

$$\begin{pmatrix} 0 & 0 & 1 & 0 & 0 \\ 0 & 0 & 0 & 1 & 0 \\ 0 & 0 & 0 & 0 & 1 \\ 0 & 0 & 0 & 0 & 0 \\ 0 & 0 & 0 & 0 & 0 \end{pmatrix}$$

is an example (there are many). No, since the eigenspace E_0 must have dimension 2.

6.2.3.   Let $\mathcal{B}_1 = \{(1,0,0),(0,3,1),(0,0,12)\}$ and $\mathcal{B}_2 = \{(0,0,1),(0,1,0),(1,0,0)\}$. Then

$$[A]_{\mathcal{B}_1} = [B]_{\mathcal{B}_2} = \begin{pmatrix} 0 & 0 & 0 \\ 1 & 0 & 0 \\ 0 & 1 & 0 \end{pmatrix}$$

6.2.5. $v_1 = (0,-3/4,1/2), v_2 = (0,1/2,0), v_3 = (1,0,0)$.

6.2.7.   There are many solutions, one is

$$\left\{ \begin{pmatrix} 1 \\ 0 \\ 0 \end{pmatrix}, \begin{pmatrix} 0 \\ 2 \\ 2 \end{pmatrix}, \begin{pmatrix} 4 \\ 6 \\ 2 \end{pmatrix} \right\}$$

6.2.9. T must have $n = dim(V)$ distinct eigenvalues.

6.2.11. $\begin{pmatrix} 1 & -1 \\ 0 & 1 \end{pmatrix}$ is one possible C.

SECTION 6.3

6.3.1.   Let $\{w_1, w_2, \ldots, w_m\}$ be a basis for W_1. Applying the basis-extension theorem, there exist $v_{m+1}, v_{m+2}, \ldots, v_n \in V$ so that $\{w_1, \ldots, w_m, v_{m+1}, \ldots, v_n\}$ is a basis for V. Set $W_2 = span\{v_{m+1}, v_{m+2}, \ldots, v_n\}$. Then $V = W_1 \oplus W_2$ as required. Note that W_2 is *not* unique.

6.3.3. W is T_A-invariant since

$$T_A \begin{pmatrix} 1 \\ 0 \end{pmatrix} = \begin{pmatrix} 1 \\ 0 \end{pmatrix}$$

Any one-dimensional T_A-invariant subspace of $\mathbf{F}^2$ must be an eigenspace. As 1 is the only eigenvalue of T_A, W is the only one-dimensional T_A-invariant subspace of $\mathbf{F}^2$.

6.3.5.   If E_c is the eigenspace of $T : V \to V$ with associated eigenvalue c, then $T(v) = cv \in E_c$ for all $v \in E_c$. Thus E_c is T-invariant. The restriction of T to E_c is the operator on E_c which mulitplies every vector by c; that is, $T|_{E_c} = cI|_{E_c}$.

6.3.7.

$$\begin{pmatrix} 0 & 1 \\ -1 & 0 \end{pmatrix}$$

gives an example. No, no, since such an operator must have a eigenvalue.

6.3.9. Let $v \in \mathbf{R}^n$. Then $v + T(v) \in E_1$ since $T(v + T(v)) = T(v) + T^2(v) = T(v) + v$. Similarly, $v - T(v) \in E_{-1}$. Since $2v = (v + T(v)) + (v - T(v))$, we find that $\mathbf{R}^n = E_1 + E_{-1}$. As $E_1 \cap E_{-1} = \{\vec{0}\}$, the conclusion that $\mathbf{R}^n = E_1 \oplus E_{-1}$ follows.

6.3.11. Suppose $W_i = span\{w_i\}$. Since W_i is T-invariant, necessarily w_i is an eigenvector for T. Thus $\{w_1, w_2, \ldots, w_n\}$ is a basis for V of eigenvectors for T; this shows that T is diagonalizable.

SECTION 6.4

6.4.1. (a) The characteristic polynmomial is $(X - 1)(X^2 + 1)$, so this matrix does not have a real Jordan form.

(b) $\begin{pmatrix} 3 & 0 & 0 \\ 1 & 3 & 0 \\ 0 & 0 & 3 \end{pmatrix}$

(c) $\begin{pmatrix} 1 & 0 & 0 \\ 0 & 1 & 0 \\ 0 & 0 & 0 \end{pmatrix}$

6.4.3. There is a basis $\mathcal{B}$ for which $[T]_{\mathcal{B}}$ is a Jordan-form matrix J. Let D be the diagonal entries of J and set $N = J - D$. These matrices work.

6.4.5. (a) Yes (b) No (c) Yes

6.4.7. (a) $P = \begin{pmatrix} 1 & 1 \\ 2 & 1 \end{pmatrix}$ and $P^{-1}MP = \begin{pmatrix} 3 & 0 \\ 1 & 3 \end{pmatrix}$

(b) $P = \begin{pmatrix} 0 & 0 & 1 & 1 \\ 0 & 0 & 1 & -1 \\ 0 & 1 & 0 & 0 \\ 1/5 & 0 & 0 & 0 \end{pmatrix}$ and $P^{-1}MP = \begin{pmatrix} 2 & 0 & 0 & 0 \\ 1 & 2 & 0 & 0 \\ 0 & 0 & 3 & 0 \\ 0 & 0 & 0 & 1 \end{pmatrix}$

(c) $P = \begin{pmatrix} 1 & -2 & 4 \\ 0 & 1 & -4 \\ 0 & 0 & 1 \end{pmatrix}$ and $P^{-1}MP = \begin{pmatrix} 2 & 0 & 0 \\ 1 & 2 & 0 \\ 0 & 1 & 2 \end{pmatrix}$

6.4.9. Note that if A has Jordan form J, then A is similar to J, and consequently, $f(A)$ is similar to $f(J)$. So, the eigenvalues of A and J are the same, and the eigenvalues of $f(A)$ and $f(J)$ are the same. Thus, the eigenvalues of A are precisely the diagonal entries of J and the eigenvalues of $f(A)$ are precisely the diagonal entries of the (necessarily triangular matrix) $f(J)$. If c is a diagonal entry of J, evidently $f(c)$ is the corresponding diagonal entry of $f(J)$. From this the result follows.

SECTION 6.5

6.5.1.   (a) The characteristic polynomial is $(X - 4)^2(X - 1)$. The primary decomposition is $\mathbf{R}^3 = W_1 \oplus W_2$ where

$$W_1 = span\left\{\begin{pmatrix} 9 \\ 0 \\ 7 \end{pmatrix}, \begin{pmatrix} 0 \\ 3 \\ 1 \end{pmatrix}\right\}$$

corresponds to the primary factor $(X - 4)^2$ and

$$W_2 = span\left\{\begin{pmatrix} 0 \\ 0 \\ 1 \end{pmatrix}\right\}$$

corresponds to the factor $(X - 1)$.
(b) The characteristic polynomial is $(X^2 + 1)(X - 1)^2$. The primary decomposition is $\mathbf{R}^4 = W_1 \oplus W_2$ where

$$W_1 = span\left\{\begin{pmatrix} 1 \\ -3 \\ 0 \\ 1 \end{pmatrix}, \begin{pmatrix} 1 \\ -1 \\ 1 \\ 0 \end{pmatrix}\right\}$$

corresponds to the factor $(X^2 + 1)$ and

$$W_2 = span\left\{\begin{pmatrix} 0 \\ 0 \\ 1 \\ 0 \end{pmatrix}, \begin{pmatrix} 0 \\ 0 \\ 0 \\ 1 \end{pmatrix}\right\}$$

corresponds to the factor $(X - 1)^2$.
(c) The characteristic polynomial is $(X - 3)^3$. Consequently the primary decomposition is $\mathbf{R}^3 = \mathbf{R}^3$.

6.5.3.   (a) If $M_T(X)$ factors into powers of two or more distinct primes, then a summand in the primary decomposition will be a nontrivial T-invariant subspace.
(b) Suppose that $M_T(X) = p(X)^r$ where $p(X)$ is a prime polynomial and $r > 1$. There must be some $v \in V$ with $p(T)^{r-1}(v) \neq \vec{0}$ as otherwise the minimal polynomial of T would be a factor of $p(X)^{r-1}$. Consider the nonzero subspace $W = p(T)^{r-1}(V) \subseteq V$. W is T-invariant as $T(p(T)^{r-1}(v)) = p(T)^{r-1}(T(v)) \in W$ for all elements $p(T)^{r-1}(v)$ of W. Clearly $W \neq V$ since the minimal polynomial of T on W is $p(X)$.

6.5.5.   (a) Given in Prob. 6.5.3. (b).
(b) Suppose that V has primary decomposition $V = W_1 \oplus W_2 \oplus \cdots \oplus W_m$. If $deg(g(X)) < deg(M_T(X))$, then there is some prime factor $p_i(X)$ of $M_T(X)$

with $p_i(X)^{r_i}$ dividing $M_T(X)$ but $p_i(X)^{r_i}$ not dividing $g(X)$. Choose $v_i \in W_i$ [the primary summand of V corresponding to $p_i(X)$] with $p_i(T)^r(v_i) \neq \vec{0}$. Then, $g(T)(v_i) \neq \vec{0}$ since $M_{T,v_i}(X) = p_i(X)^{r_i}$. Setting $v = v_1 + v_2 + \cdots + v_m$, we find that $g(T)(v) = g(T)(v_1) + g(T)(v_2) + \cdots g(T)(v_m) \neq \vec{0}$, using the fact that $g(T)(v_i) \in W_i$ together with the directness of the sum $V = W_1 \oplus W_2 \oplus \cdots \oplus W_m$. Since this vector v works for all $g(X)$, we are done.

6.5.7. X^n if the matrix is $n \times n$.

6.5.9. Since $T^2 - 3T = 0$, the minimal polynomial must be a factor of $X^2 - 3X = X(X - 3)$. Thus $M_T(X)$ is one of X, $X - 3$, or $X(X - 3)$. In particular the minimal polynomial factors into distinct linear factors. By Corollary 6.5.8 T, is diagonalizable.

SECTION 6.6

6.6.1. (a) $span \left\{ \begin{pmatrix} 1 \\ 3 \\ 2 \end{pmatrix}, \begin{pmatrix} 13 \\ 6 \\ 13 \end{pmatrix} \right\}$

(b) $\mathbf{F}^3$

(c) $span \left\{ \begin{pmatrix} 1 \\ 3 \\ 2 \end{pmatrix} \right\}$

6.6.3. (a) $X - 6$ (b) $X^2 - 3X - 1$ (c) $X^3 - 2X^2 - 4X - 1$

6.6.5. If every vector in V is a T-cyclic vector, then V contains no nontrivial T-invariant subspaces. Now apply Prob. 6.5.3 to see that $M_T(X)$ is a power of a prime polynomial.

SECTION 6.7

6.7.1. (a) $\begin{pmatrix} 0 & 0 & 3 \\ 1 & 0 & 9 \\ 0 & 1 & 3 \end{pmatrix}$ (b) $\begin{pmatrix} 0 & 0 & 45 \\ 1 & 0 & -39 \\ 0 & 1 & 11 \end{pmatrix}$ (c) $\begin{pmatrix} 0 & -15 & 0 \\ 1 & 8 & 0 \\ 0 & 0 & 3 \end{pmatrix}$

6.7.3. There are three possible real rational forms, the invariant factors are: $(X^2 + 1)^3$; $(X^2 + 1)^2$, $(X^2 + 1)$; and $(X^2 + 1), (X^2 + 1), (X^2 + 1)$. There are nine possible Jordan forms over an algebraically closed field. (These nine come from the combinations of the three Jordan forms for each primary summand corresponding to the two eigenvalues i and $-i$.)

6.7.5. If A and B are similar over $\mathbf{F}$, then they clearly are similar over L. Conversely, if A and B are similar over L, then their rational forms over L must be the same. However, the rational form is uniquely determined, and thus their rational forms over $\mathbf{F}$ must be the same (as L matrices) as their L rational forms. Consequently, A and B have the same rational form over $\mathbf{F}$ and thus are similar over $\mathbf{F}$.

6.7.7. $M_A(X) = X^2(X - 1)^2$, $C_A(X) = X^4(X - 1)^4$; the eigenvalues are 0 and 1, each with algebraic multiplicity 4, the geometric multiplicity of 0 is 2 and of 1 is 3. The invariant factors are: $X^2(X-1)^2$, $X^2(X-1)$, $X-1$. It follows that the rational form of A is

$$\begin{pmatrix} 0 & 0 & 0 & 0 \\ 1 & 0 & 0 & 0 \\ 0 & 1 & 0 & -1 \\ 0 & 0 & 1 & 2 \end{pmatrix} \oplus \begin{pmatrix} 0 & 0 & 0 \\ 1 & 0 & 0 \\ 0 & 1 & 1 \end{pmatrix} \oplus (1)$$

SECTION 6.8

6.8.1. (a) $\begin{pmatrix} 1 & 0 & 0 \\ 0 & X-1 & 0 \\ 0 & 0 & (X-1)^2 \end{pmatrix}$ (b) $\begin{pmatrix} 1 & 0 \\ 0 & X^2-4X-5 \end{pmatrix}$

(c) $\begin{pmatrix} 1 & 0 & 0 \\ 0 & 1 & 0 \\ 0 & 0 & X^3-5X^2-6X \end{pmatrix}$

6.8.3. Precisely when the original matrix is rI_n for some $r \in \mathbf{F}$.

SECTION 6.9

6.9.1 Let $V = Z_1 \oplus Z_2 \oplus \ldots \oplus Z_s$ be a cyclic decomposition for T. Then $M_T(X) = M_{T|Z_1}(X)$. But Z_1 is cyclic, so it has a cyclic vector v. Necessarily, $M_{T,v}(X) = M_{T|Z_1}(X) = M_T(X)$, as required.

6.9.3. Let $T = T_A$ where

$$A = \begin{pmatrix} 0 & 1 \\ -1 & 0 \end{pmatrix} \oplus \begin{pmatrix} 0 & 2 \\ -1 & 0 \end{pmatrix}$$

Then, $C_T(X) = (X^2+1)(X^2+2)$ and necessarily $C_T(X) = M_T(X)$. Consequently, $\mathbf{F}^4$ is T-cyclic. Such a decomposition does not contradict the cyclic-decomposition theorem since the minimal polynomial of T restricted to V_1 is not the minimal polynomial of T on all of $\mathbf{F}^4$.

6.9.5. Observe that as A is similar to its rational form R, A^t is similar to R^t. The matrices R and R^t have the same invariant factors, since R^t is obtained from R by transposing each companion matrix block of R. Thus R and R^t are similar, and consequently, A and A^t are similar.

SECTION 7.1

7.1.1. (a) $W(1, x, e^x) = e^x \neq 0$.
(b) $W(1, x, x^2, \sin x, x\sin x) = 6 + 2\cos^2 x \neq 0$.
(c) $W(\sin x, \cos x, e^x) = -e^x \neq 0$.

7.1.3. (a) Yes (b) No (c) No Lemma 7.1.2 does not apply since $\{-1, 0, 1\} \subset \mathbf{R}$ does not contain an open interval.

SECTION 7.2

7.2.1. (a) $\begin{pmatrix} e^{3t} & 0 \\ 0 & e^{4t} \end{pmatrix}$ (b) $\begin{pmatrix} 1 & t \\ 0 & 1 \end{pmatrix}$ (c) $\begin{pmatrix} e^t & 2te^t \\ 0 & e^t \end{pmatrix}$

7.2.3. (a)

$$e^{At} = \frac{1}{2} \begin{pmatrix} e^{4t} + e^{-2t} & e^{4t} - e^{-2t} \\ e^{4t} - e^{-2t} & e^{4t} + e^{-2t} \end{pmatrix}$$

and the solutions to $F'(t) = AF(t)$ are given by

$$F(t) = \begin{pmatrix} (A+B)e^{4t} + (A-B)e^{2t} \\ (A+B)e^{4t} - (A-B)e^{2t} \end{pmatrix} \quad \text{where} \quad A, B \in \mathbf{R}$$

(b)

$$e^{At} = e^t \begin{pmatrix} 1-t & -t \\ t & 1+t \end{pmatrix}$$

and the solutions to $F'(t) = AF(t)$ are given by

$$F(t) = e^t \begin{pmatrix} A(1-t) - Bt \\ At + B(1+t) \end{pmatrix} \quad \text{where} \quad A, B \in \mathbf{R}$$

7.2.5. For example, let

$$A = \begin{pmatrix} 0 & 1 \\ 0 & 0 \end{pmatrix} \quad \text{and} \quad B = \begin{pmatrix} 0 & 0 \\ 1 & 0 \end{pmatrix}$$

Then

$$e^A = \begin{pmatrix} 1 & 1 \\ 0 & 1 \end{pmatrix}, \quad e^B = \begin{pmatrix} 1 & 0 \\ 1 & 1 \end{pmatrix} \quad \text{and} \quad e^{A+B} = \frac{1}{2} \begin{pmatrix} e + e^{-1} & e - e^{-1} \\ e - e^{-1} & e + e^{-1} \end{pmatrix}$$

The reason this occurs for matrices and not for real numbers is that matrix multiplication does not commute.

7.2.7. (a) $y = c_1 e^{-2x} + c_2 x e^{-2x}$.
(b) $y = c_1 e^{2x} + c_2 e^{-3x}$.
(c) $y = c_1 e^{-x} + c_2 x e^{-x} + c_3 e^x$.

SECTION 7.3

7.3.1. $H_4 = 16X^4 - 48X^2 + 12$, $H_5 = 32X^5 - 160X^3 + 120X$.
7.3.3 (a) $(3/5)X$ (c) $1 + X + X^2$

SECTION 7.4

7.4.1. (a) $\pi - 2\sin x - \sin(2x) - (2/3)\sin(3x) - \cdots - (2/n)\sin(nx) \cdots$
(b) $1/2 - (1/2)\cos(2x)$
7.4.3. For example, consider the sequence $1, 1 + X, 1 + X + (1/2)X^2, \ldots,$
$1 + X + \cdots + (1/n!)X^n = f_n, \ldots$. Note that $|\int_0^1 (f_n - f_m)^2 \, dX| \leq (1/n!) + (1/(n+1)!) + \cdots + (1/m!) < (1/n!)e$, so we find that the sequence is a Cauchy sequence. The sequence converges to the function e^X, which is not a polynomial function.

SECTION 8.1

8.1.1. (a) $f_1(x, y, z) = x - z$, $f_2(x, y, z) = z$, $f_3(x, y, z) = -x + y + z$.
(b) In the standard basis $e_2^*(x, y, z) = y$, but $e_2 = v_3$ and $T_B(e_3) = f_3^* \neq e_2^*$.
8.1.3. (a) If $W_1 \subseteq W_2$ and $f \in W_2^0$, then $f(w) = \vec{0}$ for all $w \in W_2$. But then, $f(w) = \vec{0}$ for all $w \in W_1$ as well. Consequently, $f \in W_1^0$.
(b) Let $\{w_1, w_2, \ldots, w_s\}$ be a basis for W. Applying the basis-extension theorem, we have a basis $\{w_1, w_2, \ldots, w_s, v_{s+1}, \ldots, v_n\}$ for V. We denote by $\{w_1^*, \ldots, w_s^*, v_{s+1}^*, v_n^*\}$ the corresponding dual basis for V^*. The result follows by checking that $\{v_{s+1}^*, \ldots, v_n^*\}$ is a basis for W^0.

8.1.5. Let $\{f_1, \ldots, f_n\}$ be a basis for V^*. We denote by $\{f_1^*, \ldots, f_n^*\}$ the corresponding dual basis for V^{**}. Consider the basis $\{v_1, \ldots, v_n\}$ for V defined by $v_i = Q^{-1}(f_i^*)$ where $Q : V \to V^{**}$ is the natural isomorphism. We claim that our original basis $\{f_1, \ldots, f_n\}$ is the dual basis of $\{v_1, \ldots, v_n\}$. To see this, note that $f_i(v_j) = f_i^*(f_j) = v_i^{**}(f_j)$ which is 1 if $i = j$ and is 0 otherwise. This proves the result.

SECTION 8.2

8.2.1 By definition $V \otimes W$ is the vector space $ML(V^*, W^*; \mathbf{F})$. We define $T : ML(V^*, W^*; \mathbf{F}) \to ML(V, W; \mathbf{F})^*$ by specifying the images under T of the generators $v \otimes w$ of $V \otimes W$. (Here, $v \in V$ and $w \in W$.) For this we set $T(v \otimes w)$ to be the functional on $ML(V, W; \mathbf{F})$ defined by

$$T(v \otimes w)\left(\sum a_{ij} f_i \otimes g_j\right) = \sum a_{ij} f_i(v_i) g_j(w_j)$$

T is checked to be an isomorphsim by seeing that T sends a basis of $V \otimes W$ to a basis of $ML(V, W; \mathbf{F})^*$.
8.2.3. The standard ordered basis for $V \otimes V$ is the basis

$$\{(1, 0) \otimes (1, 0), (1, 0) \otimes (0, 1), (0, 1) \otimes (1, 0), (0, 1) \otimes (0, 1)\}$$

and the coordinates of $(1, 1) \otimes (-1, 2)$ are $(-1, 2, -1, 2)$ with respect to this basis.

A Short Bibliography

There is an abundance of elementary linear algebra texts which introduce students to basic matrix operations and their applications. The reader should have no difficulty locating several dozen at any University library, or perhaps can pick one up cheaply at a nearby garage sale. This bibliography lists a few alternative sources of the advanced material included in this text.

Advanced Linear Algebra:

1. Curtis, C. W.: *Linear Algebra, an Introductory Approach*, Allyn and Bacon Inc., Boston, Mass., 1974.

2. Finkbeiner, D. T.: *Introduction to Matrices and Linear Transformations* 3d ed., W. H. Freeman and Co., San Francisco, Calif., 1978.

3. Gantmacher, F. R.: *Theory of Matrices*, Chelsea, New York, 1959.

4. Hoffman, K. and R. Kunze: *Linear Algebra*, Prentice Hall, Englewood Cliffs, N.J., 1971.

5. Jones, B.: *Linear Algebra*, Holden Day Inc., San Francisco, Calif., 1973.

6. Halmos, P. R.: *Finite-Dimensional Vector Spaces*, Van Nostrand-Reinhold, Princeton, N.J., 1958.

Applied Linear Algebra

7. Noble, B., and J. W. Daniel: *Applied Linear Algebra* 2d ed., Prentice Hall, Englewood Cliffs, N.J., 1977.

8. Strang, G.: *Linear Algebra and Its Applicatons* 2d ed., Academic Press, New York, 1980.

Algebra with sections on Linear Algebra

9. Birkhoff, G., and S. MacLane: *A Survey of Modern Algebra*, Macmillan, New York, 1953.

10. Herstein, I. N.: *Topics in Algebra*, Xerox College Publishing, Lexington Mass., 1975.

11. Jacobson, N.: *Basic Algebra* I 2d ed., W. H. Freeman and Co., New York, New York, 1985.

12. Lang, S.: *Algebra*, Addison-Wesley, Reading, Mass., 1970.

Advanced Calculus and Analysis

13. Marsden, J., and A. Tromba: *Vector Calculus*, W. H. Freeman and Co., New York, 1988.

14. Rudin, W.: *Real and Complex Analysis*, McGraw-Hill, New York, 1974.

15. Spivak, M.: *Calculus on Manifolds, A Modern Approach to Theorems of Advanced Calculus*, W. A. Benjamin, New York, 1965.

16. Williamson, R., R. Crowell, and H. Trotter: *Calculus of Vector Functions*, Prentice-Hall, Englewood Cliffs, N.J., 1968.

Differential Equations

17. Boyce, W. E., and R. C. Di Prima, *Introduction to Differential Equations*, John Wiley, New York, 1970.

18. Coddington, E. A., and N. Levinson, *Theory of Ordinary Differential Equations*, McGraw-Hill, New York, 1955.

INDEX